Probability

Odds in Favor of *A*

$$\text{Odds}(A) = \frac{P(A)}{1 - P(A)} \qquad P(A) \text{ is the probability that } A \text{ will occur}$$

Probability That Even

$P(A \cup B) = P(A)$ … $= 0$

Probability That Even

$P(A \cap B) = P(A)$ … $= P(B)$

Expected Value of a

$\mu = E(X) = p_1x_1$

Variance of a Rando

$\sigma^2 = \text{Var}(X) = p$

Standard Deviation o

$$\sigma = \sqrt{\text{Var}(X)}$$

Binomial Probability

$P(k \text{ successes}) =$

$\mu = np$ and $\sigma =$

Statistics

Ungrouped Data

$$\bar{x} = \frac{\sum x_i}{n} \qquad s = \sqrt{\frac{\sum (x_i - \bar{x})^2}{n - 1}}$$

Grouped Data

$$\bar{x} = \frac{\sum f_i x_i}{n} \qquad s = \sqrt{\frac{\sum f_i (x_i - \bar{x})^2}{n - 1}}$$

FINITE MATHEMATICS FOR COLLEGE STUDENTS

Bernard J. Rice
Carroll M. Schleppi
Jerry D. Strange
University of Dayton

Brooks/Cole Publishing Company
Pacific Grove, California

Brooks/Cole Publishing Company
A Division of Wadsworth, Inc.

Printed in the United States of America

10 9 8 7 6 5 4 3 2 1

Library of Congress Cataloging in Publication Data

Rice, Bernard J.
Finite mathematics for college students / Bernard J. Rice, Carroll M. Schleppi, Jerry D. Strange.
p. cm.
Includes index.
ISBN 0-534-17172-9 : $32.00
1. Mathematics. I. Schleppi, Carroll M., 1941– . II. Strange, Jerry D. III. Title.
QA39.2.R5 1992
510—dc20 91-43315
CIP

ISBN 0-534-17172-9

Photo Credits
Chapter 1: Grant Heilman; Chapters 2 and 3: Lee Hocker; Chapter 4: David S. Strickler/ Monkmeyer Press Photo Service; Chapter 5: David Madison; Chapter 6: Lee Hocker; Chapter 7: Focus on Sports; Chapter 8: Lee Hocker; Chapter 9: David Powers/Stock Boston.

Sponsoring Editor: *Paula-Christy Heighton*
Project Development Editor: *Sue Ewing*
Marketing Representative: *Steve Simmons*
Editorial Assistant: *Lainie Giuliano, Carol Ann Benedict*
Production: *Julie Kranhold, ExLibris; Joan Marsh*
Manuscript Editor: *Judith Abrahms*
Interior Design: *John Edeen*
Cover Design: *Lisa Berman*
Cover Photo: *Jan Bindas Studio, Boston*
Cover Artist: *Judith Larzelere*
Interior Illustration: *Carl Brown*
Photo Researcher: *Stuart Kenter*
Typesetting: *Syntax International*
Cover Printing: *The Lehigh Press*
Printing and Binding: *R. R. Donnelley & Sons Company*

PREFACE

Finite Mathematics for College Students is written to accommodate the growing number of students who wish, and need, to learn about some of the applications of modern mathematics. Traditionally, college algebra was the standard fare for all students seeking to fulfill a mathematics requirement. But, today, with the wealth of interesting and useful topics available, which can be understood with minimal mathematical maturity, courses in finite mathematics are becoming a viable option. Such finite mathematics courses are usually designed to improve the student's ability to "see" how mathematics is used to represent physical situations. A corollary objective is to make the student aware of the fact that mathematics is an important tool for solving a wide variety of problems. The goal of this text is to help the student appreciate and understand, through the use of examples and applications, some sophisticated but accessible modern finite mathematics. After completing this book, the student will have an idea of the mathematical language and tools used in a wide variety of situations.

We have written *Finite Mathematics* to be reader-friendly, by using an intuitive approach to applications, wherever possible, in order to alleviate "math anxiety." Because of the diversity of interests of students, we have tried to choose applications that are within the experience level of a typical college student. In most cases, we have chosen examples that a student can follow and understand without long preliminary explanations of the application. This helps to avoid the situation in which the application is so involved and difficult that the mathematical idea is lost. We have sought to make the applications interesting without being overwhelming.

Our central theme, as outlined in Chapter 1, is the concept of mathematical modeling, but for the most part this is informal. Linearity is the common thread of the first three chapters beginning with elementary linear models in Chapter 1, continuing with systems of equations in Chapter 2, and, finally, an introduction to linear programming with a discussion of the simplex method in Chapter 3. Chapter 2 also includes a discussion of matrix algebra. Chapter 4 delves into sets and counting and builds the foundation for probabilistic and statistical models presented in Chapters 5 and 6. Chapter 7, on decision theory, uses expected value and Markov chains to show how mathematics is used to help us make rational decisions. Chapter 8 explores the mathematics of finance and discusses the most relevant finance formulas and terminology. Our final Chapter 9 is on Boolean algebra and shows how mathematical abstraction can be used to make certain situations more understandable. The approach used in this is intended to give a flavor of how mathematics is developed.

Finite Mathematics for College Students assumes the student has taken the usual high school algebra courses. As such, this book does not attempt to reinforce algebraic manipulative skills, rather it builds on the background already achieved in high school. Throughout the book, "Comments" and "Warnings" advise the reader of historic points, interpretations, and pitfalls.

There is more material in this text than even the most ambitious instructor could cover in a one-semester course. We suggest that Chapters 1 through 3 be made a basis for the beginning of any course. The discussion of Leontief economic models and the simplex method for linear programming are optional. Chapter 4, on sets and counting, is important to all of finite mathematics and should not be omitted. Chapters 5 through 7 form a unit in which probability is the common thread, although the material on decision theory could be easily omitted. Chapters 8 and 9 each stand alone and may be chosen at the option of the instructor to round out a course.

A word about calculators and computers: We feel that the inclusion of calculators and computers is an important aspect of a modern finite mathematics course, and the current generation of calculators, which performs matrix and statistical operations, are timely. Thus we give some hints on the use of calculators but, for the most part, we have not tied the text material to any particular calculator.

Finite Mathematics for College Students has benefited from several rounds of thoughtful reviews by the following colleagues in the mathematical community: Frederick J. Carter, St. Mary's University, San Antonio, Texas; Kenneth Davis, Albion College, Albion, Mississippi; Doug Dokken, University of St. Thomas, St. Paul, Minnesota; Robert Eicken, Illinois Central College, East Peoria, Illinois; Bennette Harris, University of Wisconsin at Whitewater; Kevin Hastings, Knox College, Galesburg, Illinois; Eric Heinz, Catonsville Community College, Catonsville, Maryland; and Byron McAllister, Montana State University, Bozeman, Montana.

We wish to take this opportunity to thank our editor, Paula-Christy Heighton, our production editor Joan Marsh, Sue Ewing and the entire staff at Brooks Cole for their continuing excellent cooperation and helpful advice during the lengthy process of writing, rewriting, and finally publishing this book.

Bernard J. Rice
Carroll M. Schleppi
Jerry D. Strange
February 1992

CONTENTS

CHAPTER 1 **LINEAR EQUATIONS AND FUNCTIONS** 1

1.1 Linear Equations 2
1.2 Applications of Linear Equations 7
1.3 Functions 13
1.4 Graphing 21
1.5 Linear Functions 31
1.6 Methods of Describing a Line 39

Review Exercises 46

CHAPTER 2 **SYSTEMS OF LINEAR EQUATIONS AND MATRICES** 49

2.1 Linear Systems 50
2.2 Higher-Order Systems: Matrix Notation 61
2.3 Matrix Algebra: Addition and Scalar Multiplication 73
2.4 Matrix Algebra: Multiplication 82
2.5 Solving Systems of Equations by Matrix Inversion 89
2.6 Leontief Models 99

Review Exercises 108

CHAPTER 3 **INEQUALITIES** 111

3.1 Linear Inequalities 112
3.2 Linear Inequalities in Two Unknowns 116
3.3 Systems of Linear Inequalities 121
3.4 Linear Programming 126
3.5 The Simplex Method 134
3.6 Nonstandard Linear Programming Problems 147

Review Exercises 157

CHAPTER 4 **SETS AND COUNTING** 159

4.1 Sets 160
4.2 More Unions and Intersections 169
4.3 Counting Permutations 179
4.4 Counting Combinations 186
4.5 Binomial Expansions 191
Review Exercises 196

CHAPTER 5 **PROBABILITY** 197

5.1 Probability Model 198
5.2 Calculating Probability 205
5.3 The Addition Rule 214
5.4 The Multiplication Rule 221
5.5 Stochastic Processes 230
5.6 Bayes' Formula 236
Review Exercises 241

CHAPTER 6 **PROBABILITY AND STATISTICS** 243

6.1 Random Variables and Probability Distributions 244
6.2 Expected Value 253
6.3 Variance and Standard Deviation 260
6.4 The Binomial Probability Distribution 268
6.5 The Normal Probability Distribution 276
6.6 Normal Approximation of the Binomial Distribution 287
6.7 Statistics 292
Review Exercises 302

CHAPTER 7 **MARKOV CHAINS AND GAME THEORY** 305

7.1 The Transition Matrix 306
7.2 Regular Markov Chains 315
7.3 Decision Making: Game Theory 324
7.4 Non-strictly Determined Games: Mixed Strategies 330
Review Exercises 340

CHAPTER 8 **THE MATHEMATICS OF FINANCE** 343

8.1 Simple Interest 344
8.2 Compound Interest 350
8.3 Arithmetic Sequences and Annual Percentage Rate 358
8.4 Geometric Sequences 366
8.5 Annuities and Mortgages 371
Review Exercises 380

CHAPTER 9 **BOOLEAN ALGEBRA** 383
9.1 Algebra of Switches 384
9.2 Boolean Algebra: An Abstract System 395
9.3 The Algebra of Statements 401
9.4 Computer Logic Circuits 408
9.5 The Algebra of Sets 413
Review Exercises 417

ANSWERS TO EXERCISES A-1

APPENDIX/TABLES A-33

INDEX I-1

CHAPTER 1

LINEAR EQUATIONS AND FUNCTIONS

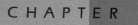

Crop irrigation is a valuable technique for raising food in arid regions. However, the process requires large quantities of water. Suppose that drinking water contains 0.5% salt, that seawater contains 3.8% salt, and that irrigation water can contain 1% salt. Since drinking-quality water is scarce, we would like to use as little of it as possible, so we might mix drinking water and seawater to make irrigation water. The question is, "How much seawater should we add to drinking water to produce irrigation water?" Later in this chapter we will show that we should mix about 15 gallons of seawater and 85 gallons of drinking water to produce each 100 gallons of irrigation water. The solution of this problem involves the use of linear equations.

A team of civil engineers is planning to survey a new roadbed through the Rocky Mountains. Roadways that rise more than 15 feet for every 100 feet of horizontal run are considered dangerous, especially for large trucks, so the team would like to avoid slopes that are this steep. The slope of a roadway is defined in terms of the properties of a straight line.

If you are interested in solving problems such as these, you need to know about linear equations. In this chapter we will examine the ways in which the geometric concept of a straight line and its algebraic counterpart, the linear equation, are used to solve a variety of physical problems. In particular, we will show how linear equations are used to solve some familiar problems in business, engineering, and science. In the course of this development, we will occasionally discuss some quantities that are nonlinear, but only to contrast their properties with those of linearity.

1.1 Linear Equations

Recall from algebra that symbols such as x, s, t, and so on are used to represent numbers. Symbols such as x, s, and t that may take on various values are called **variables.** Any grouping of numbers and symbols in which the arithmetic operations of addition, subtraction, multiplication, or division are used to combine the numbers and symbols is called an **algebraic expression.** The set of values that may be assigned to a variable is called the **domain** of the variable. Sometimes the domain of a variable may be any real number; sometimes it may be restricted to some subset of the real numbers. For instance, the domain of the variable in the expression $1/x$ includes all values of x except $x = 0$, since division by 0 is not permitted. The domain is also called the set of **permissible values** of the variable.

An **equation** is a mathematical statement that two expressions are equal. An equation is called a *conditional equation* if it is a true statement for a limited number of values of the unknown quantity. For instance, let x represent a real number in the equation $x + 3 = 7$. Then $x + 3 = 7$ is a conditional equation, because 4 is the only value of x for which the equation is true. We say that $x = 4$ **satisfies** the equation or is a **solution of the equation.** More generally, a solution of an equation is any number in the domain of the variable that makes the equation a true statement. The set of all solutions of an equation is called the **solution set.** For the equation $x + 3 = 7$, the solution set consists of a single number, 4. We can *check* to see whether a particular number is a solution by substituting that number into the original equation and determining whether the resulting statement is true. Thus we can check that 4 is a solution to $x + 3 = 7$ by substituting 4 for x and noting that $4 + 3 = 7$ is a true statement. It is always wise to check your possible solutions.

The domain of the variable must be clearly understood when a variable is used to represent physical quantities. In some instances, the domain of a variable might be limited rather arbitrarily to sets of numbers such as the natural numbers, the integers, the rational numbers, or the real numbers. However, in many cases, the domain of the variable is determined from the nature of the problem. For example, in a problem in which the variable is the number of shirts a retailer expects to sell this year, we would want to limit the permissible values of the variable to nonnegative integers. Further, we might want to limit the domain of the variable to less than some number: for example, the greatest number of shirts the retailer *could* reasonably expect to sell this year. In any case, the domain must be understood. For our discussions, independent of any real-life problem, we will assume, unless otherwise stated, that *the permissible values of the variable are the real numbers.* With this in mind, we see that the equation $x^2 = 4$ has two solutions, 2 and -2, in its solution set, but the equation $x^2 = -4$ has no solutions, because the square of a real number cannot be negative. The equation $x^2 = 2$ has two real solutions, $\sqrt{2}$ and $-\sqrt{2}$, but has no solutions if the permissible values are limited to the rational numbers. In some cases the statement is true for all permissible values of the variable. Such statements are called **identities.** The equation $x + 3x = 4x$ is an example of an identity.

A special type of equation, and the center of our attention in this book, is the kind that can be expressed in the form

$$ax + b = 0$$

where x is a variable and a and b are constants, $a \neq 0$. Such an equation is said to be *linear in the variable* x. The solution set of the linear equation consists of one element, $-b/a$.

Two equations are said to be **equivalent** if they have the same solution set. Thus, the equations $x + 3 = 7$ and $x + 2 = 6$ are equivalent, since each has 4 as its only solution. The equations $x = 2$ and $x^2 = 4$ are *not* equivalent, because the solution to the first one is 2 and the second has two solutions, 2 and -2.

To **solve** an equation means to find all of its possible solutions. But how do we proceed? In some cases, the solution is obvious. For example, if the equation is in the form $2x = 6$, it is obvious that $x = 3$ is the solution. If the solution is not obvious, we will want to manipulate the given equation so we can replace it with an equivalent equation whose solution is obvious. There are three *allowable manipulations* for transforming an equation so that the resulting equation will be equivalent.

ALLOWABLE MANIPULATIONS

When the following allowable operations are performed on an equation, the result is an equivalent equation.

- **Substitution:** A quantity may be substituted for an equal quantity. For example, $2x + 5x$ may be replaced by $7x$, and $4 + 10$ by 14, in the equation $2x + 5x + 4 + 10 = 0$, which then becomes $7x + 14 = 0$.
- **Addition/Subtraction:** The same quantity may be added to or subtracted from both sides of an equation. For example, if 5 is subtracted from both sides of the equation $2x + 5 = x - 7$, we get the equivalent equation $2x = x - 12$. This may be further changed by subtracting x from both sides to obtain $x = -12$.
- **Multiplication/Division:** Both sides of an equation may be multiplied or divided by the same *nonzero* quantity. For example, both sides of the equation $3x = 9x - 18$ may be divided by 3 to yield the equivalent equation $x = 3x - 6$.

The next two examples show how these manipulations are used to solve linear equations in one variable. The sequence in which the manipulations are performed is not unique but, of course, the solution set is unique. In a general way, the aim of the manipulations is to isolate the variable on one side of the equation so that the solutions are obvious.

Example 1 Solve $3x + 12 = 48$

Solution

$$3x + 12 = 48$$

$$3x = 36 \qquad \text{Subtracting 12 from both sides}$$

$$x = 12 \qquad \text{Dividing both sides by 3}$$

To check this solution, let $x = 12$ in the original equation. We see that

$$3(12) + 12 = 48$$

is a true statement, so the solution is $x = 12$. ■

Example 2 Solve $2(x + 3) = 7x$

Solution

$$2(x + 3) = 7x$$

$$2x + 6 = 7x \qquad \text{Expanding the left side}$$

$$6 = 5x \qquad \text{Subtracting } 2x \text{ from both sides}$$

$$x = \frac{6}{5} \qquad \text{Dividing both sides by 5}$$

Check by substituting into each side of the original equation. Substituting into the left-hand side, $2(\frac{6}{5} + 3) = 2(\frac{21}{5}) = \frac{42}{5}$; substituting into the right-hand side, $7(\frac{6}{5}) = \frac{42}{5}$. This shows that $x = \frac{6}{5}$ is the solution. ∎

Sometimes the variable occurs in the denominator of one or more terms in the equation. In this case, we usually multiply both sides of the equation by the lowest common denominator of the fractional terms in order to eliminate the fractions. If the resulting equation is linear—that is, of the form $ax + b = 0$—it can then be solved as usual.

WARNING: *Any value of the variable that produces a zero denominator cannot be a solution.*

Example 3 Solve the equation $\frac{1}{x} + \frac{3}{2x} = \frac{3}{x-5}$

Solution Notice that x may not take on the values 0 or 5, because both of these values produce a zero in the denominator. To eliminate the fractions, we multiply each term on both sides of the equality by $2x(x - 5)$, the lowest common denominator of the fractional terms.

$$\frac{1}{x} \cdot 2x(x-5) + \frac{3}{2x} \cdot 2x(x-5) = \frac{3}{x-5} \cdot 2x(x-5)$$

$$\begin{aligned} 2(x-5) + 3(x-5) &= 3(2x) && \text{Cancellation law} \\ 2x - 10 + 3x - 15 &= 6x && \text{Expanding} \\ -x &= 25 && \text{Collecting like terms} \\ x &= -25 && \text{Multiplying by } -1 \end{aligned}$$

Recall that the possible solution can be any real number except 5 or 0, so -25 is a possible solution. Checking $x = -25$, we have $\frac{1}{-25} + \frac{3}{2(-25)} = \frac{3}{-25-5}$, or $-\frac{2}{50} - \frac{3}{50} = -\frac{3}{30}$. ∎

Frequently equations have more than one variable quantity and we find it necessary to solve the equation for one variable in terms of the others. If the equation is linear in form, the manipulations described above may be used to find the solution. The examples that follow are typical. Remember that the goal is to isolate the variable of interest on one side of the equality.

Example 4 Degrees Fahrenheit (°F) and degrees Celsius (°C) are related by the formula $F = \frac{9}{5}C + 32$. This formula gives the temperature in °F if the temperature in °C is known. Write a formula that will give the temperature in °C if the temperature in °F is known.

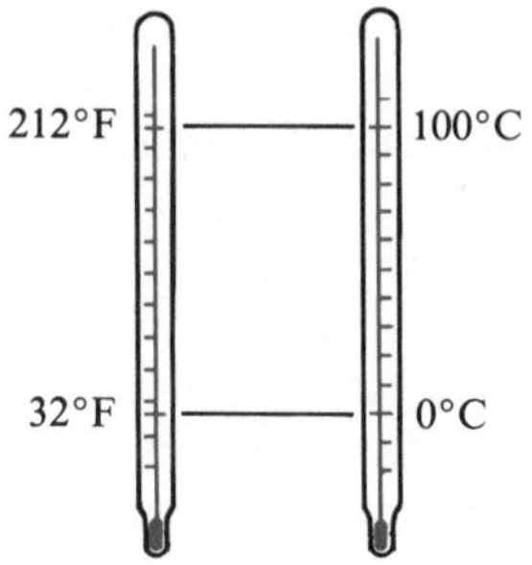

Solution Subtracting 32 from both sides of $F = \frac{9}{5}C + 32$, we obtain

$$\frac{9}{5}C = F - 32$$

Multiplying both sides by $\frac{5}{9}$ gives

$$C = \frac{5}{9}(F - 32)$$

■

Example 5 The expression

$$S = \frac{a - rL}{1 - r}$$

occurs in the study of geometric progressions. Solve for r. (This formula is used in the financial world to calculate interest payments on annuities and mortgages.)

Solution Multiplying both sides of the expression by $1 - r$, we get

$S(1 - r) = a - rL$	Canceling $1 - r$ on the right-hand side
$S - Sr = a - rL$	Expanding the left-hand side
$rL - Sr = a - S$	Adding $rL - S$ to both sides to isolate terms involving r
$r(L - S) = a - S$	Factoring
$r = \dfrac{a - S}{L - S}$	Dividing by $L - S$

■

Exercises Section 1.1

In Exercises 1–25 solve the given equation. Remember that a zero denominator is undefined.

1. $2x + 5 = 2$
2. $6x - 1 = -x - 3$
3. $3(2x - 1) - x = 4 - (x + 3)$
4. $3(x - 2) + 7(2x + 4) = -(x + 1)$
5. $3x + 5 = 4(x - 2)$
6. $3y + 7(4 - y) = 3(5 - 2y)$
7. $\dfrac{4a}{3} - 5a + 2 = \dfrac{a}{2} - 1$

8. $\frac{3x}{5} + 4 = \frac{x}{2} + 5$

9. $\frac{2x + 1}{3} + 16 = 3x$

10. $\frac{x + 1}{3} + \frac{x + 2}{7} = 5$

11. $\frac{2}{c} + \frac{3}{c} = 10$

12. $\frac{4}{y} - 3 = \frac{5}{2y}$

13. $\frac{3x}{x + 5} - 4 = 0$

14. $\frac{4}{x + 2} + 4 = 0$

15. $\frac{3}{6y + 2} = \frac{4}{7y + 3}$

16. $\frac{2}{x + 3} = \frac{5}{2x - 1}$

17. $\frac{2}{x - 2} - \frac{3}{x + 5} = \frac{10}{(x - 2)(x + 5)}$

18. $\frac{1}{(2x + 3)(x - 1)} = \frac{1}{2x + 3} + \frac{1}{x - 1}$

19. $\frac{5}{5m - 11} = \frac{3}{m - 5} - \frac{4}{2m - 3}$

20. $\frac{2}{x + 3} + \frac{3}{x - 4} = \frac{5x + 6}{(x + 3)(x - 4)}$

21. $(x + 7)(x - 1) = (x + 9)^2$

22. $(x - 3)(x + 4) = x^2 - 1$

23. $2x^2 + 1 = (2x - 1)(x + 3)$

24. $3x(x - 1) = (3x - 7)(x + 1)$

25. $x(x^2 + 3) = 5 + x^3$

26. Solve for x: $\frac{1}{x - a} + \frac{1}{x + a} = \frac{a}{x - a}$

27. Solve for x: $\frac{p + q}{x + 1} = \frac{p - q}{x - 1}$

28. The expression $l = a + (n - 1)d$ appears in a discussion of arithmetic progressions. Solve for d.

29. The formula $\frac{1}{R} = \frac{1}{R_1} + \frac{1}{R_2}$ occurs in electric-circuit analysis. Solve for R.

30. The simple-interest formula is $A = P(1 + i \cdot n)$. Solve for n.

31. The compound-interest formula is $A = P(1 + r)^n$. Solve for P.

32. The area of a trapezoid with bases a and b and height h is given by the formula $A = \frac{1}{2}h(a + b)$. Solve for b.

33. The velocity of an object may be written as $v^2 = v_0^2 + 2as$, where a is the acceleration, v_0 is the initial velocity, and s is the displacement. Solve for a.

34. The length of a metal rod whose temperature changes from T_0 to T_1 is given by

$$L = L_0[1 + \alpha(T_1 - T_0)]$$

Solve for T_1.

35. A formula for depreciating equipment is $D = \frac{nd}{n + 2}$. Solve for n.

1.2 APPLICATIONS OF LINEAR EQUATIONS

In this section we show how to use algebraic symbols to express real-world problems. We use symbols, such as letters of the alphabet, to represent physical quantities and then express relationships between the quantities in terms of these symbols. The representation of a real-world problem by mathematical symbols is referred to as *mathematical modeling*.

Example 1 Some simple mathematical models:

(a) If we let w represent the width of a rectangle, l its length, and A its area, then the fact that "the area of a rectangle is the product of the length and the width" is written as $A = l \cdot w$ or $A = lw$.

(b) If you pay 8% tax on the cost of a car, then, using x to represent the cost of the car, the tax you pay is written algebraically as $0.08x$.

(c) The distance traveled in miles, d, is equal to the product of the speed in miles per hour, v, and the time in hours, t. Symbolically: $d = vt$.

(d) The sum of two numbers is 12. If x is one of the numbers, then $12 - x$ is the other.

(e) If we represent a two-digit number by yx, then the *value* of the number is $10y + x$.

(f) If x represents a number, then $x + 4$ represents a number that is 4 greater than x; $5x$, a number that is 5 times x; and $5x + 4$, a number that is 4 greater than 5 times x.

(g) If it takes t hours to do a job, then $1/t$ is the fraction of the job that can be done in 1 hour. ■

Example 1 shows how algebraic language is used to construct mathematical models of physical situations. Mathematical models frequently involve equations that express the relationships between various quantities. There are no set rules to follow in establishing a mathematical model; however, the following guidelines are offered to provide a systematic approach to the problem.

GUIDELINES FOR WRITING A MATHEMATICAL MODEL ■

1. Read the conditions. Read the given problem carefully to determine precisely what is known and what you are trying to find. If possible, draw a picture of the situation. This frequently helps you clarify the statement of the problem.
2. Identify the variable and constant quantities. Assign letters to the variables. Begin your solution with phrases such as "Let $x =$ the speed," or "Let t be the time."
3. Form an equation. Use the given information to write a relationship between the constants and variables of the problem. This is your *mathematical model.*
4. Solve the equation. Here you must use the permissible operations outlined in Section 1.1.
5. Check your solution. Solutions should be checked to ensure that they satisfy the requirements of the original problem. For instance, -2 would not be a possible answer for one dimension of a rectangle.

Example 2 Michael pays $278 for a calculator and a typewriter. If the calculator costs $64 less than the typewriter, how much does he pay for each item?

Solution We approach the solution to this problem by using the suggested guidelines for writing a mathematical model of the situation.

1. **Read the conditions:** From the reading of the problem we know that the total cost of the calculator and the typewriter is $278 and that the calculator costs $64 less than the typewriter. The problem is to find the cost of each item.
2. **Identify the variable and constant quantities:** We identify the unknowns (variables) as the costs of the items. Constant in the problem are the

total cost, which is \$278, and the difference between the costs of the two items, which is \$64. We begin with "Let $x =$ the cost in dollars of the typewriter." Then, since the calculator costs \$64 less than the typewriter, we express its cost in dollars as $x - 64$.

3. **Form an equation:** Since we know the total cost of the two items, we have

$$\text{Cost of the typewriter} + \text{Cost of the calculator} = \text{Total cost}$$

4. **Solve the equation:** Substituting the symbolic expressions for the cost of the typewriter and the cost of the calculator (expressed in step 2) into the indicated equation, we obtain

$$x + (x - 64) = 278$$

$2x = 278 + 64$	Adding 64 to both sides and collecting like terms
$2x = 342$	Summing 278 and 64
$x = 171$	Dividing both sides by 2

Hence, the typewriter costs \$171 and the calculator costs \$171 − \$64 = \$107.

5. **Check the solution:** To check our solution, we note that \$171 − \$107 = \$64 (that is, the cost of the calculator is \$64 less than that of the typewriter) and that \$171 + \$107 = \$278 (that is, the total cost of the two items is \$278). ∎

Example 3 Coastal Air's Flight 402 left New York City for Los Angeles at 10:00 P.M., cruising at 400 mph. Flight 611 left New York for the same destination at 11:30 P.M., cruising at 550 mph. How long will it take Flight 611 to overtake Flight 402?

Figure 1.1

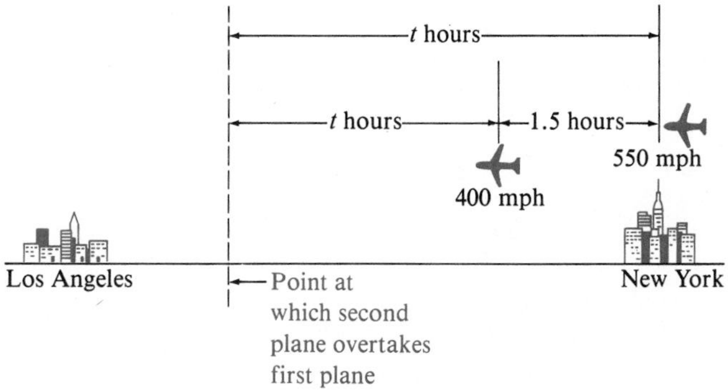

Solution Let t be the time required for Flight 611 to overtake Flight 402. At the time of rendezvous, Flight 402 will have been in the air $t + 1.5$ hr. A formula that gives the relationship between distance, velocity, and time is $d = vt$, as given in

Example 1, step c. We use this formula to write

$$400(t + 1.5) = \text{distance traveled by Flight 402}$$
$$550t = \text{distance traveled by Flight 611}$$

Since the distances traveled by the two planes are equal when Flight 611 overtakes Flight 402, we have

$$\boxed{\text{Distance traveled by Flight 402}} = \boxed{\text{Distance traveled by Flight 611}}$$

Consequently, the desired mathematical model is

$$\begin{aligned} 400(t + 1.5) &= 550t \\ 400t + 600 &= 550t && \text{Expanding the left-hand side} \\ 150t &= 600 && \text{Subtracting } 400t \text{ from both sides} \\ t &= 4 && \text{Dividing by 150} \end{aligned}$$

It takes 4 hr for Flight 611 to overtake Flight 402. ■

Example 4 Suppose that you are in charge of an orange-juice stand at Coney Island and that you have two types of orange juice: one is 55% pulp and the other is 15% pulp. How much of each must you mix to get 75 gallons of orange juice that is 40% pulp?

Solution Let n be the number of gallons of the 55% variety that will be in the final mixture. This means $75 - n$ will represent the number of gallons of the 15% variety. The amount of pulp from the two mixtures is $0.55n$ and $0.15(75 - n)$, and the total amount of pulp in the new 40% mixture is to be $0.40(75)$. Thus,

$$\boxed{\text{Amount of pulp from 55\% mixture}} + \boxed{\text{Amount of pulp from 15\% mixture}} = \boxed{\text{Amount of pulp in 40\% mixture}}$$

In symbols:

$$\begin{aligned} 0.55n + 0.15(75 - n) &= 0.40(75) \\ 0.55n + 11.25 - 0.15n &= 30.00 && \text{Expanding} \\ 0.40n &= 18.75 && \text{Subtracting 11.25 from both sides and collecting like terms} \\ n &= \frac{18.75}{0.40} && \text{Dividing by 0.40} \\ n &= 46.9 \end{aligned}$$

This means that you must mix in 46.9 gallons of the 55% mixture and $75 - 46.9 = 28.1$ gallons of the 15% mixture. Note the implicit physical limitations on this problem. For example, suppose you had only 40 gallons of the higher-pulp variety. Then you would either have to modify your requirement for a 40% mix or adjust the total amount of 75 gallons. ■

Example 5 The Mideast Agricultural Corporation wants to reduce the salt content of seawater to a level at which it can be used for irrigation by combining it with "pure" drinking water. What percentage of seawater (3.8% salt) should they combine with drinking water (0.5% salt) to get irrigation water with 1% salt?

Solution For simplicity, consider a 100-gal mixture. Let x = the number of gallons of seawater in the mixture; then $100 - x$ is the number of gallons of drinking water. The amount of salt contributed from the seawater is $0.038x$ and the amount of salt from the drinking water is $0.005(100 - x)$. The total amount of salt in the desired mixture is $0.01(100)$. Hence, we have the following literal equation:

$$\boxed{\begin{array}{c}\text{Amount of salt}\\\text{from seawater}\end{array}} + \boxed{\begin{array}{c}\text{Amount of salt}\\\text{from drinking water}\end{array}} = \boxed{\begin{array}{c}\text{Amount of salt}\\\text{in irrigation water}\end{array}}$$

After substituting the appropriate symbolic expressions, this becomes

$$\begin{aligned} 0.038x + 0.005(100 - x) &= 0.01(100) \\ 38x + 5(100 - x) &= 10(100) && \text{Multiplying by 1000 to clear decimals} \\ 38x + 500 - 5x &= 1000 && \text{Expanding} \\ 33x &= 500 && \text{Subtracting 500 from both sides and collecting like terms} \\ x &= 15.15 && \text{Dividing by 33} \end{aligned}$$

Hence, there should be about 15 gallons of seawater and 85 gallons of drinking water, or, on a percentage basis, 15% seawater and 85% drinking water in the mixture. ■

Example 6 The Beechwood Swim Club manager needs to fill a pool in 8 hr, and she knows that the main water line will take 12 hr to fill the pool. What should be the minimum rating of an auxiliary hose, in "hours to fill the pool," to get the pool filled on time if both the main line and the auxiliary hose are used?

Solution Let x = the minimum rating of the auxiliary hose. Then consider how much of the pool each hose can fill by itself in 1 hr.

$$\frac{1}{12} = \text{part of the pool filled by main line in 1 hr}$$

$$\frac{1}{x} = \text{part of the pool filled by auxiliary hose in 1 hr}$$

$$\frac{1}{8} = \text{part of the pool that must be filled in 1 hr by both hoses}$$

Since the part of the pool that can be filled in 1 hr is the sum of the capacities of the main and auxiliary lines, we have

$$\boxed{\begin{array}{c}\text{Amount filled in}\\\text{1 hr by main line}\end{array}} + \boxed{\begin{array}{c}\text{Amount filled in}\\\text{1 hr by auxiliary hose}\end{array}} = \boxed{\begin{array}{c}\text{Amount filled in}\\\text{1 hr by both}\end{array}}$$

Symbolically, the mathematical model is:

$$\frac{1}{12} + \frac{1}{x} = \frac{1}{8}$$

$$2x + 24 = 3x \qquad \text{Multiplying by } 24x \text{ to eliminate fractions}$$

$$x = 24 \qquad \text{Subtracting } 2x \text{ from both sides}$$

Thus, the auxiliary hose (when operating alone) must be able to fill the pool in 24 hours or less. ■

EXERCISES SECTION 1.2

In Exercises 1–20, express each of the given sentences or phrases symbolically.

1. The area of a triangle, A, equals one-half the product of the base, b, times the height, h.

2. The area, A, of a circle equals the product of π times the square of the radius, r.

3. The perimeter, P, of a square is 4 times the side length, s.

4. The perimeter, P, of a triangle is the sum of the lengths of the sides, a, b, and c.

5. Acceleration, a, equals velocity, v, divided by time, t.

6. Voltage, V, equals the current in the circuit, I, times the resistance of the circuit, R.

7. A number, x, is 8 less than another number, y.

8. A number, x, is 3 times the square of another number, y.

9. A number, a, is 5 times the sum of another number, b, plus three.

10. A number, a, is 3 more than 5 times another number, b.

11. Twenty percent of the amount by which a number, x, exceeds 12,000.

12. The sum of two numbers, x and y, divided by 2.

13. The average of three numbers, x, y, and z.

14. The portion of a job done in 1 day if it takes x days to do the entire job.

15. Interest, i, equals principal, p, times rate, r.

16. The product of the sum and the difference of two numbers, x and y.

17. Twice the sum of two numbers, x and y.

18. The area of a triangle, A, whose base is twice its height.

19. The surface area, A, of a cube of side s.

20. The volume, V, of a rectangular box whose dimensions are l, w, and h.

In Exercises 21–38, write a mathematical model for each problem and then solve the problem for the indicated variable.

21. The cost of a car is its price, P, plus 7% sales tax. If Sue has \$9000 to spend on the car, what price should she be considering?

22. Jim makes a weekly base salary of \$200 plus one-fourth of his total weekly salary. How much does he make per week?

23. Find three consecutive even integers whose sum is 312.

24. The total monetary value of 73 coins, all nickels and dimes, is \$5.75. How many nickels are there?

25. Randall can run a 7-min mile. He starts 0.1 mi ahead of Brent, who can run a 6-min mile. How far does Brent, the faster jogger, have to run to catch up to Randall, the slower jogger?

26. The average of the salaries of Jane, Jim, and Nancy is \$25,000 per year. If Jane earns \$3000 more than Nancy and Jim earns half as much as Jane, how much does each earn?

27. A new Buick that averages 22 mi/gal of gasoline for city driving and 28 mi/gal for highway driving uses 20 gal of gasoline on a 500-mi trip. How much of the driving was done in the city?

28. Hazel wants to winterize her car. She finds that her 10-qt radiator is filled with a mixture that is 30% antifreeze. How much of the fluid should she drain and replace with pure antifreeze to double the strength of the mixture?

29. A pilot flew an airplane with the wind for 1 hr and returned the same distance against the wind in 2 hr. If the cruising (air) speed of the plane is 400 mph, what is the velocity of the wind?

30. Suppose that in taking a vacation in your car you average 60 mph outside the city limits and 30 mph within the city limits. If you take 5 hr to travel 270 mi, how much of the time did you spend inside cities and how much outside?

31. The distance around the Indianapolis Speedway track is 2.5 mi. If one driver averages 170 mph and another averages 165 mph, in how much time will the first driver "lap" the second?

32. Adam and his friend together can build a house in 20 days. Alone, his friend can build it in 30 days. Based on this information, how long would it take Adam to build the house alone?

33. Central State University's football team scored 44 points by scoring equal numbers of touchdowns (6 points), extra points (1 point), and field goals (3 points), and half as many safeties (2 points). How many of each scoring play did the team make?

34. PetroOne's oil tanker is being filled by two pipelines, which could fill it separately in 12 hr and 16 hr, respectively. How long will it take to fill the tanker?

35. A girl weighing 100 lb sits on one end of a 20-ft teeterboard; a boy weighing 150 lb sits on the other end. Where must the fulcrum be for the two to be balanced?

36. Tom has a 20-lb mixture that is one-fourth cement and three-fourths sand. How much of a mixture that is half cement and half sand must be added to the 20-lb mixture to produce a mixture that is one-third cement?

37. A car leaves Cincinnati for Toledo at 55 mph. Thirty minutes later a truck leaves Toledo for Cincinnati at 45 mph. The two cities are 206 mi apart and both vehicles are using Interstate 75. How many hours must the truck travel before it meets the car?

38. Jane can overhaul an engine in 20 hr and Fred can do the same job in 30 hr. If they both work together for a number of hours and then Jane finishes the job by herself in 5 hr, how many hours did they work together?

1.3 Functions

In this section we develop the important idea of **function** and explore its relationship to our ability to understand the physical world around us. The idea of function is not unique to mathematics, but in fact is just the formalization of the familiar idea of pairing numbers together in which one number is related to a cause and the other to an effect. Perhaps our most common use of pairing numbers is related to money. For instance, we pair the amount of money we are paid to do a job with the number of hours we must work at the job. This idea of pairing one number with another is at the heart of mathematics.

Functional Pairing

Suppose it costs \$5 to rent a videotape of a movie. We understand that for one tape the rental fee is \$5, for two tapes it is \$10, for three tapes it is \$15, and for n tapes it is $5n$ dollars. This suggests a rule that pairs one number, n, the number of videotapes, with another number, C, the total cost; that is, $C = 5n$, where C

and n are positive integers. This simple example illustrates an important idea in mathematics; namely, that of pairing a number y with a number x according to some specified **rule of correspondence.** For instance, you can get a y for each x from the rule given by the formula $y = 5x + 2$. In this case, if $x = 0$, $y = 2$; if $x = 1$, $y = 7$; if $x = \pi$, $y = 5\pi + 2$; and so on for any real value of x. Sometimes the set of replacement values for x must be restricted; for instance, $x = 2$ may not be used in the formula $y = \dfrac{1}{x-2}$ since the formula is undefined for this value. (Why?)

A rule of correspondence need not be expressed as a formula, but may be given by a table of values. The following table pairs a number, x, the number of golf lessons, with a second number, y, the total cost in dollars for x lessons.

x	1	2	3	4
y	30	55	75	80

Here the rule of correspondence is the table itself. Notice that the table can also be represented by the set of ordered pairs $\{(1, 30), (2, 55), (3, 75), (4, 80)\}$. By **ordered pairs** we mean that the x-value is given first and the y-value second.

If we let X represent the set of values that we can assign to x, and Y the corresponding set of values obtained by the rule of correspondence f, then the general idea of what is called a **functional pairing** is as shown in Figure 1.2. The figure represents the transformation of the number x by the rule f into the number y. A functional pairing is *unique,* which means that only one value of y is paired with each value of x.

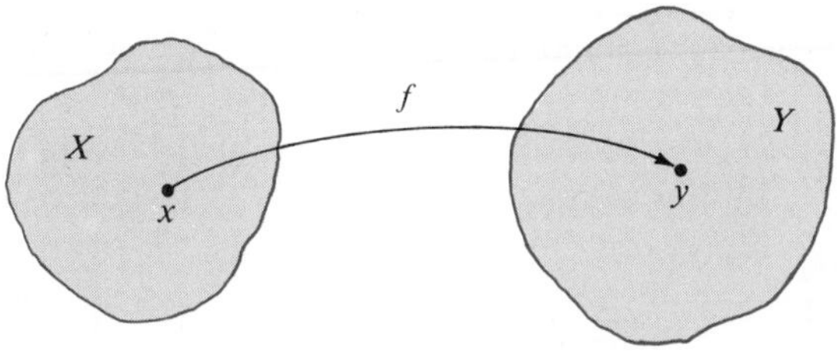

Figure 1.2

We summarize the above discussion in the following definition. Notice that this definition defines three concepts: function, domain, and range.

DEFINITION Function

A correspondence f that assigns to each x in X exactly one element y in Y is called a **function.** The set X is called the **domain** of the function and the set of the corresponding values of y is called the **range** of the function. We say that "y is a function of x."

This definition applies to rules of correspondence given both by formulas and by sets of ordered pairs of numbers. We wish to explain the concept of function as it applies to a set of ordered pairs of numbers of the form (x, y). For a set of ordered pairs (x, y) to define a function, no two distinct pairs may have the same first element. This follows from the fact that a function assigns, to each x in X, a unique y in Y. Since we usually think of obtaining values of y from values of x, we emphasize this dependency by calling x the **independent variable** and y the **dependent variable.** The last two points in the comments that follow will help clarify this idea.

- A formula such as $y = \pm\sqrt{x}$ does *not* define y as a function of x, because it assigns two values of y to each nonzero value of x. For example, the values of $+2$ and -2 are assigned to y for $x = 4$. (However, we note that $y = \sqrt{x}$ or $y = -\sqrt{x}$ does define a function.)
- The expression $y = 8$ defines a function, because y has the value 8 for every value for x. The definition does not require that y have a different value for each x, but only that y have a *single* value for each x.
- The expression $x = 5$ is not a function, because many values of y correspond to $x = 5$. In fact, since y does not occur in the expression, any value of y may be chosen as a "corresponding" value of x.
- The set $\{(2, 3), (-1, 4), (0, -5), (3, 4)\}$ is a function, because no two distinct pairs have the same first element and different second elements. The domain is $\{2, -1, 0, 3\}$ and the range is $\{3, 4, -5\}$.
- The set $\{(-3, 4), (2, 5), (2, -6), (9, 7)\}$ is not a function, because two of the pairs—namely, $(2, 5)$ and $(2, -6)$—have the same first element and different second elements. Therefore, two different values of y are assigned to the same value of x.

The domain of a function can be limited or restricted in an arbitrary manner. For example, the domain of the function whose defining equation is $y = 5x + 2$ may be arbitrarily limited to the positive integers, in which case the range is limited to the integers greater than 2. So-called "restricted" domains are common in functions that represent physical relationships involving discrete numbers of people, cars, dollars, and so on; in this case, the domain is usually restricted to the set of positive integers. However, just as with the permissible values of the variable in Section 1.1, *if the domain is not specified, we assume that it consists of all real numbers for which the rule of correspondence generates a real number.*

Example 1 The equation $y = x^2 + 5$ defines a function, because each value of x determines only one value of y. The domain consists of all real numbers, and the range consists of only those real numbers greater than or equal to 5, since the smallest value of x^2 is 0. ∎

Example 2 The expression $y = \sqrt{x}$ defines a function whose domain and range are both the set of nonnegative real numbers. Notice that x cannot be a negative number in this case, because the square root of a negative number is not a real number. Also, notice that the value of a radical such as $\sqrt{4}$ is 2, not ± 2, since *only the positive value is given by the radical.* ∎

Example 3 Find the range for $y = 2x - 3$, if the domain is $-1 \leq x \leq 4$.

Solution We note that the smallest range-value will correspond to $x = -1$ and the largest range-value will correspond to $x = 4$. Therefore, the range is the set of real numbers in the interval $-5 \leq y \leq 5$. ∎

Example 4 Find the domain and range of the function $y = \dfrac{4}{x - 3}$.

Solution The only limitation on x is that $x - 3$ cannot equal 0. Since division by zero is not permissible, we must exclude 3 as a member of the domain; thus, the domain consists of all the reals except $x = 3$. To find the range we solve for x and note any restrictions on y. Thus,

$$y = \frac{4}{x - 3}$$

$$xy - 3y = 4 \qquad \text{Multiplying both sides by } x - 3$$

$$xy = 3y + 4 \qquad \text{Adding } 3y \text{ to both sides}$$

$$x = \frac{3y + 4}{y} \qquad \text{Dividing both sides by } y$$

From this, we see that the only limitation on y is that it cannot equal zero. Therefore, the range consists of all real numbers except zero. ∎

Example 5 Functions can be given by tables of ordered pairs of numbers, but only if the ordered pairs are unique and no two distinct pairs have the same first element. The following table defines a function:

x	-2	-1	0	1	2
y	0	1	2	0	5

In this case, the domain is $\{-2, -1, 0, 1, 2\}$ and the range is $\{0, 1, 2, 5\}$. The table is the rule of correspondence for this function. ∎

In many physical situations the domain of the function is restricted by the nature of the problem. For instance, the height above ground of a ball thrown upward from ground level with an initial velocity of 32 ft/sec is described by

$$h = 32t - 16t^2$$

where t is the elapsed time in seconds and h is the vertical height of the ball in feet. The ball will strike the ground when $h = 0$; that is, when $32t - 16t^2 = 0$. Solving this equation for t yields $t = 2$, so the ball strikes the ground 2 sec after it is thrown

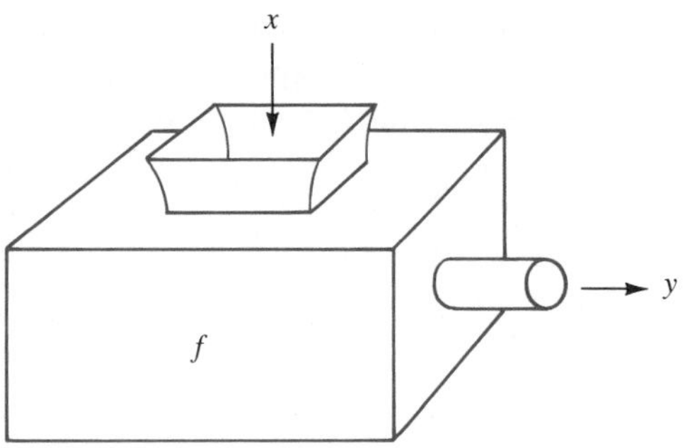

Figure 1.3

upward. Thus, the domain of this function is $0 \le t \le 2$, because the equation has no relevance to the physical problem outside of this interval.

A useful analogy compares a function to a machine that has an input and an output, as in Figure 1.3. When an element, x, enters the machine, it is transformed by the machine (function) into a new element, y. The set of elements that can be put into the machine represents the domain of the function and the set of outputs represents the range.

Example 6 An ice-vending machine dispenses a 5-lb bag of ice for \$1, a 10-lb bag for \$2, and a 15-lb bag for \$3. This machine represents a function in which the input to the machine (domain) is restricted to \$1, \$2, and \$3. The corresponding output (range) is then 5 lb of ice, 10 lb of ice, and 15 lb of ice. (See Figure 1.4.)

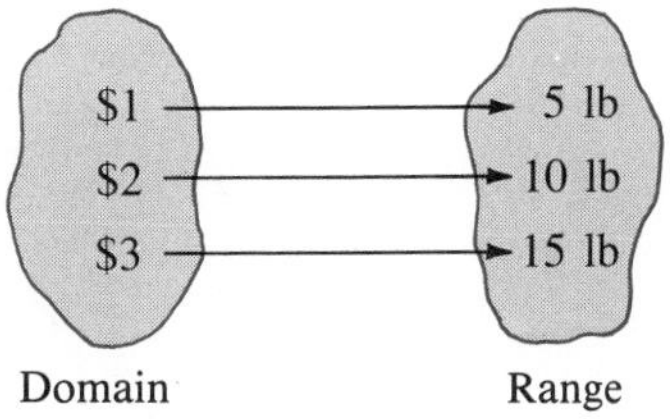

Figure 1.4

■

Functional Notation

Functions are customarily denoted by single letters such as f, F, g, G, h, or H. If x is an element in the domain of f, the corresponding value of the range element is denoted by $f(x)$, which is read "f of x" or "the function evaluated at x." A function whose rule of correspondence is $y = 5x + 2$ can also be written as $f(x) = 5x + 2$. To find the number in the range associated with a given numerical value of x, merely replace the letter x, wherever it occurs in the expression for $f(x)$, by the number from the domain. For example, if $f(x) = 5x + 2$, then $f(2) = 5 \cdot 2 + 2 = 12$, and $f(-4) = 5 \cdot (-4) + 2 = -18$.

WARNING: *The symbol f(x) does NOT mean the product of f times x. It is a symbol for the range element corresponding to the domain element x.*

It is sometimes helpful to think of x as representing a blank in the formula, in which a number or expression is to be placed. If, for instance, $f(x) = x^2 + 3x$, we may think of the formula in the following form:

$$f(\quad) = (\quad)^2 + 3(\quad)$$

Then any number or expression can be placed in the blanks. For example, $f(a^2)$ is

$$f(a^2) = (a^2)^2 + 3(a^2) = a^4 + 3a^2$$

COMMENT: *The domain element, x, is often called the* ***argument*** *of f.*

Example 7 Let $f(x) = x^2 + 3x$
Evaluate:

(a) f at $x = 2$
(b) $f(a) + f(2)$
(c) f at $x = a + 2$

Solution

(a) The value of f at $x = 2$ is denoted by $f(2)$. We substitute 2 for x to obtain $f(2) = 2^2 + 3 \cdot 2 = 10$.
(b) The value of f at $x = a$ is $f(a) = a^2 + 3a$ and the value of f at $x = 2$ is $f(2) = 10$. Therefore,

$$f(a) + f(2) = a^2 + 3a + 10$$

(c) In functional notation, this is $f(a + 2)$, which means we are to substitute $a + 2$ for x. Thus,

$$\begin{aligned} f(a + 2) &= (a + 2)^2 + 3(a + 2) \\ &= a^2 + 4a + 4 + 3a + 6 \\ &= a^2 + 7a + 10 \end{aligned}$$

■

WARNING: *A common mistake is to assume that $f(a + 2)$ is equal to $f(a) + f(2)$. By comparing the results in (b) and (c) of the previous example, you can see that $f(a + 2) \neq f(a) + f(2)$.*

Functions can be combined through the operations of addition, subtraction, multiplication, and division, as shown in the next example.

Example 8 Suppose $f(x) = 2x$ and $g(x) = x + 3$. Find the sum, difference, product, and quotient of these two functions.

Solution

Sum: $f(x) + g(x) = 2x + (x + 3) = 3x + 3$

Difference: $f(x) - g(x) = 2x - (x + 3) = x - 3$

Product: $f(x) \cdot g(x) = 2x \cdot (x + 3) = 2x^2 + 6x$

Quotient: $\dfrac{f(x)}{g(x)} = \dfrac{2x}{x+3}$ ■

Example 9 If $f(x) = 2x + 3$, find and expand the product $f(x + 1) \cdot f(5x)$.

Solution Notice that $f(x + 1) = 2(x + 1) + 3 = 2x + 5$ and $f(5x) = 2(5x) + 3 = 10x + 3$. Therefore,

$$f(x + 1) \cdot f(5x) = (2x + 5)(10x + 3) = 20x^2 + 56x + 15$$ ■

Composition

The previous examples show that functions can be combined by arithmetic operations. Another way to combine functions is called **composition.** In a composition of functions, the range of one function is used as the domain of the other function.

DEFINITION Composition of Functions

Let f and g be functions such that the range of g is in the domain of f. Then **$f(g(x))$** is called the **composition of f with g.** Similarly, if the range of f is in the domain of g, then **$g(f(x))$** is called the **composition of g with f.**

Example 10 Given $f(x) = 2x$ and $g(x) = x + 3$, find the compositions of f with g and g with f.

Solution The composition of f with g is

$$f(g(x)) = f(x + 3) = 2(x + 3) = 2x + 6$$

The composition of g with f is

$$g(f(x)) = g(2x) = 2x + 3$$

Notice that the two compositions are not the same; that is, $f(g(x)) \neq g(f(x))$. ■

WARNING: *The composition of two functions, f and g, should not be confused with the product $f(x) \cdot g(x)$. Using the two functions of the previous example,*

$$f(x) \cdot g(x) = 2x \cdot (x + 3) = 2x^2 + 6x$$

which is not the same as either of the two compositions.

Example 11 Sally and Mel start the Hot Tee's Shirt Company to sell T-shirts at the World Series. The profit made by the company is given by $P(n) = 4n - 8000$, where n represents the number of T-shirts sold. The number of T-shirts sold is in turn given by $n(x) = \dfrac{x}{4} - 2500$, where x is the attendance at the stadium. Write the company's profit as a function of attendance.

Solution The range of $n(x) = \frac{x}{4} - 2500$, the number of T-shirts sold, is the domain of $P(n) = 4n - 8000$, the profit. Hot Tee's profit, as a function of attendance, is the composition of P with n; that is,

$$P(n(x)) = P\left(\frac{x}{4} - 2500\right) = 4\left(\frac{x}{4} - 2500\right) - 8000 = x - 18{,}000$$

■

EXERCISES SECTION 1.3

Which of the expressions in Exercises 1–8 define y as a function of x?

1. $y = 2x + 5$

2. $y = x^2$

3. $y = 10$

4. $y < 3x$

5. $y^3 = x$

6. $y^2 = x^3$

7. $y^2 = 5x$

8. $y = \frac{1}{x}$

Which of the sets and tables in Exercises 9–12 define y as a function of x?

9. $\{(2, 3), (-1, 4), (3, 0), (0, 4)\}$

10. $\{(-1, 0), (2, 3), (2, -2)\}$

11.

x	1	1	2	3
y	2	3	4	7

12.

x	2	7	8
y	3	9	3

In Exercises 13–24, find the domain and range of the given function.

13. $y = 2x$

14. $y = -x$

15. $y = 4 - 3x$

16. $y = 3t^2 + 5$

17. $y = \sqrt{-x}$

18. $y = \frac{1}{x^2}$

19. $y = x^{1/3}$

20. $y = 1/x$

21. $y = \frac{4}{x + 3}$

22. $f(x) = \frac{5 - x}{x - 2}$

23. $\{(0, 1), (2, -1), (5, 7)\}$

24. $\{(1, 2), (3, 5), (7, 1), (12, -2)\}$

25. Find the range of $y = 2x + 3$ if the domain is restricted to $-1 \le x \le 4$.

26. Find the range of $y = -x + 1$ if the domain is restricted to $0 \le x \le 5$.

27. Given that $2 \le y \le 7$ is the range of $y = \frac{1}{3}x - \frac{1}{2}$, find the corresponding domain interval.

28. Given that $-1 \le y \le 3$ is the range of $y = 5x + 4$, find the corresponding domain interval.

29. Suppose $f(x) = 3x + 1$. Compute the following:
(a) $f(3)$ (b) $f(\pi)$ (c) $f(z)$
(d) $f(x - h)$ (e) $f(x) - f(h)$
(f) The domain number whose corresponding range number is 10.
(g) The range of the function.

30. Suppose $G(t) = t^2 - 2t + 1$. Compute the following:
(a) $G(2)$ (b) $G(-1)$ (c) $G(x^2)$
(d) $G(a + b)$ (e) $G(a) + G(b)$ (f) $G(\sqrt{t})$

31. Suppose the function H is defined by the set $\{(2, 3), (5, 9), (7, -3)\}$. Compute the following:
(a) $H(2)$ (b) $H(7)$
(c) $H(3)$ (d) The range of H
(e) The domain of H
(f) Compare $H(2 + 5)$ to $H(2) + H(5)$.

32. Given that $f(x) = 3x - 2$, find and expand $f(x + 3) \cdot f(2x)$.

33. Given that $g(x) = 2x - 7$, find and expand $g(x - 4) \cdot g(x - 1)$.

34. Given that $h(t) = t^2$, find and expand $h(t + 2) \cdot h(\sqrt{t})$.

35. Given that $F(x) = 3x^2$, find and expand $F(x) \cdot F(x - 2)$.

In Exercises 36–40, find $f(x) + g(x), f(x) - g(x)$, $f(x) \cdot g(x)$, and $f(x)/g(x)$.

36. $f(x) = 2x,\ g(x) = 5x$

37. $f(x) = x - 3,\ g(x) = x + 3$

38. $f(x) = g(x) = x$

39. $f(x) = x^2 - 1,\ g(x) = x + 1$

40. $f(x) = x^2,\ g(x) = x^3$

In Exercises 41–48, find $f(g(x))$ and $g(f(x))$.

41. $f(x) = x - 5,\ g(x) = 3 - 2x$

42. $f(x) = 2x + 3,\ g(x) = 5x - 1$

43. $f(x) = 3x + 4,\ g(x) = 2x$

44. $f(x) = \dfrac{1}{x+1},\ g(x) = x + 1$

45. $f(x) = -2x,\ g(x) = -x/2$

46. $f(x) = 6 - 3x,\ g(x) = -\dfrac{1}{3}x + 2$

47. $f(x) = \dfrac{1}{2x+1},\ g(x) = \dfrac{1-x}{2x}$

48. $f(x) = \dfrac{1}{x^2+1},\ g(x) = \dfrac{1}{x}$

1.4 GRAPHING

In this section we will see how to identify a function with its geometric counterpart, called the **graph of the function.** At its most fundamental level, a function is conceptualized as a set of ordered pairs of numbers (x, y), where the word "ordered" is necessary to distinguish between the pairing (3, 7), which means $x = 3$, $y = 7$, and a (totally different) pairing (7, 3), which means $x = 7$, $y = 3$.

To graph a function, therefore, we need a system sensitive to this ordered-pair structure. Recall from algebra that the "number line" associates single numbers with points on a line; the system used to graph functions associates ordered pairs of numbers with points in a plane. This system is called the **rectangular coordinate system.**

COMMENT: *The rectangular coordinate system is also referred to as the* ***Cartesian coordinate system,*** *after the French mathematician René Descartes (1596–1650), who was one of the first mathematicians to make a connection between ordered pairs of numbers and points in the plane.*

To construct the rectangular coordinate system, draw two mutually perpendicular number lines, causing them to intersect at zero on each line, as shown in Figure 1.5. The horizontal line is usually called the ***x*-axis,** the vertical line the ***y*-axis,** and the intersection the **origin.** The two axes, considered together, are called the **coordinate axes;** they divide the plane into four zones, or **quadrants.** The upper right quadrant is called the first quadrant, and the others are numbered consecutively counterclockwise, as in Figure 1.5. The axes themselves are not considered to be in any quadrant.

Points in the plane are located with respect to the origin and the axes. The horizontal displacement of a point to the right or left of the y-axis is called the *x-coordinate,* or **abscissa,** of the point. Values of x to the right of the y-axis are

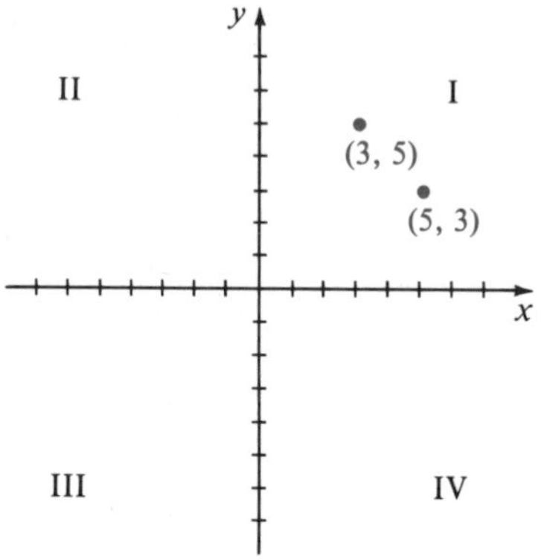

Figure 1.5

considered *positive*; those to the left, *negative*. Similarly, the vertical displacement above or below the x-axis is called the *y-coordinate*, or **ordinate,** of the point. Points above the x-axis have positive ordinates; those below, negative. When the horizontal displacement is zero, the point is on the y-axis. Similarly, when the vertical displacement is zero, the point is on the x-axis. Together, the abscissa and the ordinate of a point are called the **coordinates** of the point; they are written as the ordered pair (x, y), in which the abscissa is written as the first number. To be precise, we should not call the point an ordered pair, but it is common to blur the distinction and say "the point (x, y)" instead of "the point whose coordinates are (x, y)."

We see that a point (x, y) lies:

- In quadrant I if both coordinates are positive.
- In quadrant II, if the x-coordinate is negative and the y-coordinate is positive.
- In quadrant III if both coordinates are negative.
- In quadrant IV if the x-coordinate is positive and the y-coordinate is negative.

Notice the impact of order. The ordered pair (3, 5), for example, represents a point that is displaced 3 units to the right of the origin and 5 units above it; the ordered pair (5, 3) represents a point 5 units to the right of the origin and 3 units above it. (See Figure 1.5.) Thus, with the rectangular coordinate system, we can identify a point in the plane with any ordered pair. This point is called the **graph** of the ordered pair.

Example 1 Locate the points $P(-1, 2)$, $Q(2, 3)$, $R(-3, -4)$, $S(3, -5)$, and $T(\pi, 0)$ in the plane.

Solution $P(-1, 2)$ is in quadrant II, because the x-coordinate is negative and the y-coordinate is positive.

$Q(2, 3)$ is in quadrant I, because both coordinates are positive.

$R(-3, -4)$ is in quadrant III, because both coordinates are negative.

$S(3, -5)$ is in quadrant IV, because the x-coordinate is positive and the y-coordinate is negative.

$T(\pi, 0)$ is not in any quadrant, but lies on the positive x-axis.

The points are plotted in Figure 1.6.

Figure 1.6

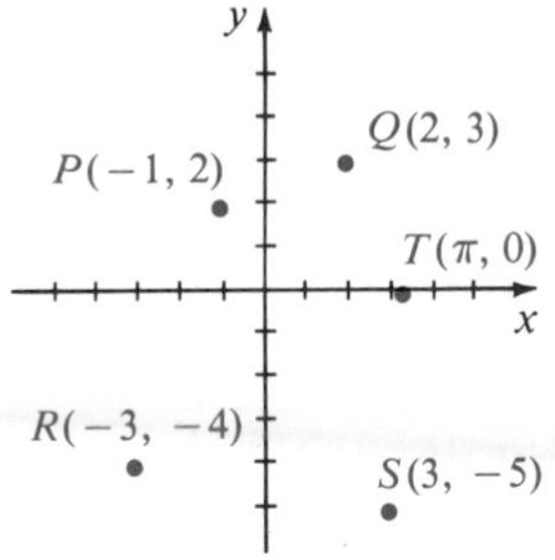

■

CALCULATOR COMMENT: *Calculators are now available that will display a rectangular coordinate system in which points may be plotted. The coordinate system can be altered by adjusting the scale on the axes. This is done by defining a minimum and maximum for each axis. The point* $(2, -3)$ *is shown below for two different scale factors. The figure on the left shows the axes for* $-10 \leq x \leq 10$ *and* $-10 \leq y \leq 10$*. The same point is displayed in the figure on the right, only in this case the axes are* $-1 \leq x \leq 4$ *and* $-4 \leq y \leq 1$.

The screen shown in the figure is that of a TI-81, but other brands will give similar displays.

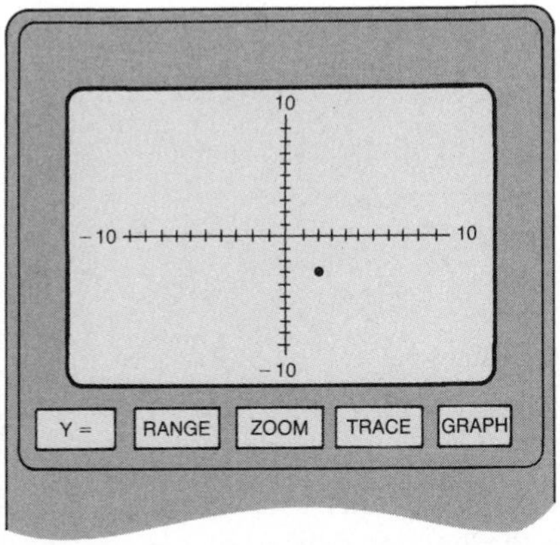

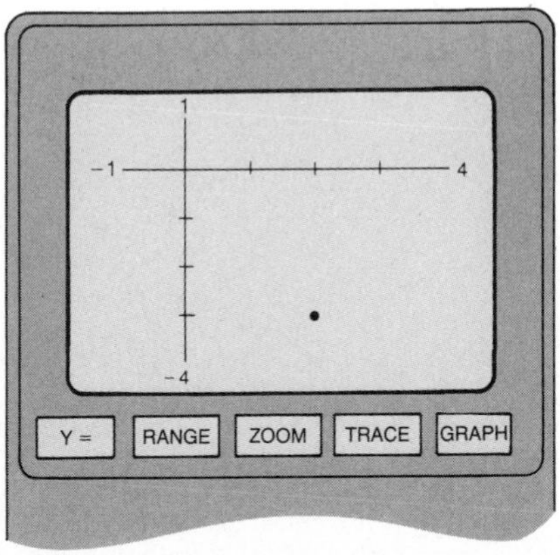

When an entire set of ordered pairs is plotted, the corresponding set of points in a plane is called the graph of the set. If the set of ordered pairs defines a function, then the points in the plane that represent the set are called the **graph of a function.** Traditionally, the domain corresponds to the abscissas and the range corresponds to the ordinates.

Example 2 Graph the function given by $\{(-4, -2), (-2, -1), (0, 0), (2, 1), (4, 2)\}$. Indicate any pattern exhibited by the points of the graph.

Solution The graph is shown in Figure 1.7. The points appear to fall along a straight line. (However, do not connect the points, because this function has a restricted domain.)

Figure 1.7

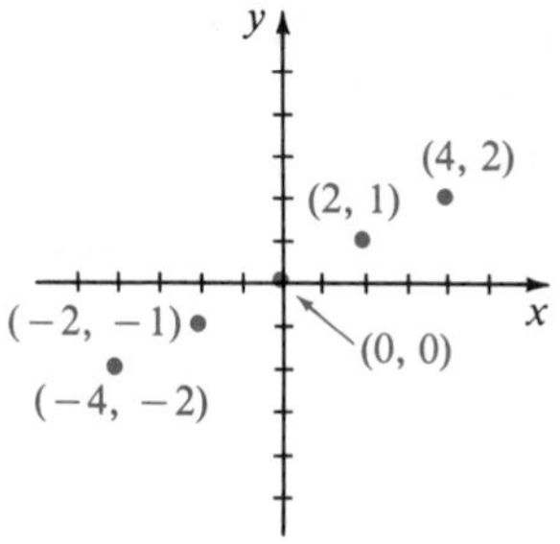

Example 3 Graph the function g whose domain is $\{0, 1, 4, 9\}$ and whose rule of correspondence is $g(x) = -\sqrt{x}$.

Solution From the domain and the rule of correspondence we can find all the ordered pairs that constitute this function. Thus,

$$g = \{(0, 0), (1, -1), (4, -2), (9, -3)\}$$

The graph of g is shown in Figure 1.8.

Figure 1.8

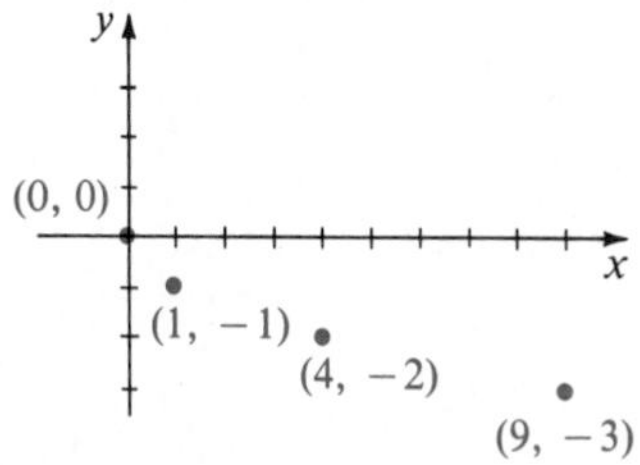

The domain of a function is usually some interval of real numbers, or even the complete set of real numbers. To graph such functions, we plot a few selected points and then connect these points with a smooth curve. As an illustration of this technique, consider the function defined by $y = x^2$. The domain of this function is the complete set of real numbers. By assigning some convenient integer values to x, we get the following table of values:

x	-3	-2	-1	0	1	2	3
y	9	4	1	0	1	4	9

Next, we plot these ordered pairs in the plane, as shown in Figure 1.9(a). The graph of $y = x^2$ is then obtained by connecting the points with a smooth curve, as shown in Figure 1.9(b). By connecting the selected points with a smooth curve, we are representing the fact that the domain of the function is the complete set of real numbers and not just the points that were plotted.

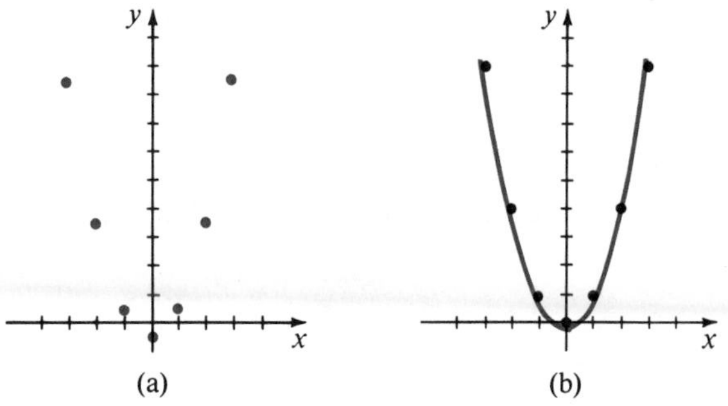

Figure 1.9

For most functions, only a segment of the graph near the origin can reasonably be plotted. That segment of the graph is meant to convey an idea of what the shape of the graph is over a more extended interval.

Example 4 Graph the function $y = \sqrt{4 - x}$.

Solution We note that the domain of this function consists of all real numbers x for which $4 - x \geq 0$, that is, $x \leq 4$. A brief table of values gives the four points shown in Figure 1.10.

x	4	2	0	-2
y	0	$\sqrt{2}$	2	$\sqrt{6}$

The graph of $y = \sqrt{4 - x}$ is then obtained by connecting these four points with a smooth curve.

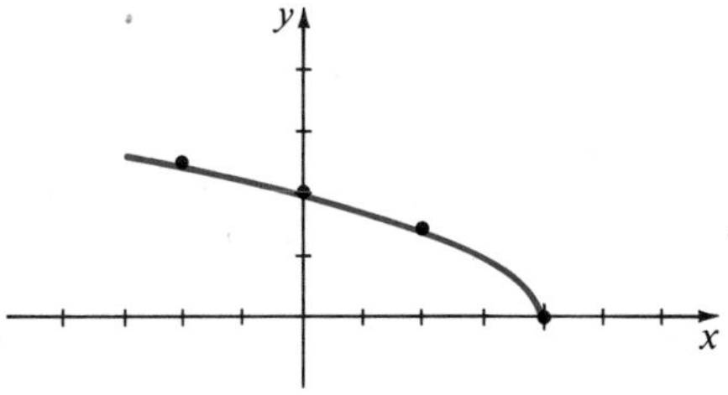

Figure 1.10

■

Example 5 Graph the function defined by $y = 3$.

Solution This is the set of points in the plane whose ordinate, y, is 3. There are no limitations on x. The graph is shown in Figure 1.11.

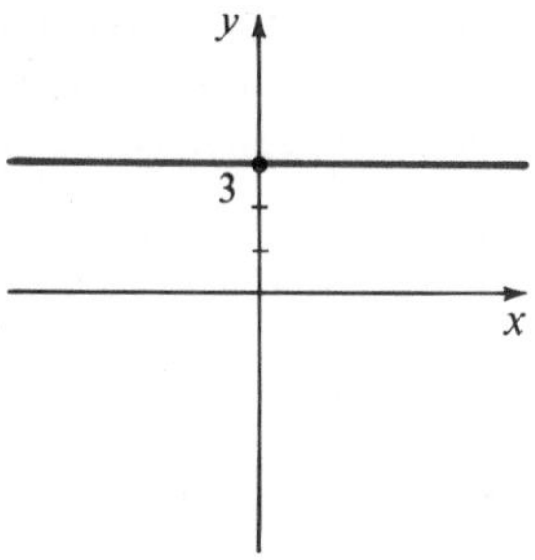

Figure 1.11

■

CALCULATOR COMMENT: *A graphing calculator can be used to display the graphs of functions. Simply enter the formula for the function as required for a specified brand and push the graph key. The following diagram shows the display of the graph of* $y = \sqrt{4 + x}$ *with* $-8 \le x \le 8$ *and* $-5 \le y \le 5$.

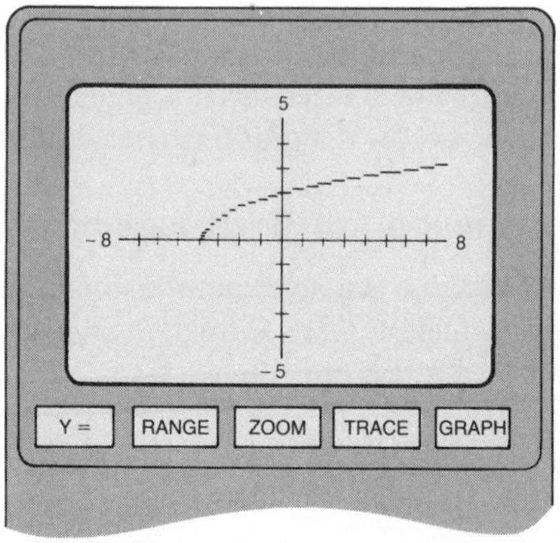

In business, the relationship between the price of an item and the number of items purchased by the consumer (in a fixed period of time) is called the **demand function.** The relation between the price and the number of items supplied by the producer is called the **supply function.** If n is the number of items and p is the price, we can express either a supply or demand function in the form

$$p = f(n)$$

Typical supply and demand curves are shown in the next example.

Example 6

(a) Draw the graph of the demand function given by $p = 8000 - \frac{1}{2}n^2$ for $0 \le n \le 100$.

(b) Draw the supply curve corresponding to $p = 100 + 5\sqrt{n}$ for $0 \le n \le 900$.

Solution Some convenient values of n are used to compute the values in the tables. Although the graphs of supply and demand curves consist of sets of discrete points, it is common practice to show a smooth curve through these points, as in Figure 1.12.

$p = 8000 - \frac{1}{2}n^2$

n	p
0	8000
20	7800
40	7200
60	6200
80	4800
100	3000

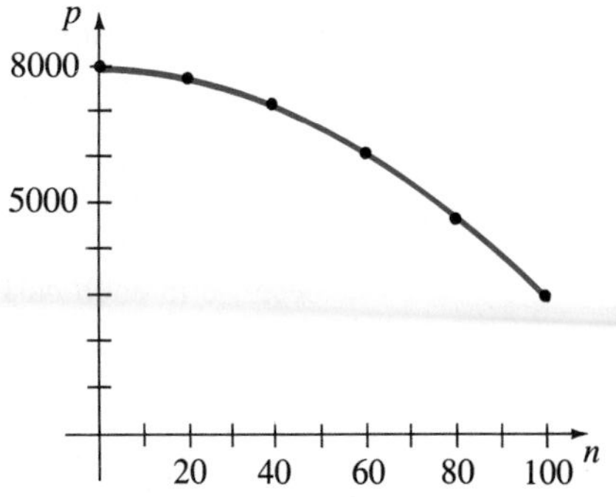

(a) Demand curve

Figure 1.12

$p = 100 + 5\sqrt{n}$

n	p
0	100
100	150
400	200
900	250

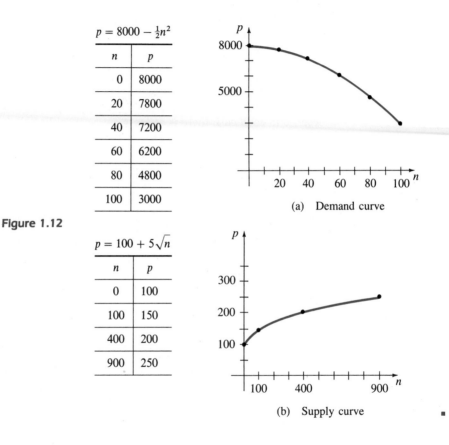

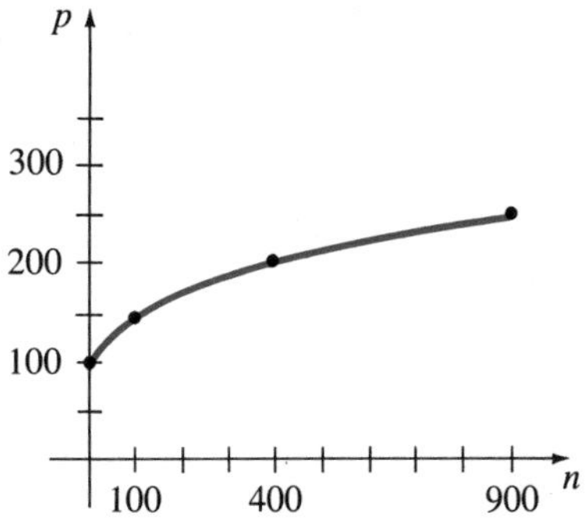

(b) Supply curve

■

The supply and demand curves shown in Figure 1.12 characterize the conditions in a free market: the demand curve shows that as the price decreases, the demand for the item will increase; the supply curve shows that as the price increases, the supply will also increase.

Using a Graph to Define a Function

A graph can be used to define a function; however, care must be taken to ensure that a particular graph actually represents a function. For a graph to define a function, there must be a unique value of y for any x in the domain of the function.

A simple test, called the **vertical line test,** can tell you whether a graph defines a function.

> **DEFINITION Vertical Line Test**
>
> If a vertical line will intersect a graph at more than one point for any value of x, the graph does NOT define y as a function of x.

Each of the graphs in Figures 1.13(a) and (b) defines y as a function of x. On the other hand, the graphs in Figures 1.13(c) and (d) do not define functions, because in each case there is at least one vertical line that intersects the graph in two or more points, indicating that there is more than one value of y for some value of x.

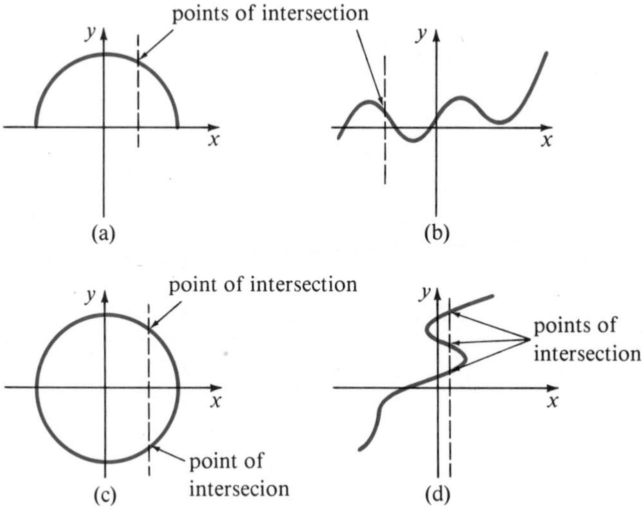

Figure 1.13

Graphs of Inequalities

The use of graphs is not limited to lines that represent functions or relations. Entire regions of the plane are sometimes used as graphs. The next example shows how a region of the plane is used to graph an inequality.

Example 7 Graph the set of points whose abscissas are greater than -1 and whose ordinates are less than or equal to 4.

Solution This set is described by a pair of inequalities, $x > -1, y \leq 4$. The shaded region in Figure 1.14 is the desired graph. Note that the solid line is a part of the region, but the broken line is not.

Figure 1.14

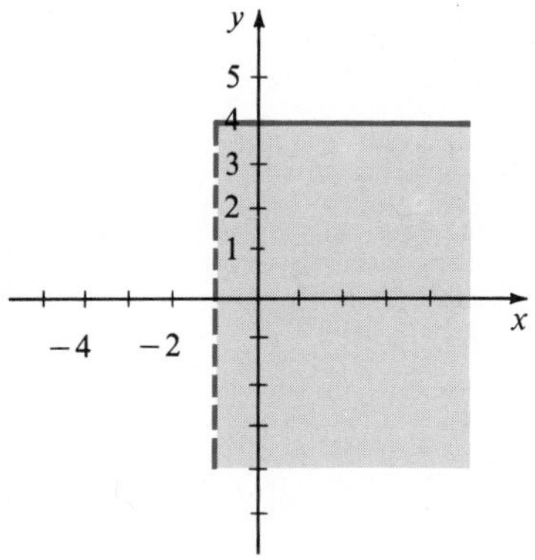

EXERCISES SECTION 1.4

In Exercises 1–6, locate the given sets of points in the plane and see if you can determine any geometric pattern exhibited by the graph.

1. $(1, 2), (2, 4), (\pi, 2\pi), (-3, -6)$

2. $(-1, -2), (-3, -6), (2, -4), (1, -2)$

3. $(2, 2), (-2, -2), (-2, 2), (2, -2)$

4. $(0, 1), (1, 0), (-1, 0), (0, -1)$

5. $(0, 2), (2, 0), (-2, 0), (0, -2), (\sqrt{2}, \sqrt{2}), (-\sqrt{2}, -\sqrt{2}), (-\sqrt{2}, \sqrt{2}), (\sqrt{2}, -\sqrt{2})$

6. $(0, 0), (-1, 1), (-2, 2), (-3, 3), (1, 1), (2, 2), (3, 3)$

In Exercises 7–16, graph the set of points defined by each given equation or inequality.

7. $x = 1$

8. $y = 2$

9. $y \le 1$

10. $x > -1$

11. $x > 2$

12. $y \le x$ and $y \ge 0$

13. $x < 2$ and $y < 1$

14. The points common to $x \ge 0$, $y \ge 0$, and $x < 1$, $y < 1$.

15. The points common to $x > -1$, $y < -1$, and $x \le -1$, $y \ge -1$.

16. All the points in the plane that are at a distance of 2 from the origin.

In Exercises 17–20, indicate which of the graphs define functions.

17. (a)

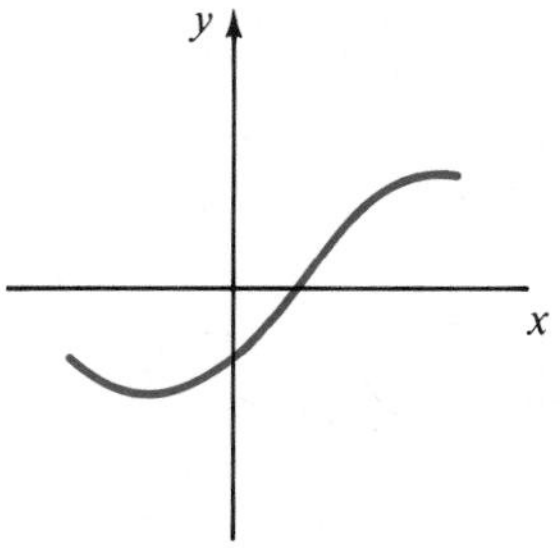

(b)

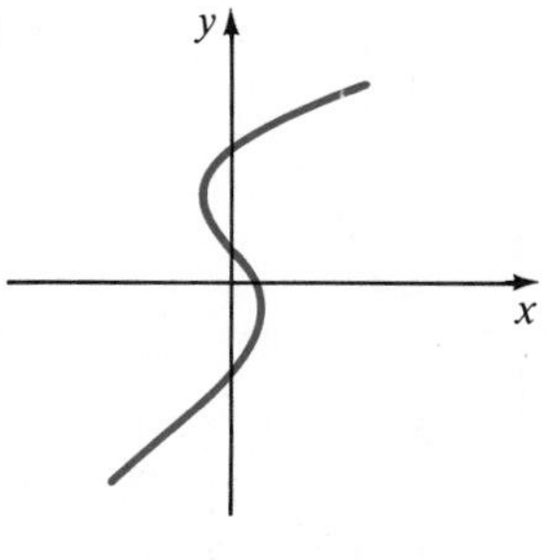

18. (a)

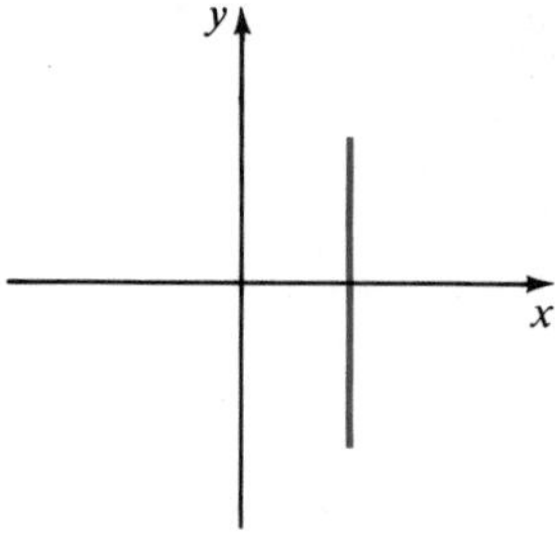

(b)

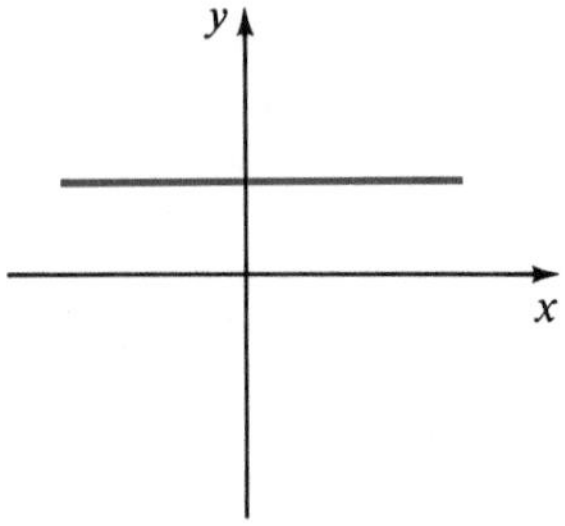

19. (a)

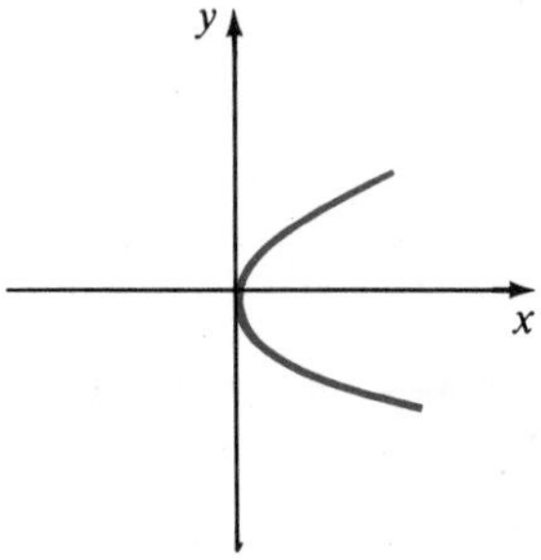

(b)

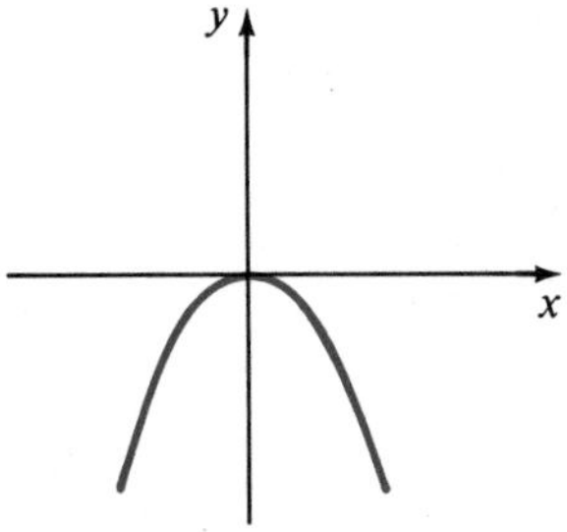

20. (a)

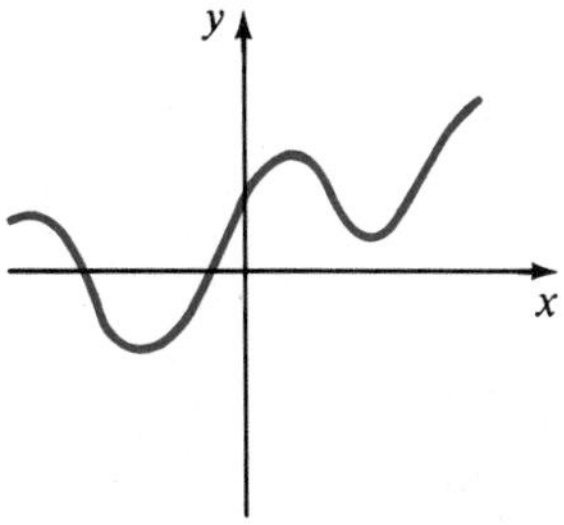

(b)

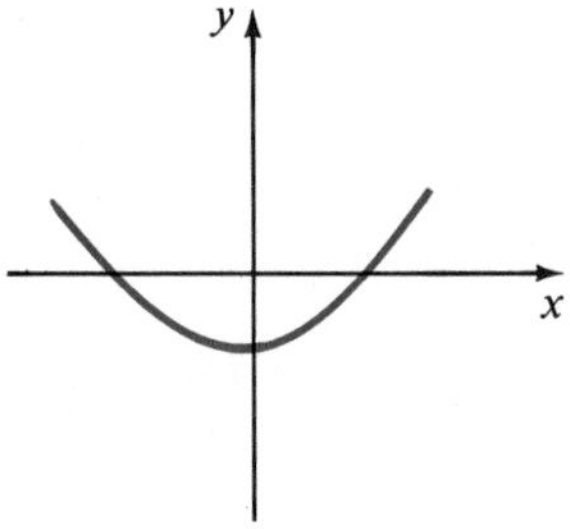

In Exercises 21–28, sketch the graph of each given function. Give the domain and range.

21. $f(x) = 3x + 5$

22. $y = 2 - \frac{1}{2}x$

23. $y = x^3$

24. $f(x) = -x^2$

25. $z = t^2 + 4$

26. $i = 3 - r^2$

27. $f(x) = \sqrt{x}$

28. $v = \sqrt{-x}$

In Exercises 29–32, sketch the graph of each given function on the indicated domain.

29. $y = -2x + 6, 0 \leq x \leq 3$

30. $y = -3x + 10, -1 \leq x \leq 2$

31. $f(x) = 0.5x + 10, 0 \leq x \leq 8$

32. $g(x) = 0.1x + 2, 0 \leq x \leq 10$

33. A biologist finds that the work w, in ergs, done by the leg muscle of a frog varies with the distance s over which it moves in centimeters, according to $w = \sqrt[3]{2s}$. Show this relationship graphically.

34. The path of a certain projectile is described by the function $h = 100x - 2x^2$, where h is the vertical height in feet and x is the horizontal displacement in feet. Show this graphically.

35. A Fax machine is to be serviced once a month. If it is not serviced, the cost of repairs is estimated by the company accountant to be \$20 plus 5 times the square of the number of months the machine goes unserviced. Express the cost of repairs as a function of the number of months the machine goes unserviced, and draw the graph.

36. A typical demand curve in economics is given by $D = \frac{50}{x + 10}$. Sketch this function for $0 \leq x \leq 10$.

37. A typical supply function in economics is given by $S = \frac{1}{2}x^2 + 20$. Sketch this function for $0 \leq x \leq 6$.

38. A traffic engineer uses the following table to describe the functional relationship between the safe speed at which a car can round a curve and the degree of the curve. Draw the graph of safe speed as a function of the degree of the curve.

Degrees of curve	5	10	15	20	25	30	35
Safe Speed (mph)	70	68	66	63	58	50	34

39. A thermocouple is a device that generates a voltage, in millivolts, when its two ends are kept at different temperatures. In laboratory experiments, the cold end of the thermocouple is usually kept at 0°C while the temperature of the other end varies. The table below represents the results of a physics lab experiment in which the output voltage of the thermocouple was recorded for various thermocouple temperatures. Draw the graph of voltage vs. temperature. Use the graph to estimate the temperature that produces a voltage of 5 mv.

T (°C)	0	50	100	150	200	250
V (mv)	0	2.1	4.1	5.9	7.0	7.8

40. The following table shows men's height and corresponding normal weights in pounds. Draw the graph of weight vs. height.

Ht	5′2″	5′4″	5′6″	5′8″	5′10″	6′0″	6′2″	6′4″
Wt	130	136	144	152	161	170	184	196

41. Profits of a new advertising firm are predicted at

$$P(t) = 2550t^2 - 6782$$

where t is the time in years and P is the profit in dollars.

(a) Find the profit after 2 years.

(b) Sketch $P(t)$ from $t = 0$ to $t = 4$ years.

(c) How many years from now will the company show a profit?

42. A physiologist is interested in how pulse rate increases during a planned 5-min exercise program. The subject's pulse rate is measured before the exercise program begins and then at 1-min intervals during the program.

t (min)	0	1	2	3	4	5
p (beats/min)	58	73	98	118	128	133

(a) Draw the graph of pulse rate versus time.

(b) Make an estimate of when the subject's pulse rate will be 90 beats/min.

Graphing Calculator Exercises

If you have a graphing calculator, use it to display the following line segments with the given endpoints. Choose a scale for the axes that will show the entire line segment and will give a clear indication of the endpoints.

1. $(1, 2), (5, 4)$

2. $(-3, 4), (4, -2)$

3. $(0, -0.5), (0.1, 0.3)$

4. $(-1, 1), (0.3, 0.2)$

5. $(-2, 15), (1, 3)$

6. $(0, 20), (-4, 10)$

7. $(13, 25), (18, 30)$

8. $(-8, 15), (-15, -7)$

Use a graphing calculator to display the graph of the following functions. Be sure to choose an appropriate scale.

9. $y = x^2$

10. $y = x - x^2$

11. $y = \sqrt{x + 1}$

12. $y = x^2 - x - 6$

13. $y = 10 + 2x$

14. $y = 3 - x$

15. $y = x^3 - 2x^2 + 5$

16. $y = 5 - x + x^3$

17. $y = \dfrac{1}{x - 2}$

18. $y = \dfrac{2}{x + 1}$

1.5 Linear Functions

Earlier in this chapter we discussed equations of the form

$$ax + b = 0, \quad a \neq 0 \qquad \textbf{(1.1a)}$$

which is the *general form* of a linear equation in one variable. In this section, we expand this discussion to include linear equations in two variables. The general form of a **linear equation** in two variables x and y is

$$Ax + By = C \tag{1.1b}$$

where A and B cannot *both* be zero. If $B \neq 0$, we can solve this equation for y, so that $y = \frac{-A}{B}x + \frac{C}{B}$. If we replace y with $f(x)$, this linear equation can be expressed as a **linear function,**

$$f(x) = \frac{-A}{B}x + \frac{C}{B}$$

or as

$$f(x) = ax + b \tag{1.2}$$

where a and b are constants. If $B = 0$ in Equation 1.1b, then $Ax + By = C$ reduces to $Ax + C = 0$, which is a linear equation in one variable, x, but not a linear function.

The phrases *linear function* and *linear equation in two variables* often mean the same thing. In fact, the words *function* and *equation* are used interchangeably for any equation that can be solved for y in terms of x. This, of course, implies that y appears in the equation or, in other words, that the coefficient of y is not zero. We will use the phrase *linear function* if we are stressing the functional concept; otherwise, we will use the phrase *linear equation.*

Example 1

(a) The equations

$$y = x, \qquad y = 2x - 5, \qquad 3x + 2y = 7, \qquad y = 0, \qquad x = 3$$

are examples of linear equations. We note that all but $x = 3$ (why?) are also linear functions.

(b) Equations and functions that are not linear are said to be *nonlinear*. Examples of nonlinear equations are

$$y = \sqrt{x}, \qquad y = x^2 - 3x, \qquad y = \frac{1}{x}$$

■

Any pair of numbers x and y for which the equation $Ax + By = C$ is satisfied is called a **solution** of the equation. (Compare this definition to that of an equation in one unknown in Section 1.1.) For example, the pair $x = 3$, $y = -2$ is a solution to the linear equation $4x + y = 10$, because $4(3) + (-2) = 10$. Other solutions are $x = 0$, $y = 10$, and $x = 1$, $y = 6$. (Check each pair for yourself.) In fact, there are an infinite number of such pairs that satisfy the equation $4x + y = 10$.

COMMENT: *Notice that a linear equation in one variable, such as $2x - 8 = 0$, has a unique solution, $x = 4$, whereas a linear equation in two variables, such as $4x + y = 10$, has infinitely many solutions. This is a major distinction between a linear equation in one variable and a linear equation in two variables.*

Example 2 List some of the solutions of the linear equation $2x + y = 2$.

Solution The usual way of finding solutions of an equation in two variables is to assign values to x and calculate the corresponding values of y. For the given equation, when $x = 0$, $y = 2$; when $x = 1$, $y = 0$; and so on. Some of the solutions are shown in the table. To list them all is not possible, since the solution consists of infinitely many ordered pairs.

x	0	1	-1	2	-2
y	2	0	4	-2	6

■

COMMENT: If $a \neq 0$, then the domain and range of $f(x) = ax + b$ are both the complete set of real numbers. If $a = 0$, then $f(x) = b$; in this case, the domain is still the entire set of real numbers (the independent variable is not restrained in any way), but the range is $f(x) = b$.

To graph a linear equation or function, we compute some of the ordered-pair solutions, plot the corresponding points, and connect them with a smooth curve.

Example 3 Graph the linear equation $2x + y = 2$.

Solution The table of values from Example 2 may be used here to obtain Figure 1.15.

Figure 1.15

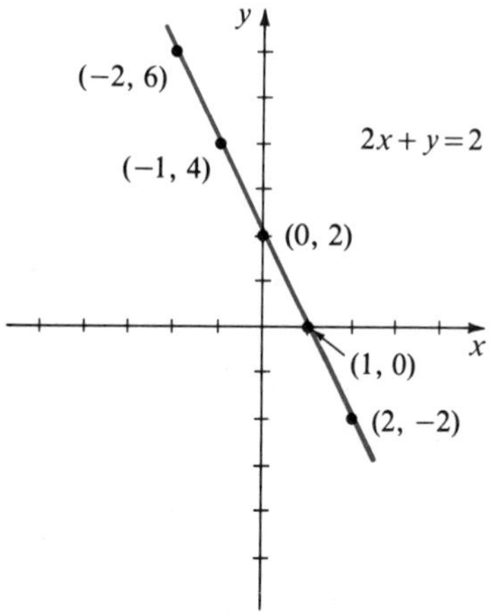

■

Figure 1.15 suggests that the graph of the linear equation $2x + y = 2$ is a straight line. More generally, we can show that **the graph of every linear equation is a straight line.** In fact, the terms *linear equation* and *linear function* are chosen to suggest the geometric nature of their graphs.

Since the graph of every linear equation is a straight line, and since a straight line is completely determined by two distinct points, we draw the graph of a linear equation by locating any two points on its graph and then drawing a straight line through the points. Plotting any more points is redundant! Often the two easiest points to find are those where the graph intercepts the x- and y-axes, called the **x-intercept** and the **y-intercept,** respectively.

x- and y-Intercepts for Functions of Two Variables

- To find the x-intercept, let $y = 0$ and solve for x.
- To find the y-intercept, let $x = 0$ and solve for y.

Example 4 Sketch the graph of $y = 2x - 4$.

Solution The two intercepts are obtained as follows:

- Letting $y = 0$ gives $2x - 4 = 0$, or $x = 2$.
- Letting $x = 0$ gives $y = -4$.

Thus, the graph passes through (2, 0) on the x-axis and (0, -4) on the y-axis. A straight line is then drawn through these points, as shown in Figure 1.16.

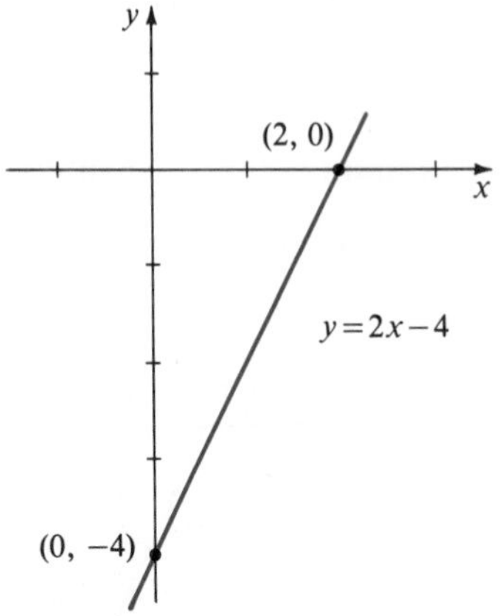

Figure 1.16

Example 5 A warehouse has 60,000 ft^2 of usable storage space. Frigidaire leases the space to store refrigerators and stoves. Each refrigerator requires 6 ft^2 and each stove requires 5 ft^2. Write an equation that describes the number of each kind of appliance that can be stored in the warehouse if it is filled to capacity.

Solution Since the total space used for refrigerators and that used for stoves must add up to 60,000 ft^2, the basic model for this problem is

$$\boxed{\begin{matrix}\text{Total space}\\ \text{for}\\ \text{refrigerators}\end{matrix}} + \boxed{\begin{matrix}\text{Total space}\\ \text{for}\\ \text{stoves}\end{matrix}} = \boxed{60{,}000}$$

Now let x be the number of refrigerators and y the number of stoves. Since each refrigerator requires 6 ft^2, the space required for x refrigerators is $6x$ ft^2. Similarly, the space required for y stoves is $5y$ ft^2. Substituting $6x$ and $5y$ into our basic model, we have the linear equation

$$6x + 5y = 60{,}000$$

Notice that there is an implied limitation: x and y may not be negative. Since $x \geq 0$ and $y \geq 0$, the graph is restricted to the first quadrant.

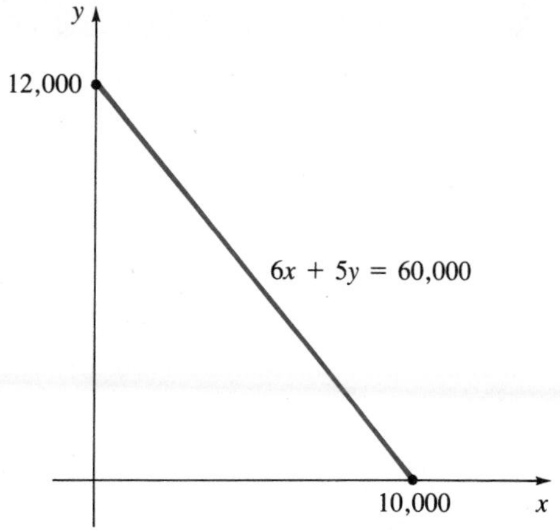

■

SLOPE OF A STRAIGHT LINE

If you were asked to describe a straight line, such as that in Figure 1.16, you might give its intercepts and you might say that the line rises from left to right; you might also give some information on the "steepness" of the rise. For instance, you might observe that the line in Figure 1.16 rises 2 units for every 1 unit moved to the right. This description of steepness is at the core of the concept of the **slope of a straight line.**

The slope of a nonvertical straight line is defined as the ratio of the vertical "rise" of the line to the corresponding horizontal "run"; that is,

$$\text{Slope} = \frac{\text{vertical rise}}{\text{horizontal run}}$$

Applying this definition to the line segment P_1P_2 in Figure 1.17, the slope of the

Figure 1.17

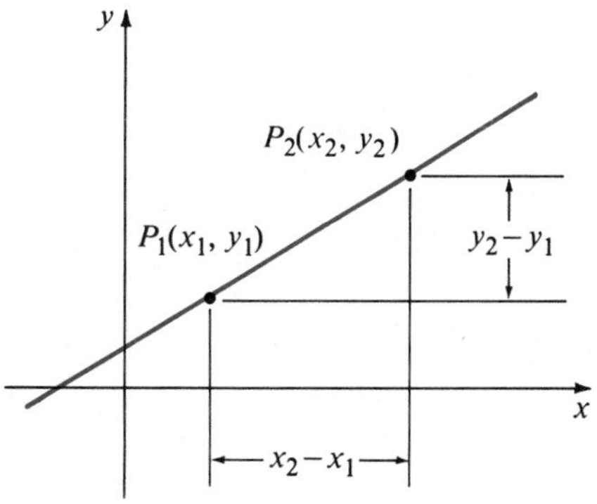

line is defined as follows:

DEFINITION Slope of a Straight Line

The slope of a straight line that passes through $P_1(x_1, y_1)$ and $P_2(x_2, y_2)$ is denoted by m and defined by

$$m = \frac{y_2 - y_1}{x_2 - x_1}, \qquad x_1 \neq x_2 \tag{1.3}$$

where $y_2 - y_1$ is the vertical distance between the points and $x_2 - x_1$ is the horizontal distance between the points.

Example 6 Find the slope of the straight line passing through the points $(-5, 1)$ and $(2, -3)$. (See Figure 1.18.)

Figure 1.18

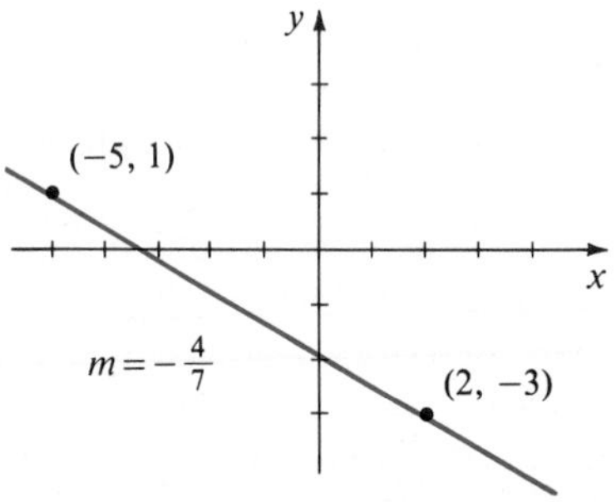

Solution Letting $(x_1, y_1) = (-5, 1)$ and $(x_2, y_2) = (2, -3)$ and using Equation 1.3, we obtain the slope as follows:

$$m = \frac{y_2 - y_1}{x_2 - x_1} = \frac{-3 - 1}{2 - (-5)} = \frac{-4}{7} = -\frac{4}{7}$$

If we interchange the labels of the given points and let $(x_1, y_1) = (2, -3)$ and $(x_2, y_2) = (-5, 1)$, the result is

$$m = \frac{y_2 - y_1}{x_2 - x_1} = \frac{1 - (-3)}{-5 - 2} = \frac{4}{-7} = -\frac{4}{7}$$

Hence, the order in which we label the given points is immaterial. We interpret a slope of $-\frac{4}{7}$ to mean that for every 7 units moved to the right, the straight line moves down 4 units, or for every 7 units moved to the left, the straight line moves up 4 units. ■

In working with straight lines, the following generalizations about slopes are useful:

- The slope of a straight line is positive if, as you follow the curve from left to right, you move up relative to the x-axis. (See Figure 1.19(a).)
- The slope of a straight line parallel to the x-axis is zero, because the rise is zero for *any* run. (See Figure 1.19(b).)
- The slope of a straight line is negative if, as you follow the curve from left to right, you move down relative to the x-axis. In Figure 1.19(c), we see that when the run is positive, the rise is negative, and when the run is negative, the rise is positive. The ratio of rise to run (that is, the slope) in either case must be negative.
- The slope of a straight line parallel to the y-axis is undefined, because the run is zero for *any* rise. Therefore, to use Equation 1.3 we would have to divide by zero, and this is an undefinable operation. (See Figure 1.19(d).)
- Parallel straight lines have equal slopes.
- Perpendicular straight lines have slopes that are negative reciprocals. That is, $m_1 = -\dfrac{1}{m_2}$ $(m_2 \neq 0)$.

Figure 1.19

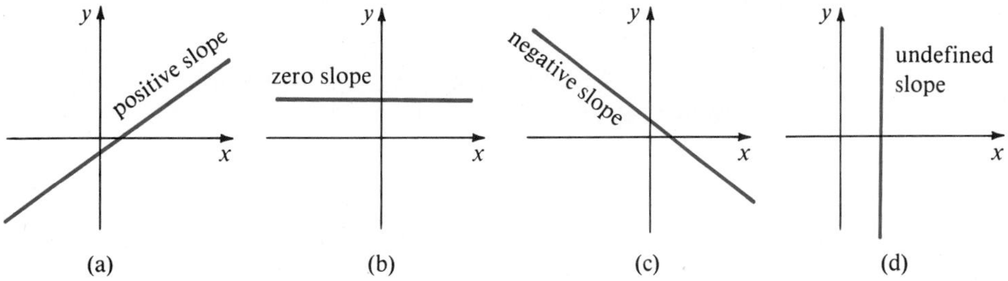

Example 7

(a) In Figure 1.20, find the slope of the line passing through (2, 1) and (−4, 6).

Figure 1.20

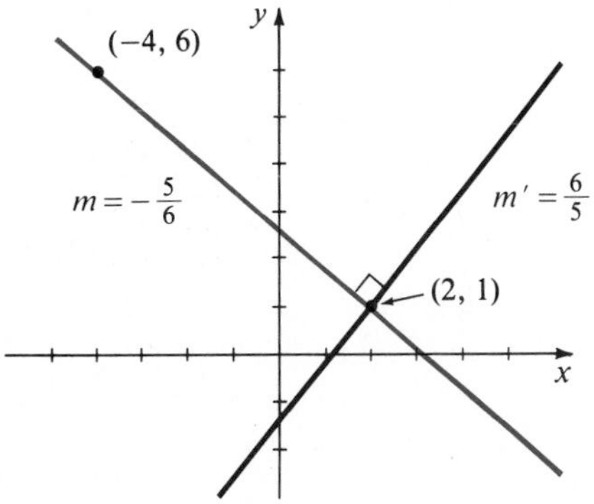

(b) Find the slope of the line drawn perpendicular to the given line at (2, 1).

Solution

(a) The slope m of the given line is

$$m = \frac{y_2 - y_1}{x_2 - x_1} = \frac{6 - 1}{-4 - 2} = -\frac{5}{6}$$

(b) The slope m' of a line perpendicular to the given line is then the negative reciprocal of m; that is,

$$m' = -\frac{1}{m} = -\frac{1}{-\frac{5}{6}} = \frac{6}{5}$$

■

The concept of slope is defined as the ratio of the *change* of two quantities. Such ratios exist in many areas where mathematics is applied. The slope expresses the ratio of a change in y to the corresponding change in x, so we also refer to this ratio as the *rate of change of y with respect to x.*

Example 8 The average velocity v of an object is defined as the distance traveled divided by the corresponding elapsed time. If a car travels 100 mi in 2.5 hr, what is the average velocity of the car?

Solution Since the distance traveled is 100 mi and the elapsed time is 2.5 hr, the average velocity of the car is

$$v = \frac{100}{2.5} = 40 \text{ mph}$$

■

EXERCISES SECTION 1.5

In Exercises 1–12, sketch the graph of each given linear equation.

1. $x + y = 1$

2. $x - y = 1$

3. $x = 5$

4. $y = -2$

5. $2x - y = 5$

6. $3x + 2y = 5$

7. $4x - y + 1 = 0$

8. $x = 7 - y$

9. $3y - 4x = 4$

10. $y = \frac{1}{2}x - 3$

11. $y = \frac{x - 2}{3}$

12. $15(x + y) = 10$

In Exercises 13–20, sketch the straight line through each pair of points and compute the slope.

13. (1, 2), (5, 4)

14. (−5, 2), (3, 7)

15. (−1, −1), (3, −6)

16. (7, 3), (0, 5)

17. (−2, −3), (−5, −7)

18. (3, −2), (7, 6)

19. (−2, 3), (5, 3)

20. $\left(\frac{1}{2}, \frac{1}{2}\right), \left(\frac{1}{2}, -\frac{2}{5}\right)$

In Exercises 21–24, sketch the line passing through each given point with the given slope.

21. (2, 5), $m = \frac{1}{2}$

22. (−1, −3), $m = 3$

23. $(5, -2)$, $m = -7$

24. $(3, 4)$, $m = -\frac{2}{5}$

25. Find the slope of a line drawn perpendicular to the line through $(-3, 0)$ and $(5, -4)$.

26. Find the slope of a line drawn perpendicular to the line through $(1, 2)$ and $(-2, -5)$.

27. A warning on an aluminum ladder says that it is unsafe to use the ladder if it is inclined at a slope that is less than 3. Find the slope of a 25-ft ladder that is leaning against a building with its foot 12 ft from the building. Does the ladder meet the safety requirement?

28. The acceleration of an atomic particle is defined as the change in velocity divided by the corresponding change in time. If the initial velocity of a particle is 25 cm/sec and 10 seconds later its velocity is 40 cm/sec, what is its acceleration? In what units is acceleration expressed?

29. The marketing director of MicroComp estimates sales for the first six months of the year by the equation $S = 25{,}000N + 200{,}000$, where N is the number of months since the first of the year and S is sales in dollars. Plot the sales curve of the company for the first six months.

30. A surveyor finds that the elevation h of a roadway varies with the distance x from a reference point, according to the formula $h = 0.1x + 1.5$. Draw the elevation of the roadway as a function of x.

31. A sociologist studying population density uses the U. S. census data to compute the yearly population growth of selected cities for the period 1980–1990. If the 1980 population of Denver was 492,000 and the 1990 population was 468,000, what was the yearly growth rate for Denver during this period?

32. The Consumer Price Index (CPI) is a measure of the nation's inflation rate and is given as a percentage. An economist estimates the CPI for the coming year by the equation $CPI = 0.15t + 4.8$, where t is time in months. Draw a graph showing the CPI for the next twelve months.

33. A GM truck plant has a 750,000-ft^2 lot where trucks are stored as they come off the production line, waiting to be transported to dealers. One-ton trucks require an area of 150 ft^2 and 2.5-ton trucks require 250 ft^2. (a) Write a linear equation that describes the number of each kind of truck that can be stored on the lot if it is filled to capacity. (b) Draw the graph. (c) If there are 1500 2.5-ton trucks in the lot, how many 1-ton trucks will the lot hold?

34. A C5-A transport airplane has 900 ft^2 of cargo space. The Army wishes to airlift jeeps (75 ft^2) and troop carriers (150 ft^2). (a) What equation describes the number of each kind of vehicle that can be loaded? (b) Draw the graph. (c) If 6 jeeps are loaded, how many troop carriers can be loaded?

35. Using the definition of the slope of a line and properties of similar triangles, show that the product of the slopes of perpendicular lines is -1.

36. Show that the composition of a linear function with a linear function is linear.

1.6 METHODS OF DESCRIBING A LINE

In the previous section, we examined some basic characteristics of straight lines in the plane. We now examine the characteristic form of linear equations and the relationship between the form of an equation and its graph. Consider the straight line that passes through $(2, 1)$ with slope $\frac{1}{3}$ as shown in Figure 1.21. Let $P(x, y)$ be any other point on the line. By definition, the slope of the line can be written $(y - 1)/(x - 2)$, which must be equal to the given slope. Thus,

$$\frac{y-1}{x-2} = \frac{1}{3}$$

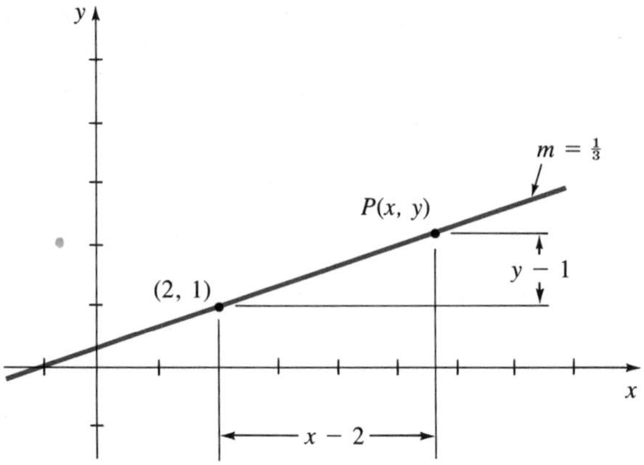

Figure 1.21

Multiplying both sides by $x - 2$, we get

$$y - 1 = \frac{1}{3}(x - 2)$$

After further simplification, we get

$$3y - x = 1$$

You should understand that any one of these three forms is the equation of the line.

Whereas every straight line can be represented by a linear equation, certain forms of representation are more useful than others. Two important forms of a straight line are the **point-slope form** and the **slope-intercept form.** The point-slope form is a generalization of the example above. To derive this form of a straight line, assume that we are given one point on the line $P(x_1, y_1)$ and that the line has slope m as shown in Figure 1.22. Choose another point on the line, represented by $P(x, y)$. By the same reasoning as above, we have

$$\frac{y - y_1}{x - x_1} = m$$

Multiplying both sides by $x - x_1$ yields the point-slope form.

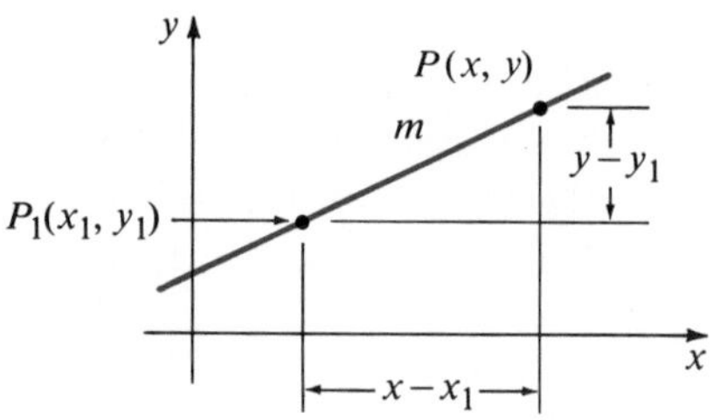

Figure 1.22

DEFINITION The Point-Slope Form of a Straight Line

The equation of a straight line passing through a fixed point (x_1, y_1) with slope m is

$$y - y_1 = m(x - x_1) \tag{1.4}$$

COMMENT: *Notice that since* any *fixed point* (x_1, y_1) *may be used in Equation 1.4, the equations will* appear different *for various choices of the fixed point, but the resulting equations will be equivalent.*

Example 1 Write the equation of the line that passes through (2, 1) and (−1, 0).

Solution First we compute the slope of the line. By the definition of slope, we get

$$m = \frac{1 - 0}{2 - (-1)} = \frac{1}{3}$$

- The point-slope form of the line passing through (2, 1) with a slope of $\frac{1}{3}$ is

$$y - 1 = \frac{1}{3}(x - 2)$$

  which can be expressed in the general form of a linear equation as $x - 3y + 1 = 0$.
- The point-slope form of the line using (−1, 0) as the fixed point is

$$y - 0 = \frac{1}{3}(x + 1)$$

  which also can be expressed in the general form $x - 3y + 1 = 0$. Therefore, even though different fixed points were used in the two cases, the two forms are equivalent. ■

If the point used in the point-slope form is the y-intercept $(0, b)$, we have a special case of the point-slope form called the *slope-intercept form.* Substituting $x_1 = 0$ and $y_1 = b$ into the point-slope form yields

$$\frac{y - b}{x - 0} = m$$

Solving for y, we have the following equation:

DEFINITION The Slope-Intercept Form of a Straight Line

The equation of a straight line passing through $(0, b)$ with slope m is

$$y = mx + b \tag{1.5}$$

COMMENT: *When an equation is in slope-intercept form, the constant on the right-hand side is the y-intercept and the coefficient of x is the slope. Given any linear equation, a few simple manipulations will result in the slope-intercept form. Simply solve for y in terms of x.*

Example 2 Rearrange the linear equation $3x + 2y = -5$ into slope-intercept form and draw its graph.

Solution To rearrange into slope-intercept form, we solve for y. Thus,

$$2y = -3x - 5$$

and

$$y = -\frac{3}{2}x + \left(-\frac{5}{2}\right)$$

from which we recognize the slope as $-\frac{3}{2}$ and the y-intercept as $-\frac{5}{2}$. (See Figure 1.23.)

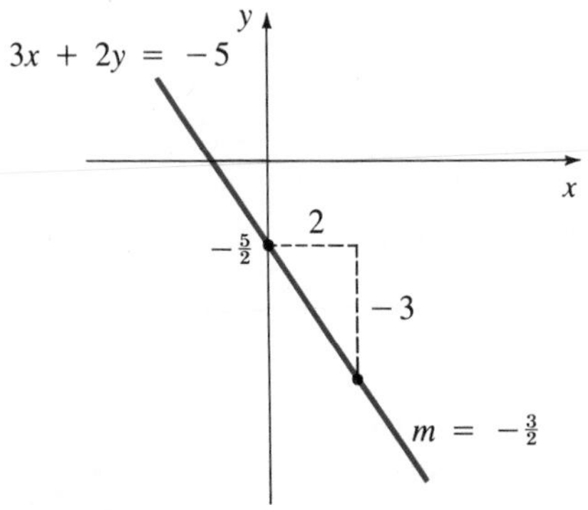

Figure 1.23

Example 3 Suppose the annual cost of depreciation and insurance for your car is a fixed amount of \$1800 and the cost of fuel and repair is \$0.12 per mile. Write the annual cost of operating your car as a linear function of miles driven. Draw a graph of this function for the first 25,000 miles.

Solution Let C = total annual cost and m = total miles driven. The cost of driving each mile is \$0.12, so the variable cost of driving m miles is $0.12m$ dollars. This is called a variable cost because it depends upon how many miles you drive during the year. Also, there is a fixed cost of \$1800 for depreciation and insurance. You pay this cost whether you drive the car or not. The total cost of operating the car is, then, the sum of the fixed and variable costs. In terms of C and m, we have the linear function

$$C = 1800 + 0.12m$$

The graph of the cost function for $0 \le m \le 25{,}000$ miles is shown in the figure. Notice that the fixed cost is the C-intercept and the cost per mile is the slope of the graph.

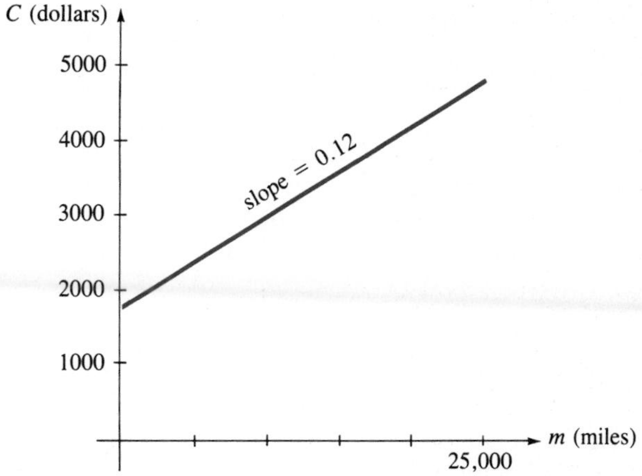

■

Example 4 Write the equation of the line perpendicular to $x - 3y = 7$ at the point $(4, -1)$.

Figure 1.24

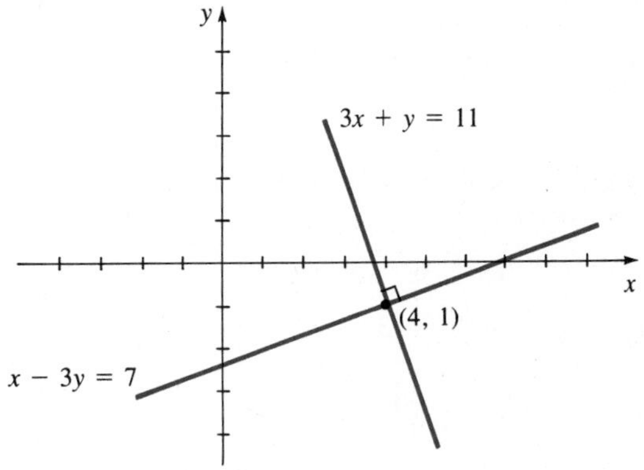

Solution First we rewrite the given equation in slope-intercept form; that is,

$$y = \frac{1}{3}x - \frac{7}{3}$$

From this form of the given equation we see that the slope of its graph is $m = \frac{1}{3}$. Denoting the slope of the perpendicular line by m', we have

$$m' = -\frac{1}{m} = -\frac{1}{\frac{1}{3}} = -3$$

Finally, using the point-slope form of a straight line, the equation of the perpendicular line is

$$y - (-1) = -3(x - 4)$$
$$y + 1 = -3x + 12$$
$$y = -3x + 11$$

The general form of the line is, then, $3x + y - 11 = 0$. ■

The general relationship between equations of lines that are perpendicular to each other is given in Exercise 34 of the exercise set below. Before looking at that exercise, see whether you can find any relationship between a given equation and that of the line perpendicular to it.

EXERCISES SECTION 1.6

In Exercises 1–4, draw the line passing through each given point with the given slope and then determine its equation.

1. $(2, 5)$, $m = \frac{1}{2}$ **2.** $(-1, -3)$, $m = 3$

3. $(5, -2)$, $m = -7$ **4.** $(3, 4)$, $m = -\frac{2}{5}$

In Exercises 5–10, draw the line through the given points and then find its equation.

5. $(1, 3)$, $(6, 2)$ **6.** $(2, 5)$, $(-3, -7)$

7. $(-1, -1)$, $(1, 2)$ **8.** $(0, 0)$, $(3, -2)$

9. $(0, 2)$, $(-5, 0)$ **10.** $\left(\frac{1}{2}, \frac{1}{3}\right)$, $\left(-\frac{1}{2}, \frac{1}{3}\right)$

In Exercises 11–22, find the slope and y-intercept of each equation and then draw the line.

11. $2x - 3y = 5$ **12.** $3x + 4y = 0$

13. $x + y = 2$ **14.** $5x + 2y = -3$

15. $4y - 2x + 8 = 0$ **16.** $-5x - y - 2 = 0$

17. $2y = x + 5$ **18.** $3x + 6 = 2y$

19. $y = 5$ **20.** $x + y = 1$

21. $x - 5y + 7 = 0$ **22.** $y = 1 + x$

In Exercises 23–32, write the equation of the line perpendicular to each given line at the indicated point.

23. $3x + 2y = 7$ at $(1, 2)$

24. $x + 3y = 11$ at $(2, 3)$

25. $x - y = 2$ at $(5, 3)$

26. $2x - 5y = 2$ at $(-4, -2)$

27. $2y - 3x = 1$ at $(1, 2)$

28. $y = x$ at $(0, 0)$

29. $5x - 7y = 0$ at $(0, 0)$

30. $5x - 7y = 3$ at $(2, 1)$

31. $-x + 3y = 5$ at $(-2, 1)$

32. $-2y - x = 3$ at $(-1, -1)$

33. Show that the equation of the line with x-intercept $(a, 0)$ and y-intercept $(0, b)$ may be written in the form

$$\frac{x}{a} + \frac{y}{b} = 1$$

This form of a line is known as the *intercept form* of a line.

(a) Write the equation $x + 3y = 6$ in intercept form.

(b) Write the equation $2x - y = 5$ in intercept form.

34. Show that the graph of $Bx - Ay = C_2$ is perpendicular to that of $Ax + By = C_1$. *Hint:* Show that the slopes are negative reciprocals.

35. The current I in a resistor is a linear function of the applied voltage V. Find the equation relating the current to the voltage if the current is $\frac{1}{2}$ amp when the voltage is 6 volts and $\frac{2}{3}$ amp when the voltage is 8 volts.

36. In an experiment to determine the coefficient of friction, it is found that a 10-lb block has a frictional force of 3 lb and a 25-lb block has a frictional force of 7.5 lb. What is the equation relating frictional force to weight if the frictional force is a linear function of weight?

37. *Linear depreciation* is one of several methods approved by the Internal Revenue Service for depreciating business property. If the original cost of the property is C dollars and if it is depreciated linearly over N years, its value, V, remaining at the end of n years is given by

$$V = C - \frac{C}{N}n$$

Find the value after 5 years of a typewriter whose initial cost of \$300 is to be depreciated over 20 years.

38. If you borrow P dollars at the simple interest rate, i, the annual interest is $P \cdot i$. Hence, the amount, A, owed at the end of n years is given by

$$A = P(1 + i \cdot n)$$

Find the amount you owe after 5 years if you borrow \$2000 at 8% simple interest.

39. Mr. Smith wants to borrow \$4000 to buy a used car. He wishes to pay off the loan with monthly payments stretching over 3 years. If he is charged 12% simple interest and if he computes the monthly payments by dividing the total amount due in 3 years by 36, how much will Mr. Smith have to pay each month?

40. A manufacturer of fountain pens can expect to sell 21,000 felt-tip pens if he charges \$.80 per pen, but only 10,000 if he raises the price to \$1.00. Assuming that the relationship is linear, find the equation of the line relating the number of pens to their price. How many felt-tip pens can he sell if he charges \$.90?

41. Express degrees Celsius as a linear function of degrees Fahrenheit if 0°C corresponds to 32°F and 100°C corresponds to 212°F.

42. The pressure at a point below the surface of a body of water is given by $P = 0.04D + 14.7$, where P is the pressure in psi and D is the depth in inches. Draw the graph of the relationship.

43. Johnson and Moore's Appliance Repair charges \$45 for a service call to repair an air conditioner, plus \$8 for each 15-min period. (a) Write the linear equation for the charges in terms of the number of 15-min periods. (b) What is the charge if the repair job takes a total of $1\frac{1}{4}$ hours to complete? (c) Sketch the graph of this linear equation.

44. Apex Truck Rental charges \$59.95 to rent a van, plus an additional charge of \$.39 for each mile. (a) Set up the linear equation relating the total rental charge to the number of miles driven. (b) Draw the graph of this linear equation.

45. The East-West Telephone Co. offers its customers a local-call plan. With this plan the local-call charge is a linear function of the number of local calls made during a billing period. (a) Determine the equation relating the charge to the number of calls if the charge is \$26.50 for 35 calls and \$31.50 for 45. (b) What is the base charge if you don't make any calls?

46. The number of cold drinks sold at the Little League park is a linear function of the temperature. Determine the equation that relates the number of cold drinks sold to the temperature in degrees Fahrenheit, if 125 drinks were sold when the temperature was 65°F and 200 drinks when the temperature was 80°F.

Important Words and Phrases

Part of getting started in a new area of study is developing an understanding of its language so you can communicate with other people. Just as in learning to speak a foreign language you must develop a vocabulary in order to communicate, in mathematics you must learn the meanings of important words and phrases if you expect to understand and be understood. Before going on to Chapter 2, you should take time to make sure you know the meanings of the following words and phrases that were discussed in this chapter:

abscissa
argument of a function
Cartesian coordinate system
composition of functions
dependent variable
domain of a function
function
graph of a function
independent variable
linear equation
linear function
ordered pair of numbers
ordinate
point-slope form of a straight line
range of a function
slope of a straight line
slope-intercept form of a straight line
slopes of perpendicular lines
solution of an equation

REVIEW EXERCISES CHAPTER 1

In Exercises 1–14, solve the indicated equations.

1. $3x - 2 = 5$

2. $2 - 5x = 8$

3. $7(3 - 2x) = 10$

4. $x + 2 = 4x + 13$

5. $2x - 3(x - 5) = 12$

6. $7 + 3(2x - 4) - 2x = 5x$

7. $\frac{3x}{2} + 2(3x - 5) = \frac{x}{3}$

8. $5x - \frac{x + 6}{2} = \frac{2x - 9}{5}$

9. $x - 3a(6 - 7x) = 19ax$

10. $a(bx + d) - 3x = c$

11. $\frac{2}{x} + \frac{3}{2x} = 5$

12. $1 + \frac{x}{x + 3} = \frac{2(x - 3)}{x}$

13. $\frac{2}{2y - 3} = \frac{5}{y + 4}$

14. $\frac{1}{x + 2} = \frac{1}{x - 1}$

15. The kinetic energy of a charged particle is given by the formula $E = c^2(m - m_0)$. Solve for m_0.

16. A banker knows that the future value, S, of money that earns simple interest is given by the formula $S = P(1 + rt)$, where P is the amount invested, r is the rate of simple interest, and t is the time over which the money is invested. Solve this formula for t.

17. The total monetary value of 84 nickels and dimes is \$5.55. How many nickels and how many dimes are there?

18. Two recliner chairs are offered as a set at a price of \$787. If the price of one of the chairs is 3 times that of the other, what is the price of each chair?

19. A car moving 50 mph crosses the starting line of the Daytona Speedway. Two minutes later, a second car moving in the same direction as the first crosses the starting line at 60 mph. How long will it take the second car to overtake the first? (Note: This is obviously not a race.)

20. A druggist mixes a 6% acid solution and a 2% acid solution together to yield 10 liters of a 5% acid solution. How much of each must the druggist use?

In Exercises, 21–28, tell whether each given equation or table defines a function.

21. $y = 3x$

22. $y = x^4$

23. $y = x \pm 5$

24. $y < 2$

25. $y = 1/x^2$

26. $y = x - x^{-1}$

27.

x	2	−1	4	7
y	3	0	7	−3

28.

x	0	5	7	8	10	5
y	0	3	4	2	-1	-3

In Exercises 29–32, find the range and domain.

29. $y = 7 + 2x$

30. $y = -3x + 12$

31. $g(x) = \sqrt{x - 3}$

32. $h(x) = \dfrac{3}{x + 5}$

In Exercises 33–36, find $f(g(x))$ and $g(f(x))$.

33. $f(x) = 3x - 1$, $g(x) = 4 - x$

34. $f(x) = 2x + 5$, $g(x) = \frac{1}{2}(x - 5)$

35. $f(x) = x + 3$, $g(x) = 1/x$

36. $f(x) = \sqrt{x}$, $g(x) = 5 - 3x$

In Exercises 37–40, draw the graph of each given function.

37. $y = 4 - 2x$

38. $F(x) = \sqrt{2 - x}$

39. $y = 1 - x^2$, for $-1 \le x \le 2$

40. $y = 3(5 + 0.5x)$, for $0 \le x \le 10$

In Exercises 41–42, sketch the straight line determined by the pairs of points and compute the slope of the line.

41. $(-1, 3)$, $(0, -2)$

42. $(2, -7)$, $(-1, -4)$

43. Draw the straight line passing through $(-1, 2)$ with a slope of 5/2.

44. Draw the straight line passing through $(0, -3)$ with a slope of 4.

45. Find the slope of a straight line that is perpendicular to the line through $(5, 4)$ and $(-1, -1)$.

46. Find the slope of a straight line that is perpendicular to the line through $(-3, 0)$ and $(-2, 5)$.

47. Determine the equation of the line passing through the point $(-5, 2)$ with slope 1/3.

48. Determine the equation of the line passing through $(1, 3)$ and $(-1, -2)$.

49. Find the equation of the line perpendicular to $x - 2y = 5$ at $(0, 5/2)$.

50. Find the equation of the line perpendicular to $5x - 2y = 4$ at $(2, 3)$.

51. The velocity of an object 4 sec after it is released is 25 cm/sec and 38 cm/sec at 10 sec after release. Find the velocity-time equation of the object if velocity is known to be a linear function of time.

52. The force required to stretch a spring is a linear function of the displacement of the spring. What is the force-displacement equation for a spring if it is 15 in. long when no force is applied and 18 in. long when a 2-lb force is applied?

53. In surveying for a new coal mine, a geologist drills two holes 500 ft apart. One hole contacts the vein at a depth of 225 ft and the other at a depth of 315 ft. Find the slope of the vein of coal, assuming that the depth of the vein is a linear function of the distance between the two holes.

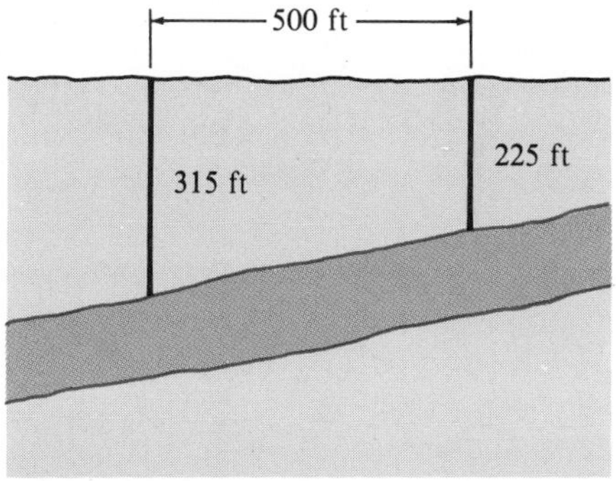

54. During a 5-min exercise workout a person's pulse rate varies with time in minutes according to $P = 15t + 58$.
(a) Sketch pulse rate as a function of time.
(b) Give the person's pulse rate at the beginning of the workout.

55. The controller of a new advertising agency estimates that the company's profit during the first 5 years will be $P(t) = 0.8(2t - 3)$, where P is the profit in millions of dollars and t is the time in years.
(a) Sketch P as a function of time for $0 \le t \le 5$ years.
(b) When will the agency begin earning a profit?

56. The cost of renting a car is \$20 plus \$0.15 for each mile driven.
(a) Express this cost as a functional relation.
(b) Graph the cost vs. the mileage driven for the first 300 miles.

57. The admission price to a concert in an 850-seat theater is \$18.

(a) Given that the cost of the performance is \$7500, write a formula showing profit as a function of paid admissions, n.

(b) Sketch the graph of this function for $0 \leq n \leq 850$.

(c) Use the graph to determine how many tickets must be sold for the theater to break even.

58. A political analyst finds that the number of registered independents voting for Republican candidates seems to vary with the inflation rate according to the function

$$D = 100{,}000(1 + 0.3i)$$

where D is the number of independents voting Republican and i is the inflation rate in percentage points. Sketch the graph of this function for $0 \leq i \leq 10\%$.

CHAPTER 2

SYSTEMS OF LINEAR EQUATIONS AND MATRICES

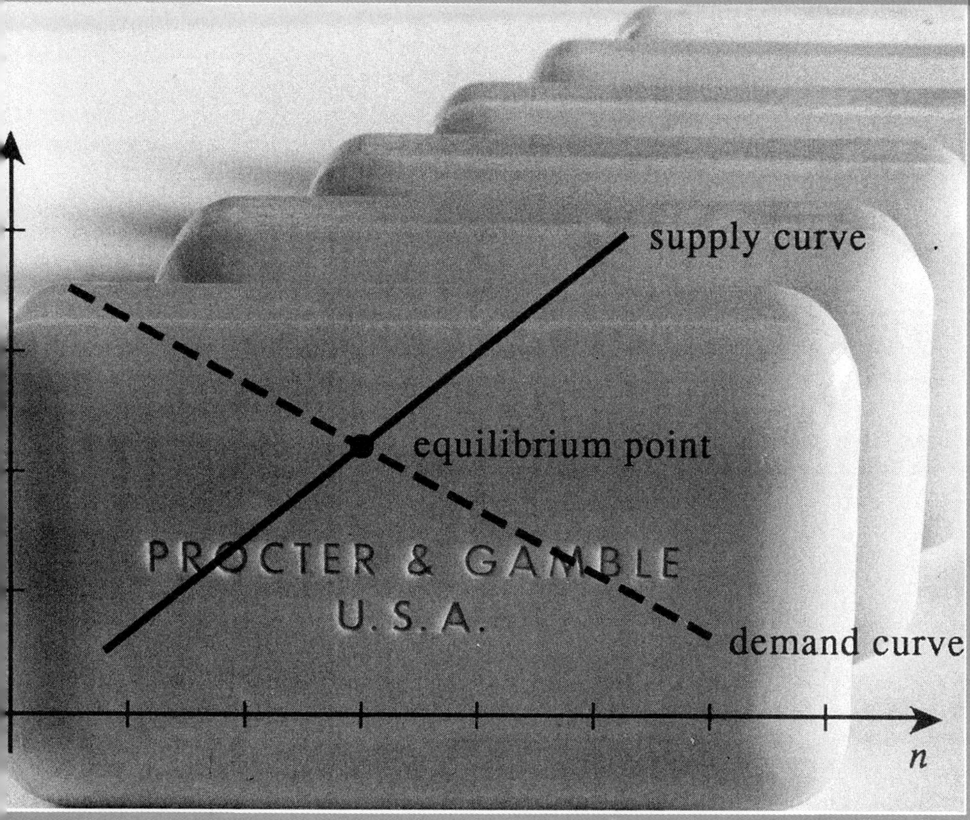

When a company such as Procter & Gamble introduces a new household product, a decision must be made as to how much to charge the consumer. Obviously, the price must be greater than the cost of producing the item or the company will not be interested in manufacturing it. On the other hand, if the price is too high, consumers may be unable or unwilling to buy it. These tensions between market forces will cause either a shortage or a surplus, neither of which is desirable. Therefore, in pricing an item, the company wants to price it low enough to attract the consumer, but high enough to make a reasonable profit. One approach to resolving this dilemma involves the use of systems of linear equations to assist in the decision-making process.

In this chapter we will extend our use of linear equations to include working with two or more linear equations at the same time. When considered together, two or more equations are called a **system of equations.** The solution of a system of linear equations is related to the solution of a linear equation in two variables, which you will recall has infinitely many solutions. For example, solutions of $3x + y = 5$ include the ordered pairs (1, 2), (2, -1), (3, -4), (0, 5), and so on. Similarly, solutions of $x - y = 7$ include (8, 1), (-1, -8), (10, 3) (3, -4), and so on. Inspection of these two solution sets reveals that, of the given solutions, only (3, -4) satisfies both equations. We say that (3, -4) is the common, or simultaneous, **solution of the system** of two equations in the unknowns x and y.

You can get an idea of what it means to solve a system of equations by graphing each equation of a system on the same set of coordinates. Figure 2.1 shows the graphs of $3x + y = 5$ and $x - y = 7$. Of interest is the fact that the ordered pair (3, -4) is the point of intersection of the two lines. The observation that the solution of a system of two linear equations corresponds to the intersection of the two lines associated with the equations is a fundamental idea, and eventually leads to a more general discussion of systems and to efficient and systematic methods of solving them.

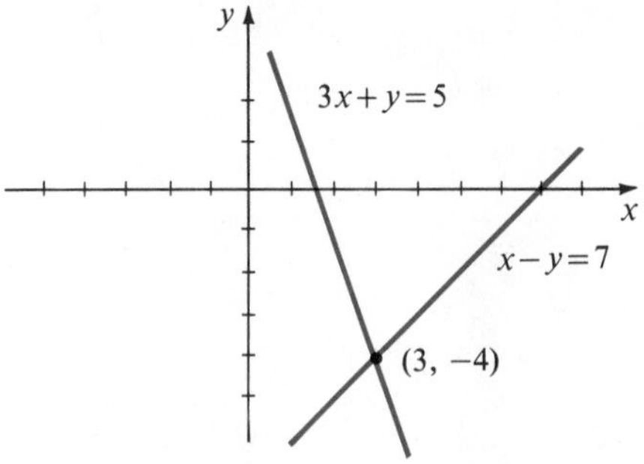

Figure 2.1

2.1 LINEAR SYSTEMS

A system of equations is referred to as a **linear system** if all the equations in the system are linear. A system that is not linear is called *nonlinear*. For example, the system

$$
\begin{aligned}
3x + y &= 5 \\
x - y &= 7
\end{aligned}
$$

is a linear system, whereas the system

$$
\begin{aligned}
x + y &= 10 \\
x^2 - 13y &= 4
\end{aligned}
$$

is nonlinear because one of the equations is a quadratic equation. The scope of this book is restricted to solving linear systems.

The method we use to solve a system of linear equations is based on the assumption that a system can be algebraically manipulated to yield an equivalent system in which the solution is more or less obvious, just as we did in solving equations. The idea is to algebraically alter the given system, without disturbing the solution, until one of the equations involves only one unknown. Before we show how this is done, a few preliminary ideas need to be defined.

Two systems of equations are **equivalent systems** if both have precisely the same simultaneous solutions. Thus, the systems

$$
\text{I}\begin{cases} 2x - y = 5 \\ x - 2y = 4 \end{cases} \quad \text{and} \quad \text{II}\begin{cases} x + y = 1 \\ 3x - y = 7 \end{cases}
$$

are equivalent systems, because for both systems, $x = 2$, $y = -1$ is the only solution. Figure 2.2(a) shows the graph of System I and Figure 2.2(b) the graph of System II. The point of intersection is the same in both cases. Notice that the two systems are equivalent even though they do not have a common equation.

Figure 2.2

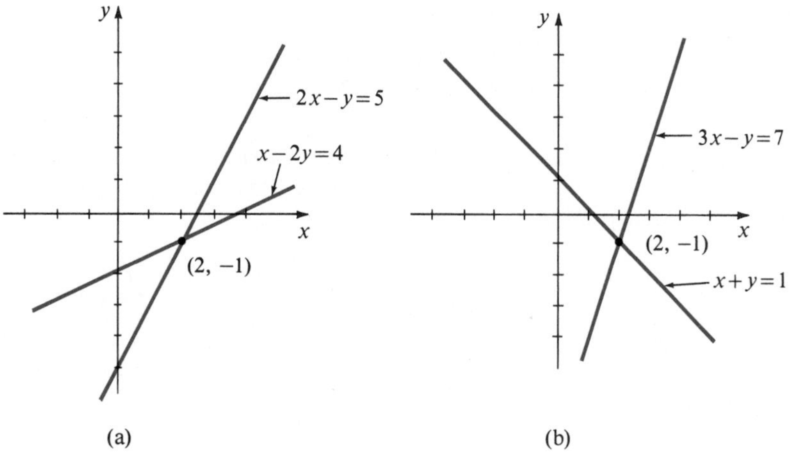

An operation that transforms a system of equations into an equivalent system is said to be allowable.

ALLOWABLE OPERATIONS FOR SYSTEMS OF EQUATIONS ■

1. The position of any two equations can be changed. Thus, the systems

$$\begin{cases} 3x + 5y = 6 \\ x - 3y = 4 \end{cases} \quad \text{and} \quad \begin{cases} x - 3y = 4 \\ 3x + 5y = 6 \end{cases}$$

are equivalent systems. Whereas this operation admittedly seems trivial, it needs to be mentioned because we will later use it in a more extended sense.

2. Any equation of a system may be replaced by a nonzero multiple of itself. Thus, the systems

$$\text{I}\begin{cases} x - 5y = 3 \\ 2x + 3y = 1 \end{cases} \quad \text{II}\begin{cases} x - 5y = 3 \\ 4x + 6y = 2 \end{cases} \quad \text{III}\begin{cases} x - 5y = 3 \\ -6x - 9y = -3 \end{cases}$$

are equivalent, since the upper equation is the same in all three and the lower equation in each system is a constant multiple of the others.

3. Any equation of a system may be replaced by the sum of that equation added to a nonzero multiple of another equation in the system. For example, the systems

$$\text{I}\begin{cases} 3x + 5y = 2 \\ x - 3y = 1 \end{cases} \quad \text{and} \quad \text{II}\begin{cases} 3x + 5y = 2 \\ 4x + 2y = 3 \end{cases}$$

are equivalent systems because the upper equation in each is the same and the lower equation in System II is obtained by adding together the equations in System I.

COMMENTS:

a. *These operations are also allowable in combination. For example, the systems*

$$\text{I}\begin{cases} 2x - y = 5 \\ x - 2y = 4 \end{cases} \quad \text{II}\begin{cases} x + y = 1 \\ 2x - y = 5 \end{cases}$$

are equivalent because:

- *The first equation of I is the second equation of II.*
- *The first equation of II is obtained by subtracting the second equation in I from the first equation in I.*

b. *These operations are* transitive *in nature; that is, if System I is equivalent to System II and System II is equivalent to System III, then System I is equivalent to System III. For example, the systems*

$$\text{II}\begin{cases} x + y = 1 \\ 2x - y = 5 \end{cases} \quad \textit{and} \quad \text{III}\begin{cases} x + y = 1 \\ 4x + y = 7 \end{cases}$$

are equivalent because their upper equations are the same and the lower equation of III was obtained from II by adding twice the upper equation of II to the lower equation of II. Thus, Systems I and III are equivalent.

ELIMINATION OF A VARIABLE

The three allowable operations are used with a technique called **Gauss-Jordan elimination** to solve systems of linear equations. The key to this technique is to multiply each of two separate equations in the system by just the right constant, such that after the two equations are added, one of the variables vanishes. That is, its coefficient becomes zero. In this way, the number of variables in the resulting equation is reduced by one. If there are initially two equations and two variables, this process will yield two equations, one of which will have only one unknown. The following example illustrates the process:

Example 1 Use Gauss-Jordan elimination to solve the system

$$\begin{aligned} 3x + 2y &= 5 \\ x - 2y &= 3 \end{aligned}$$

Sketch a figure showing the solution.

Solution First examine the system to see if any obvious allowable operations can be used to eliminate one of the variables. In this case, the y variable can be eliminated by simply adding the two equations. By replacing the second equation with this sum, we get the equivalent system

$$\begin{aligned} 3x + 2y &= 5 \\ 4x &= 8 \end{aligned}$$

Multiplying the second equation by $\frac{1}{4}$ yields the equivalent system

$$\begin{aligned} 3x + 2y &= 5 \\ x &= 2 \end{aligned}$$

Next, add -3 times the second equation to the first equation and replace the first equation with this sum. The system is now

$$\begin{aligned} 2y &= -1 \\ x &= 2 \end{aligned}$$

Finally, multiply the first equation by $\frac{1}{2}$. The resulting system of equations represents the solution to the system.

$$\begin{aligned} y &= -\frac{1}{2} \\ x &= 2 \end{aligned}$$

As always, check the solution by substituting it into the original system of equations. Thus,

$$3(2) + 2\left(-\frac{1}{2}\right) = 5$$

and

$$2 - 2\left(-\frac{1}{2}\right) = 3$$

Figure 2.3 shows a graph of the given system.

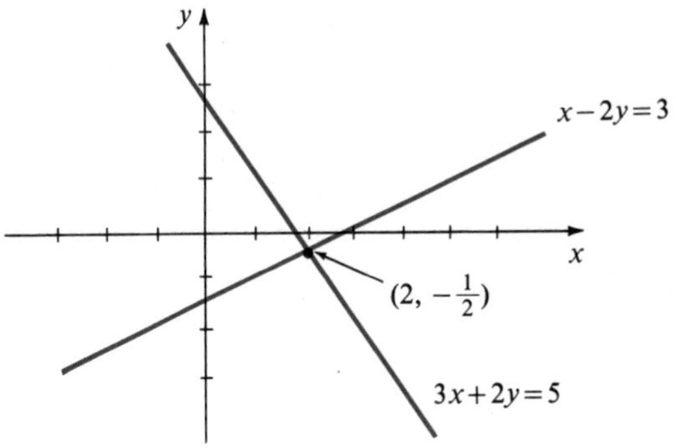

Figure 2.3

Example 2 Use Gauss-Jordan elimination to solve the system

$$\begin{aligned} 3x + 2y &= -4 \\ 4x + 9y &= 1 \end{aligned}$$

Solution Neither of the variables can be eliminated by merely adding the two equations. However, x can be eliminated by subtracting 3 times the second equation from 4 times the first equation. Replacing the second equation of the system by this difference, we have the equivalent system

$$\begin{aligned} 3x + 2y &= -4 \\ -19y &= -19 \end{aligned}$$

Divide the second equation by -19. Then subtract 2 times the second equation from the first equation. Replacing the first equation with the difference we have

$$\begin{aligned} 3x &= -6 \\ y &= 1 \end{aligned}$$

Finally, divide the first equation by 3 to obtain the solution,

$$\begin{aligned} x &= -2 \\ y &= 1 \end{aligned}$$

Checking this solution, we have $3(-2) + 2(1) = -4$ and $4(-2) + 9(1) = 1$. ▪

Solutions of Linear Systems

Since every linear system presented so far has had precisely one solution, you may think that this is always the case. However, as the next example shows, a linear system can have infinitely many solutions, or no solution.

Example 3 Find the solution to each of the following linear systems:

$$\text{I}\begin{cases} x + y = 2 \\ x + 2y = 3 \end{cases} \qquad \text{II}\begin{cases} x - 2y = 6 \\ 2x - 4y = 12 \end{cases} \qquad \text{III}\begin{cases} 2x + y = 4 \\ 6x + 3y = -6 \end{cases}$$

Solution Figure 2.4 shows a sketch of each of the three systems.

The two straight lines of System I cross at (1, 1). Thus, the solution is $x = 1$, $y = 1$, which can easily be checked by direct substitution. Since the equations of System II have the same intercepts, the two lines coincide, and the solution set is the infinite set of ordered pairs given by the coordinates of the points on the line. In System III, the lines do not intersect; that is, they are parallel. Hence, the system has no solution.

Figure 2.4

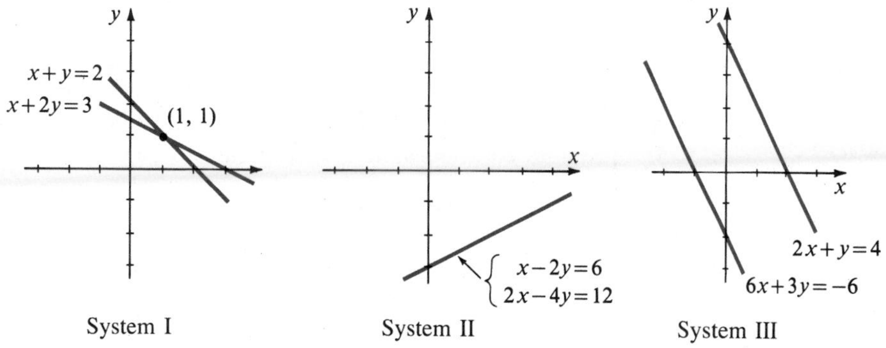

■

The same conclusion can be obtained by algebraic methods if we proceed as follows:

System I Multiplying the first equation by -1 and adding the result to the second equation, we have

$$\begin{aligned} x + y &= 2 \\ y &= 1 \end{aligned}$$

Then multiply the second equation by -1 and add the result to the first equation to obtain

$$\begin{aligned} x &= 1 \\ y &= 1 \end{aligned}$$

System II Multiplying the first equation by -2, adding it to the second equation, and replacing the second equation yields

$$\begin{aligned} x - 2y &= 6 \\ 0 &= 0 \end{aligned}$$

Since $0 = 0$ is true for all values of x and y, we conclude that this system has infinitely many solutions. Particular solutions can be obtained by assigning y a value, and determining x from $x - 2y = 6$. For example, if $y = -1$, then $x - 2(-1) = 6$, so $x = 4$. Thus, $(4, -1)$ is one of the infinitely many solutions.

System III Multiplying the first equation by -3, adding it to the second equation, and replacing the second equation yields

$$\begin{aligned} 2x + y &= 4 \\ 0 &= -18 \end{aligned}$$

Since $0 = -18$ can never be true, we conclude that there is no solution to this system.

Example 3 shows the only three possibilities for linear systems of two equations in two unknowns. Since the graph of each of the equations is a straight line, one of the following statements must be true:

1. The lines intersect at one point. (The system is said to be **consistent.**)
2. The lines are parallel. (The system is said to be **inconsistent.**)
3. The lines coincide. (The system is said to be **dependent.**)

More generally, linear systems are said to be consistent, inconsistent, or dependent as their simultaneous solution sets consist of a single element, of no elements, or of infinitely many elements. *Caution:* This terminology applies only to *linear* systems.

Applications Involving Systems of Linear Models

The next examples show how systems of linear equations might arise in solving physical problems. The hints given in Section 1.2 as to how to construct linear models apply here as well.

Example 4 The graph of $ax^2 + by = 1$ passes through the points (1, 4) and (2, 7). Determine the values of a and b for this quadratic function.

Solution At first glance this would appear to be a *nonlinear* problem, because the graph of $ax^2 + by = 1$ is a parabola and not a straight line. But, as you will see, the solution of the problem involves the use of a system of linear equations. Since both (1, 4) and (2, 7) must satisfy the given quadratic equation, we can write the following system of linear equations in a and b:

$$\begin{aligned} &(1) \quad a + 4b = 1 \\ &(2) \quad 4a + 7b = 1 \end{aligned}$$

To solve this system, multiply (1) by -4, add the result to (2) and replace (2) with this sum. Thus:

$$\begin{aligned} &(1) \quad a + 4b = 1 \\ &(2) \quad -9b = -3 \end{aligned}$$

From (2), we get $b = \frac{1}{3}$ and then $a = -\frac{1}{3}$. After checking the solution, we determine that the desired equation is

$$-\frac{1}{3}x^2 + \frac{1}{3}y = 1$$

or

$$x^2 - y = -3$$

The graph of this function is shown in Figure 2.5.

Figure 2.5

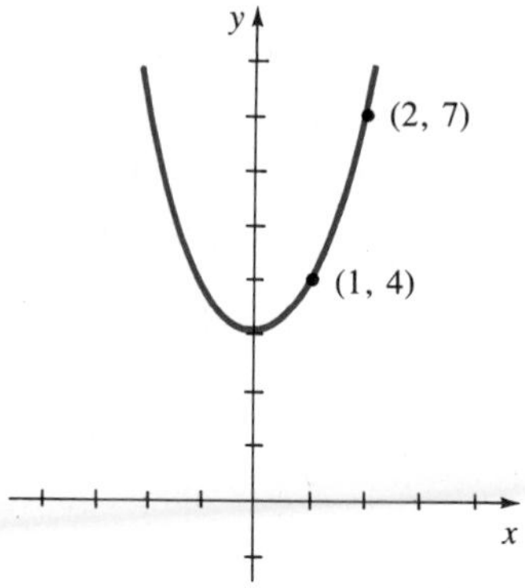

In business, the relationship between the price of an item, p, and the number of items, n, purchased by the consumer in a fixed period is called a **demand function.** The relationship between the price, p, and the number of items supplied by the producer, n, is called a **supply function.** For many business applications, we assume that supply and demand functions are linear. Figure 2.6 shows typical linear supply and demand curves drawn on the same coordinate axes. Although the graphs of supply and demand functions consist of sets of discrete points, it is common practice to treat them as continuous functions and draw their graphs as smooth curves.

Figure 2.6

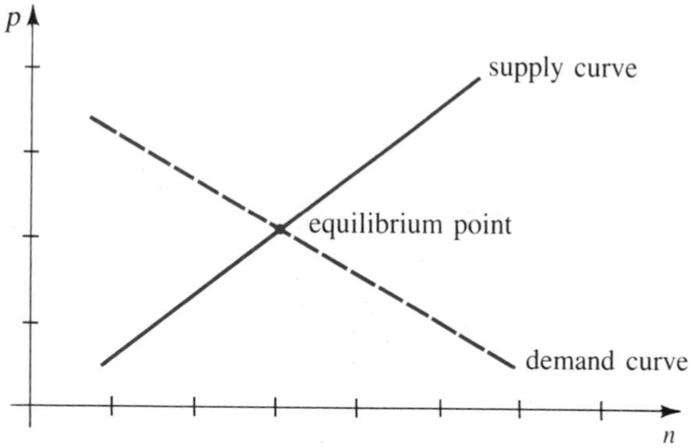

The graph shows that producers are much more willing to supply an item priced high and, conversely, consumers are more willing to buy when prices are low. This tension between the two principal market forces causes either a surplus of the item or a shortage, both undesirable. Market equilibrium results when there is no surplus or shortage. This equilibrium is represented by the point at which the supply and demand curves meet. The coordinates of the point give the equilibrium quantity and price.

Example 5 Find the equilibrium quantity and price for a VCR market in which a group of wholesalers will buy 200 VCRs if the price is \$150 and 50 if the price is \$400, while the manufacturer will supply 50 units at \$100 and 200 units at \$300. Assume linear supply and demand functions.

Solution Letting x be the number of items produced or supplied and p the price, we have the following chart:

Demand:	$x = 200,$	$p = 150$	Supply:	$x = 50,$	$p = 100$
	$x = 50,$	$p = 400$		$x = 200,$	$p = 300$

The equation of the demand line is obtained by using the point-slope form of a straight line.

$$\frac{p - 150}{x - 200} = \frac{400 - 150}{50 - 200} = -\frac{5}{3}$$

which may be written $3p + 5x = 1450$.

Similarly, the supply function is

$$\frac{p - 100}{x - 50} = \frac{300 - 100}{200 - 50} = \frac{4}{3}$$

which may be written $3p - 4x = 100$.

We solve the equations $3p + 5x = 1450$ and $3p - 4x = 100$ simultaneously by taking the negative of the second equation and adding it to the first:

$$9x = 1350$$
$$x = 150$$

which is the equilibrium number of VCRs to be supplied (and demanded). The corresponding price is obtained by substituting 150 for x in either of the two equations to obtain $p = \$233$, which is the equilibrium price. ▪

Example 6 A chemist has two acid solutions, one containing 10% acid and the other 4% acid. See Figure 2.7. How many liters of each solution are needed to make 200 liters of a solution that is 6% acid?

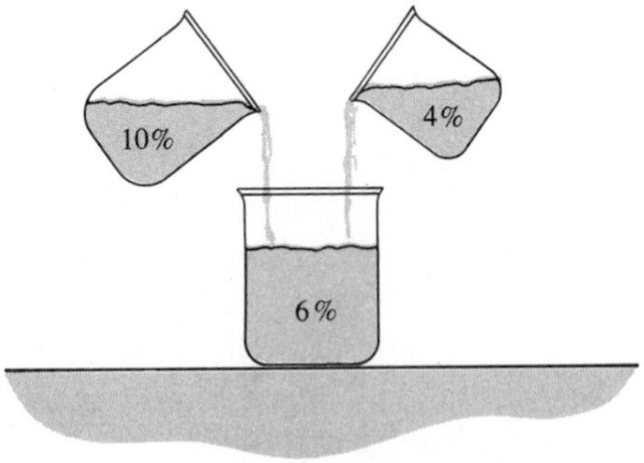

Figure 2.7

Solution Let x = the number of liters of 10% acid solution and let y = the number of liters of 4% acid solution. Since the amount of the 10% solution plus the amount of the 4% solution must equal 200 liters, we have the following equation:

$$(1)\quad x + y = 200$$

A second equation is generated by observing that the amount of acid in x liters of the 10% solution is $0.10x$, the amount of acid in y liters of the 4% solution is $0.04y$, and the amount of acid in the final 6% solution is $200(0.06) = 12$. Since the amount of acid from the two sources must equal the amount of acid in the final solution, we write the following equation:

$$(2)\quad 0.10x + 0.04y = 12$$

The solution to this system of equations, using Gauss-Jordan elimination, is

$$x = 66\tfrac{2}{3}\text{ liters}$$

and

$$y = 133\tfrac{1}{3}\text{ liters}$$ ∎

EXERCISES SECTION 2.1

Solve the systems of equations in Exercises 1–19 by the Gauss-Jordan elimination method. Indicate whether the system is consistent, inconsistent, or dependent. In each case, draw the graph of the system and check the solution.

1. $2x - 4y = 2$
   $-2x + y = 4$

2. $x + 5y = 4$
   $x + 3y = 6$

3. $3x + 4y = 23$
   $x - 3y = -1$

4. $x + y = 4$
   $x - y = 2$

5. $x - 4y = 8$
   $2x - 8y = 16$

6. $3x + 2y - 6 = 0$
   $x - 3y - 3 = 0$

7. $3x + 4y = 10$
   $6x + 8y = -2$

8. $3x + 5y = 2$
   $6x + 10y = 4$

9. $3x - y = 2$
   $2x + y = 6$

10. $2x + 3y = 7$
    $6x - y = 1$

11. $3x + y = 0$
    $2x - 2y = 2$

12. $x + 5y = 0$
    $3x + 15y = 3$

13. $x + 7y - 7 = 0$
    $8x - 7y - 3 = 0$

14. $3x + 5y = 4$
    $\frac{3}{2}x + \frac{5}{2}y = 2$

15. $3x + 5y = 5$
    $x + 4y = 11$

16. $4x - 7y = 29$
    $6x + 5y = -3$

17. $9z - 13w + 3 = 0$
    $6z - 7w = 3$

18. $5x + 4y = 0$
    $3x + 5y + 13 = 0$

19. $2x + 3y = -10$
    $3x - 2y = -2$

In each of the remaining exercises in this set, write a system of equations and then use Gauss-Jordan elimination to solve it.

20. The velocity-time equation of car A is $v = 3t + 20$ ft/sec, where t is the time in seconds. If the velocity-time equation for car B is $v = 6t + 5$ ft/sec, at what time do both cars have the same velocity? What is this velocity?

21. A company buys a total of 25 grinders from two different companies. It buys 3 more Brand A grinders than Brand B. The total number of each brand of grinder can be found by solving the system of equations

$$A + B = 25$$
$$A - B = 3$$

Can you explain how these equations are derived? Solve for A and B.

22. The sum of the tens digit and the units digit of a two-digit number is 11. The value of the number is 7 less than 30 times the units digit. Find the number.

23. The sum of two numbers is 12 and the difference is 4. Find the two numbers.

24. Some 20-lb containers and 30-lb containers are used to load 160 lb of salt. The number of 20-lb containers is one more than twice the number of 30-lb containers. How many of each kind of container are used?

25. The graph of $ax + by^2 = 2$ passes through the points $(4, -2)$ and $(-2, 2)$. Determine the coefficients a and b.

26. The resistance of a wire is given by $R = aT + K$, where a and K are constants and T is the temperature in degrees Celsius. Find a and K if $R =$ 25 ohms when $T = 10°\text{C}$ and $R = 30$ ohms when $T = 100°\text{C}$.

27. A boat travels 60 mi upstream in 10 hr. If it returns in 8 hr, find the rate of the current and of the boat. Assume that both rates are constant during the trip.

28. A truck enters a freeway traveling 50 mph. One hour later, a car enters the freeway by the same ramp, traveling 80 mph. How long does it take for the car to overtake the truck?

29. A technician needs 10 lb of an alloy that is 60% lead and 40% zinc by weight. If she has an alloy that is 80% lead and 20% zinc, and another that is 50% lead and 50% zinc, how many pounds of each must she use to obtain the desired alloy?

30. The lifeguard at a swimming pool knows that two water pipes running at the same time can fill the pool in 6 hr. If both pipes run for 2 hr and the first is then shut off, the second pipe will take 5 hr more to fill the pool. How long would it take each pipe to fill the pool by itself?

31. A collection of nickels and dimes has a total value of \$2.50. If there are 35 coins in all, how many dimes are there?

32. An airplane flying with a tailwind takes 2 hr to make a trip of 1240 mi. If the return trip against the wind takes 2.5 hr, what is the speed of the airplane relative to the air? What is the wind speed?

33. Two computers can complete a series of calculations in 10 min if they are used together. After 6 min, one of the computers breaks down and the other takes 9 min to finish the job. How long would it take each computer to do the series of calculations working alone?

34. An economy car that averages 45 mpg can go 120 mi farther on a tank of gas than can a larger luxury model that averages 15 mpg. If the total capacity of both tanks together is 40 gal, what is the size of each tank?

35. The total number of registered Democrats and Republicans in a certain district is 100,000. In the last election 52,000 voters turned out; tabulations showed that 50% of the Democrats and 60% of the Republicans voted. Determine the number of registered Democrats and the number of registered Republicans.

36. A manufacturer will supply an item as follows: 10 if the price is \$50; 40 if it is \$100. The corresponding demand is 250 if the price is \$25, and 50 if it is \$125. Assume that the supply and demand functions are linear.
(a) Write the supply function.
(b) Write the demand function.
(c) Determine the equilibrium amount and price.

37. A group of wholesalers will buy 100 television sets if the price is \$400 each, and 150 if it is \$250 each. The manufacturer is willing to supply 75 sets at a price of \$300 each, or 150 sets at \$350 each. Write the linear supply and demand functions and determine the equilibrium values.

38. A video store rents movies at \$3 per movie. Its monthly revenue function is therefore $R(x) = 3x$, where x is the number of movies rented in a month's time. Its cost function is $C(x) = 1600 + x$. Graph the revenue and cost functions on the same axes. What does the point of intersection represent? What are the values at that point?

2.2 HIGHER-ORDER SYSTEMS: MATRIX NOTATION

In this section we will consider a variety of systems of linear equations. To study these systems efficiently we use rectangular arrays of numbers, called *matrices,* to represent systems of linear equations. Initially, our use of matrices is limited to representing systems of linear equations, but in the next section, these arrays will take on a life independent of their application to systems of equations.

DEFINITION Matrix

A **matrix** is a rectangular array of numbers. A matrix is denoted by enclosing the rectangular array with brackets.

MATRICES

A political science student asked the students in class to indicate their gender and their political preference. The results were as follows: of the 15 females, 7 were Republicans, 5 were Democrats, and 3 were independents; of the 17 males, 8 were Republicans, 7 were Democrats, and 2 were independents. To present this information to the class, the student arranged the numbers into the following rectangular array:

$$\begin{array}{l} \\ \text{Female} \\ \text{Male} \end{array} \begin{array}{c} \begin{array}{ccc} \text{Rep.} & \text{Dem.} & \text{Ind.} \end{array} \\ \begin{bmatrix} 7 & 5 & 3 \\ 8 & 7 & 2 \end{bmatrix} \end{array}$$

Rectangular arrays like this one are used extensively to present information and have many important applications. Some examples of matrices are

(a) $\begin{bmatrix} 2 & -1 \end{bmatrix}$ **(b)** $\begin{bmatrix} 5 & -3 \\ 2 & 1 \end{bmatrix}$

(c) $\begin{bmatrix} x \\ y \\ z \end{bmatrix}$ **(d)** $\begin{bmatrix} 2 & 1 & 1 & 0 & 12 \\ 1 & 2 & 0 & 1 & 18 \\ 3 & 2 & 0 & 0 & 2 \end{bmatrix}$

The **dimensions** of a matrix are indicated by giving the number of *rows* followed by the number of *columns.* Thus, an $m \times n$ (read "m by n") matrix is a rectangular array with m rows and n columns. Of the matrices above: (a) is 1×2, (b) is 2×2, (c) is 3×1, and (d) is 3×5.

A **square matrix** has the same number of rows and columns. In this case the number of rows (or columns) is called the **order** of the matrix. Thus,

$$\begin{bmatrix} 2 & 1 \\ 0 & -3 \end{bmatrix} \quad \text{and} \quad \begin{bmatrix} 0 & 5 & -2 \\ 2 & 3 & 0 \\ -10 & 7 & 9 \end{bmatrix}$$

are square matrices of orders 2 and 3, respectively.

COMMENT: *A matrix has no numerical value; it is simply a device that is used to list or catalog numbers. In fact, our interest in matrices in this section is as an efficient means of listing and manipulating the coefficients of a system of equations.*

Higher-Order Linear Systems

An equation of the form $ax + by + cz = d$ is called a linear equation in the three variables x, y, and z, where a, b, c, and d are constants. A solution to this equation is an ordered triple of numbers corresponding to x, y, and z that make the equation a true statement. For example, $(1, -1, 3)$ is a solution to the equation $2x - 3y + z = 8$. Of course, there are many other ordered triples that are also solutions to this equation. But, as with linear equations in two unknowns, not all these solutions are of interest to us.

A system of three linear equations in three unknowns will be of the form

$$\begin{aligned} a_1x + b_1y + c_1z &= d_1 \\ a_2x + b_2y + c_2z &= d_2 \\ a_3x + b_3y + c_3z &= d_3 \end{aligned}$$

A *solution* of this system is any ordered triple that simultaneously satisfies each equation in the system. For example, the ordered triple $(1, -1, 2)$ is a solution of the system

$$\begin{aligned} x - y - z &= 0 \\ 2x + y + z &= 3 \\ x + y + 2z &= 4 \end{aligned}$$

You can verify this by substituting $x = 1$, $y = -1$, and $z = 2$ into each of the equations.

To *solve* a system means to find all possible solutions. The technique of Gauss-Jordan elimination that we used for systems of two equations in two unknowns is easily extended to higher-order systems. Thus, we will use the same allowable operations of adding multiples of equations to one another to obtain equivalent systems, eventually obtaining a system in which the solution is obvious. For third-order systems in which the variables are x, y, and z, our aim is to reduce the given system to an equivalent system in which the values of x, y, and z are clearly stated.

NOTATION

Since the manipulations required to solve a system of three equations are more involved than those for systems of two equations, we introduce a special notation to help us keep track of the process. First, label each equation in the system. For instance, we could number the following system as shown:

$$\begin{aligned} ①\quad & 2x + 3y - z = 1 \\ ②\quad & x - 3y + 5z = -1 \\ ③\quad & 3x + 2y + z = 0 \end{aligned}$$

Then the notation

$$a① + ② \to ②$$

means "Multiply Equation ① by a, add it to Equation ②, and replace Equation ② with this sum." Notice that this transformation affects *only* Equation ②, while Equations ① and ③ are unaltered. We will use this notation in the following example.

Example 1 Use Gauss-Jordan elimination to solve the system

$$\begin{aligned} ①\quad & x - 6y + 3z = -2 \\ ②\quad & 2x - 3y + z = -2 \\ ③\quad & 3x + 3y - 2z = 2 \end{aligned}$$

Solution

$$\left.\begin{aligned} x - 6y + 3z &= -2 \\ 2x - 3y + z &= -2 \\ 3x + 3y - 2z &= 2 \end{aligned}\right\} \xrightarrow{-2① + ② \to ②}$$

$$\left.\begin{aligned} x - 6y + 3z &= -2 \\ 9y - 5z &= 2 \\ 3x + 3y - 2z &= 2 \end{aligned}\right\} \xrightarrow{-3① + ③ \to ③}$$

$$\left.\begin{aligned} x - 6y + 3z &= -2 \\ 9y - 5z &= 2 \\ 21y - 11z &= 8 \end{aligned}\right\} \xrightarrow{-\frac{7}{3}② + ③ \to ③}$$

$$\left.\begin{aligned} x - 6y + 3z &= -2 \\ 9y - 5z &= 2 \\ \frac{2}{3}z &= \frac{10}{3} \end{aligned}\right\} \xrightarrow{\frac{3}{2}③ \to ③}$$

$$\left.\begin{aligned} x - 6y + 3z &= -2 \\ 9y - 5z &= 2 \\ z &= 5 \end{aligned}\right\} \xrightarrow{5③ + ② \to ②}$$

$$\left.\begin{aligned} x - 6y + 3z &= -2 \\ 9y \quad\quad &= 27 \\ z &= 5 \end{aligned}\right\} \xrightarrow{\frac{1}{9}② \to ②}$$

$$\left.\begin{aligned} x - 6y + 3z &= -2 \\ y \quad\quad &= 3 \\ z &= 5 \end{aligned}\right\} \xrightarrow{-3③ + ① \to ①}$$

$$\left.\begin{aligned} x - 6y \quad\quad &= -17 \\ y \quad\quad &= 3 \\ z &= 5 \end{aligned}\right\} \xrightarrow{6② + ① \to ①}$$

$$\begin{aligned} x \quad\quad\quad &= 1 \\ y \quad\quad &= 3 \\ z &= 5 \end{aligned}$$

Hence, the solution is the ordered triple, (1, 3, 5).

It is always a good idea to check your solution in the original system of equations. Thus we have

$$\begin{aligned} 1 - 6(3) + 3(5) &= -2 \\ 2(1) - 3(3) + 5 &= -2 \\ 3(1) + 3(3) - 2(5) &= 2 \end{aligned}$$

■

The Augmented Matrix

The effort required to reduce a linear system to triangular form can be simplified if we avoid the needless rewriting of variables. We easily eliminate the variable names by using a matrix to keep track of the coefficients of a system. For example, the system in Example 1 can be represented by the matrix

$$\left[\begin{array}{rrr|r} 1 & -6 & 3 & -2 \\ 2 & -3 & 1 & -2 \\ 3 & 3 & -2 & 2 \end{array}\right]$$

This matrix is called the *matrix of the system, or the* **augmented matrix.** We have inserted the dashed line to emphasize the matrix of the coefficients of the variables, which alone is called the *coefficient matrix*. Notice that *the form of the augmented matrix assumes that each equation in the system is written in the form $ax + by + cz = d$*. If an equation of a system is not given in this form, it must be so written before it can be used in the augmented matrix. Be sure to include the zero coefficients.

Once the matrix of a system of equations is written, we treat the rows of the matrix just as we treat the equations they represent. Of course, we must make a mental note that the numbers in the first column represent the coefficients of x; those in the second column, the coefficients of y; and those in the third column, the coefficients of z.

A system of equations written in matrix form is solved using the same operations as in the Gauss-Jordan elimination. The use of **equivalent matrices** corresponds to equivalent systems of equations.

Operations on Augmented Matrices—Row Operations

Two augmented matrices are equivalent if

1. The position of any two rows is interchanged.
2. Any row is replaced by a nonzero multiple of itself.
3. Any row is replaced by the sum of that row added to a nonzero multiple of another row.

The following examples show how Gauss-Jordan elimination is used with augmented matrix notation to solve systems of equations.

Example 2 Use augmented matrix notation to solve

$$\begin{aligned} 3x + 2y &= 10 \\ x - 3y &= -4 \end{aligned}$$

Solution This system is represented by the augmented matrix

$$\left[\begin{array}{cc|c} 3 & 2 & 10 \\ 1 & -3 & -4 \end{array}\right]$$

The following steps yield the desired form.

$$\left[\begin{array}{cc|c} 3 & 2 & 10 \\ 1 & -3 & -4 \end{array}\right] \xrightarrow{\textcircled{1} \leftrightarrow \textcircled{2}} \left[\begin{array}{cc|c} 1 & -3 & -4 \\ 3 & 2 & 10 \end{array}\right] \xrightarrow{-3\textcircled{1} + \textcircled{2} \to \textcircled{2}}$$

$$\left[\begin{array}{cc|c} 1 & -3 & -4 \\ 0 & 11 & 22 \end{array}\right] \xrightarrow{\frac{1}{11}\textcircled{2} \to \textcircled{2}} \left[\begin{array}{cc|c} 1 & -3 & -4 \\ 0 & 1 & 2 \end{array}\right] \xrightarrow{3\textcircled{2} + \textcircled{1} \to \textcircled{1}}$$

$$\left[\begin{array}{cc|c} 1 & 0 & 2 \\ 0 & 1 & 2 \end{array}\right]$$

So the solution is $x = 2$, $y = 2$. Checking, we see

$$\begin{aligned} 3(2) + 2(2) &= 10 \\ 2 - 3(2) &= -4 \end{aligned}$$

∎

Notice the form of the coefficient matrix in the final step. The entries on the main diagonal are one and all the other entries are zero. This form of the coefficient matrix is called the *identity matrix*.

DEFINITION Identity Matrix

The $n \times n$ square matrix I with ones on the main diagonal and zeroes elsewhere is called the **identity matrix of order n.**

The identity matrix of order 2 is $\begin{bmatrix} 1 & 0 \\ 0 & 1 \end{bmatrix}$.

The identity matrix of order 3 is $\begin{bmatrix} 1 & 0 & 0 \\ 0 & 1 & 0 \\ 0 & 0 & 1 \end{bmatrix}$.

The reason we take special notice of the identity matrix at this point is that it can be shown that a system of equations has a unique solution when the identity matrix for the coefficient matrix can be obtained by some sequence of row operations. Further, the solution of the system is obvious when the coefficient matrix is in this form. Therefore, our strategy is to apply row operations until either the identity matrix is obtained or it is clear that no such matrix is possible.

Notice in Example 2, in the second step, that row 1 remained unchanged although we used it to operate on row 2. In this case, we call row 1 the *operative row* of the operation. In matrices of three or more rows, we can perform more than one operation at a time *provided* that the operative row is the same for all uses of the operation. This is shown in the next example.

Example 3 Use matrix notation to solve the system

$$\begin{aligned} x - 6y + 3z &= -2 \\ 2x - 3y + z &= -2 \\ 3x + 3y - 2z &= 2 \end{aligned}$$

This is the same system that was solved in Example 1 of this section.

Solution The given system is represented by the augmented matrix

$$\left[\begin{array}{rrr|r} 1 & -6 & 3 & -2 \\ 2 & -3 & 1 & -2 \\ 3 & 3 & -2 & 2 \end{array}\right]$$

The following steps yield the desired form.

$$\left[\begin{array}{rrr|r} 1 & -6 & 3 & -2 \\ 2 & -3 & 1 & -2 \\ 3 & 3 & -2 & 2 \end{array}\right] \xrightarrow[-3① + ③ \to ③]{-2① + ② \to ②} \left[\begin{array}{rrr|r} 1 & -6 & 3 & -2 \\ 0 & 9 & -5 & 2 \\ 0 & 21 & -11 & 8 \end{array}\right]$$

$$\xrightarrow{7② - 3③ \to ③} \left[\begin{array}{rrr|r} 1 & -6 & 3 & -2 \\ 0 & 9 & -5 & 2 \\ 0 & 0 & -2 & -10 \end{array}\right]$$

$$\xrightarrow{-\frac{1}{2}③ \to ③} \left[\begin{array}{ccc|c} 1 & -6 & 3 & -2 \\ 0 & 9 & -5 & 2 \\ 0 & 0 & 1 & 5 \end{array}\right]$$

$$\xrightarrow[-3③ + ① \to ①]{5③ + ② \to ②} \left[\begin{array}{ccc|c} 1 & -6 & 0 & -17 \\ 0 & 9 & 0 & 27 \\ 0 & 0 & 1 & 5 \end{array}\right]$$

$$\xrightarrow{\frac{1}{9}② \to ②} \left[\begin{array}{ccc|c} 1 & -6 & 0 & -17 \\ 0 & 1 & 0 & 3 \\ 0 & 0 & 1 & 5 \end{array}\right]$$

$$\xrightarrow{6② + ① \to ①} \left[\begin{array}{ccc|c} 1 & 0 & 0 & 1 \\ 0 & 1 & 0 & 3 \\ 0 & 0 & 1 & 5 \end{array}\right]$$

So the solution is (1, 3, 5), which agrees with the result in Example 1. ▪

COMMENT: *The matrix method described here for systems of three equations can be used for linear systems of any order. There are some systems of four equations in the exercise set at the end of this section. The process for four equations is the same as for systems of three equations; there is simply more of it.*

Inconsistent and Dependent Systems

In some cases, when the row operations are used to transform an augmented matrix, the coefficient matrix cannot be reduced to the identity matrix, but will have one or more rows that are all zeroes. This situation indicates that the system does not have a unique solution. Example 4 illustrates a system that has no solution.

Example 4 Solve the system

$$\begin{aligned} x - y + z &= 8 \\ 2x + y - z &= -2 \\ -x - 5y + 5z &= 10 \end{aligned}$$

Solution The augmented matrix of the system is

$$\left[\begin{array}{ccc|c} 1 & -1 & 1 & 8 \\ 2 & 1 & -1 & -2 \\ -1 & -5 & 5 & 10 \end{array}\right]$$

The reduction of this matrix follows.

$$\left[\begin{array}{rrr|r} 1 & -1 & 1 & 8 \\ 2 & 1 & -1 & -2 \\ -1 & -5 & 5 & 10 \end{array}\right] \xrightarrow[\text{①} + \text{③} \to \text{③}]{-2\text{①} + \text{②} \to \text{②}} \left[\begin{array}{rrr|r} 1 & -1 & 1 & 8 \\ 0 & 3 & -3 & -18 \\ 0 & -6 & 6 & 18 \end{array}\right]$$

$$\xrightarrow{2\text{②} + \text{③} \to \text{③}} \left[\begin{array}{rrr|r} 1 & -1 & 1 & 8 \\ 0 & 3 & -3 & -18 \\ 0 & 0 & 0 & -18 \end{array}\right]$$

The last row signifies that $0(x) + 0(y) + 0(z) = -18$. Since this is impossible for any values of x, y, and z, there is no solution to this system of equations. A system of equations that has no solutions is called **inconsistent.** ■

A **dependent** system has an infinite number of solutions. The next example shows how this is identified if augmented matrix notation is used to represent the system.

Example 5 Solve the system

$$\begin{aligned} x + y - z &= -1 \\ 2x - y + z &= 2 \\ x - 5y + 5z &= 7 \end{aligned}$$

Solution The given system is represented by the augmented matrix

$$\left[\begin{array}{rrr|r} 1 & 1 & -1 & -1 \\ 2 & -1 & 1 & 2 \\ 1 & -5 & 5 & 7 \end{array}\right]$$

The following steps lead to the solution.

$$\left[\begin{array}{rrr|r} 1 & 1 & -1 & -1 \\ 2 & -1 & 1 & 2 \\ 1 & -5 & 5 & 7 \end{array}\right] \xrightarrow[-1\text{①} + \text{③} \to \text{③}]{-2\text{①} + \text{②} \to \text{②}} \left[\begin{array}{rrr|r} 1 & 1 & -1 & -1 \\ 0 & -3 & 3 & 4 \\ 0 & -6 & 6 & 8 \end{array}\right]$$

$$\xrightarrow{-2\text{②} + \text{③} \to \text{③}} \left[\begin{array}{rrr|r} 1 & 1 & -1 & -1 \\ 0 & -3 & 3 & 4 \\ 0 & 0 & 0 & 0 \end{array}\right]$$

The third row, containing only zeroes, signifies that $0(x) + 0(y) + 0(z) = 0$. This is true for all values of x, y, and z, which means the system is dependent. The augmented matrix shows that x, y, and z are related only by the equations represented by the first two rows of the matrix. This system is now a system involving three variables and two equations; that is, a system with more variables than equations. ■

SYSTEMS IN WHICH THE NUMBER OF VARIABLES IS NOT EQUAL TO THE NUMBER OF EQUATIONS

Systems of equations in which the number of unknowns is different from the number of equations arise quite often. They are handled in the same way as those in which the number of unknowns and equations is the same. We look again at Example 5. By virtue of the matrix operations on the rows, the matrix

$$\left[\begin{array}{rrr|r} 1 & 1 & -1 & -1 \\ 2 & -1 & 1 & 2 \\ 1 & -5 & 5 & 7 \end{array}\right]$$

is equivalent to

$$\left[\begin{array}{rrr|r} 1 & 1 & -1 & -1 \\ 0 & -3 & 3 & 4 \\ 0 & 0 & 0 & 0 \end{array}\right]$$

This means that the given system of three equations in three unknowns,

$$\begin{aligned} x + y - z &= -1 \\ 2x - y + z &= 2 \\ x - 5y + 5z &= 7 \end{aligned}$$

represented by the first matrix, is equivalent to a system of two equations in three unknowns,

$$\begin{aligned} x + y - z &= -1 \\ -3y + 3z &= 4 \end{aligned}$$

represented by the second matrix. To solve this system of equations we continue with the row operations until we obtain ones on the main diagonal. Therefore, we have

$$\xrightarrow{-\frac{1}{3}② \to ②} \left[\begin{array}{rrr|r} 1 & 1 & -1 & -1 \\ 0 & 1 & -1 & -\frac{4}{3} \\ 0 & 0 & 0 & 0 \end{array}\right]$$

$$\begin{aligned} x + y - z &= -1 \\ y - z &= -\frac{4}{3} \end{aligned}$$

The solutions are then defined by

$$z = t, \text{ where } t \text{ is any number}$$

$$y = t - \frac{4}{3}$$

$$x = -1 + t - y = -1 + t - \left(t - \frac{4}{3}\right) = \frac{1}{3}$$

or by the ordered triple $(\frac{1}{3}, t - \frac{4}{3}, t)$.

Thus, if the number of unknowns and equations is different, we proceed as in the earlier cases by writing the augmented matrix followed by Gauss-Jordan elimination. The purpose of the elimination process is to obtain zeroes in the first column below the first element, to obtain zeroes in the second column below the second element, and so on, and to have the nonzero elements on the diagonal equal to 1. At this point, all solutions can be written in terms of one or more real parameters.

Example 6 Solve the system

$$\begin{aligned} x - y - z + 2w &= 2 \\ 2x - y - z + 2w &= 1 \\ 3x - 2y - 5z + 7w &= 9 \end{aligned}$$

Solution The augmented matrix that represents this system is

$$\left[\begin{array}{rrrr|r} 1 & -1 & -1 & 2 & 2 \\ 2 & -1 & -1 & 2 & 1 \\ 3 & -2 & -5 & 7 & 9 \end{array}\right]$$

The following steps lead to the desired equivalent matrix.

$$\left[\begin{array}{rrrr|r} 1 & -1 & -1 & 2 & 2 \\ 2 & -1 & -1 & 2 & 1 \\ 3 & -2 & -5 & 7 & 9 \end{array}\right] \xrightarrow[-3① + ③ \to ③]{-2① + ② \to ②}$$

$$\left[\begin{array}{rrrr|r} 1 & -1 & -1 & 2 & 2 \\ 0 & 1 & 1 & -2 & -3 \\ 0 & 1 & -2 & 1 & 3 \end{array}\right] \xrightarrow{-1② + ③ \to ③}$$

$$\left[\begin{array}{rrrr|r} 1 & -1 & -1 & 2 & 2 \\ 0 & 1 & 1 & -2 & -3 \\ 0 & 0 & -3 & 3 & 6 \end{array}\right] \xrightarrow{-\frac{1}{3}③ \to ③}$$

$$\left[\begin{array}{rrrr|r} 1 & -1 & -1 & 2 & 2 \\ 0 & 1 & 1 & -2 & -3 \\ 0 & 0 & 1 & -1 & -2 \end{array}\right]$$

This last matrix represents the three equations

$$\begin{aligned} x - y - z + 2w &= 2 \\ y + z - 2w &= -3 \\ z - w &= -2 \end{aligned}$$

Letting w be replaced by the arbitrary value t, we obtain from the third of these equations

$$z = -2 + t$$

Using this value for z in the second equation,

$$\begin{aligned} y &= -3 - (-2 + t) + 2t \\ &= -1 + t \end{aligned}$$

And finally, using these values for w, y, and z in the first equation, we have

$$\begin{aligned} x &= 2 + (-1 + t) + (-2 + t) - 2t \\ &= -1 \end{aligned}$$

In ordered quadruple form, the solution to the given system is

$$(-1, t - 1, t - 2, t)$$

■

COMPUTER COMMENT: *Computer software is available for solving systems of equations. A number of computer algebra systems (CAS), with names such as* MAPLE, Derive, *and* Mathematica, *are now available for solving systems of linear equations. Although the syntax used in each CAS is unique to that system, the following example, using* MAPLE, *is typical in the way it represents a system of equations. To use* MAPLE *to solve the system*

$$\begin{aligned} x - 6y + 3z &= -2 \\ 2x - 3y + z &= -2 \\ 3x + 3y - 2z &= 2 \end{aligned}$$

type the following statement:

```
solve({x-6*y+3*z=-2, 2*x-3*y+z=-2, 3*x+3*y-2*z=2}, {x,y,z});
```

When this is entered, the solution

$$x = 1, \quad y = 3, \quad z = 5$$

appears on the screen.

Exercises Section 2.2

In Exercises 1–30, use augmented matrix notation and row operations to determine whether each system is consistent, inconsistent, or dependent. If it is consistent or dependent, give the solution.

1. $x + 2y = 10$, $x - 2y = -6$

2. $5x + 2y = 6$, $3x + 4y = 12$

3. $3x - y = 5$, $x + 4y = 6$

4. $x - 2y + z = 3$, $2x + y - z = 7$, $3x - y + 2z = 6$

5. $8x + 5y = 4$, $5y - 3z = 1$, $12x + 5z = 6$

6. $2x + y - z = 2$, $x - y + z = 4$, $y - z = -2$

7. $3x + 5y - 3z = 31$, $2x - 3y + 2z = 13$, $5x + 2y - 5z = 20$

8. $-x + 3y - z = 0$, $2x + y + 2z = 2$, $-x + 7y - z = -2$

9. $2x - 5z = 2$, $9y - 4z = 7$, $3x - 12y = -2$

10. $x - 2y + z = 7$, $2x + 3z = 4$, $y + 2z = 1$

11. $$\begin{aligned} x + y + 2z &= 3 \\ 3x - y - z &= 1 \\ x + 5y + 9z &= 7 \end{aligned}$$

12. $$\begin{aligned} 6x - 8y + 3z &= -5 \\ 15x + 12y + 7z &= 12 \\ 9x + 20y - 4z &= 25 \end{aligned}$$

13. $$\begin{aligned} 2x + y - z &= 9 \\ x - y + z &= 0 \\ -x + 3y - 2z &= 5 \end{aligned}$$

14. $$\begin{aligned} x + 2y - z &= 1 \\ y + z &= 3 \\ x - y &= 2 \end{aligned}$$

15. $$\begin{aligned} x - y &= 2 \\ y - z &= -3 \\ 3x + 2y - 5z &= 1 \end{aligned}$$

16. $$\begin{aligned} x - 2y + 3z &= 7 \\ 2x - y - z &= -4 \\ -x + 2y - 3z &= -5 \end{aligned}$$

17. $$\begin{aligned} x - 3y + 2z &= 0 \\ 2x + y - z &= -3 \\ 10x - 2y &= -12 \end{aligned}$$

18. $$\begin{aligned} -x - 2y + z &= -7 \\ x - 3y + 4z &= 2 \\ 2x - 5y - 7z &= 5 \end{aligned}$$

19. $$\begin{aligned} x + y + z &= 3 \\ x - 2y - 3z &= 4 \\ 2x + y + 2z &= 9 \end{aligned}$$

20. $$\begin{aligned} 3x + y - 2z &= 0 \\ -3x - y + 2z &= 1 \\ -2x + 6y - 7z &= -2 \end{aligned}$$

21. $$\begin{aligned} x + 4y - 3z &= 8 \\ 2x - 3y - z &= 2 \\ x + y + z &= 10 \end{aligned}$$

22. $$\begin{aligned} x - y - z &= 4 \\ 2x + y + z &= -2 \\ 3x - 2y - z &= 1 \end{aligned}$$

23. $$\begin{aligned} x + y - z &= 1 \\ x + 2y - 2z &= 1 \end{aligned}$$

24. $$\begin{aligned} 2x - y + z &= 2 \\ x - 2y - z &= 0 \end{aligned}$$

25. $$\begin{aligned} 2x - y &= 3 \\ y + 2z &= 2 \end{aligned}$$

26. $$\begin{aligned} x + 3y + z &= 2 \\ x - y &= 1 \end{aligned}$$

27. $$\begin{aligned} x + y + z &= 2 \\ 2x + 3z &= 4 \end{aligned}$$

28. $$\begin{aligned} y - z &= 2 \\ x + 2y - 5z &= -9 \end{aligned}$$

29. $$\begin{aligned} x - 2y + z &= 7 \\ 2x + 3z &= 4 \end{aligned}$$

30. $$\begin{aligned} x - 3y + 2z &= 0 \\ 2x + y - z &= -3 \end{aligned}$$

The systems in Exercises 31–34 are consistent. Find the unique solution for each by using matrix methods.

31. $$\begin{aligned} 3x - 2y + z + w &= 10 \\ x + y - z - w &= -5 \\ 2x + y + z + 5w &= 18 \\ -x - 2y + 5z - 2w &= 5 \end{aligned}$$

32. $$\begin{aligned} 2x - 5y + z - w &= -5 \\ x + 4y - z - 2w &= -7 \\ -x - 7y + 3z + w &= -5 \\ -3x + 8y - 6z - 3w &= -6 \end{aligned}$$

33. $$\begin{aligned} 3x + y + z - w &= -2 \\ -5x + 2y - 3z - 3w &= -1 \\ 7x + 5y + 4z - w &= -1 \\ 4x - y - 2z + 3w &= 7 \end{aligned}$$

34. $$\begin{aligned} 2x + 5y - 2z + w &= -11 \\ x - 8y - z + 3w &= -18 \\ -3x - 9y + 3z + 2w &= -1 \\ 6x + 7y + 5z + 8w &= 8 \end{aligned}$$

In each of the remaining problems, obtain the system of equations from the statement of the problem.

35. The total number of nickels, dimes, and quarters in a box is 900. There are 20 more nickels than there are dimes, and the number of nickels and dimes together is 44 more than the number of quarters. How many coins of each denomination are in the box?

36. A triangle has a perimeter of 50 inches. The longest side is 3 inches longer than the next longest and 10 inches longer than the shortest. Find the length of each side.

37. A company runs three production lines that together have an output of 45 parts/hr. Twice the production of the first line is equal to the combined output of the other two lines, and the output of the second line is four parts per hour more than that of the third line. Find the production rate of each line.

38. A laboratory produces an alloy of copper, tin, and zinc that weighs 37 g. If the copper in the alloy weighs 3 g more than the zinc and the zinc weighs 8 g more than the tin, how much of each element is in the alloy?

39. A dietician wishes to blend together three commercial high-fiber foods to yield 100 grams of a high-fiber food supplement that contains 80 grams of fiber and 15 grams of carbohydrates. The three brands of high-fiber food available are: Brand A, which is 70% fiber, 10% carbohydrates, and 20% fat; Brand B, which is 90% fiber, 10% carbohydrates, and 0% fat; and Brand C, which is 80% fiber, 20% carbohydrates, and 0% fat. How much of each brand should the dietician use to blend the supplement?

40. If you have access to a computer algebra system such as *MAPLE*, *Derive*, or *Mathematica*, use it to solve some of the systems of equations in this exercise set.

2.3 MATRIX ALGEBRA: ADDITION AND SCALAR MULTIPLICATION

The concept of a matrix was introduced in the previous section in connection with systems of linear equations, where matrices were used to keep track of the equations in the systems. In fact, there are many other uses of matrices that have nothing to do with systems of equations. For instance, matrices are used extensively in business, science, the social sciences, and engineering because they make it possible to keep large quantities of data in a relatively compact, organized form. Here is a simple illustration of this idea: Suppose you own a sporting goods store that sells golf balls and you stock Titleist and ProFlight balls in white, orange, and yellow. The fact that you have 150 boxes of white Titleist, 98 boxes of orange Titleist, 79 boxes of yellow Titleist, 109 boxes of white ProFlight, 72 boxes of orange ProFlight, and 33 boxes of yellow ProFlight balls can be written in matrix form as

	White	Orange	Yellow
Titleist	150	98	79
ProFlight	109	72	33

The efficiency of matrix notation in this case is obvious. There are additional applications of matrices, but before we discuss them we must define some of the basic operations in what is called matrix algebra. For convenience we will restate the definition of a matrix.

DEFINITION Matrix

A matrix is a rectangular array of numbers.

We denote a matrix with a capital letter and display it by enclosing the array in square brackets. For example, the number of different brands and colors of golf balls mentioned at the beginning of this section can be represented by the matrix

$$A = \begin{bmatrix} 150 & 98 & 79 \\ 109 & 72 & 33 \end{bmatrix}$$

Of course, we must remember that the first row represents Titleist brand and the second row represents ProFlight brand, and that the columns are for the colors, white, orange, and yellow.

The *dimensions* (or *size*) of a matrix are indicated by giving the number of rows and the number of columns in the array. The number of rows is given first. Thus, an $m \times n$ (read "m by n") matrix has m rows and n columns.

Example 1

$$A = \begin{bmatrix} 2 & 1 & -1 \\ 3 & 0 & 4 \end{bmatrix} \text{ is a } 2 \times 3 \text{ matrix}$$

$$B = \begin{bmatrix} 1 & 0 & 3 \\ 2 & -1 & 0 \\ 4 & -2 & 1 \end{bmatrix} \text{ is a } 3 \times 3 \text{ matrix}$$

$$C = \begin{bmatrix} 3 & -2 & 5 & 0 \end{bmatrix} \text{ is a } 1 \times 4 \text{ matrix}$$

$$D = \begin{bmatrix} 8 \\ -1 \\ 5 \end{bmatrix} \text{ is a } 3 \times 1 \text{ matrix}$$

■

If a matrix has only one row, like matrix C, it is called a **row matrix;** if it has only one column, like matrix D, it is called a **column matrix.** As we noted in Section 2.2, a matrix that has the same number of rows and columns, like matrix B, is called a **square matrix.** In this case, the number of rows or columns is called the **order** of the matrix. If all entries in a matrix are zeroes, the matrix is called a **zero matrix** and is denoted by **0.** For example,

$$\mathbf{0} = \begin{bmatrix} 0 & 0 & 0 \\ 0 & 0 & 0 \end{bmatrix} \text{ is a } 2 \times 3 \text{ zero matrix}$$

To identify a specific entry, or element, within a matrix, we use a lowercase letter with a double subscript to indicate its row and column location. The first number of the subscript refers to the row location and the second to the column location. Thus, the notation a_{23} means the element in the second row and the third column of matrix A. In general, the notation a_{ij} means the element in the ith row and the jth column of matrix A. The following 3×4 matrix shows how the notation is used:

$$A = \begin{bmatrix} a_{11} & a_{12} & a_{13} & a_{14} \\ a_{21} & a_{22} & a_{23} & a_{24} \\ a_{31} & a_{32} & a_{33} & a_{34} \end{bmatrix}$$

In a square matrix, elements such as a_{11}, a_{22}, and a_{33}, which have the same row and column number, are called **main diagonal elements.**

We also write the $m \times n$ matrix A as

$$A = [a_{ij}]$$

where i is the row number, which is understood to vary from 1 to m, and j is the column number, which is understood to vary from 1 to n.

CALCULATOR COMMENT: *Many of the calculators on the market today are capable of handling matrices. A common syntax for entering the elements of a matrix requires that the row and column location of each number be identified. A typical scheme is* *shown in diagram (a). This display shows a 2 × 3 matrix A in which* $a_{11} = 1$, $a_{12} = 3$, $a_{13} = 2$, $a_{21} = 1$, $a_{22} = -1$, $a_{23} = 0$. *Once the entries have been made, the calculator will store these elements in matrix A. Diagram (b) shows the format that the calculator will probably use to display matrix A.*

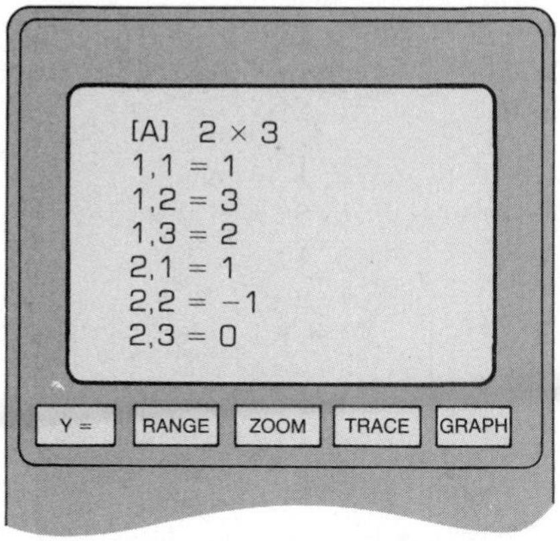

(a)

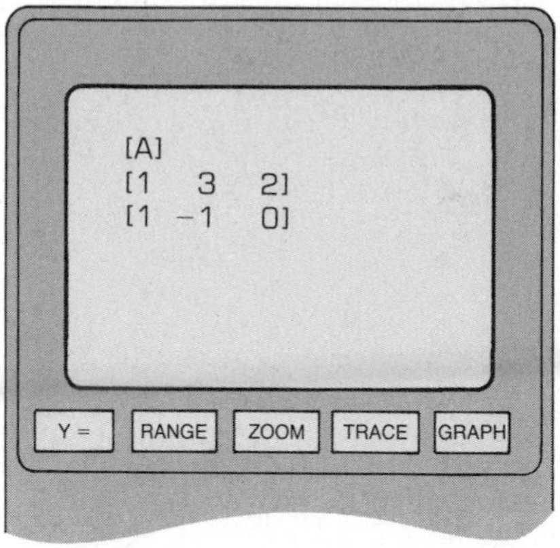

(b)

Two matrices are equal if they have the same order or dimensions and if their corresponding elements are equal. By the corresponding elements we mean elements that have the same row and column numbers. For instance, 2 and 5, −1 and 7, 4 and 1, and 0 and 3 are corresponding elements in the following matrices.

$$\begin{bmatrix} 2 & -1 \\ 0 & 4 \end{bmatrix} \quad \begin{bmatrix} 5 & 7 \\ 3 & 1 \end{bmatrix}$$

In general, if A and B are $m \times n$ matrices, then a_{11} and b_{11}, a_{12} and $b_{12}, \ldots, a_{ij}$ and b_{ij} are corresponding elements for $i = 1$ to m and $j = 1$ to n. We formalize the concept of equality of matrices in the following definition:

DEFINITION Equality of Matrices

Two matrices A and B are **equal,** written $A = B$, if they are the same size and their corresponding entries are equal.

Example 2

(a) The following matrices are equal:

$$\begin{bmatrix} 1 & 6 \\ -2 & 5 \end{bmatrix} = \begin{bmatrix} 1 & 6 \\ -2 & 5 \end{bmatrix}$$

$$\begin{bmatrix} \sqrt{4} & 2 & -1 \\ 9 & 10 & 7 \end{bmatrix} = \begin{bmatrix} 2 & 2 & -1 \\ 9 & 10 & 7 \end{bmatrix}$$

(b) The following matrices have the same dimension, but are *not* equal:

$$\begin{bmatrix} 2 & 3 & -1 \\ 9 & 0 & 7 \end{bmatrix} \neq \begin{bmatrix} 2 & 2 & -1 \\ 9 & 0 & 5 \end{bmatrix}$$

These matrices are not equal because the row 1, column 2 entries are not equal. In addition, the row 2, column 3 entries are not equal. If any of the corresponding entries are not equal, the matrices are not equal. ■

Example 3 Determine values of h and k so that the following matrices will be equal.

$$\begin{bmatrix} 3 & k-2 \\ 4 & h+1 \end{bmatrix} = \begin{bmatrix} 3 & 1 \\ 4 & 7 \end{bmatrix}$$

Solution If the two matrices are equal, the corresponding entries must be equal. This means that

$$k - 2 = 1 \qquad \text{and} \qquad h + 1 = 7$$

So

$$k = 3 \qquad \text{and} \qquad h = 6$$

are the required values of h and k. ■

Two matrices can be added together if they have the same dimensions. The sum of two matrices having the same dimensions is defined as follows:

DEFINITION Matrix Addition

If A and B are both $m \times n$ matrices, then the *sum* of A and B, written $A + B$, is obtained by adding the corresponding entries of A and B.

The definition says that if A and B are both $m \times n$ matrices and $C = A + B$, then the entries in C are formed by adding the corresponding entries in A and B; that is, $c_{11} = a_{11} + b_{11}$, $c_{12} = a_{12} + b_{12}, \ldots, c_{ij} = a_{ij} + b_{ij}$, for $i = 1$ to m and $j = 1$ to n. The process is shown in the next example.

Example 4 Find $C = A + B$ for the given matrices or explain why they may not be added.

(a) Given $A = \begin{bmatrix} 2 & 5 \\ -3 & 1 \end{bmatrix}$ and $B = \begin{bmatrix} -2 & 2 \\ 1 & 0 \end{bmatrix}$.

Since A and B are both 2×2 matrices, the desired sum is

$$C = A + B = \begin{bmatrix} 2 & 5 \\ -3 & 1 \end{bmatrix} + \begin{bmatrix} -2 & 2 \\ 1 & 0 \end{bmatrix}$$

$$= \begin{bmatrix} 2 + (-2) & 5 + 2 \\ -3 + 1 & 1 + 0 \end{bmatrix} = \begin{bmatrix} 0 & 7 \\ -2 & 1 \end{bmatrix}$$

Notice that $c_{11} = a_{11} + b_{11}$, $c_{12} = a_{12} + b_{12}$, and so on.

(b) Given $A = \begin{bmatrix} 3 & -1 \\ 7 & 12 \end{bmatrix}$ and $B = \begin{bmatrix} 4 \\ -2 \end{bmatrix}$.

These two matrices cannot be added because they are not the same size. ∎

Example 5 Suppose the sporting goods store mentioned at the beginning of this section has two branch stores, and that each branch lists its inventory of Titleist and ProFlight golf balls in a 2×3 matrix format as shown below.

$$S_1 = \begin{bmatrix} 150 & 98 & 79 \\ 109 & 72 & 33 \end{bmatrix} \qquad S_2 = \begin{bmatrix} 203 & 128 & 95 \\ 155 & 80 & 41 \end{bmatrix} \qquad S_3 = \begin{bmatrix} 98 & 76 & 69 \\ 100 & 38 & 43 \end{bmatrix}$$

The total golf-ball inventory of the three stores is then

$$T = \begin{bmatrix} 150 & 98 & 79 \\ 109 & 72 & 33 \end{bmatrix} + \begin{bmatrix} 203 & 128 & 95 \\ 155 & 80 & 41 \end{bmatrix} + \begin{bmatrix} 98 & 76 & 69 \\ 100 & 38 & 43 \end{bmatrix}$$

$$= \begin{bmatrix} 451 & 302 & 243 \\ 364 & 190 & 117 \end{bmatrix}$$

Recall that the first row represents the number of Titleist balls and the second the number of ProFlight balls, while the first column represents the number of white balls, the second the number of orange balls, and the third the number of yellow balls. We see that the total number of boxes of orange Titleist balls is 302. ∎

There are instances when we need to multiply each entry in a matrix by some constant. The operation of multiplying each element of a matrix A by a constant c is called scalar multiplication and cA is called a scalar product.

DEFINITION Scalar Multiplication

If A is an $m \times n$ matrix and c is a constant, then the matrix obtained by multiplying each entry of A by c is the **scalar product** and is denoted by cA.

Example 6 Let $A = \begin{bmatrix} 1 & 2 \\ -1 & 3 \end{bmatrix}$. Then the scalar product $2A$ is given by

$$2A = \begin{bmatrix} 2 \cdot 1 & 2 \cdot 2 \\ 2 \cdot (-1) & 2 \cdot 3 \end{bmatrix} = \begin{bmatrix} 2 & 4 \\ -2 & 6 \end{bmatrix}$$

and

$$(-1)A = \begin{bmatrix} -1 \cdot 1 & -1 \cdot 2 \\ -1 \cdot (-1) & -1 \cdot 3 \end{bmatrix} = \begin{bmatrix} -1 & -2 \\ 1 & -3 \end{bmatrix}$$

■

COMMENT: *By convention, $(-1)A$ is written $-A$. In this way, subtraction of matrices is included in the definition of matrix addition. That is,*

$$A - B = A + (-1)B$$

The following laws hold for matrix algebra whenever the matrix operation is defined. We will use these laws without proving their validity.

Laws of Matrix Algebra

Matrix Addition and Scalar Multiplication If matrices A, B, and C have the same size and c is a scalar constant, then

$$A + B = B + A$$
$$A + (B + C) = (A + B) + C$$
$$c(A + B) = cA + cB$$

Finally, we consider a matrix operation called *transposition,* which consists of interchanging the rows and columns of a matrix. This operation is used later in this book, in connection with linear programming.

DEFINITION Transpose of a Matrix

The **transpose** of a matrix A is denoted by A^t and defined as the matrix obtained by interchanging the rows and columns of A.

In transposing an $m \times n$ matrix A, the first row of A becomes the first column of A^t (that is, $a_{11} \to a_{11}, a_{12} \to a_{21}, a_{13} \to a_{31}, \ldots, a_{1n} \to a_{n1}$), the second row of A becomes the second column of A^t, and so on. This process is demonstrated in the next example. If A is an $m \times n$ matrix, then A^t is an $n \times m$ matrix.

Example 7 The transpose of

$$C = \begin{bmatrix} 2 & -1 & 4 \\ -2 & 0 & 7 \end{bmatrix} \quad \leftarrow (2 \times 3)$$

is

$$C^t = \begin{bmatrix} 2 & -2 \\ -1 & 0 \\ 4 & 7 \end{bmatrix} \quad \leftarrow (3 \times 2)$$

■

Laws of Matrix Algebra

Matrix Transposition If matrices A and B are the same size and c is a scalar constant, then

$$(A^t)^t = A$$
$$(cA)^t = cA^t$$
$$(A + B)^t = A^t + B^t$$

As we noted previously, we will use these laws without proving them.

Example 8 Show that $(A + B)^t = A^t + B^t$ for the matrices

$$A = [2 \quad 5] \qquad \text{and} \qquad B = [-1 \quad 3]$$

Solution First we note that the sum of A and B is

$$A + B = [2 \quad 5] + [-1 \quad 3] = [1 \quad 8]$$

The transpose of $A + B$ is then

$$(A + B)^t = \begin{bmatrix} 1 \\ 8 \end{bmatrix}$$

Also, the transposes of A and B are

$$A^t = \begin{bmatrix} 2 \\ 5 \end{bmatrix} \qquad \text{and} \qquad B^t = \begin{bmatrix} -1 \\ 3 \end{bmatrix}$$

So, the sum of the transposes is

$$A^t + B^t = \begin{bmatrix} 2 \\ 5 \end{bmatrix} + \begin{bmatrix} -1 \\ 3 \end{bmatrix} = \begin{bmatrix} 1 \\ 8 \end{bmatrix}$$

We have shown that $(A + B)^t = A^t + B^t$ for the two given matrices. ■

CALCULATOR COMMENT: *Calculators with matrix capability can add matrices as well as find transposes of matrices, assuming the matrices have been entered as described at the beginning of this section. For instance, if*

$$A=\begin{bmatrix}2 & -3\\ 1 & -1\end{bmatrix} \quad \text{and} \quad B=\begin{bmatrix}3 & 0\\ 5 & 1\end{bmatrix}$$

then the sum $A+B$ is displayed as shown in Figure (a) and A^t as shown in Figure (b).

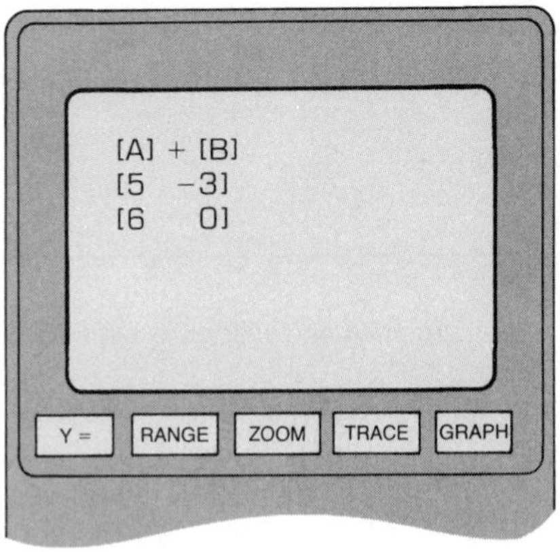

(a)

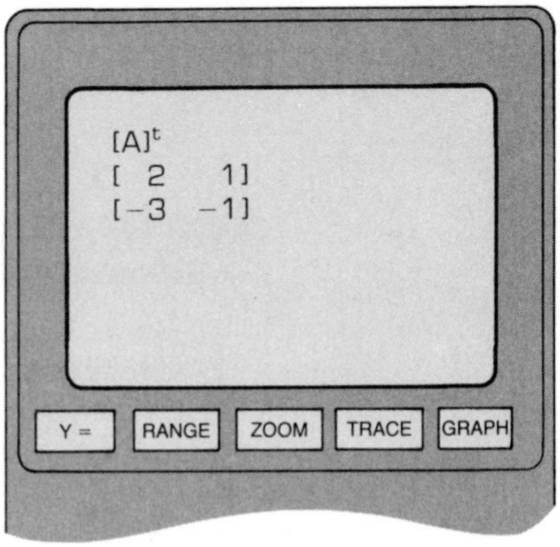

(b)

EXERCISES SECTION 2.3

In Exercises 1–4, indicate why $A \neq B$.

1. $A=\begin{bmatrix}0 & 1\\ 0 & 2\end{bmatrix}$, $B=\begin{bmatrix}0 & 1\\ 0 & 3\end{bmatrix}$

2. $A=\begin{bmatrix}1\\ 3\end{bmatrix}$, $B=\begin{bmatrix}2\\ 3\end{bmatrix}$

3. $A=\begin{bmatrix}1 & 2\\ 4 & -1\end{bmatrix}$, $B=\begin{bmatrix}1 & 2 & 0\\ 4 & -1 & 0\end{bmatrix}$

4. $A=\begin{bmatrix}1 & 1 & 1\\ 1 & 1 & 1\\ 1 & 1 & 1\end{bmatrix}$, $B=\begin{bmatrix}1 & 1\\ 1 & 1\end{bmatrix}$

In Exercises 5–10, find the values of the unknowns that will make the matrices equal.

5. $\begin{bmatrix}2 & h+2\\ 3 & 4\end{bmatrix}=\begin{bmatrix}2 & 5\\ k & 4\end{bmatrix}$

6. $\begin{bmatrix}h & -1 & 0\\ 2 & -k & 3\end{bmatrix}=\begin{bmatrix}4 & -1 & 0\\ 2 & 5 & 3\end{bmatrix}$

7. $\begin{bmatrix}2x & 3 & 7\\ 12 & 0 & y^2\end{bmatrix}=\begin{bmatrix}8 & 3 & 7\\ 12 & 0 & 16\end{bmatrix}$

8. $[x-2 \quad 1 \quad y+1 \quad 2]=[13 \quad 1 \quad 1 \quad 2]$

9. $\begin{bmatrix}h^2\\ 3k\end{bmatrix}=\begin{bmatrix}h+2\\ k^2\end{bmatrix}$

10. $\begin{bmatrix}h+12 & 3\\ -1 & 4\end{bmatrix}=\begin{bmatrix}h^2 & 3\\ -1 & 4\end{bmatrix}$

In Exercises 11–20, find $A+B$, $A-B$, and $2A$, or tell why the operation cannot be done.

11. $A=\begin{bmatrix}5 & 0\\ 1 & 1\end{bmatrix}$, $B=\begin{bmatrix}-1 & 2\\ 1 & 0\end{bmatrix}$

12. $A=\begin{bmatrix}0 & 1\\ -1 & 2\end{bmatrix}$, $B=\begin{bmatrix}2 & 1\\ 1 & 2\end{bmatrix}$

13. $A=\begin{bmatrix}1 & 2 & 3\\ 4 & 0 & 1\end{bmatrix}$, $B=\begin{bmatrix}1 & 0 & 2\\ 0 & 5 & 6\end{bmatrix}$

14. $A=\begin{bmatrix}-1\\ 5\end{bmatrix}$, $B=\begin{bmatrix}2\\ 0\end{bmatrix}$

15. $A = \begin{bmatrix} 1 \\ 0 \\ 2 \end{bmatrix}$, $B = \begin{bmatrix} 7 \\ -1 \\ 5 \end{bmatrix}$

16. $A = [1 \quad 6]$, $B = [5 \quad 2]$

17. $A = [0 \quad 3 \quad 1]$, $B = [-1 \quad 2 \quad 0]$

18. $A = [8]$, $B = [2]$

19. $A = \begin{bmatrix} 1 & 0 \\ 5 & 2 \end{bmatrix}$, $B = \begin{bmatrix} 1 \\ 2 \end{bmatrix}$

20. $A = \begin{bmatrix} 1 & 0 \\ 0 & 1 \end{bmatrix}$, $B = \begin{bmatrix} 0 & 1 \\ 1 & 0 \end{bmatrix}$

In Exercises 21–30, find the transpose of each given matrix. Verify that $(A^t)^t = A$.

21. $\begin{bmatrix} 1 & 2 \\ -1 & 5 \end{bmatrix}$

22. $\begin{bmatrix} 3 & 0 \\ -4 & 1 \end{bmatrix}$

23. $\begin{bmatrix} 2 & -1 & 0 \\ 2 & -3 & 3 \end{bmatrix}$

24. $\begin{bmatrix} 8 & 3 & 7 \\ 12 & 0 & 16 \end{bmatrix}$

25. $[-2 \quad 1 \quad 8 \quad 2]$

26. $[3 \quad 10 \quad -9]$

27. $\begin{bmatrix} 17 \\ 3 \end{bmatrix}$

28. $\begin{bmatrix} 2 \\ 13 \end{bmatrix}$

29. $\begin{bmatrix} 1 & 0 & 0 \\ 0 & 1 & 0 \\ 0 & 0 & 1 \end{bmatrix}$

30. $\begin{bmatrix} 2 & 0 \\ 0 & 2 \end{bmatrix}$

In Exercises 31–32, verify that $(A + B)^t = A^t + B^t$ for the given pairs of matrices.

31. $A = \begin{bmatrix} 5 & -2 \\ 3 & 6 \end{bmatrix}$, $B = \begin{bmatrix} 10 & 0 \\ 8 & -7 \end{bmatrix}$

32. $A = \begin{bmatrix} 1 \\ 3 \end{bmatrix}$, $B = \begin{bmatrix} 2 \\ 3 \end{bmatrix}$

33. The Softrest Chair Company produces sofas and chairs in plants operating in Dallas and Houston. Matrix A summarizes production from the two plants for January. The first row represents the production of the Dallas plant; the second row, the production of the Houston plant. The first column represents the number of sofas; the second column, the number of chairs. Matrix B represents production from the two plants for February.

$$A = \begin{bmatrix} 130 & 243 \\ 95 & 158 \end{bmatrix} \qquad B = \begin{bmatrix} 141 & 250 \\ 80 & 121 \end{bmatrix}$$

(a) How many chairs were produced by the Dallas plant in January?
(b) How many sofas were produced by the Houston plant in February?
(c) How many chairs were produced in February?
(d) What was the total production of chairs by the Dallas plant in January and February?
(e) Write a matrix representing production for the two plants for January and February.
(f) The matrix S represents shipments made during January.

$$S = \begin{bmatrix} 129 & 220 \\ 90 & 151 \end{bmatrix}$$

Use a matrix to describe the number of sofas and chairs in the inventory of the two plants at the end of January.

34. Matrix A represents the number of smokers and non-smokers participating in a psychological experiment at Good Samaritan Hospital. The first row represents participants under 40; the second row represents those over 40. The first column represents smokers; the second column, non-smokers. Matrix B represents the number of participants for the same experiment at General Hospital.

$$A = \begin{bmatrix} 78 & 55 \\ 30 & 91 \end{bmatrix} \qquad B = \begin{bmatrix} 125 & 98 \\ 89 & 201 \end{bmatrix}$$

(a) How many smokers over 40 are in the experiment at General Hospital?
(b) How many non-smokers are in the experiment at Good Samaritan?
(c) How many non-smokers over 40 are in the experiment?
(d) Write a matrix that shows the combined number of participants.
(e) Matrix C represents the number of participants who quit the experiment before it was completed.

$$C = \begin{bmatrix} 21 & 13 \\ 35 & 32 \end{bmatrix}$$

Use a matrix to show how many people completed the experiment.

Calculator Exercises

Complete Exercises 11–30 using a calculator or a computer that will do matrix operations. What response does the calculator or computer give if the operation is undefined?

2.4 MATRIX ALGEBRA: MULTIPLICATION

The Boeing airplane company builds airplanes in Seattle, Washington, and Wichita, Kansas. Suppose that matrix A represents the number of 727, 737, and 747 models built at each facility.

$$A = \begin{array}{l} \\ \text{Seattle} \\ \text{Wichita} \end{array} \begin{array}{c} \begin{array}{ccc} 727 & 737 & 747 \end{array} \\ \begin{bmatrix} 15 & 18 & 9 \\ 10 & 25 & 13 \end{bmatrix} \end{array}$$

Now suppose the cost of building each airplane is divided into the cost of airframe construction, the cost of mechanical systems, and the cost of electrical systems, and is given in the 3×3 matrix B, below. The entries are in millions of dollars (that is, 1.2 means \$1,200,000).

$$B = \begin{array}{l} \\ 727 \\ 737 \\ 747 \end{array} \begin{array}{c} \begin{array}{ccc} \text{Frame} & \text{Mech.} & \text{Elec.} \end{array} \\ \begin{bmatrix} 1.2 & 0.8 & 0.9 \\ 1.3 & 1.0 & 1.2 \\ 2.4 & 1.3 & 1.6 \end{bmatrix} \end{array}$$

The total cost of airframe construction for the 727 models built at the Seattle plant is 15×1.2—that is, the number of 727s built times the cost of airframe construction per plane. If we want the total cost of airframe construction at the Seattle plant, it would be given by the following sum:

$$15(1.2) + 18(1.3) + 9(2.4) = 63$$

That is, the total cost of airframe construction at the Seattle plant for the indicated models is \$63,000,000. Similarly, the cost of airframe construction at the Wichita plant is \$75,700,000, since

$$10(1.2) + 25(1.3) + 13(2.4) = 75.7$$

Continuing in this way, we can calculate the total cost for mechanical and electrical systems at each plant. We can then organize these six numbers that represent the total cost of airframe construction, mechanical systems, and electrical systems at the two plants into a matrix, C.

$$C = \begin{array}{l} \\ \text{Seattle} \\ \text{Wichita} \end{array} \begin{array}{c} \begin{array}{ccc} \text{Frame} & \text{Mech.} & \text{Elec.} \end{array} \\ \begin{bmatrix} 63.0 & 41.7 & 49.5 \\ 75.7 & 49.9 & 59.8 \end{bmatrix} \end{array}$$

The total cost of each operation for the three models produced at each plant is presented compactly in the 2×3 matrix C.

Our purpose in this example is to introduce the technique of **matrix multiplication.** To this end, we wish to point out that the total cost of airframe construction for all models at the Seattle plant is actually obtained by multiplying the entries in the first row of matrix A by the corresponding entries in the first column of matrix

B and then adding up the resulting products. Similarly, the total cost of mechanical systems at the Seattle plant is obtained by multiplying the entries in the first row of matrix A by the corresponding entries in the second column of matrix B and then adding up the resulting products. In fact, each entry in matrix C is obtained by multiplying the entries in a row of matrix A by the corresponding entries in a column of matrix B and then adding up the resulting products.

The process of constructing the total cost matrix C from the production matrix A and cost-per-operation matrix B in this example illustrates the process of matrix multiplication, which we now define.

DEFINITION Matrix Multiplication

Let A be an $m \times p$ matrix and B be a $p \times n$ matrix. Then the matrix product AB is the $m \times n$ matrix obtained as follows: The entry in the ith row and the jth column of AB is the sum of the products obtained by multiplying the entries in the ith row of A by the corresponding entries in the jth column of B.

Another way of stating the definition is to say that if $C = AB$, then each entry c_{ij} in product matrix C is given by

$$c_{ij} = a_{i1}b_{1j} + a_{i2}b_{2j} + \cdots + a_{ip}b_{pj}$$

COMMENT: *The definition of AB requires that the number of* columns *of A be equal to the number of* rows *of B. If the dimensions of A are $m \times p$ and the dimensions of B are $p \times n$, then the dimensions of the product matrix C are $m \times n$. That is, we have the following:*

$$A_{m \times p} \cdot B_{p \times n} = C_{m \times n}$$

Example 1 Let $A = \begin{bmatrix} 2 & 1 \\ 3 & 0 \end{bmatrix}$ and $B = \begin{bmatrix} 1 & 2 & 1 \\ 0 & 1 & 4 \end{bmatrix}$. Find AB and BA.

Solution Since there are two columns in A and two rows in B, we can form the product AB. The product matrix AB will be a 2×3 matrix. The six elements in the product $C = AB$ are generated as follows:

$$c_{11} = \begin{bmatrix} 2 & 1 \\ 3 & 0 \end{bmatrix}\begin{bmatrix} 1 & 2 & 1 \\ 0 & 1 & 4 \end{bmatrix} = 2 \cdot 1 + 1 \cdot 0 = 2$$

$$c_{12} = \begin{bmatrix} 2 & 1 \\ 3 & 0 \end{bmatrix}\begin{bmatrix} 1 & 2 & 1 \\ 0 & 1 & 4 \end{bmatrix} = 2 \cdot 2 + 1 \cdot 1 = 5$$

$$c_{13} = \begin{bmatrix} 2 & 1 \\ 3 & 0 \end{bmatrix}\begin{bmatrix} 1 & 2 & 1 \\ 0 & 1 & 4 \end{bmatrix} = 2 \cdot 1 + 1 \cdot 4 = 6$$

$$c_{21} = \begin{bmatrix} 2 & 1 \\ 3 & 0 \end{bmatrix}\begin{bmatrix} 1 & 2 & 1 \\ 0 & 1 & 4 \end{bmatrix} = 3 \cdot 1 + 0 \cdot 0 = 3$$

$$c_{22} = \begin{bmatrix} 2 & 1 \\ 3 & 0 \end{bmatrix} \begin{bmatrix} 1 & 2 & 1 \\ 0 & 1 & 4 \end{bmatrix} = 3 \cdot 2 + 0 \cdot 1 = 6$$

$$c_{23} = \begin{bmatrix} 2 & 1 \\ 3 & 0 \end{bmatrix} \begin{bmatrix} 1 & 2 & 1 \\ 0 & 1 & 4 \end{bmatrix} = 3 \cdot 1 + 0 \cdot 4 = 3$$

The product matrix is

$$C = AB = \begin{bmatrix} 2 & 5 & 6 \\ 3 & 6 & 3 \end{bmatrix}$$

The product BA cannot be obtained, because B has three columns and A has only two rows. ▪

Example 2 Let $A = [2 \quad 1 \quad -1]$ and $B = \begin{bmatrix} 1 \\ 5 \\ -2 \end{bmatrix}$. Find AB and BA.

Solution A is a 1×3 matrix and B is a 3×1 matrix. Since A has three columns and B has three rows, the product AB can be formed. AB is a 1×1 matrix:

$$AB = [2 \quad 1 \quad -1] \begin{bmatrix} 1 \\ 5 \\ -2 \end{bmatrix} = [2 \cdot 1 + 1 \cdot 5 + (-1) \cdot (-2)] = [9]$$

From the dimensions of A and B, we note that the product BA is a 3×3 matrix:

$$BA = \begin{bmatrix} 1 \\ 5 \\ -2 \end{bmatrix} [2 \quad 1 \quad -1] = \begin{bmatrix} 1 \cdot 2 & 1 \cdot 1 & 1 \cdot (-1) \\ 5 \cdot 2 & 5 \cdot 1 & 5 \cdot (-1) \\ -2 \cdot 2 & -2 \cdot 1 & -2 \cdot (-1) \end{bmatrix}$$

$$= \begin{bmatrix} 2 & 1 & -1 \\ 10 & 5 & -5 \\ -4 & -2 & 2 \end{bmatrix}$$

▪

We see from the preceding two examples that the products AB and BA of matrices A and B are not necessarily the same. In Example 1, the product BA was undefined; in Example 2, the product AB was not the same size as BA. The next example shows that even when both AB and BA are defined and of the same size, they may not be equal.

Example 3 Let $A = \begin{bmatrix} 1 & -1 \\ 2 & 0 \end{bmatrix}$ and $B = \begin{bmatrix} 0 & 1 \\ 1 & -1 \end{bmatrix}$. Find AB and BA.

Solution

$$AB = \begin{bmatrix} 0 + (-1) & 1 + 1 \\ 0 + 0 & 2 + 0 \end{bmatrix} = \begin{bmatrix} -1 & 2 \\ 0 & 2 \end{bmatrix}$$

$$BA = \begin{bmatrix} 0+2 & 0+0 \\ 1-2 & -1+0 \end{bmatrix} = \begin{bmatrix} 2 & 0 \\ -1 & -1 \end{bmatrix}$$ ■

COMMENT: *The preceding examples show that matrix multiplication is not commutative. However, matrix multiplication is associative; that is, $A(BC) = (AB)C$. Matrix multiplication is also distributive over matrix addition; that is, $A(B + C) = AB + AC$.*

CALCULATOR COMMENT: *Calculators with matrix capability can perform matrix multiplication on matrices that have been entered as described in Section 2.3. For instance, if*

$$A = \begin{bmatrix} 2 & -3 \\ 1 & -1 \end{bmatrix} \quad \text{and} \quad B = \begin{bmatrix} 3 & 0 \\ 5 & 1 \end{bmatrix}$$

then the matrix product $A \cdot B$ is displayed in the following figure.

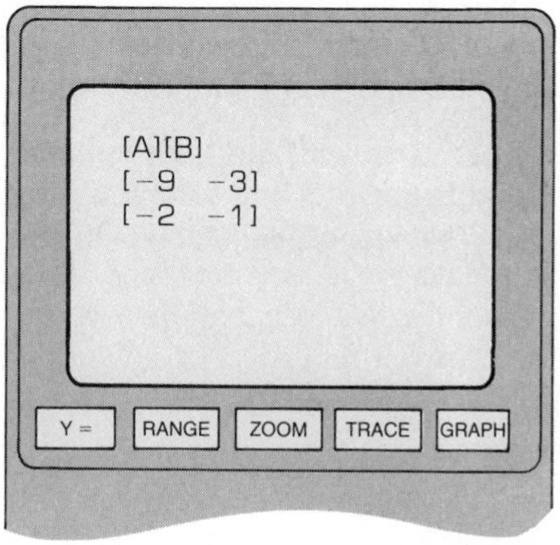

We conclude this section with a discussion of the ways in which matrix multiplication helps simplify the writing of systems of equations. Recall from Section 2.2 that we can use augmented matrices to solve systems of equations. Matrix notation can also be used to represent systems of equations in the form of matrix equations. To illustrate this notation, consider the following system of equations:

$$\begin{aligned} x - 6y + 3z &= -2 \\ 2x - 3y + z &= -2 \\ 3x + 3y - 2z &= 2 \end{aligned}$$

This is the same system solved in Examples 1 and 3 in Section 2.2. We let A be the 3×3 matrix of coefficients

$$A = \begin{bmatrix} 1 & -6 & 3 \\ 2 & -3 & 1 \\ 3 & 3 & -2 \end{bmatrix}$$

and B be the 3×1 matrix of constants

$$B = \begin{bmatrix} -2 \\ -2 \\ 2 \end{bmatrix}$$

Finally, we represent the variables x, y, and z as a 3×1 matrix:

$$X = \begin{bmatrix} x \\ y \\ z \end{bmatrix}$$

The given system of equations may then be written as a matrix equation of the form

$$AX = B$$

or

$$\begin{bmatrix} 1 & -6 & 3 \\ 2 & -3 & 1 \\ 3 & 3 & -2 \end{bmatrix} \begin{bmatrix} x \\ y \\ z \end{bmatrix} = \begin{bmatrix} -2 \\ -2 \\ 2 \end{bmatrix}$$

Solving the matrix equation $AX = B$ means finding the values of the elements in the matrix X. Any matrix X that satisfies the equation $AX = B$ is called a *solution* of the matrix equation. We know from Examples 1 and 3 in Section 2.2 that $x = 1$, $y = 3$, and $z = 5$ is a solution of the given system of equations. Consequently,

$$X = \begin{bmatrix} 1 \\ 3 \\ 5 \end{bmatrix}$$

is a solution of $AX = B$ for the given coefficient matrices. Example 4 demonstrates that this is a correct solution.

Example 4 Let $A = \begin{bmatrix} 1 & -6 & 3 \\ 2 & -3 & 1 \\ 3 & 3 & -2 \end{bmatrix}$ and $B = \begin{bmatrix} -2 \\ -2 \\ 2 \end{bmatrix}$. Show that $X = \begin{bmatrix} 1 \\ 3 \\ 5 \end{bmatrix}$ is a solution of $AX = B$.

Solution

$$\begin{bmatrix} 1 & -6 & 3 \\ 2 & -3 & 1 \\ 3 & 3 & -2 \end{bmatrix} \begin{bmatrix} 1 \\ 3 \\ 5 \end{bmatrix} = \begin{bmatrix} 1 \cdot 1 + (-6) \cdot 3 + 3 \cdot 5 \\ 2 \cdot 1 + (-3) \cdot 3 + 1 \cdot 5 \\ 3 \cdot 1 + 3 \cdot 3 + (-2) \cdot 5 \end{bmatrix} = \begin{bmatrix} -2 \\ -2 \\ 2 \end{bmatrix} = B$$

This shows that the given matrix X is a solution of $AX = B$. ▪

EXERCISES SECTION 2.4

In Exercises 1–10, find AB and BA, or tell why the product is undefined.

1. $A = \begin{bmatrix} 5 & 0 \\ 1 & 1 \end{bmatrix}$, $B = \begin{bmatrix} -1 & 2 \\ 1 & 0 \end{bmatrix}$

2. $A = \begin{bmatrix} 0 & 1 \\ -1 & 2 \end{bmatrix}$, $B = \begin{bmatrix} 2 & 1 \\ 1 & 2 \end{bmatrix}$

3. $A = \begin{bmatrix} 1 & 2 & 3 \\ 4 & 0 & 1 \end{bmatrix}$, $B = \begin{bmatrix} 1 & 0 & 2 \\ 0 & 5 & 6 \end{bmatrix}$

4. $A = \begin{bmatrix} -1 \\ 5 \end{bmatrix}$, $B = \begin{bmatrix} 2 \\ 0 \end{bmatrix}$

5. $A = \begin{bmatrix} -1 \\ 5 \end{bmatrix}$, $B = [2 \quad 0]$

6. $A = \begin{bmatrix} 2 \\ 0 \\ 1 \end{bmatrix}$, $B = [-5 \quad 5 \quad 3]$

7. $A = \begin{bmatrix} 2 & 1 \\ 1 & -1 \end{bmatrix}$, $B = \begin{bmatrix} 1 \\ 2 \end{bmatrix}$

8. $A = [3]$, $B = [2 \quad 5]$

9. $A = \begin{bmatrix} 2 & 0 & 5 \\ 6 & 1 & 4 \\ -2 & 1 & 1 \end{bmatrix}$, $B = \begin{bmatrix} 1 & 0 \\ -1 & 1 \\ 2 & 3 \end{bmatrix}$

10. $A = \begin{bmatrix} -2 & 2 \\ 1 & 3 \\ 3 & -2 \end{bmatrix}$, $B = \begin{bmatrix} 2 & -3 & 1 \\ 1 & 0 & -2 \\ -1 & 5 & 2 \end{bmatrix}$

In Exercises 11–20, write each system of equations in the form AX = B.

11. $\begin{aligned} 2x + 3y &= 7 \\ 3x + y &= 10 \end{aligned}$

12. $\begin{aligned} x + y &= -2 \\ 2x - 3y &= 15 \end{aligned}$

13. $\begin{aligned} 5x + y &= 0 \\ x - 3y &= 7 \end{aligned}$

14. $\begin{aligned} -x - 2y &= 3 \\ 6x - 5y &= -1 \end{aligned}$

15. $\begin{aligned} x + y + z &= 1 \\ 2x - y - z &= 2 \\ x - y + z &= 12 \end{aligned}$

16. $\begin{aligned} 2x - 3y - z &= 1 \\ x + y - 2z &= 0 \\ -x + y + z &= -2 \end{aligned}$

17. $\begin{aligned} x + 3y - z &= 0 \\ 4y - 2z &= 1 \\ 2x + z &= 13 \end{aligned}$

18. $\begin{aligned} x + y &= 3 \\ x - y &= -2 \\ y + 2z &= 5 \end{aligned}$

19. $\begin{aligned} r + s - 4t &= 7 \\ 2r + 3s - t &= 0 \\ 7r + 2t &= 6 \end{aligned}$

20. $\begin{aligned} 2a - 4b - 7c &= 23 \\ a - 5c &= 10 \\ 3b + c &= -1 \end{aligned}$

In Exercises 21–26, show that X is a solution of AX = B.

21. $A = \begin{bmatrix} 2 & 1 \\ 1 & -1 \end{bmatrix}$, $B = \begin{bmatrix} 1 \\ 2 \end{bmatrix}$, $X = \begin{bmatrix} 1 \\ -1 \end{bmatrix}$

22. $A = \begin{bmatrix} 5 & 2 \\ 1 & -3 \end{bmatrix}$, $B = \begin{bmatrix} 8 \\ 5 \end{bmatrix}$, $X = \begin{bmatrix} 2 \\ -1 \end{bmatrix}$

23. $A = \begin{bmatrix} 2 & 1 & 1 \\ 1 & -1 & 1 \\ 1 & -2 & 2 \end{bmatrix}$, $B = \begin{bmatrix} 6 \\ 3 \\ 5 \end{bmatrix}$, $X = \begin{bmatrix} 1 \\ 1 \\ 3 \end{bmatrix}$

24. $A = \begin{bmatrix} 2 & 2 & -3 \\ 1 & -1 & 2 \end{bmatrix}$, $B = \begin{bmatrix} 1 \\ 2 \end{bmatrix}$, $X = \begin{bmatrix} 1 \\ 1 \\ 1 \end{bmatrix}$

25. $A = \begin{bmatrix} 2 & 1 & 3 \\ 1 & -2 & -1 \end{bmatrix}$, $B = \begin{bmatrix} 3 \\ -\frac{7}{2} \end{bmatrix}$, $X = \begin{bmatrix} -\frac{1}{2} \\ 1 \\ 1 \end{bmatrix}$

26. $A = \begin{bmatrix} 1 & 1 & 0 \\ 0 & 1 & 0 \\ 0 & 0 & 0 \end{bmatrix}$, $B = \begin{bmatrix} 0 \\ 0 \\ 0 \end{bmatrix}$, $X = a\begin{bmatrix} 0 \\ 0 \\ 1 \end{bmatrix}$

27. Let $A = \begin{bmatrix} 1 & 2 \\ -1 & 0 \end{bmatrix}$, $B = \begin{bmatrix} 0 & 1 \\ 5 & 2 \end{bmatrix}$, and $C = \begin{bmatrix} 3 & 5 \\ 0 & 2 \end{bmatrix}$. Show that $A(BC) = (AB)C$. Which property does this result demonstrate?

28. Let A, B, and C be the matrices given in Exercise 27. Show that $A(B + C) = AB + AC$. Which property does this result demonstrate?

29. Let A and B be the matrices given in Exercise 27. Show that $A^2 - B^2 \neq (A - B)(A + B)$. (*Note:* $A^2 = AA$, $B^2 = BB$.)

30. Let A and B be the matrices given in Exercise 27. Show that $(A + B)^2 \neq A^2 + 2AB + B^2$.

31. Determine an appropriate formula for $(A + B)^2$ that will be valid even when $AB \neq BA$.

32. A square matrix is called a *diagonal matrix* if all entries off the main diagonal are zero. Determine a rule for multiplying diagonal matrices.

33. Let $A = \begin{bmatrix} 0 & 1 \\ 0 & 1 \end{bmatrix}$, $B = \begin{bmatrix} 1 & 1 \\ 1 & 0 \end{bmatrix}$, and $C = \begin{bmatrix} 0 & 0 \\ 1 & 0 \end{bmatrix}$. Show that $AB = AC$ even though $B \neq C$. That is, the cancellation law does not hold for matrix multiplication.

34. Let $A = \begin{bmatrix} 1 & 0 \\ 1 & 0 \end{bmatrix}$ and $D = \begin{bmatrix} 0 & 0 \\ 1 & 1 \end{bmatrix}$. Show that $AD = \mathbf{0}$ even though $A \neq \mathbf{0}$ and $D \neq \mathbf{0}$. Note that this result differs from that of a similar property of real-number multiplication.

35. The registration office at California State University lists the number of second-year men and women students majoring in engineering, business, and education. The numbers are shown in matrix A.

$A =$	Men	Women
Engr	450	102
Bus	300	269
Educ	130	207

Based on past data, the housing office expects that 0.5 of the male second-year students will choose to live on campus, 0.4 will choose to live off campus, and 0.1 will commute. Of the women, it is expected that 0.6 will choose to live on campus, 0.2 will choose to live off campus, and 0.2 will commute. These numbers are given in matrix B.

$B =$	On	Off	Comm.
Men	0.5	0.4	0.1
Women	0.6	0.2	0.2

(a) How many second-year business majors will live on campus?

(b) Construct a matrix that shows the number of second-year students, by major, who will choose to live on campus, who will choose to live off campus, and who will commute.

36. A company manufactures two types of wooden pull-toys (a dog and a cat) at two different plants. The number of each type of toy made each day at the two plants is shown in the following matrix:

	Dog	Cat
Plant 1	500	350
Plant 2	780	610

Three operations are involved in the manufacturing process: cutting, sanding, and painting. The time, in hours, required to perform each of these operations for each type of toy is shown in the next matrix.

	Cut	Sand	Paint
Dog	0.1	0.1	0.2
Cat	0.2	0.1	0.3

(a) How many hours are spent in the cutting operation at Plant 1?

(b) Construct a matrix to show the total number of hours devoted to each operation at each plant.

37. A swimming-pool manufacturer produces three styles of pools. It has orders for 500 of Style A, 2000 of Style B, and 1500 of Style C. The table below lists the number of units of raw materials the manufacturer needs for each style.

	Steel	Plastic
Style A	30	20
Style B	20	10
Style C	15	20

Steel costs \$20 per unit; plastic, \$10 per unit.

(a) Write a 1×3 matrix showing the number of units of each style ordered.

(b) Write a 3×2 matrix showing the number of units of steel and plastic needed to manufacture each style.

(c) Use matrix multiplication to determine the total number of units of each raw material needed to produce each style.

(d) Write a 2×1 matrix containing the cost per unit for each of the raw materials.

(e) Use matrix multiplication to find the total investment required to fill the orders.

Calculator Exercises

38. Complete Exercises 1–10 using a calculator or computer that will perform matrix operations. What response does the calculator or computer give when the operation is undefined?

39. Use a calculator or computer with matrix capability to complete Exercises 21–26.

2.5 SOLVING SYSTEMS OF EQUATIONS BY MATRIX INVERSION

In Section 2.4, we saw that a system of n equations can be represented by a matrix equation of the form

$$AX = B$$

where A is an $n \times n$ matrix of coefficients, B is an $n \times 1$ matrix of constants, and X is an $n \times 1$ matrix of unknowns or variables. In this section, we introduce a method for solving matrix equations of the form $AX = B$. First, we will restate the definition of the identity matrix from Section 2.2.

DEFINITION Identity Matrix

The $n \times n$ square matrix I with ones on the main diagonal and zeroes elsewhere is called the **identity matrix of order n.**

The identity matrix of order 2 is

$$I = \begin{bmatrix} 1 & 0 \\ 0 & 1 \end{bmatrix}$$

and the identity matrix of order 3 is

$$I = \begin{bmatrix} 1 & 0 & 0 \\ 0 & 1 & 0 \\ 0 & 0 & 1 \end{bmatrix}$$

The result of multiplying the $n \times n$ matrix A by the identity matrix I of order n is the matrix A, *unchanged.* For example,

$$\begin{bmatrix} 2 & -3 \\ 1 & 4 \end{bmatrix}\begin{bmatrix} 1 & 0 \\ 0 & 1 \end{bmatrix} = \begin{bmatrix} 2 & -3 \\ 1 & 4 \end{bmatrix}$$

COMMENT: *An identity matrix is always a square matrix.*

Example 1 Let $A = \begin{bmatrix} 3 & 5 \\ 2 & 7 \end{bmatrix}$. Then

$$AI = \begin{bmatrix} 3 & 5 \\ 2 & 7 \end{bmatrix}\begin{bmatrix} 1 & 0 \\ 0 & 1 \end{bmatrix} = \begin{bmatrix} 3 & 5 \\ 2 & 7 \end{bmatrix}$$

and

$$IA = \begin{bmatrix} 1 & 0 \\ 0 & 1 \end{bmatrix}\begin{bmatrix} 3 & 5 \\ 2 & 7 \end{bmatrix} = \begin{bmatrix} 3 & 5 \\ 2 & 7 \end{bmatrix}$$

■

The result obtained in Example 1 is true for all square matrices. We state this in the following property.

DEFINITION Identity Property

Let A be a square matrix of order n and let I be the identity matrix of order n. Then

$$AI = IA = A$$

A square matrix A for which a square matrix B can be found such that

$$AB = BA = I$$

is said to be *invertible*. The matrix B is called the *inverse matrix* of A, and it can be shown that such an inverse is unique.

DEFINITION Inverse Matrix

Let A be a square matrix of order n. The **inverse matrix** of A is denoted by A^{-1} and has the property that

$$AA^{-1} = A^{-1}A = I$$

COMMENT: *Since $AA^{-1} = A^{-1}A = I$, if we want to show that a given matrix B is an inverse of A, it is sufficient to show that $AB = I$.*

Example 2 Let $A = \begin{bmatrix} 2 & 1 \\ 3 & 2 \end{bmatrix}$. Show that the inverse of A is $B = \begin{bmatrix} 2 & -1 \\ -3 & 2 \end{bmatrix}$.

Solution To show that B is the inverse of A, we show that $AB = I$.

$$AB = \begin{bmatrix} 2 & 1 \\ 3 & 2 \end{bmatrix}\begin{bmatrix} 2 & -1 \\ -3 & 2 \end{bmatrix} = \begin{bmatrix} 2\cdot 2 + 1\cdot(-3) & 2\cdot(-1) + 1\cdot 2 \\ 3\cdot 2 + 2\cdot(-3) & 3\cdot(-1) + 2\cdot 2 \end{bmatrix}$$
$$= \begin{bmatrix} 1 & 0 \\ 0 & 1 \end{bmatrix} = I$$

We have shown that B is the inverse of A; we have also shown that A is the inverse of B. ▪

COMMENT: *Not every square matrix has an inverse. For instance, the matrix*

$$\begin{bmatrix} 4 & 6 \\ 2 & 3 \end{bmatrix}$$

does not have an inverse. A square matrix that has an inverse is said to be ***nonsingular;*** *one that does not have an inverse is said to be* ***singular.***

We now describe a procedure for finding the inverse matrix of any nonsingular matrix A using an augmented matrix. In this case, we augment the matrix A with the identity matrix I and write

$$[A \mid I]$$

We then perform row operations (see Section 2.2) on the augmented matrix until A is transformed into the identity matrix. The inverse matrix will then appear on the right within the augmented matrix:

$$[A \mid I] \to [I \mid A^{-1}]$$

Example 3 Let $A = \begin{bmatrix} 3 & -7 \\ 6 & 2 \end{bmatrix}$. Find A^{-1}.

Solution We form the augmented matrix

$$\left[\begin{array}{cc|cc} 3 & -7 & 1 & 0 \\ 6 & 2 & 0 & 1 \end{array}\right]$$

Next, we perform row operations on the augmented matrix until we obtain the identity matrix on the left-hand side of the dashed line. First, we multiply the first row by -2, add the result to the second row, and replace the second row with this sum. This operation is expressed by

$$\left[\begin{array}{cc|cc} 3 & -7 & 1 & 0 \\ 6 & 2 & 0 & 1 \end{array}\right] \xrightarrow{-2① + ② \to ②} \left[\begin{array}{cc|cc} 3 & -7 & 1 & 0 \\ 0 & 16 & -2 & 1 \end{array}\right]$$

To obtain a 1 in the first row of the first column, we multiply the first row by $\frac{1}{3}$.

$$\left[\begin{array}{cc|cc} 3 & -7 & 1 & 0 \\ 0 & 16 & -2 & 1 \end{array}\right] \xrightarrow{\frac{1}{3}① \to ①} \left[\begin{array}{cc|cc} 1 & -\frac{7}{3} & \frac{1}{3} & 0 \\ 0 & 16 & -2 & 1 \end{array}\right]$$

Next, we multiply the second row by $\frac{1}{16}$.

$$\left[\begin{array}{cc|cc} 1 & -\frac{7}{3} & \frac{1}{3} & 0 \\ 0 & 16 & -2 & 1 \end{array}\right] \xrightarrow{\frac{1}{16}② \to ②} \left[\begin{array}{cc|cc} 1 & -\frac{7}{3} & \frac{1}{3} & 0 \\ 0 & 1 & -\frac{1}{8} & \frac{1}{16} \end{array}\right]$$

Finally, we multiply the second row by $\frac{7}{3}$, add the result to the first row, and replace the first row with this sum.

$$\left[\begin{array}{cc|cc} 1 & -\frac{7}{3} & \frac{1}{3} & 0 \\ 0 & 1 & -\frac{1}{8} & \frac{1}{16} \end{array}\right] \xrightarrow{\frac{7}{3}② + ① \to ①} \left[\begin{array}{cc|cc} 1 & 0 & \frac{1}{24} & \frac{7}{48} \\ 0 & 1 & -\frac{1}{8} & \frac{1}{16} \end{array}\right]$$

We now have the identity matrix on the left-hand side. The inverse of A is

$$A^{-1} = \begin{bmatrix} \frac{1}{24} & \frac{7}{48} \\ -\frac{1}{8} & \frac{1}{16} \end{bmatrix}$$

We check this result by considering $A^{-1}A$:

$$A^{-1}A = \begin{bmatrix} \frac{1}{24} & \frac{7}{48} \\ -\frac{1}{8} & \frac{1}{16} \end{bmatrix} \begin{bmatrix} 3 & -7 \\ 6 & 2 \end{bmatrix} = \begin{bmatrix} 1 & 0 \\ 0 & 1 \end{bmatrix}$$ ■

COMMENT: *If it is impossible to transform the augmented matrix such that the identity matrix appears on the left-hand side, then A^{-1} does not exist.*

Example 4 Using the procedure of row operations, find A^{-1} for

$$A = \begin{bmatrix} 2 & 1 & 0 \\ 4 & 1 & -1 \\ 2 & 1 & -1 \end{bmatrix}$$

Solution We form the augmented matrix

$$\left[\begin{array}{ccc|ccc} 2 & 1 & 0 & 1 & 0 & 0 \\ 4 & 1 & -1 & 0 & 1 & 0 \\ 2 & 1 & -1 & 0 & 0 & 1 \end{array}\right]$$

Then we have the following equivalent matrices:

$$\xrightarrow{\frac{1}{2}① \to ①} \left[\begin{array}{ccc|ccc} 1 & \frac{1}{2} & 0 & \frac{1}{2} & 0 & 0 \\ 4 & 1 & -1 & 0 & 1 & 0 \\ 2 & 1 & -1 & 0 & 0 & 1 \end{array}\right]$$

$$\xrightarrow[-2① + ③ \to ③]{-4① + ② \to ②} \left[\begin{array}{ccc|ccc} 1 & \frac{1}{2} & 0 & \frac{1}{2} & 0 & 0 \\ 0 & -1 & -1 & -2 & 1 & 0 \\ 0 & 0 & -1 & -1 & 0 & 1 \end{array}\right]$$

$$\xrightarrow[-1③ \to ③]{-1② \to ②} \left[\begin{array}{ccc|ccc} 1 & \frac{1}{2} & 0 & \frac{1}{2} & 0 & 0 \\ 0 & 1 & 1 & 2 & -1 & 0 \\ 0 & 0 & 1 & 1 & 0 & -1 \end{array}\right]$$

$$\xrightarrow{-1③ + ② \to ②} \left[\begin{array}{ccc|ccc} 1 & \frac{1}{2} & 0 & \frac{1}{2} & 0 & 0 \\ 0 & 1 & 0 & 1 & -1 & 1 \\ 0 & 0 & 1 & 1 & 0 & -1 \end{array}\right]$$

$$\xrightarrow{-\frac{1}{2}② + ① \to ①} \left[\begin{array}{ccc|ccc} 1 & 0 & 0 & 0 & \frac{1}{2} & -\frac{1}{2} \\ 0 & 1 & 0 & 1 & -1 & 1 \\ 0 & 0 & 1 & 1 & 0 & -1 \end{array}\right]$$

The inverse matrix of A is then

$$A^{-1} = \begin{bmatrix} 0 & \frac{1}{2} & -\frac{1}{2} \\ 1 & -1 & 1 \\ 1 & 0 & -1 \end{bmatrix}$$

Checking this result, we have

$$A^{-1}A = \begin{bmatrix} 0 & \frac{1}{2} & -\frac{1}{2} \\ 1 & -1 & 1 \\ 1 & 0 & -1 \end{bmatrix} \begin{bmatrix} 2 & 1 & 0 \\ 4 & 1 & -1 \\ 2 & 1 & -1 \end{bmatrix} = \begin{bmatrix} 1 & 0 & 0 \\ 0 & 1 & 0 \\ 0 & 0 & 1 \end{bmatrix}$$

■

CALCULATOR COMMENT: *The inverse of a matrix can be found by using a calculator with matrix capability. To find the inverse of matrix A in Example 4, enter A as described in your owner's manual. A typical operation for A^{-1} is to enter A and then push the $\boxed{x^{-1}}$ key. In the following figure, matrix A is shown in (a) as it might be displayed on a calculator screen and A^{-1} is shown in (b).*

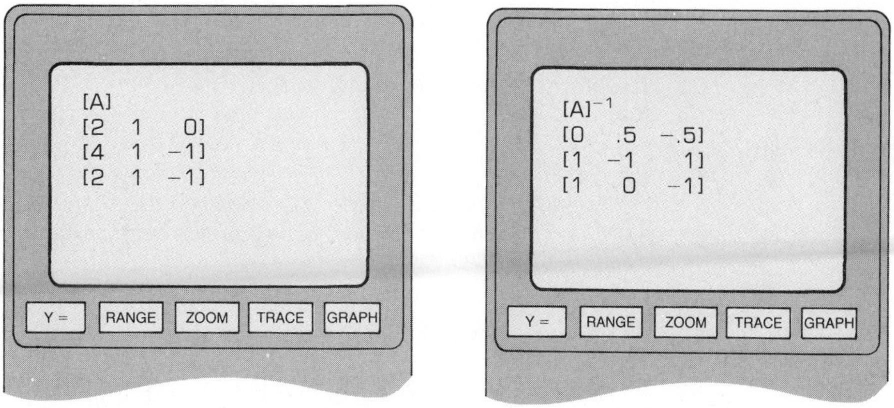

(a) (b)

We know that $AA^{-1} = I$, but this may not be obvious to you when a calculator is used to find AA^{-1}. For instance, consider the matrix A given by

$$A = \begin{bmatrix} 2 & 3 & 0 \\ 4 & 5 & 1 \\ -2 & 7 & 1 \end{bmatrix}$$

The TI-81 graphing calculator gives

$$A^{-1} = \begin{bmatrix} .09 & .14 & -.14 \\ .27 & -.09 & .09 \\ -1.73 & .91 & .09 \end{bmatrix}$$

But when the calculator is used to calculate AA^{-1}, the result is

$$AA^{-1} = \begin{bmatrix} 1 & -2 \times 10^{-13} & 0 \\ 6 \times 10^{-13} & 1 & 0 \\ -3 \times 10^{-13} & -2 \times 10^{-13} & 1 \end{bmatrix}$$

which does not appear to be the identity matrix. The problem is caused by the need for the calculator to round off the entries in A^{-1}. A nonzero entry such as -2×10^{-13} (which is very close to zero) is caused by round-off error. You should be alert to such problems when using a calculator or a computer.

One of the main uses of inverse matrices is solving a system of n equations in n unknowns. Such a system can be represented by

$$AX = B$$

where A is a square $n \times n$ matrix and X is the $n \times 1$ matrix of unknowns. If A^{-1} exists, we can multiply both sides of $AX = B$ on the left by A^{-1} to obtain

$$A^{-1}(AX) = A^{-1}B$$

The left-hand side of this equation becomes

$$A^{-1}(AX) = (A^{-1}A)X = IX = X$$

Therefore, the solution of $AX = B$ is

$$X = A^{-1}B$$

COMMENT: *When solving the matrix equation $AX = B$ for X, it is necessary to multiply by A^{-1} on the* **left** *on both sides of an equation because matrix multiplication is not commutative.*

Example 5 Solve the matrix equation $AX = B$, where

$$A = \begin{bmatrix} 2 & 1 & 0 \\ 4 & 1 & -1 \\ 2 & 1 & -1 \end{bmatrix}, \quad X = \begin{bmatrix} x_1 \\ x_2 \\ x_3 \end{bmatrix}, \quad B = \begin{bmatrix} 2 \\ -2 \\ 4 \end{bmatrix}$$

Solution Observing that A is the matrix given in Example 4, we have

$$A^{-1} = \begin{bmatrix} 0 & \frac{1}{2} & -\frac{1}{2} \\ 1 & -1 & 1 \\ 1 & 0 & -1 \end{bmatrix}$$

The solution to $AX = B$ is then

$$X = A^{-1}B = \begin{bmatrix} 0 & \frac{1}{2} & -\frac{1}{2} \\ 1 & -1 & 1 \\ 1 & 0 & -1 \end{bmatrix} \begin{bmatrix} 2 \\ -2 \\ 4 \end{bmatrix} = \begin{bmatrix} 0 - 1 - 2 \\ 2 + 2 + 4 \\ 2 + 0 - 4 \end{bmatrix} = \begin{bmatrix} -3 \\ 8 \\ -2 \end{bmatrix}$$

Thus, $x_1 = -3$, $x_2 = 8$, and $x_3 = -2$. ▪

Example 6 Solve the system

$$\begin{aligned} 2x + y &= 3 \\ 3x + 2y &= -4 \end{aligned}$$

by the method of matrix inversion.

Solution This system can be represented in the form $AX = B$ with

$$A = \begin{bmatrix} 2 & 1 \\ 3 & 2 \end{bmatrix}, \quad X = \begin{bmatrix} x \\ y \end{bmatrix}, \quad B = \begin{bmatrix} 3 \\ -4 \end{bmatrix}$$

Then, from Example 2,

$$A^{-1} = \begin{bmatrix} 2 & -1 \\ -3 & 2 \end{bmatrix}$$

Therefore,

$$X = \begin{bmatrix} 2 & -1 \\ -3 & 2 \end{bmatrix} \begin{bmatrix} 3 \\ -4 \end{bmatrix} = \begin{bmatrix} 10 \\ -17 \end{bmatrix}$$

Thus, $x = 10$ and $y = -17$ is the solution of the given system. ■

The method of matrix inversion is not a time-saver except when we are required to solve several systems of the form

$$AX = B_1, AX = B_2, \ldots, AX = B_n$$

Then, all solutions are easily obtained once A^{-1} has been found. This is more efficient than using the method of Gauss-Jordan elimination on each of the systems.

Example 7 Solve $AX = B$, if $B_1 = \begin{bmatrix} 1 \\ 0 \end{bmatrix}$, $B_2 = \begin{bmatrix} 0 \\ 1 \end{bmatrix}$, $B_3 = \begin{bmatrix} 0 \\ 0 \end{bmatrix}$, and $B_4 = \begin{bmatrix} 1 \\ 5 \end{bmatrix}$, and $A = \begin{bmatrix} 2 & 1 \\ 3 & 2 \end{bmatrix}$.

Solution As in Example 6, $A^{-1} = \begin{bmatrix} 2 & -1 \\ -3 & 2 \end{bmatrix}$, so

$$X_1 = A^{-1}\begin{bmatrix} 1 \\ 0 \end{bmatrix}, \quad X_2 = A^{-1}\begin{bmatrix} 0 \\ 1 \end{bmatrix}, \quad X_3 = A^{-1}\begin{bmatrix} 0 \\ 0 \end{bmatrix}, \quad X_4 = A^{-1}\begin{bmatrix} 1 \\ 5 \end{bmatrix}$$

From this, we obtain

$$X_1 = \begin{bmatrix} 2 \\ -3 \end{bmatrix}, \quad X_2 = \begin{bmatrix} -1 \\ 2 \end{bmatrix}, \quad X_3 = \begin{bmatrix} 0 \\ 0 \end{bmatrix}, \quad X_4 = \begin{bmatrix} -3 \\ 7 \end{bmatrix}$$ ■

Example 8 Solve the system

$$\begin{aligned} x + 2y - 3z &= 1 \\ y + 4z &= 2 \\ 3x + 4y - 17z &= -1 \end{aligned}$$

Solution The matrix of coefficients is

$$\begin{bmatrix} 1 & 2 & -3 \\ 0 & 1 & 4 \\ 3 & 4 & -17 \end{bmatrix}$$

Attempting to find the inverse, we begin with

$$\left[\begin{array}{rrr|rrr} 1 & 2 & -3 & 1 & 0 & 0 \\ 0 & 1 & 4 & 0 & 1 & 0 \\ 3 & 4 & -17 & 0 & 0 & 1 \end{array}\right]$$

Using the elementary row operation $-3① + ③ \rightarrow ③$, we get

$$\left[\begin{array}{rrr|rrr} 1 & 2 & -3 & 1 & 0 & 0 \\ 0 & 1 & 4 & 0 & 1 & 0 \\ 0 & -2 & -8 & -3 & 0 & 1 \end{array}\right]$$

Next, using $2② + ③ \rightarrow ③$, we obtain

$$\left[\begin{array}{rrr|rrr} 1 & 2 & -3 & 1 & 0 & 0 \\ 0 & 1 & 4 & 0 & 1 & 0 \\ 0 & 0 & 0 & -3 & 2 & 1 \end{array}\right]$$

This shows that the given matrix of coefficients does not have an inverse, because the bottom row of zeroes on the left-hand side prevents us from obtaining the identity matrix. However, this does not mean that the system does not have a solution. We may use Gauss-Jordan elimination on the augmented matrix, as described in Section 2.2. Thus,

$$\left[\begin{array}{rrr|r} 1 & 2 & -3 & 1 \\ 0 & 1 & 4 & 2 \\ 3 & 4 & -17 & -1 \end{array}\right] \xrightarrow{-3① + ③ \rightarrow ③} \left[\begin{array}{rrr|r} 1 & 2 & -3 & 1 \\ 0 & 1 & 4 & 2 \\ 0 & -2 & -8 & -4 \end{array}\right]$$

$$\xrightarrow{2② + ③ \rightarrow ③} \left[\begin{array}{rrr|r} 1 & 2 & -3 & 1 \\ 0 & 1 & 4 & 2 \\ 0 & 0 & 0 & 0 \end{array}\right]$$

This system is solved by setting $z = t$, $y = 2 - 4t$, and $x = 11t - 3$. ■

CALCULATOR COMMENT: *Calculators with matrix capability can solve systems of equations of the form $AX = B$ by evaluating the product $A^{-1}B$. The $n \times n$ coefficient matrix A and the $n \times 1$ constant matrix B are entered into the calculator and then $A^{-1}B$ is evaluated.*

Coded Messages, or Matrices and the Captain Midnight Secret Decoder Ring

Since the beginning of written language, people have been trying to keep the content of messages secret by encoding them. Probably one of the easiest codes is to use numbers in place of the letters of the alphabet. This technique uses a set of substitutions such as the one shown below to encode the message.

A	B	C	D	E	F	G	H	I	J	K	L	M
↓	↓	↓	↓	↓	↓	↓	↓	↓	↓	↓	↓	↓
1	2	3	4	5	6	7	8	9	10	11	12	13

N	O	P	Q	R	S	T	U	V	W	X	Y	Z
↓	↓	↓	↓	↓	↓	↓	↓	↓	↓	↓	↓	↓
14	15	16	17	18	19	20	21	22	23	24	25	26

For example, the message

I like math

is encoded as

I	L	I	K	E	M	A	T	H
↓	↓	↓	↓	↓	↓	↓	↓	↓
9	12	9	11	5	13	1	20	8

However, codes of this type are very easy to break, because A is always 1, B is always 2, and so on. We can improve the encoding process by using matrix algebra. To see how this works, we select an invertible $n \times n$ matrix A. For example, suppose A is the 3×3 matrix

$$A = \begin{bmatrix} 1 & 0 & 1 \\ 0 & -1 & 0 \\ 2 & 1 & 3 \end{bmatrix}$$

First we rewrite the coded message as the following series of 3×1 matrices:

$$B_1 = \begin{bmatrix} 9 \\ 12 \\ 9 \end{bmatrix} \qquad B_2 = \begin{bmatrix} 11 \\ 5 \\ 13 \end{bmatrix} \qquad B_3 = \begin{bmatrix} 1 \\ 20 \\ 8 \end{bmatrix}$$

We then expand the matrix products AB_1, AB_2, and AB_3 to obtain the coded message

I	L	I	K	E	M	A	T	H
↓	↓	↓	↓	↓	↓	↓	↓	↓
18	−12	57	24	−5	66	9	−20	46

Compare this with the original coded form. Notice that the first I in the message is now represented by 18 and the second I by 57. Clearly, this is a more difficult code to break.

The person receiving the coded message decodes it by rewriting it as a series of 3×1 matrices, each multiplied by A^{-1}. Here, the inverse of matrix A is

$$A^{-1} = \begin{bmatrix} 3 & -1 & -1 \\ 0 & -1 & 0 \\ -2 & 1 & 1 \end{bmatrix}$$

Thus, the receiver writes the message as

$$C_1 = \begin{bmatrix} 18 \\ -12 \\ 57 \end{bmatrix} \quad C_2 = \begin{bmatrix} 24 \\ -5 \\ 66 \end{bmatrix} \quad C_3 = \begin{bmatrix} 9 \\ -20 \\ 46 \end{bmatrix}$$

and computes the matrix products $A^{-1}C_1$, $A^{-1}C_2$, and $A^{-1}C_3$, which yields the original message.

EXERCISES SECTION 2.5

In Exercises 1–15, compute the inverse matrix if it exists.

1. $\begin{bmatrix} 0 & 1 \\ 1 & 0 \end{bmatrix}$

2. $\begin{bmatrix} 2 & 0 \\ 0 & 3 \end{bmatrix}$

3. $\begin{bmatrix} 2 & 1 \\ 1 & 0 \end{bmatrix}$

4. $\begin{bmatrix} 1 & 0 & 1 \\ 0 & 1 & 0 \\ 0 & 0 & 1 \end{bmatrix}$

5. $\begin{bmatrix} 2 & 0 & 0 \\ 0 & 3 & 0 \\ 0 & 0 & 5 \end{bmatrix}$

6. $\begin{bmatrix} 2 & 0 & 1 \\ 0 & 3 & 1 \\ -2 & 0 & 4 \end{bmatrix}$

7. $\begin{bmatrix} 1 & 0 & 1 \\ 2 & 1 & 2 \\ -1 & 0 & 0 \end{bmatrix}$

8. $\begin{bmatrix} 2 & 4 \\ 1 & 2 \end{bmatrix}$

9. $\begin{bmatrix} 1 & 0 & 2 \\ 0 & 1 & 5 \\ 0 & 0 & 0 \end{bmatrix}$

10. $\begin{bmatrix} 1 & 1 & 0 \\ 1 & 1 & 1 \\ 0 & 0 & 1 \end{bmatrix}$

11. $\begin{bmatrix} a & b \\ 0 & d \end{bmatrix}, a \cdot d \neq 0$

12. $\begin{bmatrix} 2 & 1 & 3 \\ 0 & 1 & 2 \\ 2 & 1 & 5 \end{bmatrix}$

13. $\begin{bmatrix} 2 & 4 & 3 \\ -2 & 4 & -2 \\ 1 & 0 & 1 \end{bmatrix}$

14. $\begin{bmatrix} a & 0 & 0 \\ 0 & b & 0 \\ 0 & 0 & c \end{bmatrix}, abc \neq 0$

15. $\begin{bmatrix} -1 & 2 & 2 \\ -1 & 2 & 5 \\ -2 & 2 & 10 \end{bmatrix}$

In Exercises 16–25, solve each system using the method of matrix inversion.

16. $x - y = 1$
$x + y = 2$

17. $2x + y = 4$
$x - y = 0$

18. $2x + y + 3z = 0$
$y + 2z = 1$
$2x + y + 5z = -1$
(*Hint:* See Exercise 12.)

19. $2x + z = 1$
$3y + z = 2$
$-2x + 4z = 0$
(*Hint:* See Exercise 6.)

20. $3x + 2y = 6$
$x - y = 7$

21. $x + z = 0$
$2x + y + 2z = 2$
$-x = 1$
(*Hint:* See Exercise 7.)

22. $2x + y + 3z = -2$
$y + 2z = -1$
$2x + y + 5z = -4$
(*Hint:* See Exercise 12.)

23. $-x + 2y + 2z = 0$
$-x + 2y + 5z = 0$
$-x + y + 5z = -2$
(*Hint:* See Exercise 15.)

24. $x + y - z = 1$
$x - y + z = 3$
$x + y - 2z = -2$

25. $$\begin{aligned} x + y + 5z &= 2 \\ 3x + 8y + 5z &= 0 \\ -x - 5y + 3z &= 15 \end{aligned}$$
(*Hint:* A^{-1} does not exist.)

26. Find the relation among b_1, b_2, and b_3 such that the following system has a solution.

$$\begin{aligned} x + 3y + z &= b_1 \\ -x + y - z &= b_2 \\ x - y + z &= b_3 \end{aligned}$$

27. Use a calculator or computer with matrix capability to find the inverse of each of the matrices in Exercises 1–10. In each case, evaluate AA^{-1} and observe the result. If $AA^{-1} \neq I$, explain why not.

28. Use a calculator or computer with matrix capability to solve the systems of equations in Exercises 16–25. Explain what happens when the system is inconsistent or dependent.

Encoding and Decoding Exercises

29. (a) Use the matrix A in the illustrative example to encode the message "The die is cast."
(b) The message

$$33 \; -1 \; 87 \; 23 \; -3 \; 64 \; 41 \; -14 \; 116 \; 19 \; 0 \; 38$$

was encoded using the matrix A in the illustrative example. Decode this message.

30. (a) Use the 2×2 matrix

$$M = \begin{bmatrix} 2 & 1 \\ -1 & 1 \end{bmatrix}$$

to encode the message "The eagle has landed."
(b) The message

$$27 \; -12 \; 27 \; -6 \; 51 \; 12 \; 9 \; -3 \; 50 \; -25$$

was encoded using the 2×2 matrix M. Decode this message.

31. (a) Use the 3×3 matrix

$$D = \begin{bmatrix} 2 & 1 & 0 \\ 4 & 1 & -1 \\ 2 & 1 & -1 \end{bmatrix}$$

to encode the message "The price is right."
(b) The message

$$21 \; 25 \; 9 \; 33 \; 59 \; 27 \; 45 \; 76 \; 40 \; 32 \; 60 \; 32$$

was encoded using the 3×3 matrix D. Decode.

32. (a) Use the 3×3 matrix

$$E = \begin{bmatrix} 1 & 0 & 1 \\ 2 & 1 & 2 \\ -1 & 0 & 0 \end{bmatrix}$$

to encode the message "send money."
(b) The message

$$27 \; 66 \; -5 \; 21 \; 61 \; -9 \; 14 \; 50 \; -9 \; 19 \; 38 \; -19$$

was encoded using matrix E. Decode.

2.6 LEONTIEF MODELS

An interesting use of matrices is encountered in economics in connection with the **Leontief input-output model,** named for the economist Wassily Leontief, who was awarded the Nobel prize in 1973 for his work in this area. This model is used to predict the effects on the economy of such things as changes in prices, wages, and federal subsidies. A major result of this model is the ability to express the interconnection of all phases of the economy and to analyze these forces in a coherent fashion. Before we can establish the actual model, we must make some value judgements or predictions about the economy. The fundamental idea is to divide an economy into basic sectors, such as farm products, manufactured goods, energy and fuel, and so on. Each of these will usually have many subclassifications. For example, the category of manufactured goods might be divided into subcategories such as electronics, household items, automobiles, and construction. Our discussion here will obviously concern a much-simplified version of the model used to explain the economy of a nation.

In developing a Leontief model, we attempt to determine the impact that one part of the economy will have on other parts. For example, suppose our economy has three major components that supply consumer goods—namely, the farm products industry, the manufactured-goods industry, and the energy/fuel industry. Our three industries not only supply consumer goods to purchasers in the other sectors of the economy, but supply themselves as well (i.e., the farming industry buys farm products, as well as machinery and fuel). With this in mind, suppose we make the following estimates of the needs of these three components of our economy.

ASSUMPTIONS

- To produce 10 units of farm products, the farming industry requires 5 units of farm products (farmers buy seed grain and livestock feed), 3 units of manufactured goods (farmers use machinery), and 1 unit of energy/fuel (fuel is needed to run the machinery). (A "unit" of a product is some arbitrary, but agreed-upon, amount. For instance, one unit of fuel might be a million barrels of oil.)
- To produce 8 units of manufactured goods, the manufacturing industry requires 2 units of farm products, 4 units of manufactured goods, and 1 unit of energy.
- To produce 20 units of energy, the energy/fuel industry requires 2 units of farm products, 2 units of manufactured goods, and 8 units of energy.

This information is summarized in Table 2.1. To make it possible to compare units, each number in the table is expressed in terms of the proportion of a component required for the production of one unit of each of the commodities. For instance, since 5 units of farm products are required to produce 10 units of farm goods, it will take $\frac{5}{10} = 0.5$ unit of farm goods to produce 1 unit of farm goods. This number is shown in the first row of Table 2.1 under the column labeled Farm Products. The other numbers in the table are obtained in a similar way.

Each column in Table 2.1 represents the amount of the commodities used by the indicated industry to produce *one* unit of its particular commodity. Thus, it takes 0.5 unit of farm products, 0.3 unit of manufacturing goods, and 0.1 unit of energy to produce one unit of farm product. Each row in Table 2.1 represents the production of the indicated component. For example, the manufacturing component produces 0.3 unit for the farming component, 0.5 unit for its own use, and 0.1 unit for the energy component.

TABLE 2.1 Output for One Unit

Input	Farm Products	Manuf. Goods	Energy
Farm Products	$\frac{5}{10} = 0.5$	$\frac{2}{8} = 0.25$	$\frac{2}{20} = 0.1$
Manufactured Goods	$\frac{3}{10} = 0.3$	$\frac{4}{8} = 0.5$	$\frac{2}{20} = 0.1$
Energy	$\frac{1}{10} = 0.1$	$\frac{1}{8} = 0.125$	$\frac{8}{20} = 0.4$

COMMENT: *If all segments of the economy were represented in such a table, the sum of any column would be 1. Clearly, our model is an incomplete representation of the economy. A model that does not include all the segments of an economy is called an* **open model.**

Example 1 Referring to Table 2.1,

(a) How many units of manufactured goods does it take to produce 50 units of farm products?
(b) How many units of energy are needed to produce 100 units of manufactured goods?

Solution

(a) Referring to column 1 of the table, we see that it takes 0.3 unit of manufactured goods to produce 1 unit of farm products. Thus, 50 times 0.3, or 15 units of manufactured goods are required in order to produce 50 units of farm products.
(b) Referring to column 2 of the table, we see that 1 unit of manufactured goods is produced for every 0.125 unit of energy. Thus, an input of 100 times 0.125, or 12.5 units of energy is required in order to produce 100 units of manufactured goods. ■

Example 2 Referring to Table 2.1, if the cost of energy rises, which product is most affected?

Solution A rise in energy would have the greatest impact on those areas that use the largest amounts of fuel. We notice in the third column of Table 2.1 that 0.1, 0.1, and 0.4 unit of energy are required for each unit of farm products, manufactured goods, and energy/fuel, respectively. Hence, the energy industry itself would be most affected by a rise in the cost of energy. ■

The use of a matrix to represent the input/output information of Table 2.1 is natural. Of course, when we write Table 2.1 as a matrix A, we have to remember that the columns represent input and the rows represent output.

$$A = \begin{bmatrix} 0.5 & 0.25 & 0.1 \\ 0.3 & 0.5 & 0.1 \\ 0.1 & 0.125 & 0.4 \end{bmatrix}$$

In this form, the matrix A is called the **Leontief input-output matrix** for the economy under discussion. However, the use of a matrix in economics is not limited to a shorthand description of the economy. We can use the Leontief model and our knowledge of matrix multiplication to analyze the production of the economy as well. To demonstrate this, we introduce the following quantities:

- x_1 represents the total production of farm products (that is, the total goods produced by the farm products industry).
- x_2 represents the total production of manufactured goods.
- x_3 represents the total production of energy.

The portion of the total production of each component used by the various components of the economy can then be stated in terms of the coefficients in matrix A and x_1, x_2, and x_3. Thus, for the farm industry to produce x_1 units of farm products it must produce and use $0.5x_1$ units of its own goods; for the manufacturing industry to produce x_2 units of manufactured goods, the farming industry must produce $0.25x_2$ units of its goods; and for the energy industry to produce x_3 units of energy, the farming industry must produce $0.1x_3$ units of its goods. The total goods produced by the farming industry for all three components of the economy is then given by

$$v_1 = 0.5x_1 + 0.25x_2 + 0.1x_3$$

This represents the portion of the production of the farming industry that is used internally by the three components of the economy. A similar expression can be written for each of the components. So that we can write this in matrix form, we let X be a 3×1 column matrix and V be a 3×1 column matrix. That is,

$$X = \begin{bmatrix} x_1 \\ x_2 \\ x_3 \end{bmatrix} \quad \text{and} \quad V = \begin{bmatrix} v_1 \\ v_2 \\ v_3 \end{bmatrix}$$

The matrix X is called the **production matrix.** The matrix V is called the **internal consumption matrix.** The portion of the output of the economy consumed by the farming, manufacturing, and energy industries is then given by the matrix equation $V = AX$. Specifically, in our example, this equation is

$$\begin{bmatrix} v_1 \\ v_2 \\ v_3 \end{bmatrix} = \begin{bmatrix} 0.5 & 0.25 & 0.1 \\ 0.3 & 0.5 & 0.1 \\ 0.1 & 0.125 & 0.4 \end{bmatrix} \begin{bmatrix} x_1 \\ x_2 \\ x_3 \end{bmatrix}$$

Example 3 Find the portion of total output that is required by the three components if the total output of farming is 600 units, the total output of manufacturing is 400 units, and the total output of energy is 200 units. The production matrix is

$$X = \begin{bmatrix} 600 \\ 400 \\ 200 \end{bmatrix}$$

Solution The portion of total output required by the three components of the economy is given by $V = AX$.

$$V = AX = \begin{bmatrix} 0.5 & 0.25 & 0.1 \\ 0.3 & 0.5 & 0.1 \\ 0.1 & 0.125 & 0.4 \end{bmatrix} \begin{bmatrix} 600 \\ 400 \\ 200 \end{bmatrix} = \begin{bmatrix} 420 \\ 400 \\ 190 \end{bmatrix}$$

Thus, of the 1200 units in the production matrix, the three industrial components in our model consume 1010: the farming industry uses 420 units, the manufacturing industry uses 400 units, and the energy industry uses 190 units. The difference between 1200 and 1010 is the number of units available as surplus to the other parts of the economy, such as the consumer.

If our model included all the components of the economy, the matrix product AX would give the total production of each of the components. If the components we include do not consume the total production, as in our example, then the matrix product AX is not equal to the total production. This means there will be a surplus of commodities. The surplus is the difference between the production and what is used for internal consumption by the components. We indicate the surplus in each component by the **surplus matrix,** S, which in this problem is the 3×1 column vector

$$S = \begin{bmatrix} s_1 \\ s_2 \\ s_3 \end{bmatrix}$$

We then use the matrix equation

$$S = X - V$$

to define surplus. ■

Example 4 In our example economy, use the surplus equation to determine the surplus if the total production is 600 units of farm products, 400 units of manufactured goods, and 200 units of energy.

Solution To determine the surplus in our example economy, we note from Example 3 that internal consumption by the farming sector is 420 units, by the manufacturing sector 400 units, and by the energy sector 190 units. Thus, the surplus is the difference between the production and the internal consumption, or $S = X - V$.

$$S = \begin{bmatrix} 600 \\ 400 \\ 200 \end{bmatrix} - \begin{bmatrix} 420 \\ 400 \\ 190 \end{bmatrix} = \begin{bmatrix} 180 \\ 0 \\ 10 \end{bmatrix}$$

We interpret this result to mean that the surplus available to the consumer is 180 units of farm products, 10 units of energy, and no manufactured goods. ■

If we look carefully at the equation for surplus, S, we see that it can be rewritten in terms of the input-output matrix and the production matrix. The specific steps in the process are

$S = X - V$

$S = IX - AX$ Replacing X with IX and V with AX

$S = (I - A)X$ Using the distributive law for matrix multiplication

This is known as the *technology equation* for the economy.

DEFINITION The Technology Equation

If A is the Leontief input-output matrix for an economy and X is the production matrix, then the surplus matrix S is given by

$$S = (I - A)X$$

where I is the identity matrix.

Example 5 Use the technology equation to determine the surplus in our example economy if the total production is 600 units of farm products, 400 units of manufactured goods, and 200 units of energy. This is the same problem solved in Example 4.

Solution Using the technology equation,

$$S = (I - A)X$$

where I is the identity matrix for A, we have

$$\begin{aligned} S &= \left(\begin{bmatrix} 1 & 0 & 0 \\ 0 & 1 & 0 \\ 0 & 0 & 1 \end{bmatrix} - \begin{bmatrix} 0.5 & 0.25 & 0.1 \\ 0.3 & 0.5 & 0.1 \\ 0.1 & 0.125 & 0.4 \end{bmatrix} \right) \begin{bmatrix} 600 \\ 400 \\ 200 \end{bmatrix} \\ &= \begin{bmatrix} 0.5 & -0.25 & -0.1 \\ -0.3 & 0.5 & -0.1 \\ -0.1 & -0.125 & 0.6 \end{bmatrix} \begin{bmatrix} 600 \\ 400 \\ 200 \end{bmatrix} \\ &= \begin{bmatrix} 180 \\ 0 \\ 10 \end{bmatrix} \end{aligned}$$

This agrees with the surplus matrix obtained in Example 4. ■

Sometimes we wish to determine the production matrix X that will produce a desired surplus. This is done by solving the surplus equation $S = (I - A)X$ for the production matrix X. To solve $S = (I - A)X$ for X, multiply both sides on the left by $(I - A)^{-1}$. This yields

$$\begin{aligned} S &= (I - A)X \\ (I - A)^{-1}S &= (I - A)^{-1}(I - A)X && \text{Multiplying on the left by } (I - A)^{-1} \\ (I - A)^{-1}S &= IX && (I - A)^{-1}(I - A) = I \end{aligned}$$

Finally, replacing IX with X and rearranging terms, we have

$$X = (I - A)^{-1}S$$

Example 6 Find the total production for the economy given by Table 2.1 if the surplus desired is 120 units of farm products, 20 units of manufactured goods, and 80 units of energy.

Solution In this case, $(I - A) = \begin{bmatrix} 0.5 & -0.25 & -0.1 \\ -0.3 & 0.5 & -0.1 \\ -0.1 & -0.125 & 0.6 \end{bmatrix}$ and $S = \begin{bmatrix} 120 \\ 20 \\ 80 \end{bmatrix}$.

The inverse matrix $(I - A)^{-1}$ may be found in a number of ways; we used an HP28 to obtain the following entries to the nearest hundredth.

$$(I - A)^{-1} = \begin{bmatrix} 3.29 & 1.85 & 0.86 \\ 2.17 & 3.31 & 0.91 \\ 1.00 & 1.00 & 2.00 \end{bmatrix}$$

From this it is an easy step to compute the production matrix X:

$$X = (I - A)^{-1}S = \begin{bmatrix} 3.29 & 1.85 & 0.86 \\ 2.17 & 3.31 & 0.91 \\ 1.00 & 1.00 & 2.00 \end{bmatrix} \begin{bmatrix} 120 \\ 20 \\ 80 \end{bmatrix} = \begin{bmatrix} 500 \\ 400 \\ 300 \end{bmatrix}.$$

We conclude from this computation that in order to have a surplus of 120 farm units, 20 manufacturing units, and 80 energy units, the farming industry must produce 500 units of farm products, the manufacturing industry must produce 400 units of manufactured goods and the energy industry must produce 300 units of energy/fuel. Note that the results may vary due to round-off error. ■

A Closed Leontief Model

If all output is consumed by the components of the economy in the technology matrix, the Leontief model is said to be **closed.** In this case, the surplus is by necessity equal to zero and the technology equation reduces to

$$(I - A)X = 0$$

which can also be written as $AX = X$.

Example 7 Suppose that the economy of a small country is divided into agricultural and nonagricultural products and that the Leontief input-output matrix is given by

$$A = \begin{bmatrix} 0.7 & 0.5 \\ 0.3 & 0.5 \end{bmatrix}$$

The first column of A represents the agricultural input; the second column, the nonagricultural input. Notice that in a closed model the sum of the entries in each column is 1.

(a) Write the technology equation.
(b) Solve the technology equation.
(c) Find the ratio of agricultural to nonagricultural products.

Solution

(a) In this case, the matrix represents a closed model, so we want to solve the system of equations described by $(I - A)X = 0$. Since

$$I - A = \begin{bmatrix} 0.3 & -0.5 \\ -0.3 & 0.5 \end{bmatrix}$$

the system obtained from $(I - A)X = 0$ is

$$\begin{aligned} 0.3x_1 - 0.5x_2 &= 0 \\ -0.3x_1 + 0.5x_2 &= 0 \end{aligned}$$

(b) This system has infinitely many solutions, of which x_1 and x_2 are related by

$$x_1 = \frac{5}{3}x_2$$

(c) This equation tells us the proportions of the two major economic forces in this case, where there is little or no surplus. We see that the ratio of agricultural products to nonagricultural products for this country should be 5 to 3. ■

EXERCISES SECTION 2.6

1. An economy consists of food and fuels. Suppose that the Leontief input-output matrix for this economy is

	Food	Fuel
Food	0.5	0.4
Fuel	0.3	0.4

(a) How many food units does it take to produce 50 units of fuels?
(b) If the cost of fuels rises, which is most affected, food or fuels?
(c) If the gross production is 300 units of food and 200 of fuels, find the surplus for this economy.
(d) If the desired surplus is 90 units of food and 15 of fuels, find the gross production required.

2. Matrix A is a Leontief input-output matrix for an economy in which the rows represent production of farm goods, petroleum, and manufactured goods, respectively.

$$A = \begin{bmatrix} 0.4 & 0.5 & 0.2 \\ 0.4 & 0.3 & 0.4 \\ 0.2 & 0.1 & 0.3 \end{bmatrix}$$

(a) How many units of farm products does it take to produce 100 units of manufactured goods?
(b) How many units of petroleum are needed to produce 50 units of farm products?
(c) If the cost of petroleum rises, which product is most affected?

(d) Find the portion of the total output economy that is required by the three components if the total output of farming is 700 units, the total output of manufacturing is 300 units, and the total output of petroleum is 600 units.
(e) Determine the surplus in this economy if the total production is as given in part (d).
(f) Determine the production required to yield a surplus of 80 units of farm products, 100 units of petroleum, and 10 units of manufactured goods.

3. An economy consists of manufactured goods, farm products, and fuels. Suppose that its Leontief input-output matrix is

$$\begin{array}{c} \\ \text{Goods} \\ \text{Products} \\ \text{Fuels} \end{array} \begin{array}{c} \begin{array}{ccc} \text{Goods} & \text{Prod.} & \text{Fuels} \end{array} \\ \begin{bmatrix} 0.5 & 0.3 & 0.4 \\ 0.1 & 0.4 & 0.2 \\ 0.2 & 0.1 & 0.3 \end{bmatrix} \end{array}$$

(a) If the gross production of the economy is 400 units of manufactured goods, 250 of farm products, and 150 of fuels, find the surplus for the economy.
(b) If the desired surplus is 90 units of manufactured goods, 50 units of farm products, and 15 units of fuels, find the gross production required.

4. A large international industrial conglomerate owns iron mines, steel mills, and fabricating plants. Assume that the three components of the company supply each other as well as external consumers. The internal consumption is as follows:

- To produce 10 units of iron ore requires 1 unit of iron ore, 1 unit of steel, and 3 units of fabricated products.
- To produce 10 units of steel requires 6 units of iron ore, 1 unit of steel, and 2 units of fabricated products.
- To produce 10 units of fabricated products requires no iron ore, 7 units of steel, and 2 units of fabricated products.

(a) Write the Leontief input-output matrix for these three components.
(b) Determine the surplus if the total production is 300 units of iron ore, 400 units of steel, and 400 units of fabricated products.
(c) Find the total production of the company if the desired surplus is 100 units of iron ore, 200 units of steel, and 100 units of fabricated products.

5. Consider an economy that consists of farmers, textile workers, and carpenters. Assume that each of these components needs food, clothing, and housing according to the following schedule:

- To produce 10 units of food requires 1 unit of food, 2 units of clothing, and 3 units of housing.
- To produce 10 units of clothing requires 3 units of food, 3 units of clothing, and 2 units of housing.
- To produce 10 units of housing requires 1 unit of food, 1 unit of clothing, and 2 units of housing.

(a) Write the Leontief input-output matrix for this economy.
(b) Determine the surplus in the economy if the total production is 400 units of food, 600 units of clothing, and 500 units of housing.
(c) Find the total production of the economy if the desired surplus is 100 units of food, 150 units of clothing, and 50 units of housing.

6. Suppose the economy of a small country is divided into agricultural and nonagricultural products and that the closed Leontief input-output matrix is given by

$$A = \begin{bmatrix} 0.6 & 0.3 \\ 0.4 & 0.7 \end{bmatrix}$$

The first column of A represents the agricultural input; the second column, the nonagricultural input.
(a) Write the technology equation.
(b) Solve the technology equation.
(c) Find the ratio of agricultural to nonagricultural products.

7. Suppose the economy of a country is divided into fuel and labor, farming, and manufactured goods and the closed Leontief model is given by

$$A = \begin{bmatrix} 0.3 & 0.4 & 0.2 \\ 0.3 & 0.5 & 0.6 \\ 0.4 & 0.1 & 0.2 \end{bmatrix}$$

where the first column represents input of fuel and labor; the second, farming; and the third, manufactured goods.
(a) Write the technology equation.
(b) Solve the technology equation.
(c) Indicate how the segments of the economy are related.

Important Words and Phrases

The new terminology in this chapter is largely related to the concept of a matrix. The initial application of matrices was as a bookkeeping device for solving systems of equations by Gauss-Jordan elimination and the final application was the use of matrix algebra in a model of our economy. You should know the meaning of the terms and phrases below. Sometimes the same concept is described in more than one way. For instance, "inverse of a matrix" and "matrix inversion" mean the same thing.

augmented matrix
consistent system
dependent system
equivalent matrix
equivalent system
Gauss-Jordan elimination
identity matrix
inconsistent system
inverse of a matrix
linear model
linear system
Leontief input-output model
matrix
matrix addition
matrix inversion
matrix multiplication
matrix transposition
scalar multiplication
solution of a system
system of equations
transpose of a matrix

Review Exercises Chapter 2

In Exercises 1–10, solve each of the systems by the Gauss-Jordan elimination method.

1. $x + 3y = 5$
$2x - 3y = 1$

2. $3a - b = 5$
$5a + 3b = 13$

3. $2s + 5t = -4$
$6s + 8t = 10$

4. $2x + 3y = 2$
$5x - 4y + 1 = 0$

5. $5x + 6z + 3 = 0$
$7x + 4z - 9 = 0$

6. $4a + 5c = 10$
$7a - 3c = 41$

7. $x + y + z = 1$
$x + y - z = 2$
$x - y + z = 3$

8. $x + y + z = 2$
$3x - y - 2z = 4$
$5x - 2y + 3z = -6$

9. $a + 2b - 3c = 1$
$b - c - 4 = 0$
$b + c - a = 5$

10. $2r + s + 2t = 5$
$4r - s - 3t = 1$
$8r + s - t = 5$

In Exercises 11–16, solve each system using augmented matrix notation and row operations.

11. $3x + 5y - 13 = 0$
$-x + 3y - 5 = 0$

12. $4x + y = 11$
$5x + 3y = 5$

13. $x - 2y = 1$
$x + 2y = 3$

14. $3a - 4b = 8$
$9a - 12b = 24$

15. $a - 2b + 3c = 4$
$-3a + 4b - c = -2$
$2a + b - 4c = 3$

16. $3r - 2s + t = 1$
$r + s + t = 0$
$s - 2t = 2$

In Exercises 17–20, write each system in the form $AX = B$.

17. $2x + 3y = 7$
$x - y = 8$

18. $x + 2y = 1$
$4y + 3x = 0$

19. $x + 2y + 5z = 10$
$y - z = 2$
$2x - y + z = -1$

20. $2x - y + z = -1$
$-x + 3y + 2z = 0$
$3y + z = 5$

In Exercises 21–24, find $A + B$, or tell why the operation cannot be done.

21. $A = \begin{bmatrix} 3 & 5 \\ -1 & 2 \end{bmatrix}$, $B = \begin{bmatrix} -2 & 4 & 6 \\ 0 & 1 & 5 \end{bmatrix}$

22. $A = \begin{bmatrix} 2 & 1 \\ -3 & 4 \\ 1 & 1 \end{bmatrix}$, $B = \begin{bmatrix} 0 & 7 \\ 3 & -5 \\ 2 & -2 \end{bmatrix}$

23. $A = \begin{bmatrix} 1 & 1 & 2 \\ 2 & 3 & 2 \\ 0 & 4 & -1 \end{bmatrix}$, $B = \begin{bmatrix} -3 & 7 & 6 \\ 0 & -2 & 3 \\ -2 & 7 & 1 \end{bmatrix}$

24. $A = \begin{bmatrix} 2 & 2 & 3 \\ -1 & 4 & 3 \end{bmatrix}$, $B = \begin{bmatrix} 5 & 2 & 8 \\ 3 & -2 & -6 \end{bmatrix}$

In Exercises 25–30, find AB and BA, or tell why the operation cannot be done.

25. $A = [2 \quad 3 \quad -1]$, $B = \begin{bmatrix} -1 \\ 2 \\ 4 \end{bmatrix}$

26. $A = \begin{bmatrix} 1 & 2 \\ 3 & -2 \end{bmatrix}$, $B = \begin{bmatrix} 2 & 3 & 1 \\ -3 & 0 & 5 \end{bmatrix}$

27. $A = \begin{bmatrix} 2 & -1 \\ 3 & 0 \end{bmatrix}$, $B = \begin{bmatrix} 5 & -3 \\ 1 & 7 \end{bmatrix}$

28. $A = \begin{bmatrix} 0 & 1 \\ 1 & -1 \end{bmatrix}$, $B = \begin{bmatrix} 1 & 0 \\ 0 & 0 \end{bmatrix}$

29. $A = \begin{bmatrix} 2 & 1 & -1 \\ 4 & 0 & 1 \end{bmatrix}$, $B = \begin{bmatrix} 2 & -3 \\ 1 & 2 \\ 4 & 8 \end{bmatrix}$

30. $A = \begin{bmatrix} 2 & 4 \\ 1 & 0 \\ -1 & 1 \end{bmatrix}$, $B = \begin{bmatrix} 2 & 1 & 4 \\ -3 & 2 & 8 \end{bmatrix}$

In Exercises 31–36, find the inverse of each matrix.

31. $\begin{bmatrix} 1 & 2 \\ 3 & 2 \end{bmatrix}$

32. $\begin{bmatrix} -1 & 3 \\ 4 & -2 \end{bmatrix}$

33. $\begin{bmatrix} -1 & 4 \\ 3 & 5 \end{bmatrix}$

34. $\begin{bmatrix} 5 & 3 \\ 2 & 1 \end{bmatrix}$

35. $\begin{bmatrix} 1 & 1 & -1 \\ 2 & 0 & 1 \\ 0 & -2 & 3 \end{bmatrix}$

36. $\begin{bmatrix} -2 & 2 & 0 \\ 1 & 3 & 1 \\ 3 & 1 & 1 \end{bmatrix}$

In Exercises 37–42, solve each system of equations by using the method of matrix inversion.

37. $3x + 5y - 13 = 0$
$-x + 3y - 5 = 0$

38. $4x + y = 11$
$5x + 3y = 5$

39. $x - 2y = 1$
$x + 2y = 3$

40. $3a - 4b = 8$
$9a - 12b = 24$

41. $a - 2b + 3c = 4$
$-3a + 4b - c = -2$
$2a + b - 4c = 3$

42. $3r - 2s + t = 1$
$r + s + t = 0$
$s - 2t = 2$

In Exercises 43–46, use either of the two matrix methods to solve the problem.

43. Two angles are complementary and one angle is 3° larger than the other. Find the two angles.

44. The discharge rate of two pipes together is 50 ft^3/sec. If the difference of the rates of the two pipes is 4 ft^3/sec, what is the discharge rate of each?

45. To finance a capital improvement to its plant, a company borrows $750,000, part at 9% and part at 11%. If the interest on the two loans amounts to $80,000, what is the amount of each loan?

46. A boat has enough fuel to cruise at 7 mph for 5 hr. How far upstream can the boat go and be able to return to the starting point if the current is 2 mph?

47. The Volksauto Company produces cars and trucks at two plants in Detroit and Anaheim. In matrix A, which represents last week's production, row 1 represents the number of cars; row 2, the number of trucks. Column 1 represents the production of the Detroit plant; column 2, the production of the Anaheim plant. Matrix B represents this week's total production for both plants.

$$A = \begin{bmatrix} 223 & 361 \\ 125 & 185 \end{bmatrix} \qquad B = \begin{bmatrix} 182 & 147 \\ 143 & 94 \end{bmatrix}$$

(a) How many cars were produced last week?
(b) How many vehicles were produced altogether by the Anaheim plant last week?
(c) Write a matrix representing production for last week and this week.
(d) Matrix C represents shipments made from the two plants at the end of this week.

$$C = \begin{bmatrix} 400 & 500 \\ 250 & 270 \end{bmatrix}$$

Use a matrix to show how many cars and trucks remain to be shipped.

48. A tennis-racket manufacturer produces three models. It has orders for 1500 of the Conner model, 2500 of the Becker model, and 2000 of the Evert model. The table below lists the number of units of raw materials the manufacturer needs for each model.

	Graphite	Plastic
Conner	40	25
Becker	30	20
Evert	10	40

The cost of graphite is \$0.20 per unit; the cost of plastic, \$0.12 per unit.

(a) Write a 1×3 matrix showing the number of rackets on order for each model.

(b) Write a 3×2 matrix showing the raw materials used for each model.

(c) Use matrix multiplication to determine the total number of units of each raw material needed.

(d) Write a 2×1 matrix showing the cost per unit for each of the raw materials.

(e) Use matrix multiplication to find the total investment needed to fill the orders.

49. Matrix A is a Leontief input-output matrix for an economy in which the rows represent production of farm goods, petroleum, and manufactured goods.

$$A = \begin{bmatrix} 0.5 & 0.35 & 0.25 \\ 0.3 & 0.2 & 0.4 \\ 0.1 & 0.2 & 0.3 \end{bmatrix}$$

(a) How many units of farm products does it take to produce 100 units of petroleum?

(b) How many units of petroleum are needed to produce 200 units of manufactured goods?

(c) If the cost of petroleum rises, which product is most affected?

(d) Find the portion of the total output economy that is required by the three components if the total output of farming is 550 units, that of manufacturing is 500 units, and that of petroleum is 300 units.

(e) Determine the surplus in this economy if the total production is as given in part (d).

(f) Determine the production required to yield a surplus of 50 units of farm products, 200 units of petroleum, and 100 units of manufactured goods.

CHAPTER 3

INEQUALITIES

The Kellogg Company makes and distributes to food retailers many different cereal products such as Corn Flakes, Nutri-Grain, and Rice Krispies. Clearly, one of the goals of the company should be to maximize the profit that it makes by selling its products. Profit is essentially the money that is left when the cost of the product is subtracted from the revenue earned by selling it. To maximize profit means to make the largest profit we can, subject to certain restrictions, or constraints. Maximizing profit is ordinarily a difficult process because of the many factors that affect cost and revenue, such as the price that can be charged in a competitive market, the cost of raw materials, labor costs, and interest rates on borrowed money. As you will learn later in this chapter, a mathematical process called **linear programming** is used by companies such as Kellogg to help them maximize profit. At the heart of linear programming is the study of systems of inequalities.

In this chapter, we will extend our study of linearity from statements of equality to statements that involve inequalities, that is, expressions that involve "less than" or "greater than" comparisons. While the algebraic manipulations involved in solving inequalities parallel those used to solve equations, the types of problems considered are substantially different.

3.1 LINEAR INEQUALITIES

In the first section of this book we used the basic idea of an inequality to describe **intervals** of real numbers. For instance, the inequality $x > 2$ represents the real numbers that are greater than 2, and $x \leq 0$ represents the real numbers that are less than or equal to zero. In the same way, the statement "Twice x plus 5 is less than 3 plus x" is written symbolically as the inequality $2x + 5 < 3 + x$. This statement is called a **linear inequality,** because the terms involving the unknown are all linear. An inequality is an *open mathematical sentence,* which means that it may be true for a subset of the real numbers. In this sense, the terminology that describes equations can also be used to describe inequalities. The set of values for which an inequality is true is called its *solution set,* and we say we have *solved* an inequality when we have found its solution set. Two inequalities are *equivalent* if they have the same solution set. Of course, the possible values of the variable might be restricted, either by choice or by the nature of the application, but unless there is some specific exception, the possible values of the variable will be the set of real numbers.

The procedure used to solve an inequality also resembles that used to solve an equation. To solve an inequality we use certain allowable algebraic manipulations to form a chain of equivalent inequalities, with the objective of isolating the variable on one side of the inequality. The allowable manipulations that permit us to do this are similar to the operations used for solving equations. Compare the following operations for inequalities with those for solving equations (see Section 1.1).

ALLOWABLE OPERATIONS FOR INEQUALITIES ■

- **Substitution** A quantity may be substituted for an equal quantity. For example, $2x + 5x < 7 - 5$ is equivalent to $7x < 2$.
- **Addition/Subtraction** The same quantity may be added to or subtracted from both sides of an inequality. For example, $x < y$ is equivalent to $x + 3 < y + 3$.
- **Multiplication/Division** Both sides of an inequality may be multiplied or divided by the same *positive* quantity. For example, $x < y$ is equivalent to $2x < 2y$. If both sides of an inequality are multiplied by the same *negative* quantity, *the inequality sign is reversed.* For example, $-2 < 5$, but when we multiply by -3, the inequality sign is reversed and the equivalent inequality is $(-3)(-2) > (-3)(5)$, or $6 > -15$. In the same way, $x > y$ is equivalent to $-3x < -3y$.

Example 1 Solve the inequality $x - 2 < 3$.

Solution Adding 2 to each side of the inequality, we have

$$x - 2 + 2 < 3 + 2$$

or

$$x < 5$$

Thus, the solution set consists of all real numbers less than 5. ■

Example 2 Solve the inequality $x + 1 < 3x - 2$.

Solution To get the terms involving x on one side and the constants on the other, we add $-3x$ and -1 to both sides to get the equivalent inequality:

$$-2x < -3$$

Multiplying this result by $-\frac{1}{2}$ yields

$$x > \frac{3}{2}$$

Notice that the inequality sign was reversed when the inequality was multiplied on both sides by $-\frac{1}{2}$. The solution set is the set of all real numbers greater than $\frac{3}{2}$.

■

Interval Notation

Solutions of inequalities are often written in a special notation, called **interval notation.** An **interval** is the set of real numbers between two given real numbers. The set of all real numbers greater than a and less than b is called an **open interval** and is denoted by (a, b). The points corresponding to a and b are called the **endpoints** of the interval and are not included in the open interval. If the endpoints are to be included, the interval is called a **closed interval** and is denoted by $[a, b]$. Thus, parentheses are used to denote open intervals and brackets to denote closed intervals; the distinction between open and closed is the exclusion or inclusion of the endpoints. Graphically, a filled dot is used to show an included endpoint and an open dot is used to show an excluded endpoint, as in Figure 3.1.

Figure 3.1

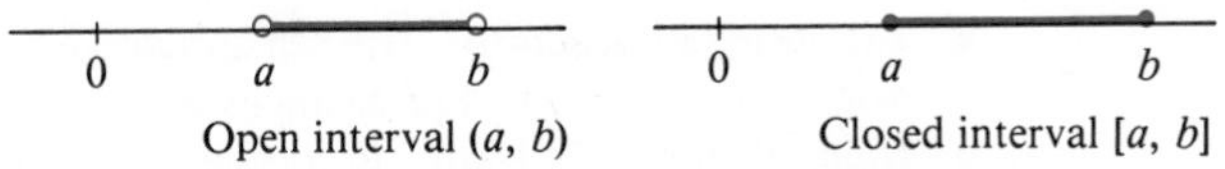

If only one of the endpoints is included in the interval, it is *half-open*. The half-open intervals $(a, b]$ and $[a, b)$ are shown in Figure 3.2.

Figure 3.2

Finally, the set of all real numbers greater than a is represented by the notation (a, ∞). The symbol ∞ (infinity) means that there is no finite number for the right-hand endpoint of the interval. (Infinity should not be thought of as a very large number, but as the idea of unboundedness.) The symbol $-\infty$ (negative infinity) is used to indicate that there is no finite number for the left-hand endpoint; therefore, the set of real numbers less than a is represented by $(-\infty, a)$. There are similar definitions for $(-\infty, a]$ and $[a, \infty)$. These intervals are shown in Figure 3.3.

Figure 3.3

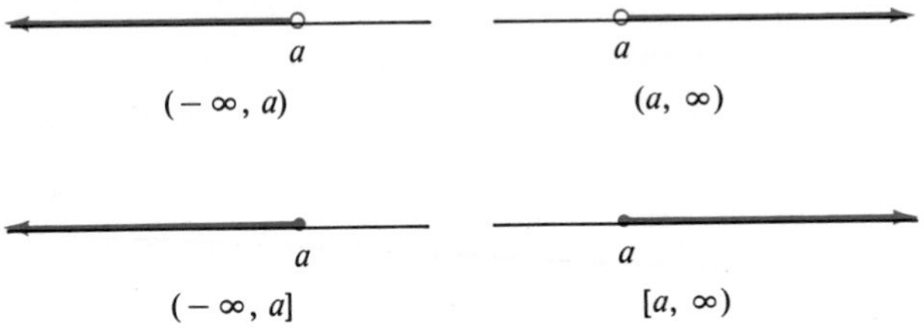

COMMENT: *The bracket notation is never used with an endpoint that is unbounded because an unbounded endpoint is assumed to be open. Thus, the notation* $(a, \infty]$ *is incorrect.*

Example 3 In Example 2, the solution to the inequality $x + 1 < 3x - 2$ was shown to be $x > \frac{3}{2}$. This solution is written in interval notation as $(\frac{3}{2}, \infty)$, and is illustrated in Figure 3.4.

Figure 3.4

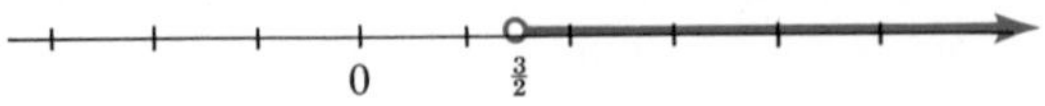

■

If we can show that $q < p$ and also that $p < r$, then it is true that $q < p < r$. The solution to an inequality of the form $q < p < r$ consists of all the real numbers that are solutions to *both* of the inequalities, $q < p$ and $p < r$. We show the solution technique in the next example.

Example 4 Solve the inequality $-2 \leq x + 3 \leq 10$.

Solution Any solution of the given inequality must be a solution of *both*

$$-2 \leq x + 3 \qquad \text{and} \qquad x + 3 \leq 10$$

Solving these two inequalities separately, we get

$$\begin{array}{r|l} -2 \leq x + 3 & x + 3 \leq 10 \\ -5 \leq x & \quad x \leq 7 \end{array}$$

Since the solution must satisfy both $-5 \leq x$ and $x \leq 7$, we conclude that the solution interval is

$$-5 \leq x \leq 7.$$

This inequality can be represented in interval notation as $[-5, 7]$. It is shown graphically in Figure 3.5.

Figure 3.5

Notice that in this problem the solution can also be obtained by subtracting 3 from each term of the given inequality. This method can always be used when the inequality is of the form

$$c \leq ax + b \leq d$$

where a, b, c, and d are constants. ■

Example 5 Solve the inequality $x - 20 < 3x - 8 \leq 4$.

Solution Any solution to the given inequality must be a solution to *both*

$$x - 20 < 3x - 8 \qquad \text{and} \qquad 3x - 8 \leq 4$$

Solving these two inequalities separately, we get the following results:

$$\begin{array}{r|l} x - 20 < 3x - 8 & 3x - 8 \leq 4 \\ x - 12 < 3x & \quad 3x \leq 12 \\ -12 < 2x & \quad\; x \leq 4 \\ -6 < x & \end{array}$$

Since the solution of the given inequality must satisfy both $-6 < x$ and $x \leq 4$, we conclude that the solution is the interval

$$-6 < x \leq 4$$

This can also be represented in interval notation as $(-6, 4]$, as in Figure 3.6.

Figure 3.6

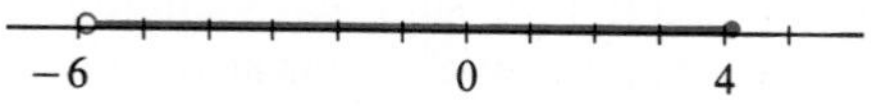

■

EXERCISES SECTION 3.1

In Exercises 1–10, sketch the graph of each interval.

1. $(-2, 3)$
2. $[-2, 3]$
3. $(-\infty, 0]$
4. $[1, \infty)$
5. $(-3, 3]$
6. $[2, 3]$
7. $\left[\frac{1}{4}, \frac{1}{2}\right]$
8. $(-1, 3)$
9. $[-3, -1]$
10. $[-5, 0)$

In Exercises 11–28, solve each inequality and sketch its graph on the number line.

11. $3x + 1 > x - 4$
12. $2 - 5x > -x$
13. $2 - 3x \leq x + 1$
14. $x \leq x + 1$
15. $\frac{2}{3}x + 5 < \frac{7}{5}x + \frac{1}{3}$
16. $3(2x - 1) + 4(x - 2) > 0$
17. $5(x - 2) - 3(x + 1) \leq 0$
18. $3(x - 1) < (x - 3) + 3$
19. $2x - 1 > 2x + 3$
20. $-4 < x - 1 \leq 4$
21. $-3 < x + 5 < 7$
22. $2 < x - 4 < 5$
23. $0 \leq 3x - 7 \leq 2$
24. $-1 < 5x + 1 \leq 6$
25. $x < 2x + 3 < 3x - 5$
26. $-2x < 2x - 3 < 2x + 7$
27. $x + 2 < 2x + 5 \leq 6x + 7$
28. $x - 1 \leq 3x + 5 < 4x - 5$
29. A national poll shows that the percentage of voters, P, expected to vote for a certain presidential candidate is 42%, with an error of $\pm 3\%$ in the poll. Express this as a double inequality.
30. The strength s of a sheet of material sufficient to hold a certain weight is given by the inequality $s + 3 \leq 2s + 1$, where s is expressed in pounds per square inch (psi). Find the range of s required to hold this weight.
31. The measurement M of a machined part for a vacuum cleaner must be greater than 9.99 inches and less than 10.02 inches. Express this as a double inequality.

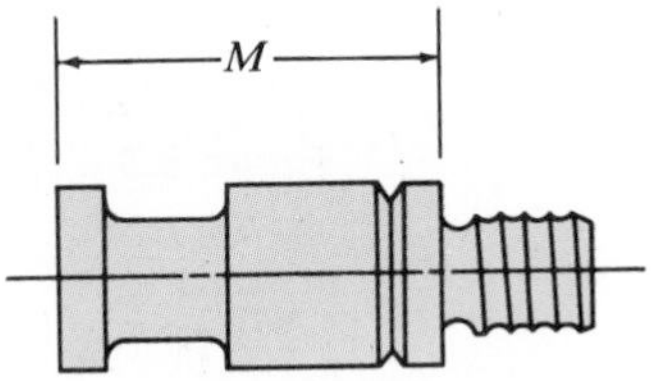

Figure 3.7

32. The number of delayed pieces of luggage per week on North-South Air is greater than 32 and less than 80. Write this as a double inequality.
33. The number of people at a recent New York Mets game was more than 25,500 and less than 30,000. Write this as a double inequality.
34. Depending on the number of births in a given month, the nursery at Natanya Hospital uses more than 180 dozen and less than 450 dozen disposable diapers per month. Write this as a double inequality.
35. Explain why $-2 > x > 2$ is incorrect.

3.2 LINEAR INEQUALITIES IN TWO UNKNOWNS

In Section 3.1, we discussed inequalities in one variable. This section extends that discussion to include inequalities in two variables. The allowable operations of substitution, addition, subtraction, multiplication, and division we established for inequalities in one variable are also allowable for inequalities in two variables.

The concept of an inequality in two variables is closely related to the corresponding idea of a linear equality in two variables. Recall that a solution to a linear equality in two variables is any ordered pair of numbers for which the equation is true. Similarly, solutions to a linear inequality in two variables are ordered pairs of numbers for which the inequality is true. For example, the linear inequality

$$x + 3y < 7$$

is true for (0, 0) (2, 1), (3, 0), and infinitely many other ordered pairs. We also observe that (1, 7), (2, 10), and (0, 3) are not solutions. Solution sets for inequalities in two variables are best described in graphical language.

THE GRAPH OF A LINEAR INEQUALITY

Consider the linear inequalities $2x + 3y < -5$ and $2x + 3y > -5$. To discover the solution sets of these two inequalities we draw the graph of the related equation

The graph of $2x + 3y = -5$ is shown in Figure 3.8 as the line L. Notice that L separates the plane into three distinct sets of points:

- The set of points that make up the line L.
- The set of points above the line L, denoted by A.
- The set of points below the line L, denoted by B.

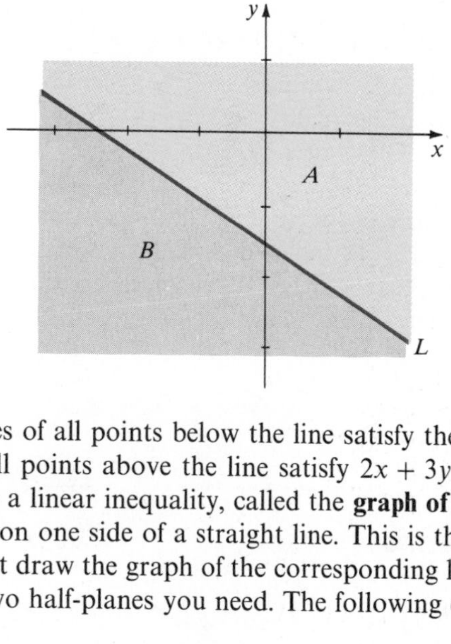

Figure 3.8

The coordinates of all points below the line satisfy the inequality $2x + 3y < -5$, and those of all points above the line satisfy $2x + 3y > -5$. Thus, the graph of all solutions to a linear inequality, called the **graph of the inequality**, corresponds to a *half-plane* on one side of a straight line. This is the clue to graphing a linear inequality: First draw the graph of the corresponding linear equation, then decide which of the two half-planes you need. The following examples illustrate the way to do this.

Example 1 Graph the solution set for the inequality $2x + y < 1$.

Solution Notice that the problem could also have been stated, "Graph the inequality $2x + y < 1$." To find the graph of $2x + y < 1$, we first draw the graph of the line $2x + y = 1$. Figure 3.9 shows it as a dashed line to indicate that it is not part of the solution set. To decide which of the two half-planes represents the given inequality, we select any "test point" (x_1, y_1), *not on the line* and substitute it into the inequality. Then, if (x_1, y_1) satisfies the inequality, this test point lies in the desired half-plane and the *entire half-plane is the graph of the inequality.* In this problem, suppose we use the origin, (0, 0) as our test point. Since $2(0) + 0 = 0 < 1$ satisfies the inequality, then the desired half-plane is below the line. The shaded half-plane in Figure 3.9 is the graph of the given inequality.

Figure 3.9

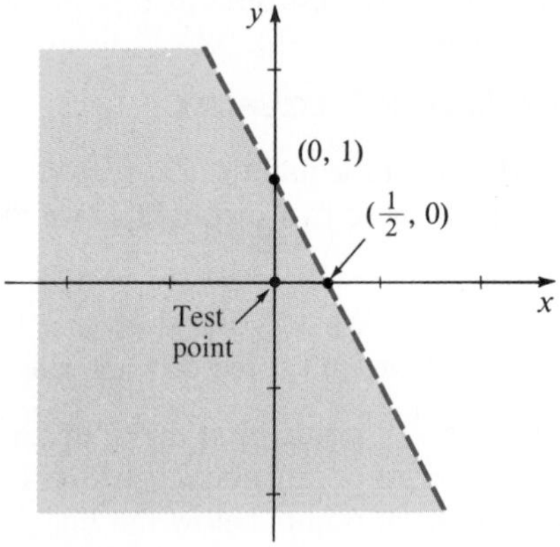

■

Example 2 Graph the solution set for the inequality $2x + y - 1 \leq x + 3y + 4$.

Solution By adding $-x - 3y + 1$ to both sides of the inequality, we obtain the equivalent inequality

$$x - 2y \leq 5$$

The graph of the related linear equation, $x - 2y = 5$, is shown as a solid line in Figure 3.10 to indicate its inclusion in the solution set. For the test point (0, 0), we find that $x - 2y = 0$, which is less than 5, and therefore all the points on the same side of the line as the origin satisfy $x - 2y \leq 5$.

Figure 3.10

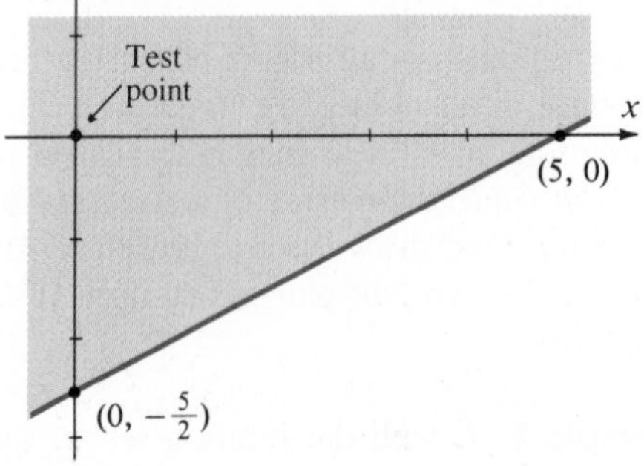

■

Example 3 Graph the solution set for the inequality $-3x - 2y + 5 \geq 5$.

Solution By subtracting 5 from both sides of the inequality, we obtain $-3x - 2y \geq 0$. The graph of the related linear equation is $-3x - 2y = 0$. It is shown in Figure 3.11 as a solid line to indicate its inclusion in the solution set. For the test point (1, 1), we see that the inequality is not true, so the points on the opposite side of the line satisfy this inequality.

Figure 3.11

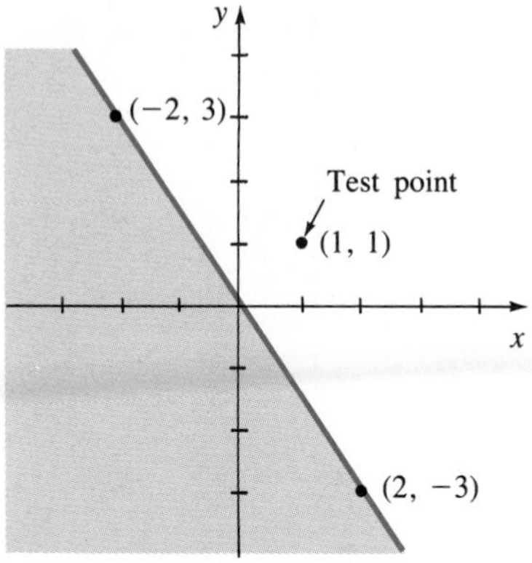

■

Example 4 Graph the solution set for the inequality $y \leq 2$.

Solution The graph consists of all the points in the plane whose y-coordinates are less than or equal to 2. The graph is shown in Figure 3.12.

Figure 3.12

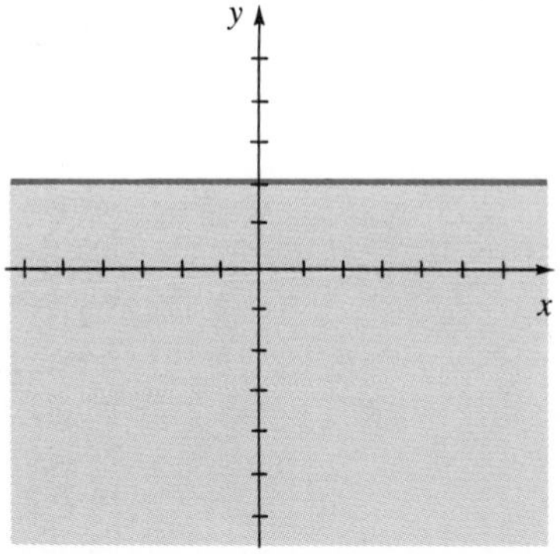

■

Example 5 A warehouse has 10,000 ft^2 of usable storage space in which to store refrigerators and air conditioners. If each refrigerator requires 6 ft^2 and each air conditioner requires 4 ft^2, write an inequality describing the number of each that can be stored in the warehouse. Draw the graph.

Solution The total area needed to store x refrigerators is $6x$ ft^2 and that for y air conditioners is $4y$ ft^2. Therefore, the sum $6x + 4y$ must be less than or equal to 10,000 ft^2; that is,

$$6x + 4y \leq 10{,}000$$

Figure 3.13 shows the graph of the line $6x + 4y = 10{,}000$. The desired half-plane lies below this line, as indicated by the shading. Notice that there are some implied restrictions on this problem owing to its physical nature. Since the number of both refrigerators and air conditioners must be greater than zero, the solution is restricted to the triangular region indicated by the dark shading.

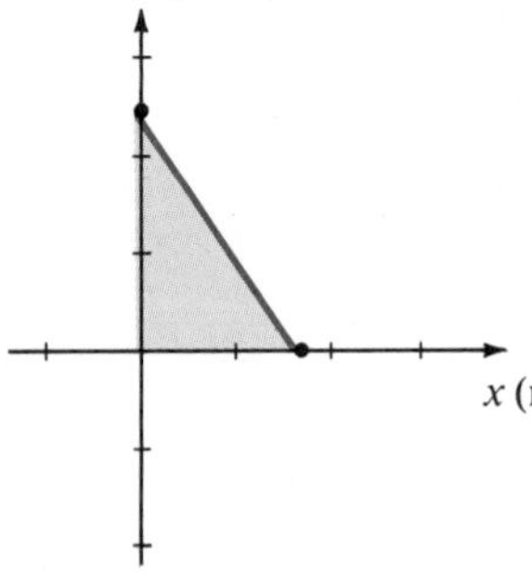

Figure 3.13

EXERCISES SECTION 3.2

Graph the solution sets for the inequalities in Exercises 1–16.

1. $x - 2y \le 3$

2. $-x + 3y \ge 2$

3. $2x + 3y \ge -2$

4. $5x + 4y \le -1$

5. $x > 5$

6. $y \ge -2$

7. $x > y$

8. $x \le y$

9. $3x - y + 7 \le 0$

10. $y < x + 4$

11. $x + y > x - y$

12. $y - x \ge y + x$

13. $2x - y - 8 > x - 2y + 8$

14. $6x - 2 < y + 1$

15. $3(x - 1) + 4x \ge 7x + y + 2(x - y)$

16. $3x + y + 2x - 3y \le 2$

17. The formula relating degrees Fahrenheit to degrees Celsius is $F = \frac{9}{5}C + 32$. For what values of C is $F < 0$?

18. The current in amperes in a certain electrical network is given by $i = 25 - 0.5t$, where t is the elapsed time in milliseconds. For what time t is $i \ge 0$?

19. A retailer wishes to purchase two types of football shoes: a premium brand for $35 per pair and a lesser-quality brand for $15 per pair. Make a sketch showing the possibilities if the capital outlay is limited to $1000.

20. In designing lighting for a given outdoor area, you are given two different kinds of light bulbs, one rated at 100 watts and the other at 200 watts. The lights will normally be on an average of 6 hr per night. Make a sketch showing the possible number of the two different kinds of light bulbs if you are limited to a cost of $1.00 per night and the cost per kilowatt hour is $0.05.

21. A paint company produces latex-base and oil-base paint. It costs $2 to produce oil-base paint and $1 to produce latex-base paint. The company wants to keep total expenditure per day to less than or equal to $1000. Make a sketch showing the possibilities of production of latex-base and oil-base paint.

22. The Jones family owns two cars: a van, which uses regular gasoline but only gets 15 miles per gallon, and a station wagon, which gets 20 miles per gallon but uses high-test premium gasoline. The cost of regular gas is $1.20 per gallon; the cost of premium is $1.35 per gallon. If the Joneses wish to budget less than $30 per week for the cost of gasoline, make a sketch of the possibilities in terms of miles driven per week.

3.3 Systems of Linear Inequalities

In Chapter 2, we considered systems of linear equations. The fundamental objective for systems of inequalities is the same as that for systems of equations—that is, to find solutions *common* to the statements in the system. The discussion here is restricted to inequalities in two unknowns, even though linear inequalities with more than two variables are possible. This restriction is made so that the solutions can be graphed.

A very simple system of inequalities is the system $x \geq 0$, $y \geq 0$. A solution to this system is any ordered pair (x, y) that satisfies both inequalities simultaneously. If we think of the points in the Cartesian plane (represented by the ordered pairs (x, y)) for which $x > 0$ and $y > 0$, then it is obvious that the first quadrant of the plane represents the solution to this system.

The Graphical Solution of a System of Inequalities

The solution set of a system of linear inequalities in two variables is usually found by using graphical procedures. To solve a system of linear inequalities in two variables, we graph all the inequalities on the same set of coordinate axes. The solution set of the system is then the region in the plane that is common to all the inequalities in the system.

Example 1 Graphically solve the system

$$x > 0, \qquad y > 0$$

Solution In Figure 3.14, the half-plane representing $x > 0$ is shown by the blue shading and the half-plane representing $y > 0$ is shown by the gray shading. The solution to the system is the overlapping region in the first quadrant.

Figure 3.14

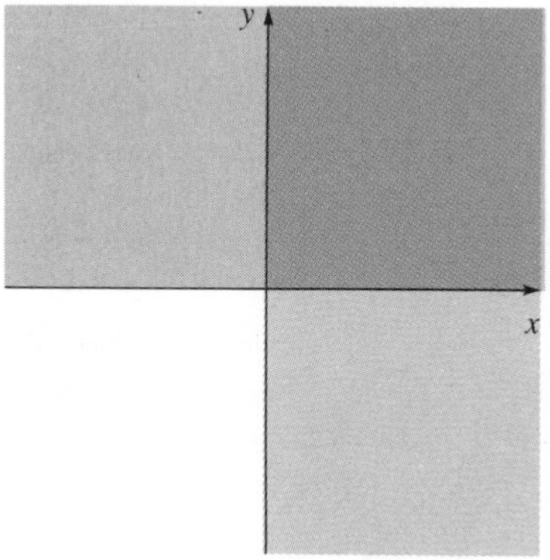

Example 2 Graphically show the solution set to the system

$$2x + y < 4$$
$$x + 2y \leq 2$$

Solution We first sketch the boundary lines for each of the half-planes. In Figure 3.15, the line $2x + y = 4$ is shown with arrows pointing toward the half-plane that contains the test point (0, 0), and is dashed because it is not a part of the region. Similarly, $x + 2y = 2$ is the boundary of the second region, and it too is shown with arrows pointing toward the region containing the test point (0, 0). Finally, the region common to these two half-planes is shown as a shaded area. This common region, along with the section of the line $x + 2y = 2$ for $x \leq 2$, is a graphical representation of the solution set of our given system. The intersection point of the two boundary lines is noted on the graph because such points are often important.

Figure 3.15

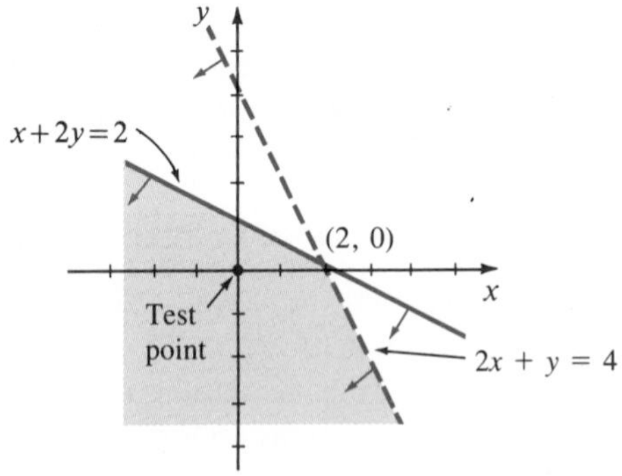

■

Example 3 Graph the solution to each of these systems:

(a) $x + y \leq 2$, $x + y \geq 4$ **(b)** $x + y \geq 2$, $x + y \geq 4$ **(c)** $x + y \geq 2$, $x + y \leq 4$

Solution For all three of these systems, the boundary lines for the two half-planes are $x + y = 2$ and $x + y = 4$, and they are parallel. But, using (0, 0) as the test point in each case, we find that the first system has no points in common, the second solution set includes the entire half-plane $x + y \geq 4$, and in the third system, the solution set is the part of the plane between the two boundary lines. (See Figure 3.16.)

Figure 3.16

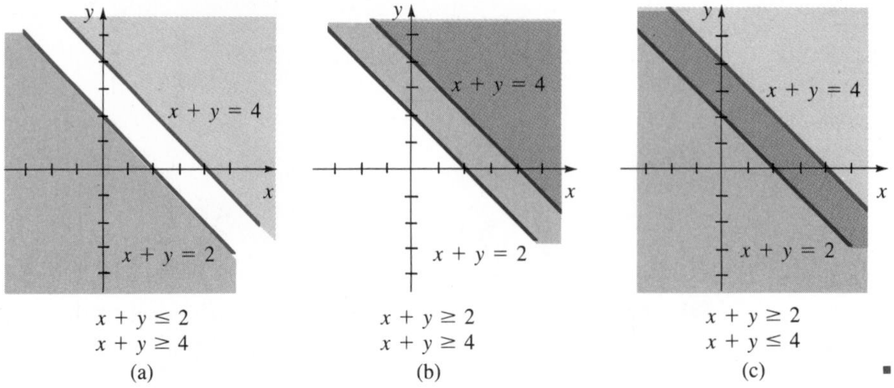

■

Example 4 Graphically solve each system:

(a) $x + y < 4$, $2x + 2y \geq 8$ **(b)** $x + y \leq 4$, $2x + 2y \geq 8$ **(c)** $x + y > 4$, $2x + 2y \geq 8$

Solution When we sketch the boundary lines for each half-plane, we recognize that the boundary lines are the same; they coincide. The test point (0, 0) in the first system indicates that there are no points that satisfy both inequalities. In the second system, we see that the line itself satisfies both inequalities. Finally, in the third system, we see that the half-plane that does not contain (0, 0) satisfies the system. (See Figure 3.17.)

Figure 3.17

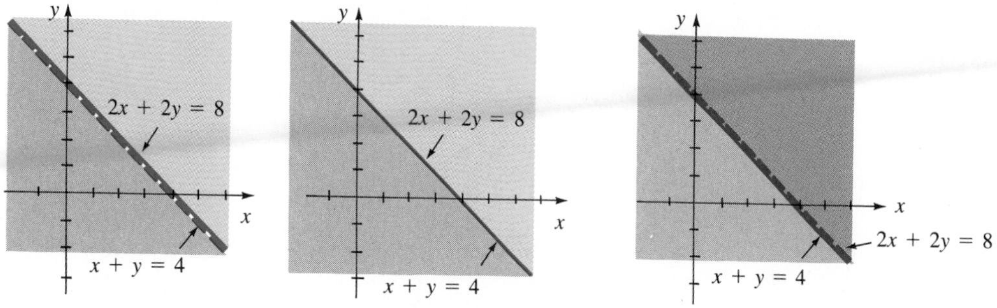

$x + y < 4$, $2x + 2y \geq 8$ (a)

$x + y \leq 4$, $2x + 2y \geq 8$ (b)

$x + y > 4$, $2x + 2y \geq 8$ (c) ■

COMMENT: *Examples 2, 3, and 4 show the possible solutions for linear systems of two inequalities in two unknowns:*

- *If the two boundary lines intersect, the solution set will be one of the four regions into which the lines divide the plane.*
- *If the two boundary lines are parallel, then the solution may be a half-plane bounded by one of the two boundary lines, or an infinite strip bounded on both sides by the boundary lines, or there may be no solution.*
- *If the two boundary lines coincide, then the solution may be a half-plane bounded by the line, or the boundary line itself, or there may be no solution.*

Example 5 Graphically solve the system

$$6x + y \leq 24$$
$$2x + 3y \leq 24$$
$$x \geq 0$$
$$y \geq 0$$

Include all points of intersection of the boundary lines.

Solution To simplify the graphing process, we note that the last two inequalities restrict us to the first quadrant. The lines $6x + y = 24$ and $2x + 3y = 24$

are shown in Figure 3.18. The two half-planes are found as usual, but the resulting solution set is restricted to the first quadrant. We also find that the point of intersection of the two lines is (3, 6). The other points of intersection are intercepts, and are found as a natural consequence of graphing the lines. The points (0, 0), (0, 8), (3, 6), and (4, 0) are called the **corner points** of the region.

Figure 3.18

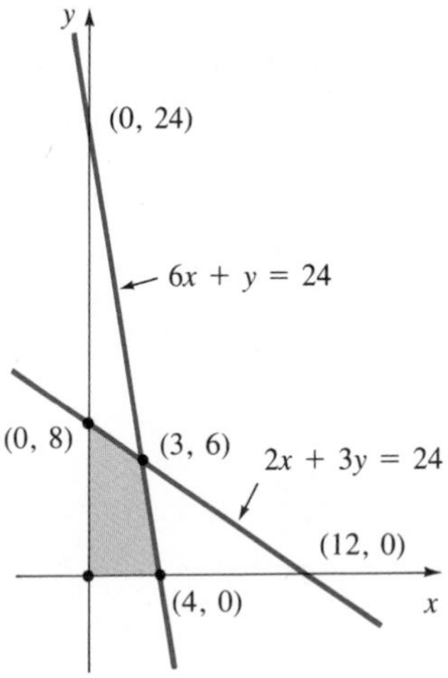

■

Example 6 Sketch the solution set for the system

$$x + 2y \geq 4$$
$$2x + y \geq 4$$
$$x \geq 0$$
$$y \geq 0$$

Solution As before, the lines $x + 2y = 4$ and $2x + y = 4$ are sketched and the point of intersection $(\frac{4}{3}, \frac{4}{3})$ is found. Also as in the previous example, we note that

Figure 3.19

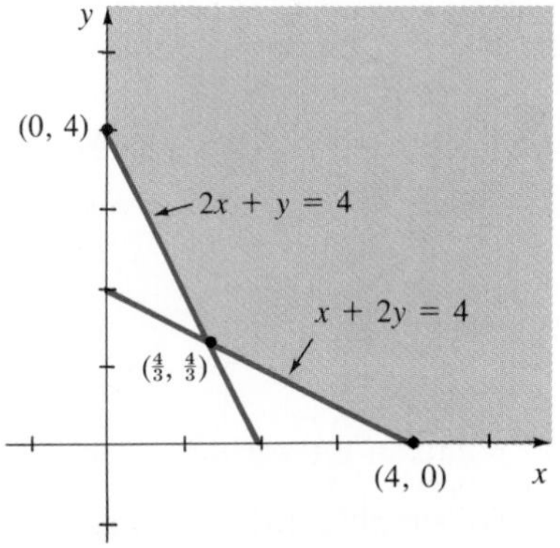

the latter two inequalities merely restrict the variables to the first quadrant. In this case, the test point (0, 0) is not in the solution set of either inequality, so the common solution set becomes the *unbounded* region shown in Figure 3.19. ■

Example 7 The Precision Brake Company manufactures brake drums and disks for Formula I–type race cars. A set of drum brakes requires 2 hr of foundry work and 4 hr of machining. A set of disk brakes requires 3 hr of foundry work and 2 hr of machining. The company operates two 8-hr shifts per day. Sketch a system of inequalities showing the possible production of drum brakes and disk brakes.

Solution Let x be the number of sets of drum brakes produced and y be the number of sets of disk brakes produced. Then $2x + 3y$ represents the total time that the foundry must be used to produce x sets of drum brakes and y sets of disk brakes. Since the foundry only operates 16 hr per day, the inequality $2x + 3y \le 16$ must be satisfied. Similarly, since the total time required for machining is $4x + 2y$, the inequality $4x + 2y \le 16$ must also be satisfied. Of course, x and y cannot be negative, so $x \ge 0$ and $y \ge 0$. The solution set (shaded region) of this system of inequalities represents the number of drum brake sets and the number of disk brake sets that can be produced in one day. The graph is shown in Figure 3.20.

Figure 3.20

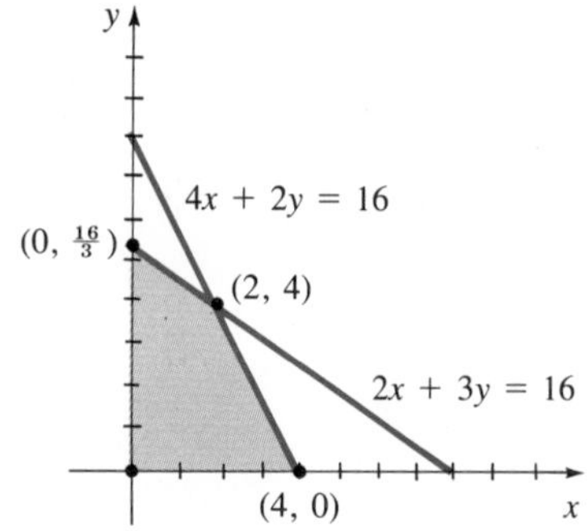

■

EXERCISES SECTION 3.3

In Exercises 1–20, solve each system of inequalities graphically.

1. $x \le 2$
$y > 6$

2. $x > -1$
$x + y > 0$

3. $x \le 0$
$y \le x$

4. $x - 4y \le 2$
$-2x + y \ge -4$

5. $2x + 5y > -4$
$4x + 3y \le 6$

6. $x + y < 4$
$x - y > 2$

7. $x \ge 0, y \ge 0$
$2x + 3y \le 6$

8. $x \ge 0, y \ge 0$
$2x + y \ge 2$

9. $y \ge 0$
$x - 3y \ge -3$
$-2x + y \ge -4$

10. $x > 2, x < 5$
$y > -1$

11. $x + 2y > 2$
$x - y < 1$
$x > 0, y > 0$

12. $2x - y > 1$
$x - y < 1$
$x > 0, y > 0$

13. $x > 0, y > 0$
$x + y < 2$
$2x + y > 1$

14. $x \geq 0, y \geq 0$
$x + y < 2$
$x + y > 1$

15. $x > 0, y > 0$
$x + y < 2$
$x + y < 1$

16. $x + y > 4$
$2x + y > 8$

17. $x + y > 4$
$2x + y < 8$
$x - y > 2$

18. $4x + 3y \leq 6$
$2x + 5y \geq -4$
$-2x + y \leq 2$

19. $4x + 3y \leq 6$
$2x + 5y > -4$
$-2x + y \leq 2$

20. $x + 2y > 4$
$x - y < 4$
$y < 5$

21. A toy company has the capacity to make 1000 toy cars and trucks per day. The number of toy cars that can be made is less than 800 and the number of toy trucks is less than 700 per day. Write a system of inequalities describing the possible combinations of toy trucks and cars the company can produce in a day. Graph the solution set.

22. Suppose you can invest \$10,000, part at 7% and part at 10%, but the amount invested at 10% cannot be more than twice what is invested at 7%. Describe the investment possibilities with a system of inequalities and draw the graph.

23. The final grades for Math 101 are due in the registrar's office 10 hr after the final exam. The professor has two graduate assistants to grade final exam papers. Sally can work no more than 5 hr and Joe can work no more than 7 hr. Write a system of inequalities describing the possibilities and draw the graph.

24. The time available to a bakery for making cakes and pies is no more than 6 hr per day. The cake-baker will work no more than 5 hr per day and the pie-baker will work no more than 4 hr per day. If x is the number of hours for baking cakes and y is the number of hours for baking pies, describe the baking schedule possibilities with a system of inequalities and draw the graph.

25. A farmer has 160 acres on which to plant soybeans or corn. Soybeans cost \$250 per acre to plant and harvest; corn costs \$150 per acre. The farmer can spend no more than \$30,000. Write a system of inequalities describing the planting possibilities and draw the graph.

26. Julie wishes to diet using a food supplement of carbohydrates and protein, which she mixes from two available products. Food A is 75% carbohydrates, 20% protein, and 5% fat. Food B is 40% carbohydrates, 50% protein, and 10% fat. Julie wants 100 grams or more of carbohydrates and 60 grams or more of protein in 200 grams or less of the supplement. Write a system of inequalities describing the possibilities and draw the graph.

3.4 LINEAR PROGRAMMING

One of the major applications of linear inequalities is in the area of business and economics, where we frequently want to maximize (profit) or minimize (cost), subject to certain limitations or constraints on capital, raw materials, worker availability, and so forth. If the profit or cost functions can be expressed in terms of linear functions, and the constraints in terms of linear inequalities, then the method that is used to maximize or minimize the quantities of interest is called **linear programming.** The phrase *linear programming* is chosen because the function and the constraining inequalities are linear and some usually limited resource is being allocated, or programmed, to optimize some quantity such as cost or profit. Thus, the word *programming* is not used here in the same way as it is in computer programming.

Problems of maximizing or minimizing (or, more generally, *optimizing*) a quantity are usually covered in calculus courses. However, the unique approach employed in linear programming allows us to study this subject without using calculus when the constraints and the function to be optimized are linear. Linear programming is a method for optimizing a linear function, called the **objective function,** subject to certain limitations in the form of linear inequalities, called **constraints.** Linear programming problems typically involve many variables, but in this section we will limit our discussion to problems involving two variables. With two variables we can use graphical techniques to help us solve the problems. Example 1 illustrates a simple linear programming problem.

Example 1 Suppose the Specialty Pen Company produces two types of pens: silver-tip and gold-tip. Because of raw-material limitations, the company can produce no more than 200 pens per day. Suppose the profit, P, on the gold-tip model is \$3 per pen and the profit on the silver-tip model is \$2 per pen. How many pens of each type should be produced per day to maximize the profit?

Solution Here, the objective function to be maximized is the profit, P. If we let x be the number of gold-tip pens to be produced in a day and y the number of silver-tip pens, then the profit is given by $P = 3x + 2y$. The linear constraints are $x \geq 0$ and $y \geq 0$, because the number of pens must be nonnegative, and $x + y \leq 200$, because of the limitation on the total daily output. The question we wish to answer is "How many of each type of pen should be produced to maximize profit?" To answer this question, we first solve the system of inequalities $x + y \leq 200$, $x \geq 0$, $y \geq 0$. In linear programming, the solution of the system of constraint inequalities is called the **feasible region** and any solution in the feasible region is called a **feasible solution.** The feasible region in this problem is the shaded region in Figure 3.21(a). Any point inside the shaded region or on its boundary is a feasible solution. Thus, $x = 50$, $y = 75$ is a feasible solution, as is $x = 100$, $y = 100$ and $x = 200$, $y = 0$.

If the values of a feasible solution are substituted into the objective function, a value for the objective function is obtained. For the objective function $P = 3x + 2y$, some typical substitutions of feasible solutions into the objective function

Figure 3.21

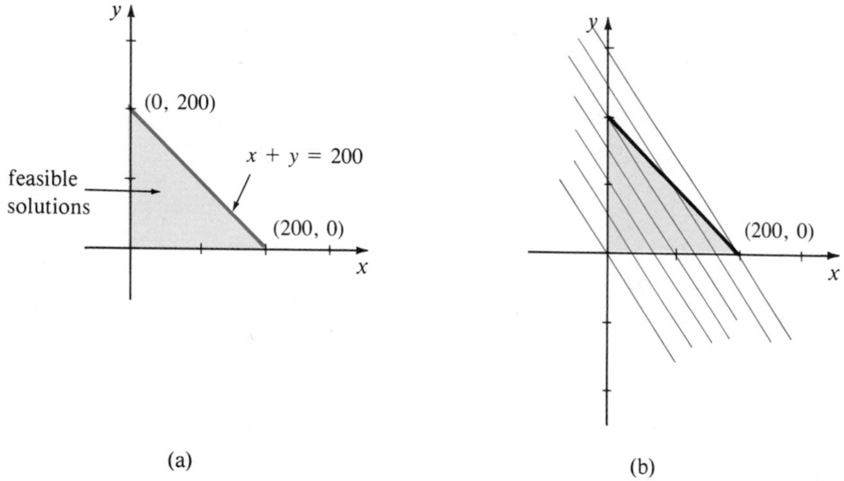

are $P = 3(100) + 2(100) = 500$, $P = 3(200) + 2(0) = 600$, and $P = 3(0) + 2(200) = 400$. A point in the feasible region that yields a maximum or minimum value for the objective function is called the **optimal solution.**

How do we know which feasible solution, if any, will give a maximum value for P? To answer this question, observe that for any particular value of P, say P_0, the objective function is $3x + 2y = P_0$, where P_0 is a constant. For each value of P_0, the graphs of the equation $3x + 2y = P_0$, or $y = \frac{1}{2}P_0 - \frac{3}{2}x$, form a family of parallel lines, some of which cut across the feasible region. (See Figure 3.21(b).) The objective function P will be at a maximum when the y-intercept of the line $y = \frac{1}{2}P_0 - \frac{3}{2}x$ has its largest value. Since the optimal solution must occur at a point in the feasible region, the line $y = \frac{1}{2}P_0 - \frac{3}{2}x$ must intersect the feasible region. Hence, we need only select the line from the family of parallel lines that has the largest y-intercept *and* intersects the feasible region. This is shown in Figure 3.21(b) to be the line with x-intercept (200, 0). Thus, the optimal solution is 200 gold-tip pens and no silver-tip pens, which yields the maximum profit of $P = \$600$. ▪

COMMENT: *The optimal solution in Example 1 is, of course, obvious without linear programming: that is, make only gold-tip pens since they yield the largest profit margin. Note, however, that linear programming gives the same solution and has the advantage that it can be used when the answer is not so obvious.*

The procedure of Example 1 shows the basis for the technique of linear programming. A solution set (feasible region) to a system of constraint inequalities is determined. Then a linear objective function is varied to produce a family of parallel lines cutting through the feasible region. Out of this family we select the line with the largest y-intercept. Typically, objective functions are of the form $P = ax + by$, where a and b are positive, so that the slope of the family of parallel lines, $y = -\frac{a}{b}x + \frac{1}{b}P$, is negative. Figures 3.22(a) and (b) show some possible situations different from that shown in Figure 3.21. In Figure 3.22(a), all points on the straight line $x + y = 200$ in the first quadrant would not only be feasible, but would also give the same maximum value to the objective function $P = x + y$.

Figure 3.22

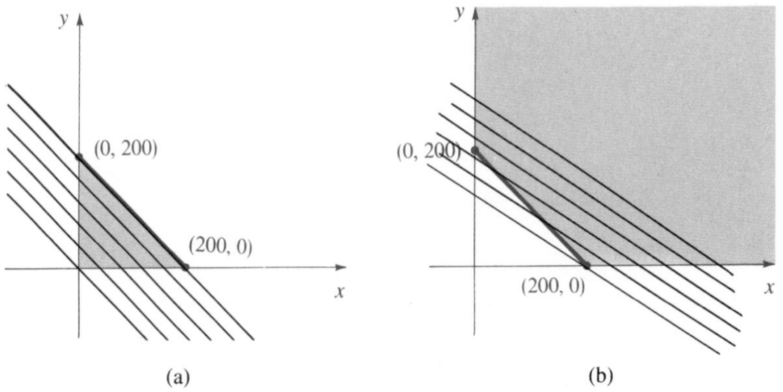

Figure 3.22(b) shows a feasible region that is unbounded and some graphs of $P = 2x + 3y$. In this case, a family of parallel lines cannot have a y-intercept with a maximum value. Hence, there is no maximum value for the feasible region shown in Figure 3.22(b). Notice, however, that there is a *minimum* value of the y-intercept, so we could *minimize* the objective function $P = 2x + 3y$ with this procedure. In that case, we would look for the feasible solution that gives the smallest y-intercept, which is the point (200, 0). The optimal solution—that is, the minimum solution—is (200, 0), which yields the minimal value of $P = 2(200) + 0 = 400$.

COMMENT: *If the feasible region obtained by graphing the system of constraint inequalities does not contain* any *points (see, for example, the system of inequalities in Example 3a of Section 3.3), then the problem has no solution. We say the problem is* unfeasible.

Figures 3.21 and 3.22 typify the graphical linear programming procedure. Although these examples are certainly insufficient to justify any conclusion, we do see that for these examples the optimal solutions occur on the boundary of the feasible region, and at a corner point of the feasible region if the solution is unique. The following theorem is proven in most elementary linear programming texts:

THEOREM Optimal Solution

If an optimal solution of a linear objective function $P = ax + by$ exists, it occurs at a corner point of the graph of the set of feasible solutions.

Because of this theorem, we may summarize the graphical method of solving a linear programming problem in two variables as follows:

LINEAR PROGRAMMING: THE GRAPHICAL METHOD ■

Assume we have a linear objective function $P = ax + by$ and a system of linear inequalities representing the constraints. Then, to find the optimal solution:

1. Find the solution set for the system of linear inequalities. This is the set of *feasible solutions.*
2. Determine the coordinates of all corner points of the feasible region.
3. Evaluate the given objective function at each of the corner points. From these, choose the largest or smallest value (*depending on the nature of the problem*) to *find the solution.*

■ **Example 2** Determine the maximum value of $P = 5x + 3y$, subject to the constraints $2x + 3y \le 12$, $3x + y \le 11$, $x \ge 0$, $y \ge 0$.

Solution Figure 3.23 shows the feasible region found by graphing $2x + 3y = 12$ and $3x + y = 11$ in the first quadrant. The point of intersection of the two lines is (3, 2). The four vertices of the feasible region are (0, 0), ($\frac{11}{3}$, 0), (3, 2) and (0, 4). We evaluate P at these four points:

$$\begin{aligned} &\text{at } (0, 0), && P = 0 \\ &\text{at } \left(\frac{11}{3}, 0\right), && P = \frac{55}{3} \\ &\text{at } (3, 2), && P = 21 \\ &\text{at } (0, 4), && P = 12 \end{aligned}$$

Of these four values, 21 is the maximum value and occurs at $x = 3$, $y = 2$.

Figure 3.23

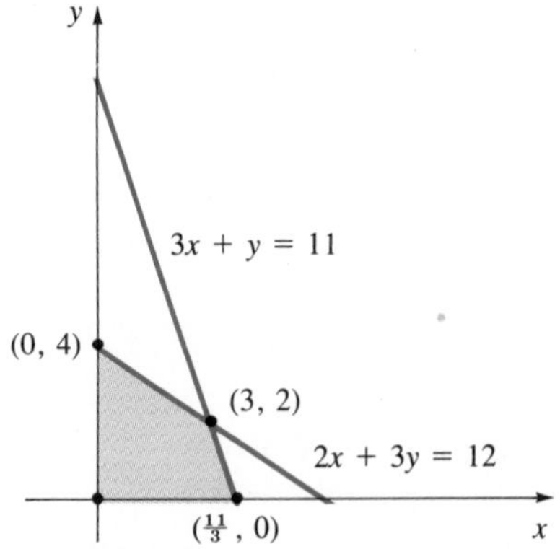

■

As we noted earlier, linear programming can also be used to minimize an objective function, subject to certain constraints. The next example considers a minimization problem.

Example 3 Minimize $K = y + 5x$ subject to $2x + 3y \geq 12$, $3x + y \geq 11$, $x \geq 0$, $y \geq 0$.

Solution Figure 3.24 shows the feasible region obtained by graphing $2x + 3y = 12$ and $3x + y = 11$. The corner points are (0, 11), (3, 2), and (6, 0).

Figure 3.24

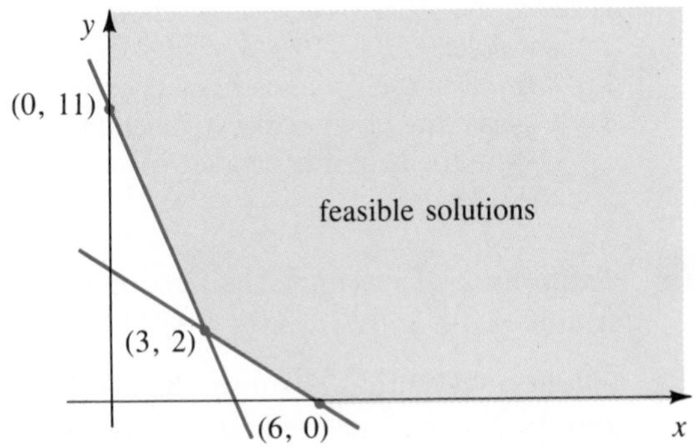

The constraint region is unbounded, but the key to the solution is still the vertices of the region. We evaluate K at the corner points:

$$\begin{aligned} &\text{at } (6, 0), \quad K = 0 + 5(6) = 30 \\ &\text{at } (3, 2), \quad K = 2 + 5(3) = 17 \\ &\text{at } (0, 11), \quad K = 11 + 5(0) = 11 \end{aligned}$$

Thus, K is minimized at (0, 11) and the minimum value is 11. ∎

The practical applications of simple linear programming problems are numerous. Establishing the constraint inequalities is usually the most difficult task.

Example 4 A small book company produces the same book in two styles, paperback and hardcover. Producing the paperback yields a profit of \$400 per day; the hardcover yields \$700 per day. The company can employ only one printing crew, which will work only 5 days per week. Paper limitations demand that the paperback be produced no more than 3 days per week, whereas the hardcover production is limited to 4 days or less per week. If you owned the company, how would you assign the printing crew?

Solution First, establish the constraints in mathematical terms. Let x equal the number of days the paperback is printed and let y equal the number of days the hardcover is printed. Then we want to maximize the profit given by $P = 400x + 700y$. The constraints are:

$x + y \leq 5$	(restriction to 5-day week)
$x \leq 3$	(restriction on paperback supplies)
$y \leq 4$	(restriction on hardcover supplies)
$x \geq 0, y \geq 0$	(number of days must be positive or zero in either case)

Figure 3.25 displays the graph of the constraint inequalities. The corner points are (0, 4), (1, 4), (3, 2), (3, 0), and (0, 0). The profit at each of these points is:

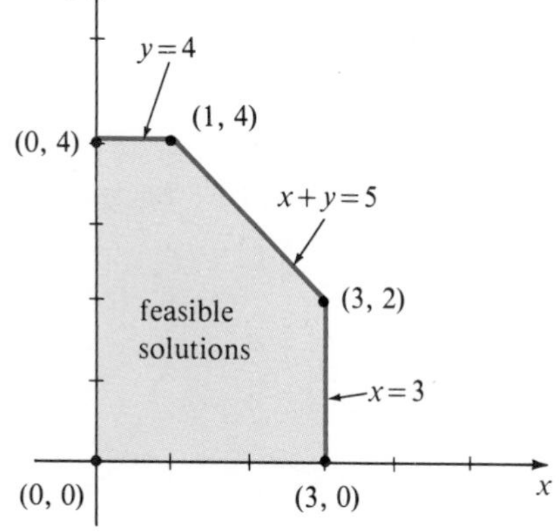

Figure 3.25

$$\begin{aligned} &\text{at } (0, 4), \quad P = 2800 \\ &\text{at } (1, 4), \quad P = 3200 \\ &\text{at } (3, 2), \quad P = 2600 \\ &\text{at } (3, 0), \quad P = 1200 \\ &\text{at } (0, 0), \quad P = 0 \end{aligned}$$

Hence the crew should devote one day to paperbacks and four to hardcovers to realize the greatest profit, \$3200. ■

Example 5 A machine shop makes both drum brakes and disk brakes. Both types require a lathe and a grinder in production. Drum brakes require 2 hr on the lathe and 4 hr on the grinder, and disk brakes require 3 hr on the lathe and 2 hr on the grinder. The company makes \$140 profit on drum brakes and \$180 on disk brakes. Assuming that the company has one lathe and one grinder and that both machines are used 16 hr per day, how many of each should the company make in order to make the largest profit?

Solution List the required times in a table.

	Lathe	Grinder
Drum brake	2 hr	4 hr
Disk brake	3 hr	2 hr

Let x equal the number of drum brakes produced, and y the number of disk brakes. Then we want to maximize the profit, $P = 140x + 180y$. The constraints are:

$$2x + 3y \leq 16 \quad \text{(restriction of lathe to 16-hour day)}$$
$$4x + 2y \leq 16 \quad \text{(restriction of grinder to 16-hour day)}$$
$$x \geq 0, \ y \geq 0 \quad \text{(number of pieces made must be positive or zero)}$$

Figure 3.26 shows the graph of the constraint inequalities. We leave it to you to show that the corner point, (2, 4), gives the maximum profit. Hence, the company makes the maximum profit by producing 2 drum brakes and 4 disk brakes. What is the maximum profit?

Figure 3.26

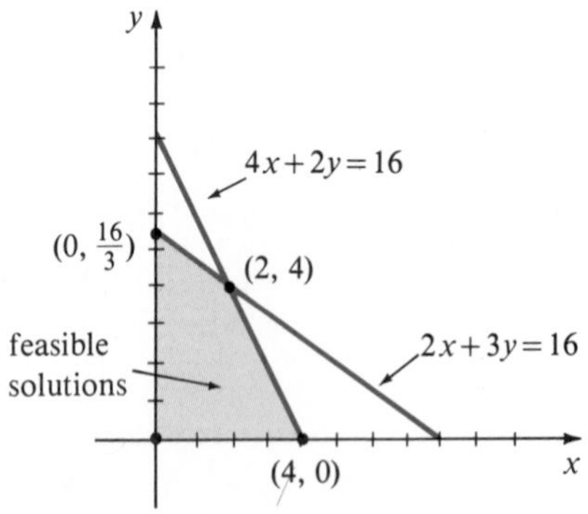

■

EXERCISES SECTION 3.4

In Exercises 1–15, graphically solve the indicated linear programming problems. In each case, assume that $x \geq 0$ and $y \geq 0$.

1. Given $y \leq 2$, $x + 2y \leq 8$, maximize $P = x + 3y$.

2. Given $x + y \leq 2$, $x + 3y \leq 4$, maximize $P = 2x + 5y$.

3. Given $x + y \leq 5$, $x + 2y \leq 8$, maximize $P = 5x + y$.

4. Given $x \leq 10$, $x + y \leq 15$, maximize $P = 3x + 2y$.

5. Given $x + 3y \leq 8$, $2x + y \leq 6$, maximize (a) $P = 2x + 3y$; (b) $P = 3x + y$.

6. Given $3x + 5y \leq 30$, $2x + y \leq 8$, maximize (a) $P = 3x + 2y$; (b) $P = 4x + 5y$.

7. Given $x + y \geq 5$, $x + 2y \geq 8$, minimize $P = 7x + 2y$.

8. Given $x + 2y \geq 3$, $x + 3y \geq 4$, minimize $P = 5x + 3y$.

9. Given $x + 3y \geq 8$, $2x + y \geq 6$, minimize (a) $P = 2x + 3y$; (b) $P = 3x + y$.

10. Given $3x + 5y \geq 30$, $2x + y \geq 8$, minimize (a) $P = 3x + 2y$; (b) $P = x + 5y$.

11. Given $x + 2y \leq 4$, $x + 2y \geq 5$, maximize $P = 5x + 3y$.

12. Given $x + 2y \leq 4$, $x + 2y \geq 2$,
(a) maximize $P = 5x + 3y$;
(b) minimize $P = 5x + 3y$.

13. Given $x + 2y \geq 2$, $2x + y \leq 4$,
(a) maximize $P = 3x + 4y$;
(b) minimize $P = 3x + 4y$.

14. Given $x + 2y \geq 2$, $2x + y \leq 4$, $x + y \leq 3$,
(a) maximize $P = 3x + 4y$;
(b) maximize $P = 5x + 2y$.

15. Given $x + 2y \geq 2$, $2x + y \leq 4$, $x + y \geq 3$,
(a) minimize $P = 3x + 4y$;
(b) minimize $P = 5x + 2y$.

A condition in the system of inequalities that does not affect the feasible region is said to be superfluous. *In Exercises 16–17, indicate any superfluous constraints.*

16. Given $x + 3y \leq 8$, $2x + y \leq 6$, $x + y \leq 10$, maximize $P = x + 3y$.

17. Given $x + 3y \geq 8$, $2x + y \geq 6$, $x + y \geq 1$, minimize $P = 2x + 5y$.

18. The owner of an office furniture store finds that the profit equation for desks and filing cabinets is $P = 5x + 3y$, where x is the number of desks and y the number of filing cabinets sold each day. How many of each need to be sold each day to maximize profit, if x and y are both nonnegative and subject to the constraints $x + \frac{1}{2}y \leq 8$, $\frac{1}{2}x + y \leq 7$.

19. A company makes a profit of \$10 on steel castings and \$8 on aluminum castings. If x is the number of steel castings, and y the number of aluminum castings, the profit equation is $P = 10x + 8y$. Company engineers have found that they must satisfy the following constraints for each day's production:

$$x \geq 0, \qquad y \geq 0,$$
$$2x + 3y \leq 120, \qquad 4x + y \leq 90$$

How many of each kind of casting should the company make each day to maximize its profit?

20. A company produces deluxe- and economy-model radios. Each economy model yields a profit of \$6, whereas each deluxe model yields \$10 profit. Both models require approximately the same amount of material to build and the supply allows for a total of 500 radios per day. The sales force reports that the demand for the economy model does not exceed 350 per day, but the deluxe radio can be sold as fast as it is made. The deluxe model requires twice as much time to build as the economy model, and if only the economy model were built, there would be enough time to produce 700 radios per day. How many economy models and deluxe models should the company construct each day to maximize its profit?

21. A broker wishes to invest \$20,000 for her client, part at 7% and part at 8.5%. The amount she invests at 8.5% cannot be more than twice the amount she invests at 7%. How much should she deposit at each rate to maximize total income? Only multiples of $100 can be invested.

22. A company produces two kinds of vacuum cleaners, with profits of $5 and $8, respectively, on the economy and deluxe models. No more than 1000 of the economy model can be sold per day. The deluxe model takes three times as long to make as the economy model and if only the economy model were made, time would allow for production of 1300 per day. Find the number of each that should be built to maximize the profit. What is the maximum profit?

23. UPS has 1400 packages to deliver overnight by air freight. The air freight company can supply large planes that will hold 200 packages at a cost of $2500 per flight, and smaller planes that will hold 100 packages at a cost of $1000 per flight. The large planes require a crew of four; the smaller planes, a crew of three. Only 36 crew members are available. How many planes of each type should be used to minimize the cost to UPS?

3.5 THE SIMPLEX METHOD

The graphical method of solving linear programming problems that was presented in Section 3.4 is limited to constraints that can be graphed in two dimensions. Hence, that method is limited to two-variable problems. Sophisticated extensions of the graphical method can be made, but most often we turn to an algebraic technique called the **simplex method.**

HISTORICAL COMMENT: *The simplex method was developed by George Dantzig in 1946, and later, with the principle of duality, by John von Neumann. Many of the early applications of this technique were for military problems; today, it is used extensively in government and private industry as well.*

The simplex method is an algebraic method, but its initial inspiration and rationale is actually a disguised use of the graphical approach, in which one systematically moves from vertex to vertex in the feasible region until the optimal solution is obtained. We will begin this section with an example that compares the simplex method with the graphical approach. Before discussing the simplex method, we will explain the concept of a basic solution of a system of linear equations.

In the development of the simplex method, we make the assumption that an inequality can be transformed into an equation by introducing a new variable, called a **slack variable.** For example, we assume that the inequality $x + 3y < 5$ can be written as $x + 3y + w_1 = 5$, where w_1 takes up the slack between the two sides of the inequality. By introducing slack variables into each of the inequality constraints on the objective function, we can convert this system of inequalities into a system of linear equations.

COMMENT: *The term* slack variable *is chosen to convey the idea that the variable is taking up the slack between the inequality and the corresponding equation. In general, if* $x \leq A$, w_1 *is a slack variable if* $x + w_1 = A$.

We explain the simplex method in the context of the linear programming problem in Example 1. We have included the graphical solution of this problem as part of the example.

Example 1 Use the simplex method to maximize the objective function $P = 3x + 2y$, with the constraints that $x \geq 0$, $y \geq 0$, $2x + y \leq 12$, and $x + 2y \leq 18$. Figure 3.27 shows the feasible region. Notice that the optimal solution occurs at (2, 8) and gives $P = 3(2) + 2(8) = 22$ as the maximum value of the objective function.

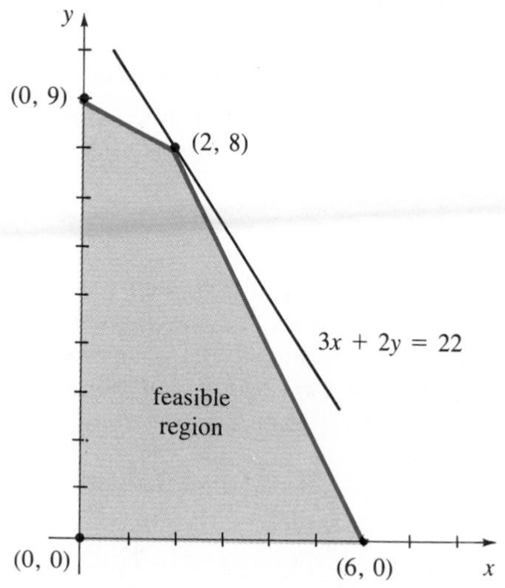

Figure 3.27

Solution As noted above, we want to introduce new variables, called slack variables, into each of the inequality constraints in order to transform them into linear equations. Calling the slack variables w_1 and w_2, we write the following equations.

$$2x + y + w_1 = 12 \quad (\text{instead of } 2x + y \leq 12)$$
$$x + 2y + w_2 = 18 \quad (\text{instead of } x + 2y \leq 18)$$

The linear programming problem that we graphically solved at the beginning of the example is now recast in the form of a problem in which we wish to maximize the objective function $P = 3x + 2y$ by finding values of x, y, w_1, and w_2 that satisfy the system of equations obtained from the inequality constraints. It is customary to write the objective function in the form $P - 3x - 2y = 0$, and then to put this equation, along with the constraint equations, into augmented matrix form. The specified augmented matrix is

$$A = \begin{array}{c} \\ \\ \\ \end{array}\begin{bmatrix} x & y & w_1 & w_2 & P & \\ 2 & 1 & 1 & 0 & 0 & 12 \\ 1 & 2 & 0 & 1 & 0 & 18 \\ -3 & -2 & 0 & 0 & 1 & 0 \end{bmatrix} \begin{array}{l} \\ \leftarrow \text{constraint} \\ \leftarrow \text{constraint} \\ \leftarrow \text{objective function} \end{array}$$

The matrix A represents the system of three equations: $2x + y + w_1 = 12$, $x + 2y + w_2 = 18$, and $-3x - 2y + P = 0$, in the five unknowns x, y, w_1, w_2, and P. This matrix is called the **initial simplex matrix.**

In matrix A the first column represents the coefficients of x; the second column, the coefficients of y; the third column, the coefficients of w_1; the fourth column, the coefficients of w_2; and the fifth column, the coefficients of P. The last column in the initial simplex matrix is the *augment,* and is often set off by a vertical line, as it was in Section 2.4. The last row is called the **objective row,** because it is the row that represents the equation involving the objective function.

We now wish to solve this system of three equations in five unknowns. However, more is involved than just getting a solution. Since the number of unknowns exceeds the number of equations, there are *infinitely many solutions* to this system. For example, $x = 0$, $y = 0$, $w_1 = 12$, $w_2 = 18$, and $P = 0$ is a solution, and many others can be obtained by letting two of the variables be fixed and solving for the remaining three. However, we do not want just *any* solution; we want the one that maximizes the value of the objective function P.

COMMENTS:

1. *The solution obtained from the initial simplex matrix for which $P = 0$ is often called the* ***initial solution,*** *because it always includes the values $x = 0$, $y = 0$. The initial solution corresponds to the value of P at the origin in the graphical method.*
2. *The infinitely many solutions of the simplex matrix correspond to the infinitely many solutions of the feasible region in the graphical method.*

The initial simplex matrix is the starting point for the simplex method. The simplex method involves the use of the elementary row operations to transform the initial simplex matrix into equivalent simplex matrices, until the maximum value of P and the corresponding values of x and y can be obtained by inspection. But how do we proceed? The key to the simplex method is the choice of what is called a **pivot element.** Then, in a manner analogous to Gauss-Jordan elimination, the pivot element is used to obtain zeroes for the other entries in that column.

The choice of a pivot element is guided by our knowledge of the graphical technique, in which the maximum value of the objective function P always occurs at a corner point of the feasible region. Since the origin is always a corner point of the feasible region, the simplex method begins there and moves systematically to other corner points of the feasible region *while increasing the value of P*. The technique is to maximize P by changing only one variable at a time until the coefficients of the variables in the objective function are zero or positive.

pivot element ↓

$$A = \left[\begin{array}{ccccc|c} \textcircled{2} & 1 & 1 & 0 & 0 & 12 \\ 1 & 2 & 0 & 1 & 0 & 18 \\ -3 & -2 & 0 & 0 & 1 & 0 \end{array}\right]$$

pivot column

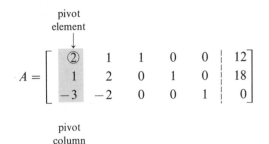

We begin by inspecting the objective (last) row of the simplex matrix A in order to identify the variable that will cause the greatest increase in P. If such a variable exists it must have a negative coefficient. Why? The entry -3 represents the coefficient of x, and -2 the coefficient of y, in the objective function, and therefore the greatest increase in the value of P is made by changing x. The column of matrix A that contains the most negative entry in the objective function row is called the **pivot column.** The pivot column is shaded in color in the matrix A. The amount by which x can be increased is limited by the two constraint rows. Figure 3.27 shows that x is limited to a value no larger than 6. This same limiting value can also be obtained from the simplex matrix by examination of the pivot column. To find the maximum value that x can have, we divide the entries in the augmented column of A by the corresponding entries in the pivot column and choose the smaller of the quotients; this entry is called the **pivot point.** The desired quotients are $\frac{12}{2} = 6$ and $\frac{18}{1} = 18$. Since $\frac{12}{2} < \frac{18}{1}$, the pivot element in this column is 2, the entry in the first row and first column. This corresponds to the corner point (6, 0) of the feasible region shown in Figure 3.27.

We use the pivot element to obtain an equivalent simplex matrix in which the pivot element is 1 and the other entries in the pivot column are 0. We make the pivot element 1 and the other elements of the pivot column 0 by using the following elementary row operations on A:

- Row 1 of A is multiplied by $\frac{1}{2}$ and becomes row 1 of B.
- Row 1 of B is then multiplied by -1 and added to row 2 of A, and this sum becomes row 2 of B.
- Row 1 of B is then multiplied by 3 and added to row 3 of A, and this sum becomes row 3 of B.

This process is called **pivoting.**

new pivot element

$$B = \left[\begin{array}{ccccc|c} 1 & \frac{1}{2} & \frac{1}{2} & 0 & 0 & 6 \\ 0 & \textcircled{\frac{3}{2}} & -\frac{1}{2} & 1 & 0 & 12 \\ 0 & -\frac{1}{2} & \frac{3}{2} & 0 & 1 & 18 \end{array}\right]$$

new pivot column

Keep in mind that each row of the original simplex matrix A represents an equation. The rows of the equivalent matrix B also represent equations. In particular, the last row of matrix B represents the objective equation

$$-\frac{1}{2}y + \frac{3}{2}w_1 + P = 18$$

Thus, we see that P now depends upon y and w_1. Since the coefficient of y is negative, any increase in y will increase the value of P. Thus, a new pivot column is determined by the number $-\frac{1}{2}$ in the objective row, which is the coefficient of y. The greatest increase in P is then obtained by changing y.

As before, we examine the elements in the pivot column of B in order to obtain a new pivot element. This is done by finding the constraints on y imposed by each of the first two rows (constraint equations). Dividing the entries in the augmented column of the matrix B by the corresponding entries in the pivot column, we obtain $6/\frac{1}{2} = 12$ and $12/\frac{3}{2} = 8$ for the first and second rows, respectively. The smallest value, 8, results from the entry of $\frac{3}{2}$, so $\frac{3}{2}$ becomes the new pivot and corresponds graphically to the selection of the corner point (2, 8) of the feasible region. (See Figure 3.27.) Using $\frac{3}{2}$ as a pivot, we use the elementary row operations to obtain a simplex matrix equivalent to B with a 1 in the pivot position and zeroes elsewhere in the pivot column. The matrix C was obtained from B as follows:

- Row 2 of B was multiplied by $\frac{2}{3}$ and became row 2 of C.
- Row 2 of C was then multiplied by $-\frac{1}{2}$ and added to row 1, and row 1 was replaced by this sum.
- Row 2 of C was then multiplied by $\frac{1}{2}$ and added to row 3, and row 3 was replaced by this sum.

$$C = \left[\begin{array}{ccccc|c} 1 & 0 & \frac{2}{3} & -\frac{1}{3} & 0 & 2 \\ 0 & 1 & -\frac{1}{3} & \frac{2}{3} & 0 & 8 \\ 0 & 0 & \frac{4}{3} & \frac{1}{3} & 1 & 22 \end{array}\right]$$

The last row of the simplex matrix C corresponds to the objective equation

$$\frac{4}{3}w_1 + \frac{1}{3}w_2 + P = 22 \quad \left(\text{or } P = 22 - \frac{4}{3}w_1 - \frac{1}{3}w_2\right)$$

which shows, by inspection (since both w_1 and w_2 are assumed to be nonnegative), that the maximum value of P is 22 and occurs when the slack variables w_1 and w_2 are 0. With $w_1 = 0$ and $w_2 = 0$, the first two rows give $x = 2$ and $y = 8$. This is the same as the corner point (2, 8) used in the graphical method.

COMMENT: *Once again, we note that when constraints with two variables are analyzed, the simplex method is the algebraic analog of the graphical technique. The choice of successive pivotal elements moved us from the vertex* (0, 0) *to the vertex* (6, 0) *and then to* (2, 8) *on the boundary of the feasible region.*

The procedure described in the above example is a sequence of steps that will lead to the optimal solution in a linear programming problem.

OUTLINE OF THE SIMPLEX METHOD ■

1. *Transform each of the constraint inequalities into a linear equation* by adding appropriate slack variables.
2. *Set up the initial simplex matrix.* This is the matrix of linear equations obtained from the constraint inequalities, augmented with the objective function in the last row.

Note: The objective function should be written with the coefficient of the variable to be optimized equal to 1, with the other variables on the same side of the equation, and with this expression equated to zero.

3. *Select a pivot column.* The pivot column is the column that contains the entry in the last row with the most negative value. (If there is a tie, either one may be used.)
4. *Select a pivot element.* Divide each positive entry in the augmented column by the corresponding entry in the pivot column. The minimum value of the positive quotients determines the pivot element. (If the quotients are equal, either one may be used. If the quotients are all zero, then no increased value of P is possible and the problem is finished. If the quotients are all negative, then there is no solution.)
5. *Perform pivoting.* Use the elementary row operations to make the pivot element 1 and all other entries in the pivot column 0. Pivoting is completed in two steps:
   (a) Divide the pivot row by the pivot element so that the pivot element is 1;
   (b) Obtain zeroes above and below the pivot element using row operations.
6. *Examine the last row.* If there are negative entries, choose a new pivot column and pivot element as described in steps 3 and 4 and compute a new equivalent simplex matrix. When the equivalent matrix that has been generated has only nonnegative quantities in the last row, the simplex method is finished and we have an optimum solution. The maximum value is determined by making the variables with positive coefficients equal to zero. The values of the other variables are obtained from the remaining rows of this simplex matrix.

Example 2 Use the simplex method to maximize $P = 3x + 5y$, subject to the constraints

$$x \geq 0, \qquad y \geq 0, \qquad x + 3y \leq 15, \qquad 2x + y \leq 10$$

Solution

Step 1 We introduce slack variables w_1 and w_2 to obtain the following system of linear constraint equalities:

$$x + 3y + w_1 = 15$$
$$2x + y + w_2 = 10$$

Step 2 The objective function is written in the form

$$P - 3x - 5y = 0$$

The augmented matrix form of this system is then

$$A = \left[\begin{array}{ccccc|c} 1 & \textcircled{3} & 1 & 0 & 0 & 15 \\ 2 & 1 & 0 & 1 & 0 & 10 \\ -3 & -5 & 0 & 0 & 1 & 0 \end{array}\right]$$

Step 3 Since the most negative value in the last row is -5 (the coefficient of y), the pivot column is the second column.

Step 4 To determine the pivot element, we divide each entry in the augmented column by the corresponding entry in the pivot column. These quotients are $\frac{15}{3} = 5$ and $\frac{10}{1} = 10$. The smaller of these is $\frac{15}{3}$, so 3 is the pivot element.

Step 5 Perform the following row operations on A:

- Multiply row 1 by $\frac{1}{3}$.
- Multiply the new row 1 by -1, add it to row 2; replace row 2 with this sum.
- Multiply the new row 1 by 5, add it to row 3; replace row 3 with this sum.

This gives a new equivalent simplex matrix:

$$B = \left[\begin{array}{cccccc|c} \frac{1}{3} & 1 & \frac{1}{3} & 0 & 0 & 5 \\ \textcircled{\frac{5}{3}} & 0 & -\frac{1}{3} & 1 & 0 & 5 \\ -\frac{4}{3} & 0 & \frac{5}{3} & 0 & 1 & 25 \end{array}\right]$$

Step 6 In the last row there is one negative entry, $-\frac{4}{3}$. Hence, column 1 is the new pivot column. The required quotients are $5/(\frac{1}{3}) = 15$ and $5/(\frac{5}{3}) = 3$. Since the smallest number is 3, the new pivot is in row 2 of column 1—that is, $\frac{5}{3}$. Using this pivot, we obtain a new equivalent simplex matrix by using the following row operations:

- Multiply row 2 by $\frac{3}{5}$ and replace row 2 with the new entries.
- Multiply the new row 2 by $-\frac{1}{3}$, add it to row 1, and replace row 1 with this sum.
- Multiply row 2 by $\frac{4}{3}$, add it to row 3, and replace row 3 with this sum.

$$C = \left[\begin{array}{cccccc|c} 0 & 1 & \frac{2}{5} & -\frac{1}{5} & 0 & 4 \\ 1 & 0 & -\frac{1}{5} & \frac{3}{5} & 0 & 3 \\ 0 & 0 & \frac{7}{5} & \frac{4}{5} & 1 & 29 \end{array}\right]$$

From this last simplex matrix, we see that all the coefficients in the last row are positive and this simplex matrix yields the following system of equations:

$$y + \frac{2}{5}w_1 - \frac{1}{5}w_2 = 4$$

$$x - \frac{1}{5}w_1 + \frac{3}{5}w_2 = 3$$

$$\frac{7}{5}w_1 + \frac{4}{5}w_2 + P = 29$$

Letting $w_1 = w_2 = 0$, we see that the maximum value $P = 29$ occurs for the values $x = 3$ and $y = 4$. ■

COMMENT: *The method outlined in the above example applies to linear programming problems in which the objective function is to be* maximized. *In Section 3.6, we will show how to use this method when a function is to be minimized.*

Example 3 Maximize the objective function $P = 2x + z$, subject to the constraints $x \geq 0$, $y \geq 0$, $z \geq 0$, $x + y + z \leq 10$, $x + 2y + z \leq 18$, $x + z \leq 10$.

Solution

Step 1 We introduce the three slack variables w_1, w_2, w_3. Then:

- Instead of $x + y + z \le 10$, we have $x + y + z + w_1 = 10$.
- Instead of $x + 2y + z \le 18$, we have $x + 2y + z + w_2 = 18$.
- Instead of $x + z \le 10$, we have $x + z + w_3 = 10$.

Step 2 The initial simplex matrix for these three equations and the objective equation $-2x - z + P = 0$ is

$$A = \left[\begin{array}{ccccccc|c} \textcircled{1} & 1 & 1 & 1 & 0 & 0 & 0 & 10 \\ 1 & 2 & 1 & 0 & 1 & 0 & 0 & 18 \\ 1 & 0 & 1 & 0 & 0 & 1 & 0 & 10 \\ -2 & 0 & -1 & 0 & 0 & 0 & 1 & 0 \end{array}\right]$$

Step 3 Observe that in the last row, -2 is the most negative value. Hence, the first column is the pivot column.

Step 4 The quotients obtained by dividing each entry in the last column by the corresponding entry in the pivot column are 10, 18, and 10. Thus we may choose either the 1 in the first row, first column, or the 1 in the third row, first column as the pivot element. We choose the 1 in the first row, first column as the pivot element.

Step 5 Using the pivot element of step 4, we obtain the equivalent matrix B:

$$B = \left[\begin{array}{ccccccc|c} 1 & 1 & 1 & 1 & 0 & 0 & 0 & 10 \\ 0 & 1 & 0 & -1 & 1 & 0 & 0 & 8 \\ 0 & -1 & 0 & -1 & 0 & 1 & 0 & 0 \\ 0 & 2 & 1 & 2 & 0 & 0 & 1 & 20 \end{array}\right]$$

Matrix B was obtained from matrix A with the element in the first row and first column as pivot element.

- The negative of row 1 was added to row 2, and this sum used to replace row 2.
- The negative of row 1 was added to row 3, and this sum used to replace row 3.
- Row 1 was multiplied by 2 and added to row 4, and this sum used to replace row 4.

Step 6 We note now that there are no negative elements in the last row, so the simplex method is finished and we can read off the solution. The maximum value is 20. The variables with nonzero coefficients in the last row are zero. Hence, y, z, and $w_1 = 0$. The value of the other variables are obtained from the remaining rows (using, of course, $y = z = w_1 = 0$). We obtain $x = 10$, $w_2 = 8$, $w_3 = 0$. Thus, the solution to the given problem is that the maximum value is obtained for $x = 10$, $y = z = 0$, and that maximum value is 20. ∎

Example 4 A bakery makes three kinds of cookies: peanut butter, raisin-nut, and chocolate chip. The cookies are sold wholesale in quantities of 100 dozen boxes for \$300, \$600, and \$800, respectively. The table below shows the units of flour, eggs, and baking soda needed to bake one unit (100 dozen boxes) of each kind of cookie. The right column in the table shows the supply of each ingredient available on any given day.

	Demand			
	Peanut Butter	Raisin Nut	Chocolate Chip	Supply Limitation
Flour	1	2	3	6
Eggs	3	8	5	14
Baking Soda	2	9	6	30

Note: We assume that an order may consist of a fractional part of 100 dozen boxes, so we are not limited to integer solutions. Maximize the revenue.

Solution Let x, y, and z be the number of units of peanut butter, raisin-nut, and chocolate chip cookies that are baked and sold in one day, and let P be the revenue from the sale of these cookies. To simplify the linear programming matrix we will write the revenue equation as

$$P = 3x + 6y + 8z$$

where the coefficients 3, 6, and 8 represent hundreds of dollars of cost. The actual revenue is then equal to $100P$.

The linear programming model for this problem is: Maximize $P = 3x + 6y + 8z$ subject to the constraints

$$\begin{aligned} x + 2y + 3z &\le 6 && \text{(limitation on flour)} \\ 3x + 8y + 5z &\le 14 && \text{(limitation on eggs)} \\ 2x + 9y + 6z &\le 30 && \text{(limitation on baking soda)} \end{aligned}$$

We now proceed with the simplex method.

Step 1 By introducing slack variables w_1, w_2, and w_3, the constraints may be written as equalities:

$$\begin{aligned} x + 2y + 3z + w_1 &= 6 \\ 3x + 8y + 5z + w_2 &= 14 \\ 2x + 9y + 6z + w_3 &= 30 \end{aligned}$$

Step 2 The equation involving the objective function is $-3x - 6y - 8z + P = 0$. We write the simplex matrix as follows:

$$A = \left[\begin{array}{ccccccc|c} 1 & 2 & \textcircled{3} & 1 & 0 & 0 & 0 & 6 \\ 3 & 8 & 5 & 0 & 1 & 0 & 0 & 14 \\ 2 & 9 & 6 & 0 & 0 & 1 & 0 & 30 \\ -3 & -6 & -8 & 0 & 0 & 0 & 1 & 0 \end{array}\right]$$

Step 3 The most negative value in the last row is -8, so the pivot column is column 3.

Step 4 In column 3, the quotients obtained by dividing the entries in the augment column by the elements of the third column are $\frac{6}{3} = 2$, $\frac{14}{5}$, and $\frac{30}{6} = 5$. The smallest of these is 2, so the pivot element is the element 3 in the first row and third column.

Step 5 We now perform pivoting with this element:

- Multiply row 1 by $\frac{1}{3}$ and make this row 1.
- Multiply new row 1 by -5 and add it to row 2; replace row 2 with this sum.
- Multiply new row 1 by -6 and add it to row 3; replace row 3 with this sum.
- Multiply new row 1 by 8 and add it to row 4; replace row 4 with this sum.

This yields the new equivalent simplex matrix:

$$B = \left[\begin{array}{ccccccc|c} \frac{1}{3} & \frac{2}{3} & 1 & \frac{1}{3} & 0 & 0 & 0 & 2 \\ \frac{4}{3} & \textcircled{\frac{14}{3}} & 0 & -\frac{5}{3} & 1 & 0 & 0 & 4 \\ 0 & 5 & 0 & -2 & 0 & 1 & 0 & 18 \\ -\frac{1}{3} & -\frac{2}{3} & 0 & \frac{8}{3} & 0 & 0 & 1 & 16 \end{array}\right]$$

The objective row has two negative numbers. Therefore, we return to step 3:

Step 3′ The most negative number in the last row is $-\frac{2}{3}$. Therefore, the second column is the new pivot column.

Step 4′ The three quotients are 3, $\frac{6}{7}$, and $\frac{18}{5}$. The smallest of these is $\frac{6}{7}$, so the element in the second row and second column is the new pivot.

Step 5′ We perform pivoting on the element $\frac{14}{3}$.

- Multiply row 2 by $\frac{3}{14}$ and make this the new row 2.
- Multiply the new row 2 by $-\frac{2}{3}$; add it to row 1; replace row 1 with this sum.
- Multiply the new row 2 by -5; add it to row 3; replace row 3 with this sum.
- Multiply the new row 2 by $\frac{2}{3}$; add it to row 4; replace row 4 with this sum.

The result is the equivalent matrix C.

$$C = \left[\begin{array}{ccccccc|c} \frac{1}{7} & 0 & 1 & \frac{4}{7} & -\frac{1}{7} & 0 & 0 & \frac{10}{7} \\ \textcircled{\frac{2}{7}} & 1 & 0 & -\frac{5}{14} & \frac{3}{14} & 0 & 0 & \frac{6}{7} \\ -\frac{10}{7} & 0 & 0 & -\frac{3}{14} & -\frac{15}{14} & 1 & 0 & \frac{96}{7} \\ -\frac{1}{7} & 0 & 0 & \frac{17}{7} & \frac{1}{7} & 0 & 1 & \frac{116}{7} \end{array}\right]$$

Step 6 We examine the last row and see that there is one remaining negative number, so we do steps 3 and 4 again.

Step 3″ Column 1 is our new pivot column.

Step 4″ We consider only the positive elements in the pivot column, so the two quotients are 10 and 3. The smallest quotient indicates that the pivot element is $\frac{2}{7}$ in row 2.

Step 5″ Perform pivoting on the element $\frac{2}{7}$. In this case we need only pivot on rows 1, 2, and 4 because the element in row 3 of that column is negative.

- Multiply row 2 by $\frac{7}{2}$.
- Multiply the new row 2 by $-\frac{1}{7}$; add it to row 1; replace row 1 with this sum.
- Multiply the new row 2 by $\frac{1}{7}$; add it to row 4; replace row 4 with this sum.

We obtain another equivalent simplex matrix:

$$D = \left[\begin{array}{ccccccc|c} 0 & -\frac{1}{2} & 1 & \frac{3}{4} & -\frac{1}{4} & 0 & 0 & 1 \\ 1 & \frac{7}{2} & 0 & -\frac{5}{4} & \frac{3}{4} & 0 & 0 & 3 \\ -\frac{10}{7} & 0 & 0 & -\frac{3}{14} & -\frac{15}{14} & 1 & 0 & \frac{96}{7} \\ 0 & \frac{1}{2} & 0 & \frac{9}{4} & \frac{1}{4} & 0 & 1 & 17 \end{array}\right]$$

Since there are no negative numbers in the objective row, the simplex method is completed and we may read the result from the final matrix. The maximum value of P is 17. The maximum revenue is obtained by baking 3 units of the peanut butter cookies, 1 unit of the chocolate chip cookies, and no raisin-nut cookies. Thus, the maximum revenue is $100(17) = \$1700$. ■

COMMENT: *The simplex method is available in many different computer software packages. One such system has the trade name of MAPLE. If your computer has MAPLE, enter "help(simplex)" to find out how to use the simplex routine.*

Because this is a basic introduction to the simplex method, we have not presented some of the troublesome aspects of the algorithm. Here are just a few:

NONUNIQUENESS Recall from the graphical technique that when the objective function $P = ax + by$ was such that $ax + by = c$ was parallel to one of the sides of the feasible region, the optimal solution was nonunique. In fact, the optimal solution occurred for any of the points along that boundary of the feasible region parallel to $ax + by = c$. When using the simplex method, the situation is not always quite so obvious. Consider, for example, the following simplex matrix:

$$A = \left[\begin{array}{ccccc|c} 1 & 1 & 1 & 0 & 0 & 18 \\ 0 & 1 & 2 & 1 & 0 & 9 \\ 0 & 0 & 3 & 0 & 1 & 36 \end{array}\right]$$

Since there are no negative elements in the last row, the optimum value has been found ($P = 36$, $x = 18$, $y = 0$). However, note that we could reach another solution by using the 1 in the second row, second column as a pivot element to obtain an equivalent simplex matrix:

$$B = \left[\begin{array}{ccccc|c} 1 & 0 & -1 & -1 & 0 & 9 \\ 0 & 1 & 2 & 1 & 0 & 9 \\ 0 & 0 & 3 & 0 & 1 & 36 \end{array}\right]$$

From this new matrix, we can read the same optimal solution ($P = 36$), but now we have $x = 9$, $y = 9$.

We can make a general statement about nonuniqueness:

COMMENT: *Nonunique solutions may occur if one of the objective function coefficients is zero. Examine the final simplex matrix. Use the column of the variable that is zero as a new pivot column and determine a new pivot element. If this pivoting operation results in a new set of values for the variables, the solution is nonunique. Note that the optimal solution itself is unaffected by this pivoting operation, because the pivoting operation does not affect the object row.*

NO SOLUTION A linear programming problem could have an unbounded solution, and hence no optimum value for P. This will occur *if a pivot column has no positive elements.* For example, consider the following simplex matrix:

$$A = \left[\begin{array}{ccccc|c} 1 & -3 & 1 & 0 & 0 & 15 \\ 2 & -1 & 0 & 1 & 0 & 10 \\ -3 & -5 & 0 & 0 & 1 & 0 \end{array}\right]$$

The pivot column is column 2 because the most negative element is -5. However, all the elements of this column are negative. We conclude that there is no maximum solution.

EXERCISES SECTION 3.5

In Exercises 1–7, set up the simplex matrix for each linear programming problem. Assume the variables are nonnegative.

1. Maximize $P = x + 2y$, subject to $3x + 2y \leq 10$ and $x + 4y \leq 15$.

2. Given $x \leq 10$, $x + y \leq 15$, maximize $P = 3x + 2y$.

3. Given $y \leq 2$, $x + 2y \leq 8$, maximize $P = x + 3y$.

4. Given $x + y \leq 2$, $x + 3y \leq 4$, maximize $P = 2x + 5y$.

5. Given $x + y \leq 5$, $x + 2y \leq 8$, maximize $P = 5x + y$.

6. Maximize $P = x + 3y + 2z$, subject to $x + y \leq 5$, $x + 2y + 6z \leq 10$, $x \leq 3$.

7. Maximize $P = 2x + 5z$, subject to $x \leq 5$, $y \leq 8$, $z \leq 10$.

In Exercises 8–12, a simplex matrix is given. Identify the pivot element.

8. $$\left[\begin{array}{ccccc|c} 2 & 1 & 1 & 0 & 0 & 12 \\ 1 & 2 & 0 & 1 & 0 & 18 \\ -3 & -2 & 0 & 0 & 1 & 20 \end{array}\right]$$

9. $$\left[\begin{array}{ccccc|c} 3 & 1 & 1 & 0 & 0 & 20 \\ 6 & 2 & 0 & 1 & 0 & 18 \\ -5 & -7 & 0 & 0 & 1 & 0 \end{array}\right]$$

10. $$\left[\begin{array}{ccccc|c} 3 & 1 & 1 & 0 & 0 & 20 \\ 6 & 2 & 0 & 1 & 0 & 18 \\ -5 & -5 & 0 & 0 & 1 & 0 \end{array}\right]$$

11. $$\left[\begin{array}{ccccc|c} 3 & 1 & 1 & 0 & 0 & 20 \\ 6 & 2 & 0 & 1 & 0 & 40 \\ -5 & -8 & 0 & 0 & 1 & 0 \end{array}\right]$$

12. $$\left[\begin{array}{ccccccc|c} 5 & 1 & 3 & 1 & 0 & 0 & 0 & 10 \\ 6 & 2 & 3 & 0 & 1 & 0 & 0 & 18 \\ 2 & 5 & 1 & 0 & 0 & 1 & 0 & 10 \\ -2 & -4 & -1 & 0 & 0 & 0 & 1 & 0 \end{array}\right]$$

In Exercises 13–18, indicate whether the given simplex matrix represents an optimal solution. If it does, give the maximum value of P and the values of the variables for which the maximum value is obtained. If not, find the next pivot, and give the next equivalent simplex matrix.

13. $$\left[\begin{array}{ccccc|c} 0 & 1 & 1 & 3 & 0 & 12 \\ 1 & 0 & 2 & 1 & 0 & 18 \\ 0 & 0 & 5 & 3 & 1 & 20 \end{array}\right]$$

14. $$\left[\begin{array}{ccccc|c} 1 & 0 & 1 & 3 & 0 & 12 \\ 0 & 1 & 2 & 1 & 0 & 18 \\ 0 & 0 & 5 & 3 & 1 & 20 \end{array}\right]$$

15. $$\left[\begin{array}{ccccc|c} 2 & 5 & 1 & 0 & 0 & 10 \\ 1 & 1 & 0 & 1 & 0 & 10 \\ -1 & -1 & 5 & 3 & 1 & 20 \end{array}\right]$$

16. $$\left[\begin{array}{ccccc|c} 2 & 0 & 1 & 3 & 0 & 40 \\ 1 & 1 & 0 & 1 & 0 & 20 \\ -1 & 0 & 5 & 3 & 1 & 20 \end{array}\right]$$

17. $$\left[\begin{array}{ccccccc|c} 1 & 0 & 0 & -1 & -3 & 4 & 0 & 10 \\ 0 & 1 & 0 & 2 & 1 & -6 & 0 & 18 \\ 0 & 0 & 1 & -2 & 0 & 1 & 0 & 10 \\ 0 & 0 & 0 & 2 & 5 & 2 & 1 & 250 \end{array}\right]$$

18. $$\left[\begin{array}{ccccccc|c} 1 & 1 & 0 & 1 & -5 & -2 & 0 & 10 \\ 0 & 2 & 0 & -3 & -1 & 0 & 0 & 18 \\ 0 & 0 & 1 & -8 & 0 & 1 & 0 & 10 \\ 0 & 4 & 0 & 2 & 7 & 1 & 1 & 0 \end{array}\right]$$

19. Maximize $P = 6x + 9y + 5z$, subject to $2x + 5y + 3z \le 20$, $x + y \le 8$, $x + 2y \le 10$.

20. Maximize $P = 32x + 40y + 24z$, subject to $x + y + 2z \le 20$, $4x + 3y + 5z \le 45$, $2x + 2y + z \le 15$.

21. Maximize $P = 5x + 3y + 3z$, subject to $x + z \le 10$, $x + 2y \le 5$, $y + 4z \le 20$.

22. Maximize $P = x + 5y + 10z$, subject to $x + y + 2z \le 10$, $x + 2y \le 15$, $y + z \le 20$.

23. Show that the solution to the linear programming problem "Maximize $P = x + 2y$, subject to $x + 2y \le 10$, $x + y \le 8$" is nonunique. Use both the graphical method and the simplex method.

24. Repeat Exercise 23 for the problem: Maximize $P = 3x + 2y$, subject to $3x + 2y \le 12$, $x + y \le 5$.

25. A company makes \$10 on steel castings and \$8 on aluminum castings. If x is the number of steel castings, and y the number of aluminum, the profit equation is $P = 10x + 8y$. Company engineers have found that they must satisfy the following constraints for each day's production:

$$x \ge 0, \qquad y \ge 0,$$
$$2x + 3y \le 120, \qquad 4x + y \le 90$$

How many of each kind of casting should the company make each day to maximize its profit?

26. A company produces deluxe and economy model radios. Each economy model yields a profit of \$6, whereas each deluxe model yields \$10 profit. Both models require approximately the same amount of material to build and the supply allows for a total of 500 radios per day. The sales force reports that the demand for the economy model does not exceed 350 per day but the deluxe radio can be sold as fast as it is made. The deluxe model requires twice as much time to build as the economy, and if only the economy model were built, there would be enough time to produce 700 radios per day. How many economy models and deluxe models should the company construct to maximize its profit?

27. A broker wishes to invest \$20,000 for her client, part at 7% and part at 8.5%. The amount she invests at 8.5% cannot be more than twice the amount she invests at 7%. How much should she deposit at each rate to maximize total income? Only multiples of \$100 can be invested.

28. A company produces two kinds of vacuum cleaners, with profits of \$5 and \$8, respectively, on the economy and deluxe models. No more than 1000 of the economy model can be made per day. The deluxe model takes three times as long to make as the economy model and if only the economy model were made, time would allow for production of 1300 per day. Find the number of each that should be built, and find the maximum profit.

29. Suppose you own a company that makes tables, desks, and cabinets. Each item has wood, steel, and plastic components. The number of units of each material required to build the furniture, and the available supply of each, are shown in the table below.

	Tables	Desks	Cabinets	Supply
Wood	2	2	3	10
Steel	1	2	1	6
Plastic	0	6	4	15

If you can make a profit of \$30 on each table, \$60 on each desk, and \$50 on each cabinet, how many of each should you build to maximize your profit?

3.6 NONSTANDARD LINEAR PROGRAMMING PROBLEMS

In Section 3.4 we were able to solve both maximization and minimization linear programming problems, but only graphically in two dimensions. The simplex method was introduced in Section 3.5 to solve linear programming problems in which:

(a) The objective function is to be maximized.
(b) All variables are nonnegative.
(c) All constraints are set so that the variables are less than or equal to ($\leq$) a nonnegative number.

Linear programming problems that satisfy these three conditions are said to be *standard* linear programming problems.

In this section we will show how the simplex method can be extended to include some **nonstandard linear programming problems,** including maximization problems with nonstandard constraints and problems in which the objective function is to be minimized. Minimizing the objective function is of practical concern, because companies are naturally interested in minimizing the cost of their operations within the constraints of various output demands or of limited access to supplies.

MAXIMIZATION PROBLEMS WITH NONSTANDARD CONSTRAINTS

A maximization problem is nonstandard if at least one of the constraints is expressed as greater than or equal to ($\geq$) a nonpositive number. (This would violate condition c.) To change a nonstandard ($\geq$) constraint into a standard ($\leq$) constraint, multiply the nonstandard constraint by -1. Recall that when we multiply an inequality by -1, the sense of the inequality is reversed; that is, $\geq$ becomes $\leq$.

For example, the nonstandard constraint $2x + 3y \geq -5$ should be replaced with $-2x - 3y \leq 5$.

COMMENT: *In the examples in this section we will use subscripted variables to indicate the generality of the method, even though many of these problems are two-dimensional and could be solved by a graphical method.*

Example 1 Maximize $P = 8x_1 + 5x_2$, subject to $2x_1 + 5x_2 \leq 60$ and $3x_1 + 4x_2 \geq -20$ and $x_1 \geq 0$ and $x_2 \geq 0$.

Solution The constraint $3x_1 + 4x_2 \geq -20$ is nonstandard, because it violates condition c. We change this into a standard constraint by multiplying through by -1 to obtain $-3x_1 - 4x_2 \leq 20$. The linear programming problem is now stated as follows:

$$\begin{aligned} &\text{Maximize } P = 8x_1 + 5x_2 \\ &\text{subject to } 2x_1 + 5x_2 \leq 60 \\ &\qquad -3x_1 - 4x_2 \leq 20 \\ &\qquad x_1 \geq 0, x_2 \geq 0 \end{aligned}$$

This is now a linear programming problem in standard form. Notice that the actual linear programming problem is unchanged; only its appearance is altered, to permit the use of the simplex method. The simplex matrix for this problem, with slack variables w_1 and w_2, is

$$A = \begin{array}{c} \begin{matrix} x_1 & x_2 & w_1 & w_2 & P & \end{matrix} \\ \left[\begin{array}{rrrrr|r} 2 & 5 & 1 & 0 & 0 & 60 \\ -3 & -4 & 0 & 1 & 0 & 20 \\ -8 & -5 & 0 & 0 & 1 & 0 \end{array}\right] \end{array}$$

The pivot column is column 1. The quotients to be compared are $\frac{60}{2}$ and $\frac{20}{-3}$. Only the first of these is positive, so 2 is the pivot element. Performing the pivoting, we obtain the equivalent simplex matrix

$$B = \left[\begin{array}{rrrrr|r} 1 & \frac{5}{2} & \frac{1}{2} & 0 & 0 & 30 \\ 0 & \frac{7}{2} & \frac{3}{2} & 1 & 0 & 110 \\ 0 & 15 & 4 & 0 & 1 & 240 \end{array}\right]$$

From matrix B, we see that the maximum value of the objective function is $P = 240$, when $x_1 = 30$, $x_2 = 0$, $w_1 = 0$, $w_2 = 110$. Only the linear programming variables need be considered so we have a maximum of $P = 240$ at $(30, 0)$. ■

Maximization Problems with a Nonpositive Variable

If one or more of the variables is less than or equal to ($\leq$) zero, instead of greater than or equal to ($\geq$) zero, make a substitution of a new variable that has the opposite sign.

Example 2 Maximize $P = 2x_1 + 3x_2$, subject to $x_1 \geq 0$, $x_2 \leq 0$ and to the constraints $2x_1 + 4x_2 \leq 60$, $5x_1 - 3x_2 \leq 200$.

Solution The nonstandard part of this problem is $x_2 \leq 0$; this violates condition b. To make this variable nonnegative, we make the substitution $x_2 = -x_3$. Then the standard linear programming problem in variables x_1 and x_3 is

$$\begin{aligned} &\text{Maximize } P = 2x_1 - 3x_3 \\ &\text{subject to } 2x_1 - 4x_3 \leq 60 \\ &\qquad\qquad 5x_1 + 3x_3 \leq 200 \\ &\qquad\qquad x_1 \geq 0, x_3 \geq 0 \end{aligned}$$

The simplex matrix, with slack variables w_1 and w_2, is

$$A = \begin{array}{c} \begin{matrix} x_1 & x_3 & w_1 & w_2 & P & \end{matrix} \\ \left[\begin{array}{rrrrr|r} 2 & -4 & 1 & 0 & 0 & 60 \\ 5 & 3 & 0 & 1 & 0 & 200 \\ -2 & 3 & 0 & 0 & 1 & 0 \end{array}\right] \end{array}$$

The pivot column is column 1, because -2 is the only negative element in the objective row. The quotients are $\frac{60}{2}$ and $\frac{200}{5}$; since 30 is the lesser of these two numbers, 2 is the pivot element. Carrying out the pivoting, we derive the new equivalent simplex matrix:

$$B = \left[\begin{array}{rrrrr|r} 1 & -2 & \frac{1}{2} & 0 & 0 & 30 \\ 0 & 13 & -\frac{5}{2} & 1 & 0 & 50 \\ 0 & -1 & 1 & 0 & 1 & 60 \end{array}\right]$$

However, the objective row still has a negative number in the last row. The second column becomes a new pivot column, with 13 the new pivot. After pivoting, we obtain

$$C = \left[\begin{array}{rrrrr|r} 1 & 0 & \frac{3}{26} & \frac{2}{13} & 0 & \frac{490}{13} \\ 0 & 1 & -\frac{5}{26} & \frac{1}{13} & 0 & \frac{50}{13} \\ 0 & 0 & \frac{21}{26} & \frac{1}{13} & 1 & \frac{830}{13} \end{array}\right]$$

The simplex method is thus completed, with a maximum of $\frac{830}{13}$. The values of the variables are $x_1 = \frac{490}{13}$, $x_3 = \frac{50}{13}$, $w_1 = 0$, $w_2 = 0$, or, in terms of the original linear programming variables, the maximum occurs at $x_1 = \frac{490}{13}$, $x_2 = -\frac{50}{13}$. ■

Minimization Problems

A linear programming problem that calls for a minimization of the objective function is nonstandard because it violates condition a. If the variables and the constraints are standard, a minimization problem can be made into a standard maximization problem by replacing the objective function with its negative. It can be shown that the minimization of P is the same as the maximization of $-P$.

Example 3 Minimize $C = -2x_1 + 3x_2$, subject to $5x_1 + 2x_2 \leq 40$, $x_1 + 2x_2 \leq 14$, $x_1 \geq 0$, and $x_2 \geq 0$.

Solution We make the substitution $C = -P$. Then $P = 2x_1 - 3x_2$. Now observe that when P is maximized, C is minimized, so that the linear programming problem takes the standard form:

$$\begin{aligned} \text{Maximize } P &= 2x_1 - 3x_2 \\ \text{subject to } 5x_1 + 2x_2 &\leq 40 \\ x_1 + 2x_2 &\leq 14 \\ x_1 \geq 0, x_2 &\geq 0 \end{aligned}$$

The simplex matrix, with slack variables w_1 and w_2, is

$$A = \left[\begin{array}{ccccc|c} 5 & 2 & 1 & 0 & 0 & 40 \\ 1 & 2 & 0 & 1 & 0 & 14 \\ -2 & 3 & 0 & 0 & 1 & 0 \end{array}\right]$$

Column 1 is the pivot column and 5 is the pivot element. Carrying out the pivoting operation, we obtain

$$B = \left[\begin{array}{ccccc|c} 1 & \frac{2}{5} & \frac{1}{5} & 0 & 0 & 8 \\ 0 & \frac{8}{5} & -\frac{1}{5} & 1 & 0 & 6 \\ 0 & \frac{19}{5} & \frac{2}{5} & 0 & 1 & 16 \end{array}\right]$$

Thus, the maximum for P is 16 at $x_1 = 8$, $x_2 = 0$. This means that the minimum for C is -16 at the same point. ■

John von Neumann, a twentieth-century mathematician, showed how to adapt the simplex method to a type of problem in which the objective function has to be minimized and the constraints are greater than or equal to ($\geq$) a nonnegative number. He did this by solving a related maximization problem called the **dual problem** of the minimization problem.

The dual problem of a linear programming problem is obtained from a given minimization problem in three steps.

DEFINITION The Dual Problem

To obtain the dual problem of a minimization problem in which the constraints are greater than or equal to ($\geq$) a nonnegative number:

1. Write the initial simplex matrix, but without slack variables.
2. Find the transpose of this matrix.
3. Interpret this last matrix in terms of a maximization problem.

An example should help to clarify the process used to write the dual problem for a minimization problem.

Example 4 Consider this linear programming problem:

$$\begin{aligned} &\text{Minimize } C = 7x_1 + 4x_2 \\ &\text{subject to } 2x_1 + x_2 \geq 10 \\ &\qquad 2x_1 + 5x_2 \geq 15 \\ &\qquad x_1 \geq 0, x_2 \geq 0 \end{aligned}$$

Find the dual problem of this minimization problem.

Solution

1. The initial matrix of constraints and objective function is given by

$$\begin{bmatrix} 2 & 1 & 10 \\ 2 & 5 & 15 \\ 7 & 4 & 0 \end{bmatrix}$$

Remember, no slack variables are used in this process.

2. The transpose of this matrix is obtained by interchanging its row and column elements. Thus, the desired transpose is

$$\begin{bmatrix} 2 & 2 & 7 \\ 1 & 5 & 4 \\ 10 & 15 & 0 \end{bmatrix}$$

3. This matrix may be interpreted as the initial simplex matrix, without slack variables, of the maximization problem:

$$\begin{aligned} &\text{Maximize } P = 10y_1 + 15y_2 \\ &\text{subject to } 2y_1 + 2y_2 \leq 7 \\ &\qquad y_1 + 5y_2 \leq 4 \\ &\qquad y_1 \geq 0, y_2 \geq 0 \end{aligned}$$

This is the dual maximization problem to the given minimization problem. ▪

COMMENT: *A minimization problem and its dual maximization problem are two different problems in two different coordinate systems. To emphasize this, we have expressed the minimization problem in* (x_1, x_2) *coordinates and the dual maximization problem in* (y_1, y_2) *coordinates. Both the minimization problem and its dual problem in Example 4 may be solved graphically, since both are problems in two dimensions. Figure 3.28 shows the solution to the minimization problem; Figure 3.29, the solution to the maximization problem. The optimal value for the minimization problem occurs at* $(\frac{35}{8}, \frac{5}{4})$. *The minimum value is* $C = 7 \cdot \frac{35}{8} + 4 \cdot \frac{5}{4} = \frac{285}{8}$. *The optimal value of the dual problem occurs at* $(\frac{27}{8}, \frac{1}{8})$. *The maximum value is* $P = 10 \cdot \frac{27}{8} + 15 \cdot \frac{1}{8} = \frac{285}{8}$.

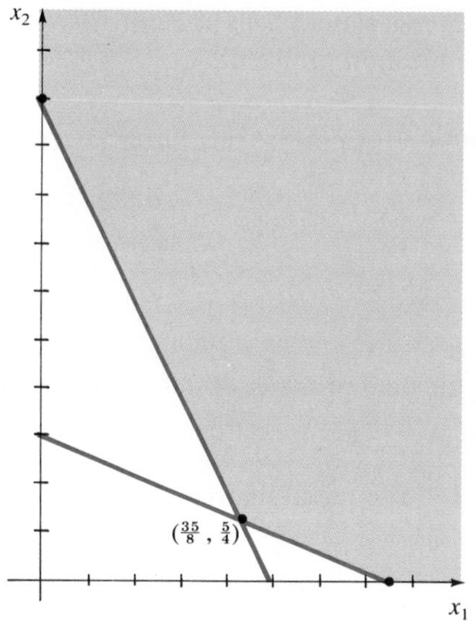

Figure 3.28

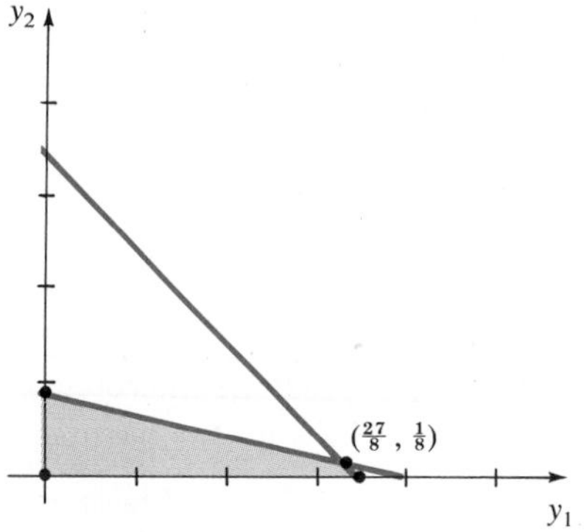

Figure 3.29

COMMENT: *Note that the optimal value for the minimization problem is the same as the optimal value for the dual maximization problem. This is always true, and is a fundamental theorem in linear programming, called the* von Neumann Duality Principle. *A formal statement of the duality principle is given at the end of this section.*

Example 5 Solve the following linear programming problem by using the simplex method with the dual of the given problem:

$$\text{Minimize } C = 7x_1 + 4x_2$$
$$\text{subject to } 2x_1 + x_2 \geq 10$$
$$2x_1 + 5x_2 \geq 15$$
$$x_1 \geq 0, x_2 \geq 0$$

Solution Note that this is the same minimization problem for which we found the dual maximization problem in Example 4. Thus, we want to solve the problem:

$$\begin{aligned} &\text{Maximize } P = 10y_1 + 15y_2 \\ &\text{subject to } 2y_1 + 2y_2 \leq 7 \\ &\qquad y_1 + 5y_2 \leq 4 \\ &\qquad y_1 \geq 0,\ y_2 \geq 0 \end{aligned}$$

The simplex matrix for this problem is

$$A = \begin{array}{c} \begin{array}{ccccc} y_1 & y_2 & w_1 & w_2 & P \end{array} \\ \left[\begin{array}{ccccc|c} 2 & 2 & 1 & 0 & 0 & 7 \\ 1 & 5 & 0 & 1 & 0 & 4 \\ -10 & -15 & 0 & 0 & 1 & 0 \end{array}\right] \end{array}$$

Column 2 is the pivot column and 5 is the pivot element. Performing the pivoting operation, we obtain

$$B = \left[\begin{array}{ccccc|c} \frac{8}{5} & 0 & 1 & -\frac{2}{5} & 0 & \frac{27}{5} \\ \frac{1}{5} & 1 & 0 & \frac{1}{5} & 0 & \frac{4}{5} \\ -7 & 0 & 0 & 3 & 1 & 12 \end{array}\right]$$

Our new pivot column is column 1, with pivot element $\frac{8}{5}$. Performing the pivoting operation, we obtain the equivalent simplex matrix C.

$$C = \left[\begin{array}{ccccc|c} 1 & 0 & \frac{5}{8} & -\frac{1}{4} & 0 & \frac{27}{8} \\ 0 & 1 & -\frac{1}{8} & \frac{1}{4} & 0 & -\frac{1}{8} \\ 0 & 0 & \frac{35}{8} & \frac{5}{4} & 1 & \frac{285}{8} \end{array}\right]$$

From this we read the optimal value of P to be $\frac{285}{8}$, which, by the von Neumann Duality Principle, is the same as the minimal value of the original minimization problem. ■

COMMENT: *Look at the entries in the bottom row in the columns corresponding to the slack variables in this last matrix of the dual problem. They are* $\frac{35}{8}$ *and* $\frac{5}{4}$*. But, from the graphical solution, we know that these are the coordinates of the point at which the minimum is obtained. Any time we solve the dual problem, the values in the last row corresponding to the coefficients of the slack variables are the coordinates of the original optimal point.*

We summarize below the two important results of our observation of the solution to the dual problem.

THEOREM Von Neumann's Duality Principle

- The optimum value of a linear programming problem is the same as the optimum value of the dual problem.
- The entries in the last row of the final matrix in the columns corresponding to the slack variables are the same as the values of the variables at the optimal point in the original problem.

COMMENT: The Duality Principle applies to the dual of a minimization problem as well as to the dual of a maximization problem. We have stated the principle in its most general form, although in this book we have considered only the dual of a minimization problem.

Example 6 A health food store gives the following information on the protein, carbohydrate, and calorie units in two recommended foods:

	Food 1	Food 2
Protein*	10	20
Carbohydrates*	30	20
Calories*	50	40

* Per gram.

Suppose the health food store says you need at least 80 units of protein and 120 units of carbohydrates in a given time period. How much of each food would you eat in order to minimize your caloric intake?

Solution Let x_1 be the amount of Food 1 and x_2 the amount of Food 2 that you will eat. Then, the inequality $10x_1 + 20x_2 \geq 80$ is the constraint that arises from the fact that we must have at least 80 grams of protein; $30x_1 + 20x_2 \geq 120$ is the constraint that arises from the carbohydrate requirement. The quantity to be minimized is the number of calories, which is given by $C = 50x_1 + 40x_2$. The initial simplex matrix of this minimization problem, without slack variables, is

$$\begin{bmatrix} 10 & 20 & 80 \\ 30 & 20 & 120 \\ 50 & 40 & 0 \end{bmatrix}$$

Without loss of generality, we may divide this matrix by 10 in order to simplify the arithmetic. The transpose of this matrix after dividing by 10 is

$$\begin{bmatrix} 1 & 3 & 5 \\ 2 & 2 & 4 \\ 8 & 12 & 0 \end{bmatrix}$$

From this we may read the dual maximization problem:

$$\begin{aligned} &\text{Maximize } P = 8y_1 + 12y_2 \\ &\text{subject to } y_1 + 3y_2 \leq 5 \\ &\qquad 2y_1 + 2y_2 \leq 4 \\ &\qquad y_1 \geq 0,\ y_2 \geq 0 \end{aligned}$$

The simplex matrix for this problem is

$$A = \begin{array}{c} \\ \\ \end{array}\left[\begin{array}{ccccc|c} y_1 & y_2 & w_1 & w_2 & P & \\ 1 & 3 & 1 & 0 & 0 & 5 \\ 2 & 2 & 0 & 1 & 0 & 4 \\ -8 & -12 & 0 & 0 & 1 & 0 \end{array}\right]$$

The pivot column is column 2 and the pivot element is the element 3 in that column. Pivoting on 3, the equivalent simplex matrix is

$$B = \left[\begin{array}{ccccc|c} \frac{1}{3} & 1 & \frac{1}{3} & 0 & 0 & \frac{5}{3} \\ \frac{4}{3} & 0 & -\frac{2}{3} & 1 & 0 & \frac{2}{3} \\ -4 & 0 & 4 & 0 & 1 & 20 \end{array}\right]$$

Now, the new pivot element is $\frac{4}{3}$ in column 1, and the equivalent matrix becomes

$$C = \left[\begin{array}{ccccc|c} 0 & 1 & \frac{1}{2} & -\frac{1}{4} & 0 & \frac{3}{2} \\ 1 & 0 & -\frac{1}{2} & \frac{3}{4} & 0 & \frac{1}{2} \\ 0 & 0 & 2 & 3 & 1 & 22 \end{array}\right]$$

The simplex method is now finished. By the duality principle, the minimum value for the calorie intake is 22 units. Recall that we divided the entries in the simplex matrix by 10, so the minimum caloric intake is $22(10) = 220$ units. The amounts of the two foods is given in the last line of the final simplex matrix in the columns for the slack variables w_1 and w_2. Hence, you should eat 2 units of Food 1 and 3 units of Food 2. ■

EXERCISES SECTION 3.6

In Exercises 1–5, each maximization problem has a nonstandard constraint. Remove the irregularity and solve the resulting problem.

1. Maximize $P = 3x_1 + 5x_2$, subject to $x_1 + 3x_2 \leq 60$, $x_1 + 5x_2 \geq -20$, $x_1 \geq 0$, $x_2 \geq 0$.

2. Maximize $P = x_1 + 2x_2$, subject to $x_1 + 4x_2 \leq 20$, $3x_1 + 5x_2 \geq -30$, $x_1 \geq 0$, $x_2 \geq 0$.

3. Maximize $P = x_1 - 2x_2$, subject to $x_1 + 3x_2 \leq 15$, $2x_1 + 7x_2 \geq -28$, $x_1 \geq 0$, $x_2 \geq 0$.

4. Maximize $P = 2x_1 - 5x_2$, subject to $x_1 + 3x_2 \geq -15$, $2x_1 + 7x_2 \geq -28$, $x_1 \geq 0$, $x_2 \geq 0$.

5. Maximize $P = 2x_1 + 4x_2$, subject to $x_1 + 3x_2 \geq -15$, $2x_1 + 7x_2 \geq -28$, $x_1 \geq 0$, $x_2 \geq 0$. Can you generalize this result?

In Exercises 6–10, one of the variables in each maximization problem is less than or equal to zero. Make a change of variables and solve the resulting problem.

6. Maximize $P = 2x_1 + x_2$, subject to $x_1 \geq 0$, $x_2 \leq 0$ and to the constraints $x_1 + 4x_2 \leq 60$, $5x_1 - 3x_2 \leq 200$.

7. Maximize $P = 5x_1 + 3x_2$, subject to $x_1 \geq 0$, $x_2 \leq 0$, $x_1 - 4x_2 \leq 40$, $5x_1 + x_2 \leq 75$.

8. Maximize $P = 5x_1 - 3x_2$, subject to $x_1 \geq 0$, $x_2 \leq 0$, $x_1 - 4x_2 \leq 40$, $5x_1 - x_2 \leq 75$.

9. Maximize $P = x_1 - 3x_2$, subject to $x_1 \geq 0$, $x_2 \leq 0$, $x_1 + 4x_2 \leq 40$, $5x_1 + 3x_2 \leq 75$.

10. Maximize $P = x_1 - 3x_2$, subject to $x_1 \leq 0$, $x_2 \leq 0$, $x_1 + 4x_2 \leq -40$, $5x_1 + 3x_2 \leq -75$.

In Exercises 11–20, write the dual of each minimization problem and solve.

11. Minimize $C = 3x_1 + 2x_2$, subject to

$$2x_1 + x_2 \geq 8$$
$$3x_1 + 5x_2 \geq 15$$
$$x_1 \geq 0,\ x_2 \geq 0$$

12. Minimize $C = 3x_1 + x_2$, subject to

$$2x_1 + x_2 \geq 8$$
$$3x_1 + 5x_2 \geq 15$$
$$x_1 \geq 0,\ x_2 \geq 0$$

13. Minimize $C = 3x_1 + 2x_2$, subject to

$$2x_1 + 3x_2 \geq 8$$
$$3x_1 - 5x_2 \geq 15$$
$$x_1 \geq 0, x_2 \geq 0$$

14. Minimize $C = x_1 + 2x_2$, subject to

$$2x_1 + x_2 \geq 8$$
$$3x_1 + 4x_2 \geq 24$$
$$x_1 \geq 0, x_2 \geq 0$$

15. Minimize $C = 3x_1 - 2x_2$, subject to

$$2x_1 - x_2 \geq 8$$
$$3x_1 + 5x_2 \geq 15$$
$$x_1 \geq 0, x_2 \geq 0$$

16. Minimize $C = x_1 + 2x_2 + x_3$, subject to

$$2x_1 + x_2 + x_3 \geq 8$$
$$3x_1 + 5x_2 + x_3 \geq 15$$
$$x_1 + 2x_2 + 3x_3 \geq 12$$
$$x_1 \geq 0, x_2 \geq 0, x_3 \geq 0$$

17. Minimize $C = x_1 + 2x_2 + x_3$, subject to

$$2x_1 + x_3 \geq 8$$
$$3x_1 + 5x_2 - x_3 \geq 15$$
$$x_1 + 2x_2 + x_3 \geq 12$$
$$x_1 \geq 0, x_2 \geq 0, x_3 \geq 0$$

18. Minimize $C = 3x_1 + x_3$, subject to

$$2x_1 - x_2 \geq 8$$
$$3x_1 + x_3 \geq 15$$
$$x_1 + 2x_2 + 3x_3 \geq 12$$
$$x_1 \geq 0, x_2 \geq 0, x_3 \geq 0$$

19. Minimize $C = x_1 + 2x_2 - x_3$, subject to

$$2x_1 + x_2 + x_3 \geq 8$$
$$3x_1 + 5x_2 - x_3 \geq 15$$
$$x_1 + 2x_2 + 3x_3 \geq 12$$
$$x_1 \geq 0, x_2 \geq 0, x_3 \geq 0$$

20. Minimize $C = 3x_1 + 2x_2 + x_3$, subject to

$$2x_1 + x_3 \geq 8$$
$$3x_1 - x_3 \geq 15$$
$$2x_2 + 3x_3 \geq 12$$
$$x_1 \geq 0, x_2 \geq 0, x_3 \geq 0$$

21. In the daily newspaper a nutritionist advises her readers about two types of cornflakes that they might buy and eat in a given week. She knows that her readers are not only interested in nutrition, but also wish to minimize their caloric intake. Suppose that in 7 days they wish to minimize their breakfast calories while getting at least 60 units of iron and 100 units of calcium. The table below lists the calories, iron, and calcium per ounce of each cereal. How much of each cereal should a reader eat in order to fulfill the nutritional requirements of iron and calcium while minimizing the number of calories consumed?

	Cereal 1	Cereal 2
Iron*	5	10
Calcium*	20	10
Calories*	50	75

* Per oz.

22. A fruit farmer has an orchard of apple and pear trees. The orchard will have an output of at least 4000 bushels in a given summer. There are at least twice as many pears picked as apples. It costs \$4 per bushel to grow and harvest the apples and \$3 per bushel for pears. How many bushels of each should be grown and harvested to minimize cost?

23. A book company has orders for at least 6000 hardback books and 10,000 paperbacks of the same title. The company has two plants. The daily production and cost per day are given in the table:

	Plant 1	Plant 2
Hardback	200	300
Paperback	500	200
Cost/day	\$600	\$500

Determine how many days each facility should work to minimize the cost.

24. A chemical engineer wants to make at least 1000 gal of industrial strength cleaner by mixing three cleaning agents. At least 9 gal of Agent 1 are used for every 7 gal of Agent 2 and there must be at least 50 gal of Agent 3. The cost of the cleaning agents is \$0.80, \$0.65, and \$0.77, respectively. How much of each agent should be used to minimize the cost of the mixture?

IMPORTANT WORDS AND PHRASES

In this chapter we have extended our concept of linearity to include inequalities, and have used these, along with our ideas of pivoting from Chapter 2, to solve some rather sophisticated problems involving maximization and minimization. These techniques involve the use of many new ideas, most of which are included in the following list:

duality principle	linear programming	pivot element
dual problem	maximization	simplex method
feasible set	minimization	slack variables
interval notation	nonstandard problems	systems of inequalities
linear inequalities	objective function	

REVIEW EXERCISES CHAPTER 3

In Exercises 1–12, solve each inequality and graph the solution on the number line.

1. $x - 4 > 2x + 3$
2. $4x \leq 8 + 3x$
3. $3x + 5 \leq 2 - 3x$
4. $3(4x - 1) \geq 15x$
5. $x - 5 < x + 1$
6. $x + 5 > x + 5$
7. $3 \leq 2x - 5 \leq 8$
8. $-1 \leq 2x - 3 \leq 7x - 8$
9. $x - 1 < 2x < 3x + 5$
10. $x - 2 < 3x < 5x - 6$
11. $x < 2x - 3 \leq x + 5$
12. $2x - 1 \leq x + 3 \leq 2x + 1$

In Exercises 13–18, graph the solution set for each inequality.

13. $2x + y - 7 \leq 0$
14. $x - 2y \geq y - 2x$
15. $5x - 3 > x + y - 2$
16. $2(x - 1) < y$
17. $2(x + y) \leq 3y + 2$
18. $3x + 2y \geq 3(5 + x)$
19. The formula relating degrees Fahrenheit to degrees Celsius is $F = \frac{9}{5}C + 32$. For which values of C is $F > 32$?
20. A retailer wishes to purchase two kinds of sweatshirts, a deluxe brand for \$25 each and a lesser-quality shirt for \$15. Make a sketch showing the possibilities if the capital outlay is limited to \$600.

In Exercises 21–24, solve the systems of inequalities graphically.

21. $x + y < 1$
    $2x - y < 5$
22. $3x - 4y \leq 5$
    $x + 2y > 2$
23. $2x + 3y \geq 6$
    $x - 2y < 8$
24. $x - 5y \leq 10$
    $3x + 5y \leq 15$
25. The Unisex Hair Design Studio makes appointments for up to 15 people per hairdresser per day. It takes $\frac{1}{2}$ hr to style a man's hair and $\frac{3}{4}$ hr to style a woman's. The operators work 8-hr days. Show

graphically the possible number of men and women clients that could be scheduled in one day for a hairdresser.

26. Marie and Kevin are planning their wedding. They have \$3000 to spend on the reception. The cost for the hall and music ranges from \$500 to \$1000. The cost for the refreshments ranges from \$1000 to \$2500. If x is the cost of the hall and music and y the cost of refreshments, show graphically the possible amounts they can spend on each for their reception.

In Exercises 27–30, solve the indicated linear programming problems graphically. Assume $x \geq 0$ and $y \geq 0$.

27. Given $x \leq 8$, $x + y \leq 12$, maximize $P = x + 2y$.

28. Given $x + y \leq 3$, $x + 2y \leq 4$, maximize $P = 5x + 2y$.

29. Given $x + y \geq 4$, $x + 2y \geq 6$, minimize $P = 3x + 5y$.

30. Given $2x + y \geq 4$, $2x + 3y \geq 6$, minimize $P = x + y$.

Solve Exercises 31 and 32 using the simplex methods of linear programming.

31. The Specialty Nut Company packages gift boxes of mixed nuts and dried fruits. The mixed nuts sell for \$7.50 per box, and the fruit for \$5.00 per box. The cost to the company for each box of nuts is \$5.50, and for each box of fruit \$4.00. The company's suppliers guarantee enough nuts, fruit, and boxes for 500 of each for each day the company is packaging. A packer takes 6 minutes to prepare a box of mixed nuts, and 10 minutes for a box of dried fruit. There are 80 worker-hours available each day to package the nuts and fruit. Determine the number of boxes of mixed nuts and of dried fruit that the company should package each day in order to maximize its profits.

32. A toymaker makes rocking horses and doll cradles to sell at crafts shows. The profit on each rocking horse is \$100; on each cradle, \$130. The rocking horse requires 10 hr of labor and the cradle requires 16. For the 16 weeks prior to a major show, the toymaker spends 30 hr per week crafting these toys. There is enough wood available for 40 toys. How many of each should be made to maximize the toymaker's profit?

In Exercises 33–38, solve each linear programming problem using the dual problem approach. Assume the variables are nonnegative.

33. Minimize $C = 7x_1 + 2x_2$, subject to

$$x_1 + x_2 \geq 5$$
$$x_1 + x_2 \geq 8$$

34. Minimize $C = 5x_1 + 3x_2$, subject to

$$x_1 + 2x_2 \geq 3$$
$$x_1 + 3x_2 \geq 4$$

35. Minimize $C = 3x_1 + 2x_2$, subject to

$$2x_1 + 3x_2 \geq 24$$
$$3x_1 + x_2 \geq 15$$

36. Minimize $C = 3x_1 + 2x_2$, subject to

$$2x_1 + 3x_2 \geq 8$$
$$3x_1 + 5x_2 \geq 15$$

37. Minimize $C = x_1 + 2x_2 + x_3$, subject to

$$2x_1 + x_2 - x_3 \geq 8$$
$$5x_1 + x_2 + x_3 \geq 15$$
$$2x_1 + x_3 \geq 12$$

38. Minimize $C = 2x_1 + 5x_3$, subject to

$$x_1 + x_3 \geq 8$$
$$3x_1 + x_2 + 3x_3 \geq 15$$
$$x_1 + 2x_2 + 2x_3 \geq 12$$

CHAPTER 4

SETS AND COUNTING

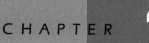

Sociologists who study population trends, bankers who evaluate investment opportunities, engineers who investigate power plants, and students who apply to graduate school—all use the concept of a set of objects even though they may not specifically use this terminology. For instance, a sociologist may talk about the set of people from a particular socioeconomic background and the banker may refer to the set of opportunities for investing money. The concept of sets, and their properties, are studied in the branch of mathematics called **set theory.** By using the mathematical concepts of set theory, we can often make studies of non-mathematical topics, such as population trends and investment opportunities, easier to understand.

4.1 SETS

In this section we will review some of the basic concepts of set theory. A set is an arbitrary collection of similar objects, whether it be numbers, people, or cattle. Set notation is a basic "language" that we sometimes use informally to describe the grouping or comparison of similar objects. In fact, set language is so basic that we hardly notice its use in everyday conversation, on the radio, or on television. The use of sets permeates essentially all scientific or quantitative discussions:

- Social scientists often talk about such things as the set of all high school dropouts, or the set of Americans who are retired.
- Engineers may discuss the set of all electrical power plants on the Ohio River, or perhaps the set of cars crossing a bridge at a given time.
- Economists may describe the set of all U.S. citizens who earn less than $10,000 per year as living at or below poverty level; they may also describe as wealthy those people who make over $1 million per year.
- A student considering graduate schools may be interested in the graduate schools offering graduate assistantships, or in those offering Ph.D. programs in Asian Studies.

Notice that in some of these cases the phrase *the set of* is omitted, even though it is understood in the context of the discussion. Sets and set notation are at the foundation of the language of mathematics. Our discussion in this first section will attempt to formalize the language of sets.

HISTORICAL COMMENT: *Sets have always been the basis of elementary mathematics, but it was Georg Cantor (1845–1918), a German mathematician, who formalized the theory of sets. Cantor's theory of sets was initially rejected by many of his contemporaries, but is accepted today as part of the foundation of modern mathematics.*

Set language is used to describe collections of similar objects, as well as the relationship between two or more sets. A **set** is an arbitrary collection of objects, whose rules of membership are well defined. The individual objects in a set are called the **elements,** or **members,** of that set. Notationally, we use uppercase letters to denote sets and lowercase letters to represent the elements of sets. The symbol $\in$ means "is a member of." Thus, $a \in A$ means that a is a member of set A. To indicate that b is not a member of A, we write $b \notin A$.

Example 1 Given that set A is the set of the first ten counting numbers, then we indicate that 2 is in this set by writing $2 \in A$. Likewise, $7 \in A$ and $10 \in A$. We indicate that -1 is *not* in A by writing $-1 \notin A$. ▪

Set Builder Notation

A set is defined by describing all its elements. In some cases, this is done by simply enclosing the members of the set in braces. For instance, $A = \{1, 3, 5, 7, 9\}$ represents the set of odd counting numbers between 0 and 10. This same set can be described in what is called **set builder notation** by stating a property of membership, instead of listing each element. Thus, the set

$$A = \{1, 3, 5, 7, 9\}$$

can also be expressed in set builder notation as

$$A = \{x \mid x \text{ is an odd integer between 0 and 10}\}$$

This is read, "A is the set of all x such that x is an odd integer between 0 and 10." Note that the vertical line $|$ is read as "such that." Set builder notation is not efficient when the number of elements in the set is small, but it is very useful for describing sets that contain a large number of elements, such as the first 1000 counting numbers. The concept of describing a set by stating its property of membership is the way we tend to speak of a set in everyday language. We talk of the members of the U. S. Senate, or the teams in the Big Ten. We seldom list the members of a set, because it is too cumbersome to do so.

Example 2 Use set builder notation to describe the set $M = \{a, e, i, o, u\}$.

Solution M is the set of vowels, so we say "M is the set of all x such that x is a vowel," and we write

$$M = \{x \mid x \text{ is a vowel}\}$$ ▪

Example 3 Enumerate the elements in $B = \{x \mid x \text{ is an even positive integer less than } 10\}$.

Solution The even positive integers less than 10 are 2, 4, 6, and 8, so B is

$$B = \{2, 4, 6, 8\}$$ ▪

Set builder notation is also appropriate when we are unsure of the precise membership of a set but do know what it takes to be a member of the set. This kind of set description is common in quiz questions, which often give the property of membership and ask for the elements of the set. For example, consider the set

$$\{x \mid x \text{ is one of the first ten presidents of the U.S.}\}$$

or the set

$$\{x \mid x \text{ is a Super Bowl winner}\}$$

In both of these cases, the property of membership precisely defines the membership of the set, but this does not necessarily mean that you could actually list the members.

Sometimes set builder notation may describe impossible properties of membership. For example,

$$\{x \mid x \text{ is a person 20 ft tall}\}$$

or

$$\{x \mid x \text{ is an integer between 6 and 7}\}$$

Clearly, there is no one who is 20 ft tall and there are no integers between 6 and 7, so these sets have no elements. A set with no elements is called the **empty set,** or the **null set,** and is denoted by $\varnothing$ or $\{\quad\}$. Keep in mind that the empty set is not the set $\{0\}$, because $\{0\}$ is a set with one element—namely, the element 0; the empty set has no elements.

COMMENT: *The property of membership must be* well defined; *that is, it must define the same set for everyone. Examples of sets for which the property of membership is not well defined are*

$$\{x \mid x \text{ is a happy person}\}$$

and

$$\{x \mid x \text{ is a distant star}\}$$

The property of membership for each of these sets is not well defined because the membership of each set is disputable.

Cardinal Number of a Set

If a set A is finite, we denote the number of elements in the set by $n(A)$, called the **cardinal number** of A. For example, since $A = \{a, b, c, d\}$ has four elements, the cardinal number of A is $n(A) = 4$. A set is said to be **finite** if its cardinal number is finite; otherwise, it is said to be **infinite.**

COMMENT: *The empty set $\varnothing$ has no elements, so its cardinal number is zero; that is,* $n(\varnothing) = 0$.

Example 4 Use set language to describe each of the given sets, and give the cardinal number of those sets that are finite:

(a) A is the set of counting numbers between 2 and 5.
(b) D is the set of counting numbers between 2 and 3.
(c) N is the set of counting numbers.
(d) $B = \{x | x \text{ is an integer and } -1 < x < 3\}$.

Solution

(a) Set A can be written as $A = \{x | x \text{ is a counting number between 2 and 5}\}$. Since A is a finite set, consisting of the two numbers 3 and 4, we can also write this set as $A = \{3, 4\}$. Both descriptions are correct. In either case, the cardinal number of A is $n(A) = 2$.
(b) There are no elements in set D, because there are no counting numbers between 2 and 3. Therefore, D is the null set, and we write $D = \varnothing$ and $n(D) = 0$.
(c) The set of all counting numbers can be written $N = \{x | x \text{ is a counting number}\}$. The set of all counting numbers is infinite.
(d) The elements of B are 0, 1, and 2, so its cardinal number is $n(B) = 3$.

Two sets M and N are said to be **equal** if they have exactly the same elements, and we write $M = N$ to describe the equality of the two sets. For example, the set $M = \{1, 3, 5, 7, 9\}$ and $N = \{x | x \text{ is an odd positive integer between 0 and 10}\}$ are equal; that is, $M = N$. The sets $A = \{1, 2, 3\}$ and $B = \{a, b, c\}$ are not equal, even though they have the same number of elements. ■

COMMENTS:

- *The order in which elements in a set are listed is irrelevant. Thus, sets A and B are equal if A has the same elements as B, but in a different order. For instance,* $\{a, b\} = \{b, a\}$.
- *Elements of a set should not be repeated in any listing. For example, suppose that a coin is tossed five times with the following outcomes: H, T, H, H, T. The set of possible outcomes is given by* $\{H, T\}$*, even though heads occurred three times and tails twice.*

Example 5

(a) If $A = \{-1, 0, 2, 5\}$ and $B = \{2, 0, -1, 5\}$, then $A = B$.
(b) If $A = \{-2, 2\}$ and $B = \{x | x^2 = 4\}$, then $A = B$. ■

SUBSETS

Sometimes we are interested in only a portion of a set. For instance, we might want to consider only the U.S. senators who are Democrats. This portion of a set is known as a subset. A **subset** of A is a set all of whose elements are also elements of set A. If $M = \{1, 2, 3, 4, 5\}$ and $N = \{1, 3, 5\}$, then N is a subset of M, because every member of N is also a member of M. We indicate that N is a subset of M by writing $N \subseteq M$. If $N \subseteq M$, but $N \neq M$, then N is a *proper* subset of M; this is denoted by $N \subset M$. We note that every set is a subset of itself; that is, $A \subseteq A$. Also, the null set, $\varnothing$, is a subset of every set, including itself.

Example 6 List all the subsets of $\{1, 2, 3\}$.

Solution The subsets are $\{1, 2, 3\}$, $\{1, 2\}$, $\{1, 3\}$, $\{2, 3\}$, $\{1\}$, $\{2\}$, $\{3\}$, and $\varnothing$. All except the first are proper subsets. (Remember, $\varnothing$ is a subset of every set.) ▪

COMMENT: *From a practical standpoint, we regularly deal with the concept of a subset. For example, a student wants to learn a subset of all knowledge, and income taxes are a subset of our total income. In one form or another, sets and subsets are more a part of our lives than we realize!*

Most discussions of sets focus on a fixed set of objects or, if not, go on endlessly until someone decides that the discussion should be properly limited to such a fixed set. If all the sets to be discussed are subsets of one set, this set is called the **universal set** and is denoted in the discussion by U. For instance, the alphabet is the universal set of all the sets of letters needed for words in the English language. Generally, the universal set varies with the situation, and changing the universal set may change the response to a given question.

Example 7 Determine the set of even counting numbers if the universal set is:

(a) $U_1 = \{2, 3, 4, 5, 6, 7\}$
(b) $U_2 = \{1, 3, 5, 7\}$

Solution Let E be the set of even counting numbers. Then

(a) For the universal set U_1, the set of even counting numbers is $E_1 = \{2, 4, 6\}$.
(b) For the universal set U_2, the set of even counting numbers is $E_2 = \varnothing$, since there are no even counting numbers in U_2. ▪

Earlier, we mentioned the subset of U.S. senators who are Democrats. Now, consider the senators who are not Democrats. That would mean the Republicans and those who belong to other political parties. We often need to look at all the elements that are in a universal set, but *not* in a given set A. Such a set is called the *complement* of A.

DEFINITION Complement of a Set

If U is the universal set for a set A, then the **complement** of A is the set consisting of those elements in U that are not in A. The complement of A is denoted by A'.

Example 8

(a) If U is the set of vowels and $A = \{a, e, o\}$, then the complement of A is $A' = \{i, u\}$.

(b) If U is the set of students taking Math 101 and C is the set of those students who pass the course, then C' is the set of students who do not pass Math 101.

(c) The complement of the universal set is the empty set—that is, $U' = \varnothing$. Also, the complement of the empty set is the universal set—that is, $\varnothing' = U$. ∎

COMMENT: *For any set A, it is easy to show that $(A')' = A$; that is, the complement of the complement is the set itself.*

VENN DIAGRAMS

Sets are often represented pictorially by using the points within a rectangular region of the plane to represent the universal set. Subsets of the universal set are then represented as circles within the rectangle. For example, in Figure 4.1, sets A and B are shown as circles within the rectangular region U. The fact that B is wholly contained in A indicates that $B \subseteq A$.

Figure 4.1

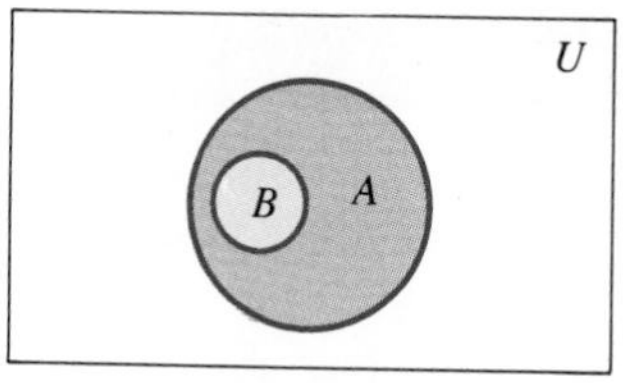

HISTORICAL COMMENT: *Diagrams like the one in Figure 4.1 are called* ***Venn diagrams,*** *after the English mathematician John Venn (1834–1923). Venn, who was an ordained priest, eventually left the priesthood to teach and study logic.*

Venn diagrams are used extensively to represent the relationship between two or more sets. In Figure 4.2(a), the sets A and B represent two sets that have no points in common. Sets with no points in common are called **disjoint sets.** The sets A and B in Figure 4.2(b) are drawn as overlapping circles to indicate that they have some points in common. However, A is not a subset of B and B is not a subset of A.

Figure 4.2

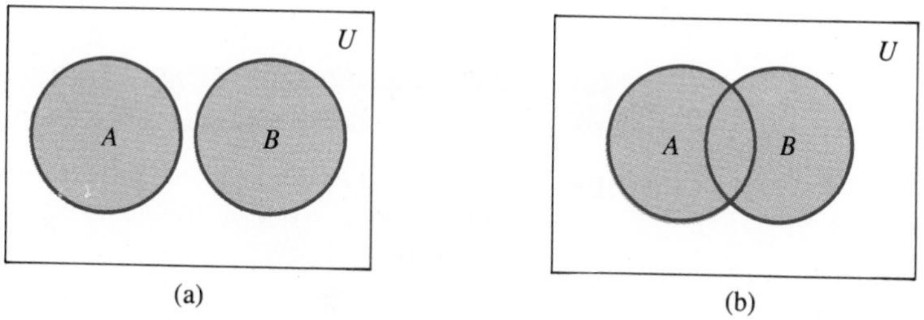

Figure 4.3 shows the Venn diagram for a set A and its complement A'.

Figure 4.3

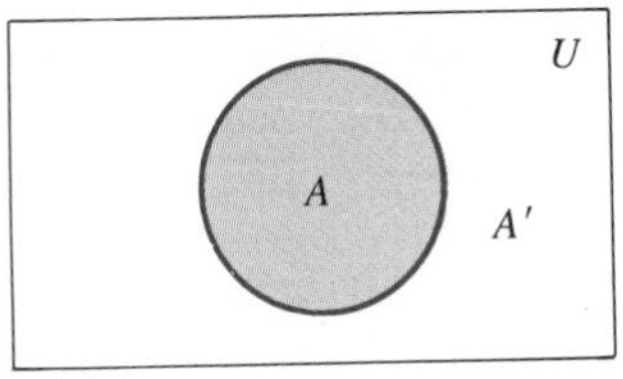

Operations on Sets

The word *operation* is used here to mean the act of doing something to a set or sets. For example, $n(A)$ refers to the operation of counting the number of elements in the set A. Finding the complement of set A is an operation on set A that yields the complementary set A'. Determining the cardinal number of a set and finding its complement are both operations on only one set. An operation that is performed on *two* sets to form a new set is called a **binary operation.** There are two basic binary set operations used to form new sets—namely, union ($\cup$) and intersection ($\cap$).

DEFINITION The Union of Two Sets

Given two sets A and B, the set of all elements that belong either to A or to B is called the **union** of A and B. The union of A and B is denoted by $A \cup B$ and defined by

$$A \cup B = \{x | x \in A \text{ or } x \in B\}$$

COMMENT: *The word* **or** *has two meanings in the English language. Our everyday use of the word* **or** *is* exclusive; *that is, when we say "red or green," we imply that either it is red or it is green, one or the other, but not both. In mathematical logic, the word* **or** *is used in the* inclusive *sense; that is, $x \in A \cup B$ means that x is in set A or in set B or in both set A and set B.*

Example 9 Given $A = \{1, 2, 3, 4\}$ and $B = \{2, 4, 6\}$, then $A \cup B = \{1, 2, 3, 4, 6\}$. Notice that 2 and 4, which are members of both sets, are only listed once in $A \cup B$. ∎

Having defined the set operation of union, we now define the set operation of intersection.

DEFINITION The Intersection of Two Sets

Given sets A and B, the set of all elements that belong to both A and B is called the **intersection** of A and B. The intersection of A and B is denoted by $A \cap B$ and defined by

$$A \cap B = \{x | x \in A \text{ and } x \in B\}$$

Example 10 If $A = \{1, 2, 3, 4\}$ and $B = \{2, 4, 6\}$, then $A \cap B = \{2, 4\}$. ■

COMMENT: *Study carefully the definitions of union and intersection. They look very similar. The only difference is in the use of the words* **and** *and* **or.** *However, this emphasizes the basic difference between the two operations. When we find the union of two sets, we include every element that is in* **either** *of those two sets. When finding the intersection, we include only those elements common to* **both** *sets.*

$A \cup B$ and $A \cap B$ are illustrated by the Venn diagrams in Figure 4.4, in which A and B are shown as overlapping circles. The elements of $A \cap B$, which in Figure 4.4(b) has been shaded more heavily for emphasis, are the points in the overlapping region.

Figure 4.4

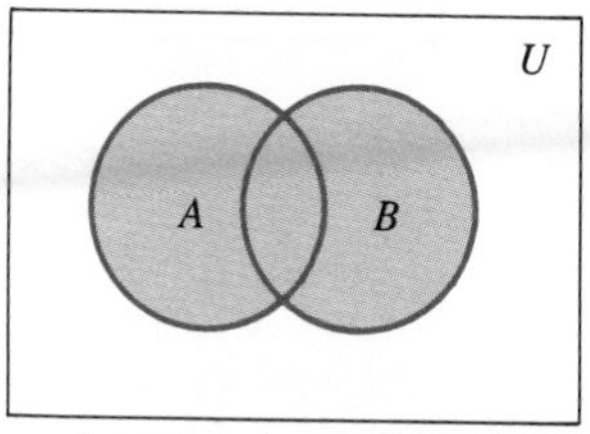

A ∪ B is the shaded area

(a)

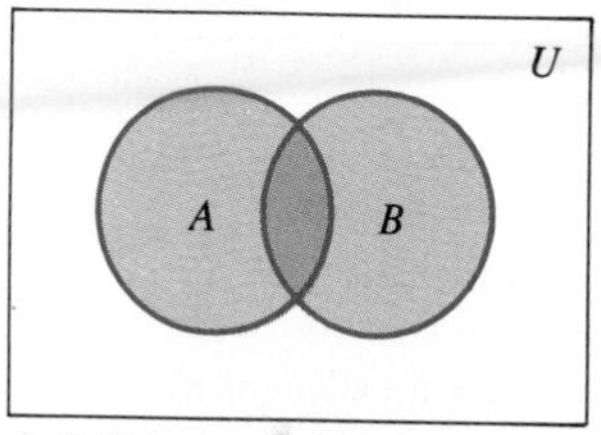

A ∩ B is the darker shaded area

(b)

Venn Diagrams of $A \cup B$ and $A \cap B$

Example 11 Given $A = \{a, b, c, e, m, p\}$ and $B = \{c, g, h, m, s\}$,

(a) Find $A \cup B$.
(b) Find $A \cap B$.

Solution

(a) $A \cup B$ consists of all elements in A or in B. Hence,

$$A \cup B = \{a, b, c, e, g, h, m, p, s\}$$

Notice that elements c and m, which are in both sets, are only listed once in $A \cup B$.

(b) The only elements that belong both to A and to B are c and m. Therefore, the intersection of the two sets is

$$A \cap B = \{c, m\}$$

■

COMMENT: *If the intersection of two sets is the empty set, the two sets are disjoint.*

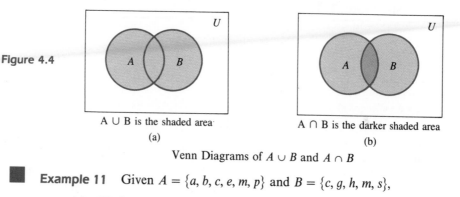

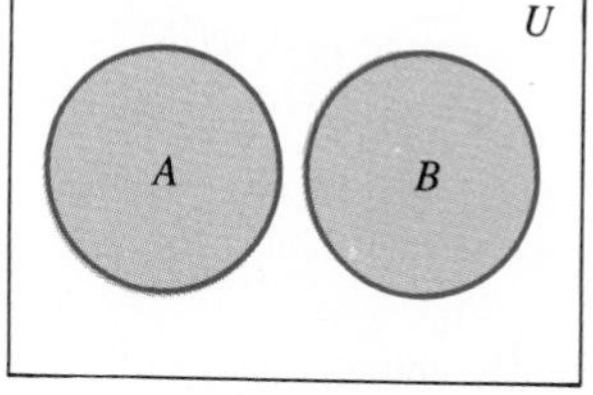

$A \cap B = \varnothing$

Example 12 Let A represent the set of males at Centerville High School, B the set of honor students, and C the set of females at Centerville High School. Then:

- $A \cap B$ represents the set of male honor students.
- $B \cap C$ represents the set of female honor students.
- $A \cup C$ represents all those attending Centerville High School. ■

COMMENT: *Note that $A \cup A' = U$ (the universal set) and $A \cap A' = \varnothing$ (the empty set).*

EXERCISES SECTION 4.1

1. Let $A = \{a, c, m, p, z\}$. Use set notation to write the following statements:
(a) c is an element of A.
(b) m is an element of A.
(c) d is not an element of A.

2. Let $A = \{-1, 2, 5, 9\}$. Use set notation to write the following statements:
(a) 2 is an element of A.
(b) 3 is not an element of A.
(c) -1 is an element of A.

3. Let $S = \{x | x \text{ is a letter of the alphabet between } b \text{ and } h\}$. Which of these statements are true or false?
(a) $d \in S$ (b) $a \in S$ (c) $m \in S$ (d) $e \in S$

4. Let $D = \{x | x \text{ is an integer and } -2 \le x \le 4\}$. Which of the following statements are true or false?
(a) $-3 \in D$ (b) $4 \in D$ (c) $0 \in D$ (d) $\frac{1}{2} \in D$

In Exercises 5–8, describe each set by listing its elements.

5. $A = \{x | x \text{ is an integer and } -2 \le x \le 5\}$

6. $S = \{x | x^2 - 9 = 0\}$

7. $B = \{x | 3x + 2 = 8\}$

8. $C = \{x | x \text{ is a letter of the alphabet that precedes } f\}$

In Exercises 9–12, use set builder notation to describe each set.

9. $S = \{3, 5, 7, 9\}$

10. $B = \{-1, 0, 1, 2\}$

11. C is the set of people in the U.S. under 25 years of age.

12. Y is the set of solutions of the equation $x^2 - 3x + 2 = 0$.

Which of the statements in Exercises 13–18, is true? Explain your answer.

13. $\{x | x + 2 = 0\} = \varnothing$

14. $\{x | 3x - 1 = 2x + 3\} = \{4\}$

15. $\{x | x^2 = 4\} = \{x | x^2 - 4 = 0\}$

16. $\{x | 2x + 7 = x - 2\} = \{x | 3x + 5 = 2x + 4\}$

17. $\{1, 3\} \subset \{x | x^2 - 3x + 2 = 0\}$

18. $\{2\} \subset \{x | x^2 = 4\}$

19. Indicate the minimum universal set of letters necessary to form the words *seven* and *element.*

20. Indicate the minimum universal set of letters necessary to form the words *amigo* and *friend.*

21. Indicate which sets are the empty set and enumerate the nonempty sets:
(a) $\{x | x \text{ is a real number and } x - 2 = 0\}$
(b) $\{x | x \text{ is an integer and } 3x + 2 = 0\}$
(c) $\{x | x \text{ is a real number and } x^2 + 1 = 0\}$
(d) $\{x | x \text{ is a real number and } x^2 - 3 = 0\}$

22. Indicate which sets are the empty set and enumerate the nonempty sets:
(a) $\{x | x \text{ is an integer and } 2x + 1 = 0\}$
(b) $\{x | x \text{ is an integer and } x^2 = 4\}$
(c) $\{x | x$ is a vowel and a letter in the word *black*$\}$
(d) $\{x | x$ is a vowel and a letter in the word *Mississippi*$\}$

23. Determine all the subsets of $\{5, 6, 8, 9\}$.

24. Determine all the subsets of $\{r, s, t\}$.

25. Suppose the universal set is $U = \{x | x \text{ is a counting number less than } 11\}$.
(a) Write the set of even counting numbers.
(b) Write the set of counting numbers larger than 5.

26. Repeat Exercise 25 with $U = \{1, 2, 3, 4, 5\}$.

Let $A = \{1, 2, 3, 5, 7, 8\}$, $B = \{1, 3, 6, 8\}$, and $C = \{2, 3, 4, 5\}$. Evaluate the operations in Exercises 27–32.

27. $A \cup B$
28. $A \cup C$
29. $A \cap B$
30. $A \cap C$
31. $B \cap C$
32. $B \cup C$

Evaluate the expressions in Exercises 33–40, if $U = \{x | x \text{ is a counting number less than } 11\}$, $A = \{2, 4, 6, 8, 10\}$, $B = \{1, 2, 3, 4\}$, and $C = \{4, 7, 9\}$.

33. A'
34. B'
35. C'
36. $A \cup B'$
37. $B \cap C'$
38. $B' \cap C'$
39. $(A \cap B)'$
40. $(A \cup B)'$

41. Let A represent the NFL football teams that won the Super Bowl between 1980 and 1990, and B those teams that were runners-up. Describe the elements in each of the following sets:
(a) A (b) B (c) $A \cup B$ (d) $A \cap B$

42. Let A represent the set of U.S. citizens who can speak a foreign language and let B be the people living in Berlin.
(a) What does $A \cap B$ represent?
(b) What does A' represent?

43. Let A represent the set of U.S. citizens whose first marriage ended in divorce and B represent the set of U.S. citizens whose second marriage ended in divorce.
(a) What does $A \cap B$ represent?
(b) What does B' represent?

44. (a) If $A \cup B = \varnothing$, what can be said about A and B? What can be said about $A \cap B$?
(b) If $A \cap B = \varnothing$, what can be said about A and B?

4.2 More Unions and Intersections

In this section, we show how to combine three or more sets by union and intersection. At first glance this may seem trivial, but remember that we initially defined the operations of union and intersection in terms of two sets; that is, they are binary operations. In extending these operations to three or more sets, we discover some fundamental mathematical properties related to the order in which the operations are applied.

Associative Properties

The operations of union and intersection are both binary operations and, as such, can be performed on only two sets at a time. So, you might ask, how can we form the union of three sets, A, B, and C? More than one approach is possible. For instance, we could form $A \cup B$ first, and then form the union of this result with C. We represent this sequence of operations by $(A \cup B) \cup C$, where the parentheses around $A \cup B$ tell us to perform that operation first. Or we could form $B \cup C$ and then form the union of this result with A. This is represented by $A \cup (B \cup C)$. Figure 4.5(a) shows $(A \cup B)$ in union with C; Figure 4.5(b) shows A in union with $(B \cup C)$. In each case, the order of taking the unions was different, but the result

Figure 4.5

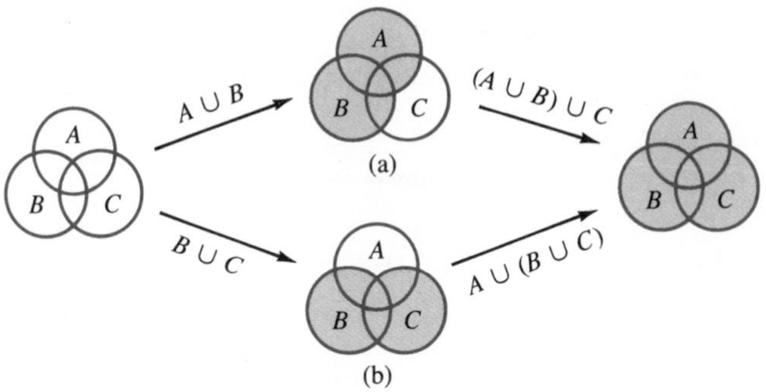

is the same! That is,

$$(A \cup B) \cup C = A \cup (B \cup C)$$

Similarly, Figure 4.6 shows that

$$(A \cap B) \cap C = A \cap (B \cap C)$$

Figure 4.6

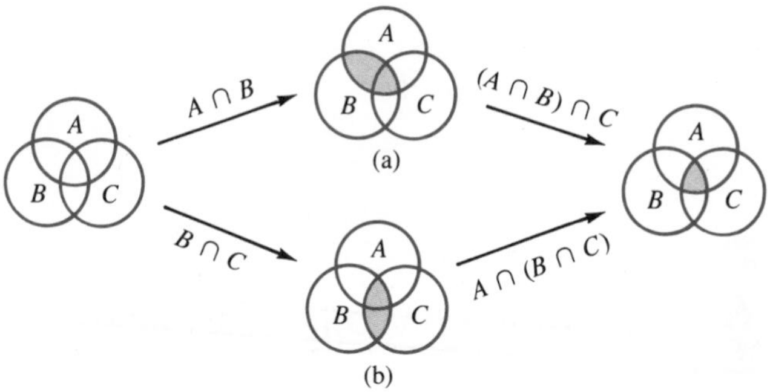

The fact that the unions of three sets may be taken in any order, and the fact that the intersections of three sets may be taken in any order, are called the **associative properties** of unions and intersections. Each of these properties says that when combining three sets with the given operation, the grouping may be done in any order. Since this is true, the parentheses are normally omitted and we simply write $A \cup B \cup C$ for the union of three sets and $A \cap B \cap C$ for the intersection of three sets.

DEFINITION Associative Properties for Union and Intersection

Given the sets A, B, and C:

$$A \cup B \cup C = A \cup (B \cup C) = (A \cup B) \cup C$$
$$A \cap B \cap C = A \cap (B \cap C) = (A \cap B) \cap C$$

Example 1

(a) $A \cap B \cap \varnothing = \varnothing$, since the intersection of the empty set with any other set is the empty set. Either we have $A \cap (B \cap \varnothing) = A \cap \varnothing = \varnothing$ or we have $(A \cap B) \cap \varnothing = \varnothing$. The order makes no difference.

(b) In performing the operations $\{3, 4, 5, 6\} \cup \{6, 7, 9, 10\} \cup \{1, 3, 5, 7, 9\}$, we may do the unions in any order. In either case, the result is $\{1, 3, 4, 5, 6, 7, 9, 10\}$. ∎

Distributive Properties

Sometimes three or more sets are operated on by union and intersection, mixing the operations. For instance, we could have the union of A and B intersected with C—that is, $(A \cup B) \cap C$; or the union of A with the intersection of B and C—that is, $A \cup (B \cap C)$. In these instances, it is important to make note of the order of operation, because, as Figure 4.7 shows, $(A \cup B) \cap C$ and $A \cup (B \cap C)$ are not the same set.

Figure 4.7

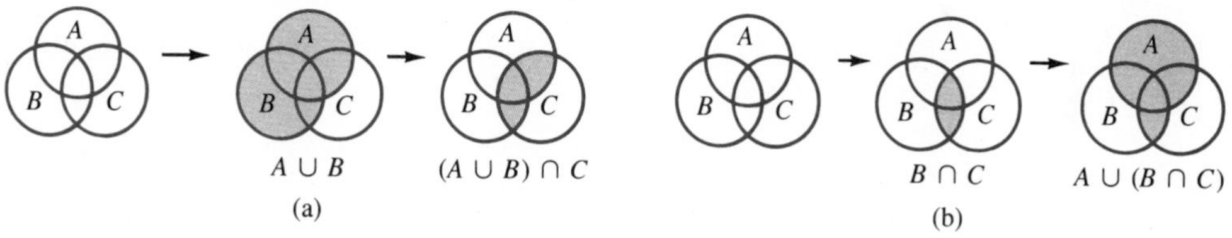

Example 2 Given $A = \{a, b, c, d\}$, $B = \{b, c, e, f\}$, and $C = \{a, c, e\}$,

(a) Find $A \cup (B \cap C)$.
(b) Find $(A \cup B) \cap C$.

Solution

(a) Performing the operation in parentheses first, we have $B \cap C = \{c, e\}$. The desired result is then the union of $B \cap C$ with A; that is,

$$A \cup (B \cap C) = \{a, b, c, d, e\}$$

(b) Performing $A \cup B$ yields $A \cup B = \{a, b, c, d, e, f\}$. The intersection of this set with C is then the desired set. Thus,

$$(A \cup B) \cap C = \{a, c, e\}$$ ∎

In algebra we learn that $a \cdot (b + c) = a \cdot b + a \cdot c$, a fact that we refer to as the distributive law of multiplication over addition. In a similar way, the set operation of union will distribute over intersection and the set operation of intersection will distribute over union. We state the distributive laws for the operations of union and intersection as follows:

DEFINITION Distributive Properties for Union and Intersection

Given sets A, B, and C:

$$A \cup (B \cap C) = (A \cup B) \cap (A \cup C)$$
$$A \cap (B \cup C) = (A \cap B) \cup (A \cap C)$$

The validity of the distributive properties is easily verified using Venn diagrams, and will be explored in the exercise set. We make use of these properties in Example 3.

Example 3 Use the distributive laws and the properties of complements to show that

$$B \cup (B' \cap C) = B \cup C$$

Solution The expression on the left side can be transformed by the distributive law to give

$$\begin{aligned} B \cup (B' \cap C) &= (B \cup B') \cap (B \cup C) \\ &= U \cap (B \cup C) && \text{Because } B \cup B' = U \\ &= B \cup C && \text{Because } U \cap A = A \end{aligned}$$

A Venn diagram illustrating this is shown in Figure 4.8. In Figure 4.8(a), $B' \cap C$ is the shaded region. The union of $B' \cap C$ with B is shown in Figure 4.8(b), but this is just the Venn diagram of $B \cup C$.

Figure 4.8

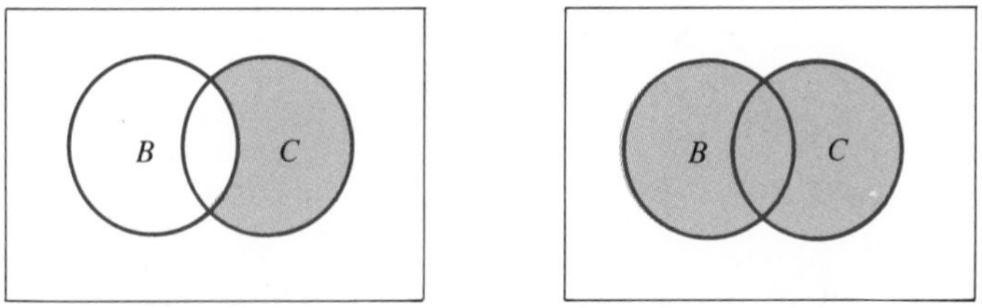

Shading is $B' \cap C$
(a)

Shading is $B \cup (B' \cap C) = B \cup C$
(b)

■

COMMENT: We note that $(A \cup B)' \neq A' \cup B'$. In Figure 4.9, the region representing $(A \cup B)'$ is shown in (a) and that for $A' \cup B'$ is shown in (b). $(A \cup B)'$ is everything except $A \cup B$, and $A' \cup B'$ is everything except $A \cap B$; that is, $(A \cap B)'$. Since the region in the Venn diagram for $(A \cup B)'$ is not the same as that for $A' \cup B'$, the two expressions are not equal.

Figure 4.9

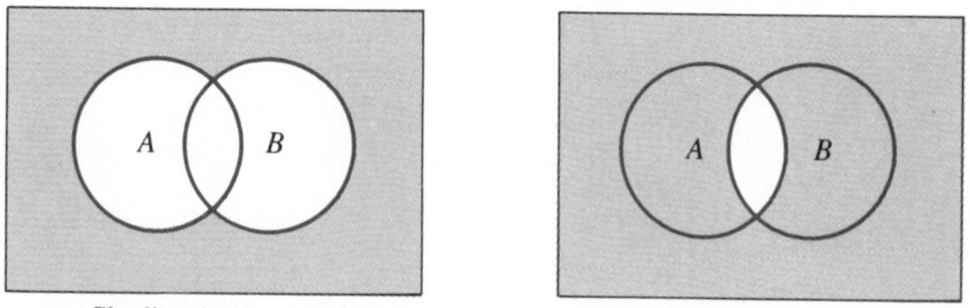

Shading is $(A \cup B)'$
(a)

Shading is $A' \cup B'$
(b)

In the process of proving that $(A \cup B)' \neq A' \cup B'$, we have uncovered an interesting property of complements. In Figure 4.9(b), we saw that the union of the complement of A and the complement of B is equal to the complement of the intersection of A and B; that is, $A' \cup B' = (A \cap B)'$. This is one of two laws of complements known as DeMorgan's laws.

DeMorgan's Laws

Given sets *A*, *B*, and *C*:

$$(A \cup B)' = A' \cap B'$$
$$(A \cap B)' = A' \cup B'$$

HISTORICAL COMMENT: *Augustus DeMorgan (1806–1871) taught at the University of London. His work in logic prepared the way for another English mathematician, George Boole. (See Chapter 9.)*

Example 4 Show that $(A \cup B)' = A' \cap B'$ is valid for $A = \{1, 2, 3, 4\}$ and $B = \{2, 3, 4, 7\}$ and $U = \{1, 2, 3, 4, 5, 6, 7, 8, 9, 10\}$.

Solution We note that $A \cup B = \{1, 2, 3, 4, 7\}$, so $(A \cup B)' = \{5, 6, 8, 9, 10\}$. We now want to show that $A' \cap B'$ is the same set. The complement of A is $A' = \{5, 6, 7, 8, 9, 10\}$; that of B is $B' = \{1, 5, 6, 8, 9, 10\}$. It follows that $A' \cap B' = \{5, 6, 8, 9, 10\}$, which is the same as the set generated by $(A \cup B)'$, so we have shown that DeMorgan's law holds for this example. ▪

Cardinal Number of a Union of Two Sets

The cardinal number of a set A was defined in Section 4.1 as the number of elements in the set and was denoted by $n(A)$. We now look at this concept as it relates to the operations of union and intersection.

Example 5 Consider the disjoint sets $A = \{a, b, c, d\}$ and $B = \{g, h\}$. Find the cardinal number of $A \cup B$.

Solution Since $A \cup B = \{a, b, c, d, g, h\}$ has six elements, its cardinal number is 6. We write this as $n(A \cup B) = 6$. ▪

Observing that $n(A) = 4$ and $n(B) = 2$, and that $6 = 4 + 2$, we might speculate that $n(A \cup B) = n(A) + n(B)$. This is, in fact, true for all finite, disjoint sets. We state this in the following rule:

If A and B are finite, disjoint sets, then

$$n(A \cup B) = n(A) + n(B) \tag{4.1}$$

If A and B are finite sets that are not disjoint, then $n(A \cup B) \neq n(A) + n(B)$. To verify this, consider $A = \{1, 3, 5\}$ and $B = \{3, 6, 7, 9\}$. The union of A and B is the set $A \cup B = \{1, 3, 5, 6, 7, 9\}$, whose cardinal number is $n(A \cup B) = 6$. The cardinal number of A is $n(A) = 3$ and that of B is $n(B) = 4$. Clearly, $6 \neq 3 + 4$, which demonstrates that $n(A \cup B) \neq n(A) + n(B)$. The reason is that the element 3 is a member of both sets and was therefore counted twice. In general, when A and B are not disjoint sets, the elements in $A \cap B$ will be counted twice, so $n(A \cap B)$ must be subtracted from $n(A) + n(B)$ to give $n(A \cup B)$. We state this in the following rule:

The Cardinal Number of $A \cup B$

If A and B are finite sets, then the cardinal number of $A \cup B$ is

$$n(A \cup B) = n(A) + n(B) - n(A \cap B) \tag{4.2}$$

Example 6 Given $A = \{0, 1, 3, 5\}$ and $B = \{2, 3, 4, 5, 6\}$, find $n(A \cup B)$ by counting the elements in $A \cup B$, then use Formula 4.2 to find $n(A \cup B)$.

Solution We note that $A \cup B = \{0, 1, 2, 3, 4, 5, 6\}$ and $A \cap B = \{3, 5\}$. Counting the elements in $A \cup B$ yields

$$n(A \cup B) = 7$$

Since $n(A) = 4$, $n(B) = 5$, and $n(A \cap B) = 2$, we have

$$n(A \cup B) = n(A) + n(B) - n(A \cap B) = 4 + 5 - 2 = 7$$

This agrees with the number obtained by counting the elements in $A \cup B$. ▪

COMMENT: *If A and B are disjoint sets, then $A \cap B = \varnothing$. Since the empty set has no elements, $n(A \cap B) = 0$. In this case, Formula 4.2 reduces to Formula 4.1.*

An application of Formula 4.2 is given in the next example. The problem is typical of the kind of problem that can be solved using this formula.

Example 7 The campus bookstore sells *Time* and *Newsweek* magazines. The manager surveyed 100 students who bought at least one of these magazines and found that 85 of the students bought *Time* and 24 bought both. Use Formula 4.2 to find out how many students bought *Newsweek*.

Solution Let X represent the set of students who bought *Time* and let Y represent the set of students who bought *Newsweek*. The set of students who bought either *Time* or *Newsweek* is then $X \cup Y$, and the set of those who bought both is $X \cap Y$. The number of students who bought either *Time* or *Newsweek* is $n(X \cup Y) = 100$, the number who bought *Time* is $n(X) = 85$, and the number who bought both is $n(X \cap Y) = 24$. Using Formula 4.2, we have

$$n(X \cup Y) = n(X) + n(Y) - n(X \cap Y)$$

Or, solving for $n(Y)$, the number of students who bought *Newsweek,* we get

$$n(Y) = n(X \cup Y) + n(X \cap Y) - n(X) = 100 + 24 - 85 = 39$$

We conclude that 39 students bought *Newsweek* magazine. Of course, you could figure this out without using Formula 4.2, but the approach used here demonstrates a procedure that can be extended to more difficult problems. ■

Instead of relying upon Formula 4.2, you can usually sketch a Venn diagram. Thus, as you can see from Figure 4.10, the number of elements in the intersection (that is, those who bought both *Time* and *Newsweek*) is initially identified. This then permits determination of the number who bought only *Time*. Finally, we are able to find the number who bought only *Newsweek* and, from this, the total number of students who bought *Newsweek*.

Figure 4.10

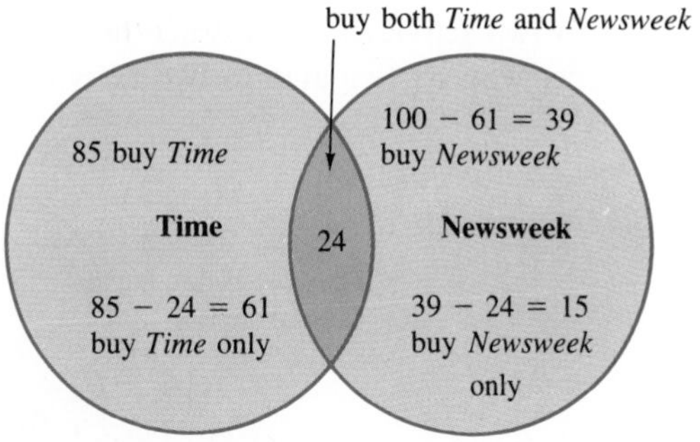

The use of a Venn diagram to relate various unions and intersections is easily extended to three sets. First we identify $A \cap B \cap C$; then, in turn, $A \cap B$, $A \cap C$, and $B \cap C$. Finally, the other regions are identified. The example below illustrates this idea.

Example 8 At a major midwestern university, 100 students were polled to determine their free-time habits. Sixty-five said they listened to music, 60 said they watched TV, and 35 said they exercised. Forty-eight said they listened to music and watched TV, 25 said they watched TV and exercised, and 20 said they listened to music and exercised. Fifteen said they did all three. Draw a Venn diagram that illustrates this situation. How many students in the survey listened to music and engaged in neither of the other two activities? How many students in the poll engaged in none of the three activities?

Solution Let M, TV, and E represent the sets of students who listened to music, watched TV, and exercised, respectively. We begin by sketching a general Venn diagram, showing the indicated sets as three overlapping circles, as in Figure 4.11(a). The last piece of information given, "Fifteen said they did all three," is identified by shading the intersection of all three sets, as in Figure 4.11(a).

Figure 4.11(a)

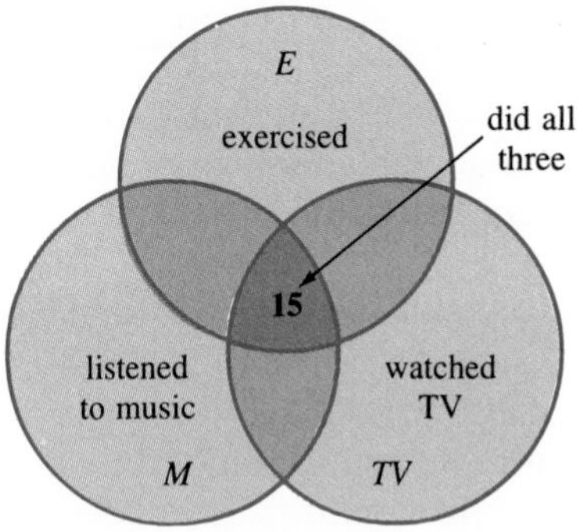

Next, we identify those who engaged in only two activities by shading the intersections of pairs of circles, as in Figure 4.11(b). To find the number who engaged in only two activities, we subtract the 15 in the intersection of all three sets from the number who participated in two activities, as shown in Figure 4.11(b).

Figure 4.11(b)

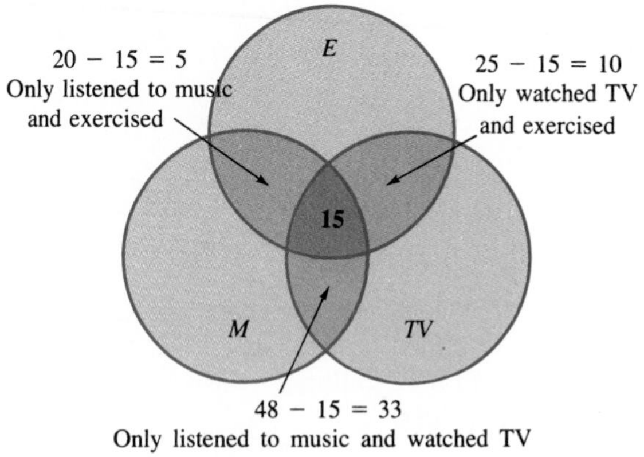

The students who engaged in only one of the activities are then identified by the region of each circle that does not intersect another circle, as shown in Figure 4.11(c). To find these numbers, subtract the sum of the numbers in the intersected regions of a circle from the total number of students represented by that circle, as shown in Figure 4.11(c).

Figure 4.11(c)

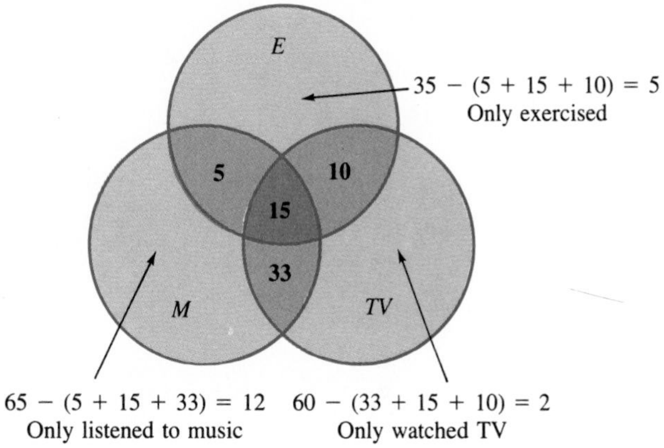

Finally, the numbers accounted for in all the parts of the Venn diagram are added together to obtain 82. There are then $100 - 82 = 18$ students who did none of the three, which is indicated by the region outside the circles shown in Figure 4.11(d).

Figure 4.11(d)

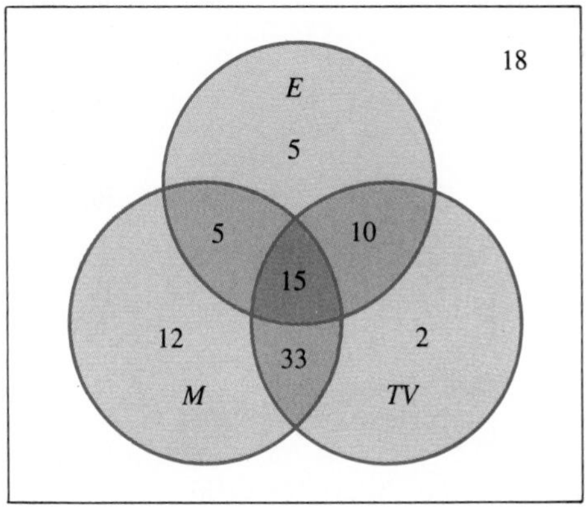

Exercises Section 4.2

Evaluate the expressions in Exercises 1–6, if $A = \{1, 2, 3, 4, 5, 6\}$, $B = \{3, 5, 7, 9\}$, *and* $C = \{4, 5, 6\}$.

1. $A \cup B \cup C$

2. $A \cap B \cap C$

3. $A \cup (B \cap C)$

4. $(A \cap B) \cup C$

5. $(A \cup B) \cap C$

6. $A \cap (B \cup C)$

7. Use Venn diagrams to show that $A \cup (B \cap C) = (A \cup B) \cap (A \cup C)$.

8. Use Venn diagrams to show that $A \cap (B \cup C) = (A \cap B) \cup (A \cap C)$.

Evaluate the expressions in Exercises 9–20, if $U = \{x | x \text{ is a counting number less than } 11\}$, $A = \{2, 4, 6, 8, 10\}$, $B = \{1, 2, 3, 4\}$, *and* $C = \{4, 7, 9\}$.

9. $A' \cup A \cup C$

10. $B' \cap \varnothing \cap U$

11. $C' \cup A \cup B$

12. $A \cup B' \cup A$

13. $(B \cap C') \cup C$
14. $B' \cap C' \cap C$
15. $(A \cap B)' \cap A$
16. $(A \cup B)' \cup A$
17. $A \cap B \cap C'$
18. $A' \cup B \cup C'$
19. $A' \cup (B \cap C)$
20. $B \cap (A' \cup C')$

In Exercises 21–24, use the distributive laws and the properties of complements to verify each statement.

21. $A \cap (A' \cup B) = A \cap B$
22. $(A \cup A') \cap B = B$
23. $(A \cup B) \cap A' = A' \cap B$
24. $A' \cup (A \cap B') = A' \cup B'$

Let X, Y, and Z be the sets shown in the figure. Shade the region that represents each set in Exercises 25–30.

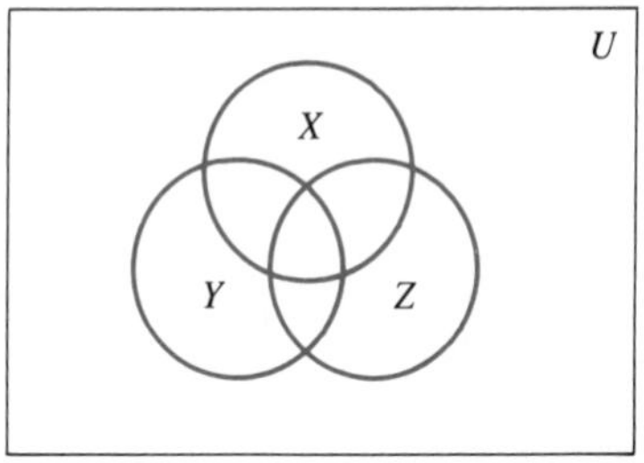

25. $X \cap (Y \cup Z)$
26. $X' \cap Y$
27. $X \cup Y'$
28. $X' \cup Z$
29. $(X \cup Y') \cap Z'$
30. $(X \cup Y)' \cap Z$
31. The Athletic Department classifies the athletes each term by sex, academic eligibility, and scholarships. Let U represent the set of all athletes at the university, and M, E, and S represent the sets of male athletes, academically eligible athletes, and athletes on scholarships, respectively. Using the set operations of complement, union, and intersection, represent each of the following:
    (a) Male athletes on scholarships.
    (b) Ineligible athletes on scholarships.
    (c) Female athletes who are either academically eligible or on scholarships.
    (d) Male athletes who are academically eligible and not on scholarships.
32. In reviewing the services and charges of various banks, Diane discovers that the banks can be grouped according to whether they charge for checking accounts, credit cards, or access to the ATMs (automatic teller machines). Let C, V, and A represent banks that charge for checking accounts, credit cards, and access to ATMs, respectively. Using the set operations of complement, union, and intersection, represent each of the following:
    (a) Banks that charge for credit cards, but not for access to ATMs.
    (b) Banks that charge for credit cards, checking accounts, and access to ATMs.
    (c) Banks that charge either for credit cards or for checking accounts.
    (d) Banks that have no charge for checking accounts, credit cards, or access to ATMs. (Can you express this in two different ways?)

In Exercises 33–36, use $n(A \cup B) = n(A) + n(B) - n(A \cap B)$ to find the cardinal number of $A \cup B$.

33. $A = \{2, 3, 5, 7\}$, $B = \{2, 4, 8\}$.
34. $A = \{b, e, o\}$, $B = \{x | x \text{ is a vowel}\}$.
35. $A = \{\text{Mary, Sally, Pete, Juan, Lou}\}$, $B = \{\text{Bill, Lou, Sam, Mary}\}$.
36. $A = \{x | x \text{ is an integer and } 0 < x < 10\}$, $B = \{x | x$ is an even counting number less than 8).
37. A survey of 1000 Ford and Chevy owners reveals that 472 own Fords and 133 own both. Use $n(A \cup B) = n(A) + n(B) - n(A \cap B)$ to calculate how many own Chevys.
38. Of 265 physics and math majors, 172 are majoring in physics, 104 are majoring in math, and some are majoring in both. Use $n(A \cup B) = n(A) + n(B) - n(A \cap B)$ to calculate how many are majoring in both physics and math.
39. Of 350 factory employees, some can run a lathe, some can run a drill press, and some can run both. If 194 can run a lathe and 189 can run a drill press, use $n(A \cup B) = n(A) + n(B) - n(A \cap B)$ to calculate how many can run both.
40. A car dealer sold 78 cars, some with air conditioning, some with a sunroof, and some with both. If 61 had air conditioning and 43 had both, use $n(A \cup B) = n(A) + n(B) - n(A \cap B)$ to calculate how many were sold with sunroofs.
41. At a midwestern university, 100 students were polled to determine who liked math and who liked

movies. Fifty said they liked math, 15 said they liked both, and the rest said they liked one or the other. How many liked movies? Use a Venn diagram to justify your result.

42. A month-long study of 100 preschool children shows that during their free play time, 75 choose water play, 65 choose to dress up, and 25 choose to paint. Forty-three selected both water play and dressing up, 20 selected both dressing up and painting, and 15 selected painting and water play. Ten chose all three. In deciding which activity to drop, compare how many children chose only water play, only to dress up, or only to paint. How many chose none of the activities?

4.3 Counting Permutations

The Counting Principle

If a coin is tossed into the air and allowed to hit the ground, there is a 50:50 chance that it will turn up heads. The theory of chance events (such as flipping a coin) is studied in a branch of mathematics called *probability*. The basic concepts of probability depend on our ability to count accurately the number of possible outcomes for a given event. In this section, we show how to count in a systematic way.

To illustrate the counting process, suppose that you manage an Astro automobile dealership. The Astro is available in orange, yellow, or blue exterior paint, with or without air conditioning, and with AM, AM/FM, or AM/FM/tape audio options. How many cars must you order if you wish to have one of each possible option on the lot? The choices can be displayed graphically in a **tree diagram,** as shown in Figure 4.12. Starting at point C, we represent the three color options

Figure 4.12

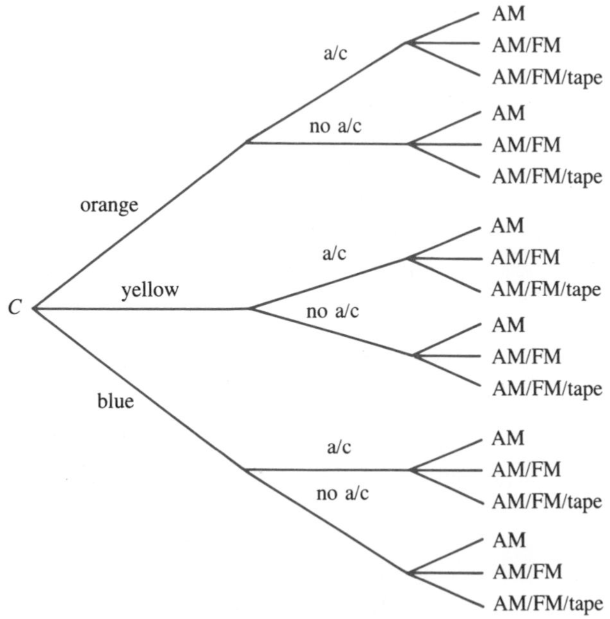

by separate lines, or branches, of the tree. The next level of branches represents the air-conditioning option, and, finally, the last level of branches represents the audio option. Each distinct path represents a different car, and you can see that there are 18 cars from which to choose.

Tree diagrams can be used to count outcomes of events whenever the events can be considered a succession of two or more independent subevents. Thus, choosing a new car can be thought of as successive events of choosing color, air conditioning, and audio equipment. Similarly, tossing three coins can be considered as tossing one coin three times.

Example 1 Draw a tree diagram showing the number of possible outcomes of tossing three coins.

Solution Since tossing three coins at the same time can be considered as tossing the same coin three times, we can readily draw the tree diagram shown in Figure 4.13. Notice that there are eight possible outcomes or paths. ■

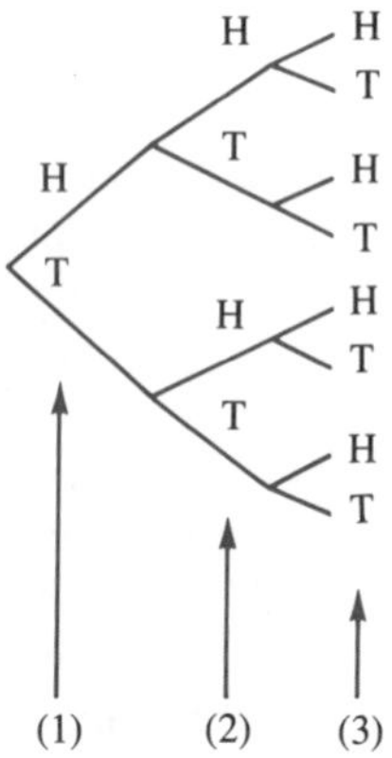

Figure 4.13

Tree diagrams can be used to count possible outcomes of events, as illustrated by Figures 4.12 and 4.13, but the process can be somewhat cumbersome. The following principle allows us to determine the number of outcomes of an event more easily than by drawing a tree diagram.

The Fundamental Principle of Counting

If one event can occur in h ways, and if after the first event, a second event can occur in k ways, then the two events can occur together in $h \cdot k$ ways.

The principle is stated for two events but can be generalized to include any finite number of successive events. Although the fundamental principle of counting can be used to find the number of possible outcomes of successive events, it does not tell us anything about the form of the outcomes.

Example 2 Use the fundamental principle of counting to determine the number of different possible types of cars, assuming that each car can have one of three exterior colors, air conditioning or no air conditioning, and one of three types of audio systems.

Solution Each different car has one color, air conditioning or no air conditioning, and one type of audio system. Since there are three color options, two air-conditioning options, and three audio options, the number of different cars is given by $3 \cdot 2 \cdot 3 = 18$. (This agrees with the number indicated in Figure 4.12.) ■

Example 3 How many different patterns are possible for the answers to a true/false test consisting of ten questions?

Solution Assuming each answer to be independent of the others, there are two possible choices for each answer. The total number of patterns is

$$2 \cdot 2 \cdot 2 \cdot 2 \cdot 2 \cdot 2 \cdot 2 \cdot 2 \cdot 2 \cdot 2 = 2^{10} = 1024$$

■

Example 4 How many five-letter "words" can be made from the letters *a*, *b*, *c*, *d*, and *e* if none of the letters can be repeated? A "word" is any arrangement of the five letters, whether it makes sense or not. Thus, *abcde* and *cbaed* are words under this definition. Notice that *abcdd* is not a valid word in this case, because *d* is repeated.

Solution It may be helpful to think of composing the words by inserting the letters into predetermined spaces or registers, much like those shown here.

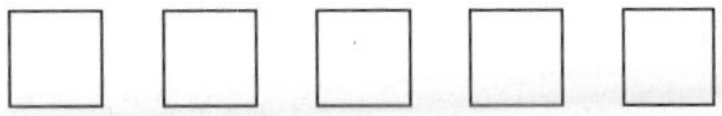

Then, if we start by filling the register on the left, there are five letters that can be placed in this space; that is, we can fill this space in five different ways. Once the first space has been filled, only four letters remain to be used in the next space, because we cannot reuse the letter we placed in the first space. Continuing in this manner, we fill each of the five spaces. The number of words that can be generated is then determined by the counting principle to be

$$5 \cdot 4 \cdot 3 \cdot 2 \cdot 1 = 120$$

Thus, there are 120 five-letter words that can be generated from the letters *a*, *b*, *c*, *d*, and *e*. Notice that if the letters could be repeated there would be 5^5 words, because then there would be five possible letters for each position. ■

The product $5 \cdot 4 \cdot 3 \cdot 2 \cdot 1$, which we encountered in Example 4, is typical of products that occur in many counting problems. This product consists of the integers from 1 to 5. The product of the integer 5 and each of the positive integers less than 5 is called **factorial 5,** or **5 factorial,** and is denoted by 5!. We defined the factorial of a positive integer *n* as follows:

DEFINITION Factorial *n*

If *n* is a positive integer, then

$$n! = n(n-1)(n-2) \cdots 3 \cdot 2 \cdot 1$$

In general, *n*! has the following property:

$$n! = n \cdot (n-1)!$$

Thus, $9! = 9 \cdot 8!$. This formula is used frequently to simplify expressions that contain factorials.

COMMENT: *Our definition of $n!$ assumes that n is a positive integer, and we noted that $n! = n(n-1)!$. If this formula is to hold for $n = 1$, we need to have $0! = 1$, because $1! = 1$ and therefore*

$$1! = 1(1-1)! = 1 \cdot 0!$$

must equal 1. Although it might seem natural to have $0! = 0$, it is necessary to have $0! = 1$, because we want the formula $n! = n(n-1)!$ to hold for all positive integers. Therefore, we make the definition:

DEFINITION

$$0! = 1$$

Example 5 Evaluate each of the following:

(a) $4!$
(b) $7!$
(c) Use (b) to evaluate $10!$
(d) $\dfrac{10!}{7!}$

Solution

(a) $4! = 4 \cdot 3 \cdot 2 \cdot 1 = 24$
(b) $7! = 7 \cdot 6 \cdot 5 \cdot 4 \cdot 3 \cdot 2 \cdot 1 = 5040$
(c) Since $10!$ is the product of 10 and all positive integers less than 10, and $7!$ is the product of 7 and all of the integers less than 7, we can write $10!$ as

$$10! = 10 \cdot 9 \cdot 8 \cdot 7! = (720)(5040) = 3{,}628{,}800$$

(d) The technique used in (c) is particularly helpful when dividing one factorial number by another. To divide $10!$ by $7!$, we write

$$\frac{10!}{7!} = \frac{10 \cdot 9 \cdot 8 \cdot 7!}{7!} = 10 \cdot 9 \cdot 8 = 720$$

■

CALCULATOR COMMENT: *Most scientific calculators can compute factorials. The value of $60!$ is found here using the* [n!] *key.*

$$60! = 8.321 \times 10^{81}$$

Factorial n is usually a second function key. Try your calculator to see how it does factorials.

Example 6 How many ways can Lou Piniella arrange nine players on a baseball team on the field if every player can play each of the nine positions?

Solution There are nine successive independent decisions to be made. The first player can be placed in any of the nine positions, the second player in any of the remaining eight positions, the third in any of the remaining seven positions, and so on. Thus, the total number of ways in which Lou can arrange the players is given by the product

$$9 \cdot 8 \cdot 7 \cdot 6 \cdot 5 \cdot 4 \cdot 3 \cdot 2 \cdot 1 = 9! = 362{,}880$$

■

An important generalization of the previous example given by the next rule is:

Arrangements in a Definite Order Without Repetition

The number of ways a set of n objects can be arranged in a definite order without repetition is equal to $n!$

Another important type of counting problem is concerned with the number of groups of k objects that can be formed from a set of n objects, where $k < n$. Suppose, for example, that you want to form a committee of two from a group of five candidates. In how many ways can this be done if the first person picked is the chairperson? Since there are five ways to pick the first person and four ways to pick the second, the fundamental counting principle assures us that there are $5 \cdot 4 = 20$ possible committees. To clarify this, let A, B, C, D, and E represent the five persons. Then the 20 committees are AB, AC, AD, AE, BA, BC, BD, BE, CA, CB, CD, CE, DA, DB, DC, DE, EA, EB, EC, and ED. The committees AB and BA are different committees because A is the chairperson of AB and B is the chairperson of BA. Sometimes the order in which elements are arranged is not important. For instance, if the two-person committee is formed without designating a chairperson, only half as many committees can be formed, because the committee consisting of A and B is the same as that consisting of B and A.

The two ways of forming committees described in the above paragraph are typical of the ways in which sets of objects are arranged; that is, arrangements in which order is important and arrangements in which order is not important. An arrangement without repetition of a set of objects in which the order of the objects *is* important is called a **permutation.** An arrangement without repetition of a set of objects in which the order of the objects *is not* important is called a **combination.**

Example 7

(a) List the permutations of the letters x, y, z.
(b) List the combinations of x, y, z.

Solution

(a) There are six permutations of x, y, z; namely, xyz, xzy, yxz, yzx, zxy, and zyx. This verifies that the number of permutations of the three letters is equal to $3! = 6$.
(b) There is only one combination of x, y, z; namely, xyz. The other five permutations of xyz are considered to be the same combination. They differ from each other only in the order of the letters. ■

PERMUTATIONS OF n OBJECTS TAKEN k AT A TIME

We have counted the number of two-person committees that can be formed from a list of five candidates, in which the first person chosen is the chairperson of the committee; in the jargon of counting, we say *we counted the number of permutations of five people taken two at a time.* By the number of permutations of n objects taken k at a time, we mean the number of possible distinct arrangements consisting of k objects. The number of permutations of n objects taken k at a time is denoted by $P_{n,k}$. To find a formula for $P_{n,k}$, we note that the first object can be chosen from any of the n objects in the set, the second from any of the $n-1$ objects remaining after the first object is chosen, the third from any of the $n-2$ objects remaining after the second has been chosen, and so on until the kth object is chosen from the $n-(k-1)$ objects remaining after the $(k-1)$th object has been chosen. Using the fundamental counting principle, we get

$$P_{n,k} = n(n-1)(n-2)\cdots(n-k+1)$$

This formula is usually modified by multiplying the right-hand side by the expression $(n-k)!/(n-k)!$. Thus,

$$P_{n,k} = \frac{n(n-1)(n-2)\cdots(n-k+1)(n-k)!}{(n-k)!}$$

You should be able to convince yourself that $n(n-1)(n-2)\cdots(n-k+1)\times(n-k)! = n!$. Therefore, the number of permutations of n objects taken k at a time is given by the following formula:

The Number of Permutations of n Objects Taken k at a Time

If n and k are positive integers and $k < n$, then the number of permutations of n objects taken k at a time is denoted by $P_{n,k}$ and given by the formula

$$P_{n,k} = \frac{n!}{(n-k)!}$$

In the special case for which $k = n$, we have, since $0! = 1$,

$$P_{n,n} = n!$$

Since $P_{n,n}$ is the number of ways n objects can be arranged in a definite order, this agrees with the result stated after Example 6.

COMMENT: *On some calculators $P_{n,k}$ is denoted by ${}_nP_k$.*

Example 8 If there are 15 entries in an art contest, in how many ways can the judge award the first, second, and third prizes?

Solution Since order is important in awarding the prizes, we want the number of permutations of 15 entries taken 3 at a time. The following computation shows that there are 2730 ways of awarding the three prizes:

$$P_{15,3} = \frac{15!}{(15-3)!} = \frac{15!}{12!} = 15 \cdot 14 \cdot 13 = 2730$$

Alternately, observe that the first prize can be awarded to any one of the 15 entries. After the first prize is awarded, there remain 14 entries that can be awarded the second prize, and when the second prize has been awarded there are 13 entries that can be awarded the third prize. So, by the counting principle, there are $15 \cdot 14 \cdot 13 = 2730$ ways to award the three prizes. ■

Example 9 In how many ways can five cards be dealt from a deck of 52 playing cards if order is important—that is, if 2H, 3H, 7S, JC, KC is different from 2H, 7S, JC, 3H, KC? (H, C, S, and D represent the four suits, hearts, clubs, spades, and diamonds.)

Solution Here, we want the number of permutations of 52 cards taken five at a time, so

$$P_{52,5} = \frac{52!}{(52-5)!} = \frac{52!}{47!} = 52 \cdot 51 \cdot 50 \cdot 49 \cdot 48 = 311{,}875{,}200$$

We note that this is not the number of distinct poker hands, since in the game of poker the order in which the cards are dealt to any one hand is not important. The number 311,875,200 is the number of ways of dealing five cards successively from a deck of 52 playing cards. ■

Example 10 Stockbrokers often send lists of "top ten" stocks to their clients by choosing randomly from a New York Stock Exchange list of 35 stocks. In how many ways can this "top ten" list be arranged for the clients?

Solution Here, we assume that order is important, since a client will notice whether, for example, General Motors is at the top or at the bottom of the list. In this case, the number of permutations of 35 stocks taken 10 at a time is

$$P_{35,10} = \frac{35!}{(35-10)!} = \frac{35!}{25!} > 6 \times 10^{14}$$

Since there are more than 6×10^{14} possible lists, the stockbroker should have no trouble generating many distinctive lists! ■

EXERCISES SECTION 4.3

(Use calculators when possible, particularly for evaluating factorials and permutations.)

1. If there are seven boys in a race, how many ways can second place be filled once first place has been decided? How many ways can third place be filled once first and second places have been decided?
2. List all permutations of A, B, C, D, taken three at a time.
3. List all permutations of the letters w, x, y, z, taken two at a time.
4. How many permutations are there of the letters in the word *angle*?

5. Find the number of permutations of seven objects taken four at a time.

6. Evaluate $P_{9,7}$.

7. How many two-digit numbers can be formed from the numbers 1, 2, 3, 4, 5, 6 if no digits can be repeated?

8. If five women enter a 10-km run, in how many ways can three of them finish first, second, and third?

9. In how many ways can ten people line up at a box office?

10. In how many ways can four political candidates be listed on the ballot?

11. How many batting orders of nine players can a baseball manager make up from a roster of 15 players?

12. Four persons enter a room in which there are seven chairs. In how many ways can they be seated? (Assume one chair per person.)

13. Find the total number of four-digit integers that can be formed from the digits 1, 2, 3, and 4. Assume that no digit is repeated in any one integer.

14. Sorority and fraternity names are selected from the 24-letter Greek alphabet. How many such two- or three-letter names can be formed if letters are not to be repeated within any name?

15. A computer can turn out "words" by randomly selecting letters from a list, even though the combination of letters might not make sense. How many such three- or four-letter words can be made with the letters of the word *slope* if letters are not repeated within a word?

16. The President of the U.S. and eight of his cabinet officers enter a room with a large oval table. If the President's spot is fixed by custom, in how many ways can the others be arranged?

17. A choreographer for eight cheerleaders wants to form a circle. If the eight positions on the circle are designated by numbers, in how many ways can they be arranged to form the circle?

18. An insurance broker has a list of six different policies for health insurance and ten different life insurance policies. In order to simplify matters for her clients she wants to narrow the list to a "top five" group of life insurance policies and a first and second choice for health insurance. In how many ways can she do this?

19. A social worker has eight clients, but she can only visit six of them each week. In how many weeks can she carry out her visits without repeating a list of weekly visits in exactly the same order?

20. The president of an auto agency has a list of eight people to fill the five vice-presidencies. The vice-presidents are not ranked by number, but from their titles everyone knows that the vice-presidents are essentially ranked, from Senior Vice-President down to Vice-President in Charge of Loans. In how many ways can the president make up this list?

21. Given that $P_{n,3} = 9P_{n,2}$, find n.

22. Given that $P_{n,2} = 56$, find n.

23. Show that $P_{20,2} = 19P_{5,2}$.

24. Show that $P_{4,1} + P_{4,2} + P_{4,3} + P_{4,4} = 4^3$.

4.4 COUNTING COMBINATIONS

In the previous section, we considered the process of counting permutations—that is, the number of *ordered* arrangements, without repetition, of k objects taken from a set of n objects, where $k \leq n$. In this section, we consider the process of counting combinations—that is, the number of *unordered* arrangements, without repetition, of k objects from a set of n objects. Some examples of unordered arrangements are:

- In the quarter finals and finals of the Miss America Contest, the top contestants are listed without regard to order.

- A hand in a game of bridge is unordered, because it makes no difference in which order the hand is dealt. Further, the hand can be shuffled without changing the nature of the hand.
- The list of guests at a party is usually unordered, because one is ordinarily not concerned with giving priority to the people as they arrive at the party.

We are interested in counting the number of combinations of n objects taken k at a time. The next two examples approach this problem by listing all of the possible combinations and then counting them.

Example 1 List and count the three-letter combinations that can be formed from w, x, y, and z.

Solution There are four three-letter combinations that can be formed from w, x, y, and z: wxy, wxz, wyz, and xyz. Of course, in considering the number of combinations, we assume that the objects being counted are not allowed to be repeated. Thus, xxy is not a valid three-letter combination of w, x, y, and z. ■

Example 2 Two new members are to be added to a university committee charged with recommending ways to improve attendance at late-afternoon classes. List and count the two-member combinations that can be added to this committee if the five eligible candidates are Sam, Pete, Ann, Lou, and Joe.

Solution We assume that order is not important, since the addition of Pete and Ann is the same as that of Ann and Pete. Consequently, we want the number of combinations of five candidates taken two at a time. The specific combinations are (Sam, Pete), (Sam, Ann), (Sam, Lou), (Sam, Joe), (Pete, Ann), (Pete, Lou), (Pete, Joe), (Ann, Lou), (Ann, Joe), and (Lou, Joe). Counting the pairs, we see that there are ten combinations of five candidates taken two at a time. ■

The process of finding the number of combinations of objects by listing all of the possible combinations is instructive, but is impractical for large numbers of objects. Therefore, we wish to develop a general formula for the number of combinations of n objects taken k at a time. To do so, we use the formula for the number of permutations of n objects taken k at a time and the fundamental principle of counting. Let $C_{n,k}$ denote the number of combinations of n objects taken k at a time. Since each of the combinations can be permuted in $P_{k,k} = k!$ ways, it follows by the definition of permutations and the fundamental principle of counting that the number of permutations of these objects is given by

$$P_{n,k} = (C_{n,k})k!$$

Solving for $C_{n,k}$, we have

$$C_{n,k} = \frac{P_{n,k}}{k!}$$

Thus, the number of combinations of n objects taken k at a time is equal to the number of permutations divided by $k!$. Replacing $P_{n,k}$ with $n!/(n-k)!$, we get the combination formula.

The Number of Combinations of n Objects Taken k at a Time

If n and k are positive integers and $k < n$, the number of combinations of n objects taken k at a time is given by the formula

$$C_{n,k} = \frac{n!}{k!(n-k)!}$$

(In the special case in which $n = k$, we have $C_{n,n} = \dfrac{n!}{n!0!} = 1$.)

COMMENT: *From the formula for $C_{n,k}$ it follows immediately that $C_{n,k} = C_{n-k,k}$.*

Example 3 From a hat with ten names in it, five are to be drawn without being replaced.

(a) In how many ways can the five names be drawn?
(b) If the first name drawn receives \$5000, the second \$4000, and so on, in how many ways can the five names be drawn?

Solution

(a) When the order in which the names are drawn is not important, we want the number of combinations of ten names taken five at a time. This is given by

$$C_{10,5} = \frac{10!}{5!(10-5)!} = \frac{10!}{5!(5)!} = \frac{10 \cdot 9 \cdot 8 \cdot 7 \cdot 6}{5 \cdot 4 \cdot 3 \cdot 2 \cdot 1} = 252$$

(b) Since the first name drawn is worth \$5000, the second \$4000, and so on, the order is important, so we want the number of permutations of the ten names taken five at a time. This is given by

$$P_{10,5} = \frac{10!}{5!} = 10 \cdot 9 \cdot 8 \cdot 7 \cdot 6 = 30{,}240$$

■

Example 4 A coin is tossed five times and the number of heads is recorded. How many ways can you get three heads in five tosses?

Solution In tossing the coin and counting the number of heads, the order is not important. So we consider

$$C_{5,3} = \frac{5!}{3!2!} = \frac{5 \cdot 4}{2 \cdot 1} = 10$$

■

Example 5 In the card game of draw poker each player is dealt five cards from a standard deck of 52 playing cards.

(a) In how many different ways can one five-card hand be dealt?
(b) In how many ways can a five-card hand containing all clubs, called a *flush,* be dealt?
(c) In how many ways can a five-card hand containing a flush of any suit be dealt?

Solution

(a) The order in which the cards are dealt is not important, so the number of five-card hands that can be dealt to one player from a deck of 52 playing cards is determined by the combination formula to be

$$C_{52,5} = \frac{52!}{5!(52-5)!} = \frac{52!}{5!(47)!} = \frac{52 \cdot 51 \cdot 50 \cdot 49 \cdot 48}{5 \cdot 4 \cdot 3 \cdot 2 \cdot 1} = 2{,}598{,}960$$

Thus, 2,598,960 distinct five-card hands can be dealt.

(b) There are 13 clubs in a deck, so the number of ways a hand can contain five clubs is

$$C_{13,5} = \frac{13!}{5!(13-5)!} = \frac{13!}{5!(8)!} = \frac{13 \cdot 12 \cdot 11 \cdot 10 \cdot 9}{5 \cdot 4 \cdot 3 \cdot 2 \cdot 1} = 1287$$

(c) To find out how many ways a flush can be dealt, we note that there are 1287 ways in which a flush of any suit can be dealt. Since there are four suits, it follows from the counting principle that there are 4(1287) = 5148 ways in which a flush can be dealt. ▪

As we found in part (c) of the previous example, we can use the counting principle in conjunction with the combination formula to count the number of possible outcomes of a given event.

Example 6 There are seven women and five men in the National Honor Society at the local high school. A committee of three women and two men is formed to plan the induction ceremony. How many ways can the committee be formed?

Solution Order is not important in forming a committee, so the problem is one of selecting *combinations* of people to serve. The three women can be chosen in 35 ways, since

$$C_{7,3} = \frac{7!}{3!(7-3)!} = \frac{7 \cdot 6 \cdot 5}{3 \cdot 2 \cdot 1} = 35$$

The two men can be chosen in ten ways, since

$$C_{5,2} = \frac{5!}{2!(5-2)!} = \frac{5 \cdot 4}{2 \cdot 1} = 10$$

Since the three women can be chosen in 35 ways and the two men in ten ways, we use the counting principle to conclude that the committee of five can be chosen in $35 \cdot 10 = 350$ ways. ▪

Example 7 A baseball team has a roster of three catchers, nine pitchers, six infielders, and seven outfielders. In how many ways can the manager arrange the nine-player starting lineup? A nine-player team consists of one catcher, one pitcher, four infielders, and three outfielders.

Solution First we calculate the number of ways each position can be filled:

$$\text{Number of ways of choosing a catcher } C_{3,1} = 3$$
$$\text{Number of ways of choosing a pitcher } C_{9,1} = 9$$
$$\text{Number of ways of choosing an infielder } C_{6,4} = 15$$
$$\text{Number of ways of choosing an outfielder } C_{7,3} = 35$$

Thus, by the counting principle, the number of possible starting lineups is

$$3 \cdot 9 \cdot 15 \cdot 35 = 14{,}175$$

EXERCISES SECTION 4.4

In Exercises 1–4, evaluate each combination.

1. $C_{4,2}$
2. $C_{5,3}$
3. $C_{7,2}$
4. $C_{9,5}$
5. How many combinations of four players can be selected from a roster of 15?
6. How many combinations of three can be selected from 12 different coins?
7. How many seven-card hands can be obtained from a deck of 52 playing cards?
8. From a committee of eight, how many subcommittees of three can be selected?
9. Six coins are tossed onto a tabletop. In how many ways can six heads appear? five heads? four heads?
10. In a ten-team basketball league, how many games must be scheduled if each team plays one game with every other team in the league? How many games must be scheduled if each team plays two games with every other team?
11. Big Al offers the following toppings for his pizzas: pepperoni, sausage, ham, onions, peppers, olives, and anchovies. How many different two-topping pizzas are available from Big Al's menu?
12. The housing office at Iowa State University is assigning 150 first-year women to 75 dormitory rooms. In how many ways can the assignments be made?
13. A computer program is designed to make arbitrary four-letter arrangements from some given word. How many four-letter combinations ending in a vowel can be made from the word *vowel*?
14. Committees are being formed from the members of your student government association. How many committees of four men and two women can be selected from ten male candidates and five female candidates?
15. At the local YMCA a picnic committee consisting of three women and two men is formed from six married couples. In how many ways can this be done?
16. At the office, a bowling team of two men and two women is selected from eight men and seven women. In how many ways can this be done?
17. A coin collector needs to select three dimes and four quarters from a collection of ten dimes and eight quarters. In how many ways can this be done?
18. A stockbroker, advising a client on a portfolio, suggests seven individual growth stocks and ten oil stocks. How many portfolios can the stockbroker develop if each portfolio is to have three growth stocks and two oil stocks?

19. Julie and Ben are preparing for a party at which beer, wine, and soft drinks will be served. Suppose there are six kinds of beer to choose from, five kinds of wine, and ten types of soft drinks. In how many ways can the drinks be selected if Julie and Ben decide to have two types of beer, two kinds of wine, and four types of soft drinks available to the guests?

20. Dan and Kristin are team-teaching a third-grade class of 23 students: thirteen girls and ten boys. Each week they choose four boys and four girls as "helpers." How many ways can Dan and Kristin choose these eight helpers?

21. Find x if $C_{x,3} = 2C_{x,2}$.

22. If $P_{n,r} = 120$ and $C_{n,r} = 20$, what are n and r?

23. An old mathematical puzzle asks how many diagonals can be drawn in a 24-sided polygon. You should be able to solve this puzzle!

4.5 BINOMIAL EXPANSIONS

Consider tossing a coin three times and recording the sequence of heads and tails. The possible results are HHH, HHT, HTH, THH, TTH, THT, HTT, and TTT. If we consider each outcome as an algebraic product of the H's and T's, the outcome HHH can be thought of as H^3, the outcomes HHT, HTH, and THH can be thought of as H^2T, and so on, as shown in Figure 4.14.

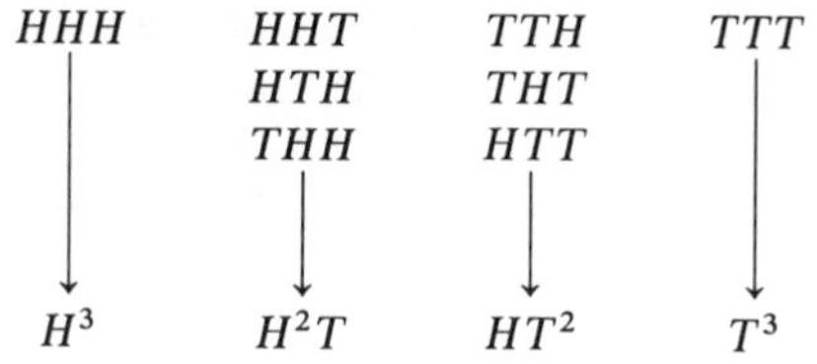

Figure 4.14

We note that these terms occur in the expansion of $(H + T)^3$; that is,

$$(H + T)^3 = H^3 + 3H^2T + 3HT^2 + T^3$$

We note also that the coefficient of each term in the expansion of $(H + T)^3$ corresponds to the number of times a particular sequence of heads and tails occurs. This relationship between the coefficients of the terms in the expansion of $(H + T)^3$ and the number of times heads or tails occurs in three tosses of a coin is closely related to our previous study of combinations and to the study of probability in the next chapter.

BINOMIAL COEFFICIENTS

A *binomial* is an indicated sum of two terms, usually written as $x + y$, but x and y may themselves be longer expressions. Thus, $3a + 2b$ and $2x^3 - 5y^2$ are considered to be binomial expressions. In this section, however, we will deal only with binomials of the form $(x + y)^n$. Binomials are often raised to powers and the indicated multiplications of the factors are carried out, or *expanded*. Hence the title of this section, **binomial expansions.** Here are the binomial expansions of $(x + y)^n$ for n from 0 to 5:

Figure 4.15

$$(x + y)^0 = 1$$
$$(x + y)^1 = x + y$$
$$(x + y)^2 = x^2 + 2xy + y^2$$
$$(x + y)^3 = x^3 + 3x^2y + 3xy^2 + y^3$$
$$(x + y)^4 = x^4 + 4x^3y + 6x^2y^2 + 4xy^3 + y^4$$
$$(x + y)^5 = x^5 + 5x^4y + 10x^3y^2 + 10x^2y^3 + 5xy^4 + y^5$$

In this section we develop a formula which is applicable to any integral power of a binomial, such as $(x + y)^9$. This important formula will allow you to easily expand any binomial of the form $(x + y)^n$ and also to determine any particular term within the expansion. The key in this discussion will be a general formula for the coefficients in the expansion, called the **binomial coefficients.** In the binomial expansions in Figure 4.15, the binomial coefficients in the expansion of $(x + y)^2$ are 1, 2, 1; in the expansion of $(x + y)^3$ they are 1, 3, 3, 1; in the expansion of $(x + y)^4$ they are 1, 4, 6, 4, 1; and for $(x + y)^5$ they are 1, 5, 10, 5, 1.

As noted earlier, the binomial coefficients of $(H + T)^3$ will tell us the number of times that a given sequence of heads and tails will occur when a coin is tossed three times. In fact, it can be shown that the binomial coefficients of $(H + T)^n$ correspond to the number of times that a given sequence of heads and tails will occur when a coin is tossed n times. Thus, the determination of these coefficients finds a significant application in counting problems. We wish to show how to evaluate the binomial coefficients without actually expanding $(H + T)^n$.

Careful observation of the expansions in Figure 4.15 reveals these interesting patterns:

1. There are always $n + 1$ terms; the first is $x^n y^0$ and the last is $x^0 y^n$.
2. Each intermediate term contains a product of $x^j y^k$, and the sum of j and k in every term is n. Thus, the product is of the form $x^{n-k} y^k$.
3. The exponent of x decreases by 1 from one term to the next, whereas the exponent of y increases by 1.
4. The coefficients follow a symmetric pattern; that is, if we follow the coefficients from either end of the expansion, we find that they match the coefficients followed from the other end (in the reverse direction).
5. It is difficult to guess the actual formula for the coefficients with such limited information. However, the coefficient of the second term is $n/1$, the coefficient of the third term is $n(n - 1)/2$, and the coefficient of the fourth term is $n(n - 1)(n - 2)/1 \cdot 2 \cdot 3$.

In a more formal treatment of taking the general power of a binomial, we would show that the coefficient of $x^{n-k} y^k$ in the expansion of $(x + y)^n$ is given by the following formula:

$$\text{Coefficient of } x^{n-k}y^k = \frac{n(n-1)(n-2)\cdots(n-k+1)}{1\cdot 2\cdot 3\cdot 4\cdot\cdots\cdot k} \quad \text{if } k \neq 0 \qquad (4.3)$$
$$= 1 \quad \text{if } k = 0$$

Formula 4.3 for the binomial coefficient may also be recognized as the number of combinations of n things taken k at a time. Thus, we may also write

$$\text{Coefficient of } x^{n-k}y^k = C_{n,k} = \frac{n!}{k!(n-k)!} \qquad (4.4)$$

The binomial coefficient depends not only on the order of the expansion, n, but also on the number of the term within that expansion, $k + 1$. Thus, the general expression for an expanded binomial consists of terms of the form

$$C_{n,k}x^{n-k}y^k = \frac{n!}{k!(n-k)!}x^{n-k}y^k, \quad \text{where } k = 0, 1, 2, \ldots, n.$$

For example, the sixth term (that is, $k = 5$) in the expansion of the binomial $(x + y)^9$ is

$$C_{9,5}x^4y^5 = \frac{9!}{4!5!}x^4y^5 = 126x^4y^5$$

Example 1 Find the coefficient of x^3y^7 in the expansion of $(x + y)^{10}$.

Solution Let A be the coefficient of x^3y^7; then the desired term of the expansion is

$$Ax^3y^7 = C_{10,7}x^3y^7 = \frac{10!}{7!3!}x^3y^7 = 120x^3y^7$$

Because of the symmetry of the binomial coefficients, A could also be evaluated as $A = C_{10,3} = 120$.

■

Example 2 Toss eight coins and record the number of heads and tails. How many ways could five heads and three tails occur?

Solution Tossing eight coins once and recording the number of heads and tails is the same mathematical problem as tossing one coin eight times and recording the number of heads and tails. In either case, there are 2^8 possible results. The number of ways in which five heads and three tails can occur is equal to the binomial coefficient of H^5T^3 in the expression $(H + T)^8$. That is,

$$C_{8,3} = \frac{8!}{5!3!} = 56$$

■

Example 3 On a true/false test of 15 questions, how many ways can you answer six false and nine true? (Assume you are just guessing at the answers.)

Solution There are 2^{15} possible sequences of answers, including none true, only one true, only two true, and so on. We are interested in the sequences that have nine true and six false answers. Or, in the notation we used for heads and tails, we are concerned with the coefficient of T^9F^6. That coefficient is $C_{15,9}$ or $C_{15,6} = 5005$. ■

Example 4 Toss one coin seven times and record the number of heads and tails. How many ways can five or more heads occur?

Solution The number of ways five or more heads can occur is the number of ways exactly five heads can occur plus the number of ways exactly six heads can occur plus the number of ways exactly seven heads can occur. That is,

$$C_{7,5} + C_{7,6} + C_{7,7} = \frac{7!}{2!5!} + \frac{7!}{1!6!} + \frac{7!}{0!7!} = 29$$ ■

Re-examine the coefficients in the binomial expansions of $(x + y)$ in Figure 4.15. Notice that when we write the coefficients $C_{n,k}$ for $k = 0, 1, 2, 3, 4, 5$ on successive rows, as shown below in Figure 4.16, we get a triangular array of numbers called **Pascal's triangle.** Pascal's triangle can be a useful device for remembering the binomial coefficients, since a recognizable pattern exists between numbers in successive rows of the triangle.

Figure 4.16

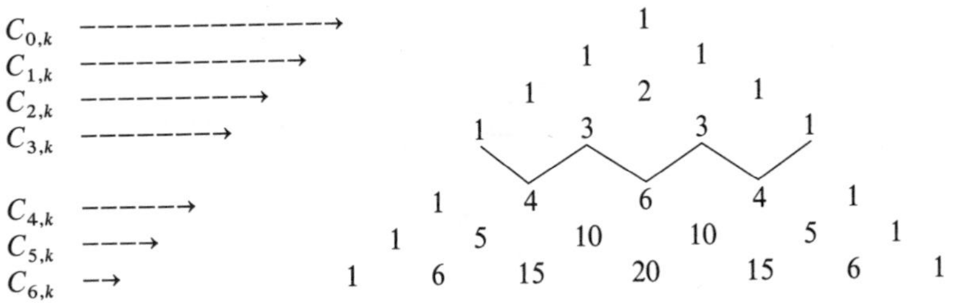

Each number interior to (between the 1's of) a row is the sum of the two numbers directly adjacent to it in the previous row. For example, the interior numbers of the fifth row are $4 = 1 + 3$, $6 = 3 + 3$, and $4 = 3 + 1$. This observation suggests the following formula:

$$C_{n+1,k+1} = C_{n,k+1} + C_{n,k}$$

COMMENT: *Using Pascal's triangle can be very cumbersome for larger values of n, for example n = 25, because in order to determine the 26th row you must first produce the previous 25 rows.*

HISTORICAL COMMENT: *This triangle was known to mathematicians long before the time of Blaise Pascal (1623–1662), the French mathematician for whom it is named.*

EXERCISES SECTION 4.5

Expand and simplify the expressions in Exercises 1–4 using the binomial formula.

1. $(x + y)^8$
2. $(a + b)^5$
3. $(H + T)^{12}$
4. $(T + H)^{10}$
5. Find the fourth term in the expansion of $(a + b)^7$.
6. Find the sixth term in the expansion of $(x + y)^{11}$.
7. Find the twelfth term in the expansion of $(H + T)^{17}$.
8. Find the fifteenth term in the expansion of $(T + H)^{20}$.
9. Find the coefficient of x^5y^3 in $(x + y)^8$.
10. Find the coefficient of H^3T^8 in $(H + T)^{11}$.
11. Find the coefficient of a^7b^2 in $(a + b)^9$.
12. Find the coefficient of T^2H^{15} in $(T + H)^{17}$.
13. $1287x^5y^8$ is a term in the expansion of $(x + y)^n$ for what value of n?
14. $55a^2b^n$ is a term in the expansion of $(a + b)^{11}$ for what value of n?
15. If you toss a coin five times, how many possible outcomes can you have with (a) five heads, (b) three heads and two tails, and (c) four tails and one head?
16. If you toss a coin nine times, how many possible outcomes can you have with (a) seven heads and two tails, (b) three heads and six tails, and (c) four heads and five tails?
17. If you toss a coin eight times, how many possible outcomes can there be with more than six heads?
18. If you toss a coin eleven times, how many possible outcomes can there be with more than seven heads?
19. If you toss a coin fifteen times, how many possible outcomes can there be with fewer than five heads?
20. If you toss a coin five times, how many possible outcomes can there be with fewer than three tails?
21. If you toss a coin fourteen times, how many possible outcomes can there be with more than five heads but fewer than nine heads?
22. If you toss a coin ten times, how many possible outcomes can there be with more than three heads but fewer than seven heads?
23. On a 20-question true/false test, how many possible ways are there to answer 12 true and 8 false? (Assume your answer are random guesses.)
24. A candy machine dispenses 15 pieces of candy, one at a time, for a quarter. The pieces are either plain chocolate or nut-filled. How many possible ways can the machine dispense 7 plain and 8 nut-filled pieces of candy to the customer?
25. Show that:
    (a) $C_{2,0} + C_{2,1} + C_{2,2} = 4$
    (b) $C_{3,0} + C_{3,1} + C_{3,2} + C_{3,3} = 8$
    (c) Show that in general:
    $$C_{n,0} + C_{n,1} + C_{n,2} + \cdots + C_{n,n} = 2^n$$
26. Using the formula for $C_{n,k}$ show that $C_{n+1,k+1} = C_{n,k+1} + C_{n,k}$.

IMPORTANT WORDS AND PHRASES CHAPTER 4

In this chapter we have explored the rich language of sets. The fundamental counting principle, together with permutations and combinations, enables us to find the cardinal numbers of some rather complicated sets in an orderly fashion. Here are some of the important words and phrases from this chapter:

associative property	counting principle	DeMorgan's laws
binomial expansions	combinations	disjoint sets
cardinal number of a set	complement of a set	distributive property

equal sets	intersection of sets	subsets
empty set	Pascal's triangle	tree diagrams
factorial	permutations	union of sets
finite set	sets	universal set
infinite set	set builder notation	Venn diagrams

REVIEW EXERCISES CHAPTER 4

For Exercises 1–10, let $U = \{1, 2, 3, 4, 5, 6, 7, 8, 9, 10\}$, $A = \{2, 3, 4\}$, $B = \{4, 5, 6\}$, *and* $C = \{2, 7, 10\}$. *Find each of the following:*

1. $A \cup B$

2. $A \cap B$

3. $(A \cup B) \cup C$

4. $(A \cup B)'$

5. $(A \cap B)'$

6. $(A \cap B) \cup C$

7. $A \cap (B \cup C)$

8. $B \cup (A \cap C)$

9. $(A \cap B)' \cup C'$

10. $(A \cap B) \cup C'$

11. In which of Exercises 1–10 may the parentheses be omitted?

12. Use set builder notation to describe the set $\{2, 4, 6, 8\}$.

13. Determine all the subsets of the set $\{1, 2, 3\}$. How many subsets does a set of n elements have?

14. Let A be the set of college students who take two or fewer mathematics courses while in college, and let B be the set of students who take one or more.
(a) What does $A \cup B$ represent?
(b) What does $A \cap B$ represent?

15. Seven members of the student council enter a room with a round table. How many different ways can they be seated at the table?

16. A stockbroker sends out a list of ten "stocks to watch" to his clients. He includes only stocks from the *Forbes* magazine Top 20, which has fifteen growth stocks and five high-income stocks. If he wants his list to contain six growth and four high-income stocks, how many different lists can he make?

17. An automobile dealer randomly invites 20 customers to his ranch who have purchased cars during the past month. During October, ten people bought model A, twelve bought model B, eight bought model C, and six bought model D. How many different groups of people can he invite to his ranch if he wants an equal number of customers for each model?

18. The local theater took a survey of the tastes of modern moviegoers with respect to some of the movies of the past. Of 150 people surveyed, 75 said they had seen *Treasure of the Sierra Madre,* 85 said they had seen *Casablanca,* and 50 said they had seen *Key Largo.* Forty-five had seen the first two, 35 had seen the second two, and 25 had seen the first and third. Fifteen said they had seen all three. Based on these numbers, how many people saw only *Key Largo*? How many people saw none of these movies?

19. What is the coefficient of x^2y^4 in the expansion of $(x + y)^6$?

20. Expand $(a + b)^{11}$.

21. If you toss a coin eight times, how many possible outcomes can there be with (a) three heads and five tails and (b) more than two heads but fewer than five heads?

22. A set of ten on/off switches controls access to a computer lab. Audra and Brian have forgotten the correct sequence of *ons* and *offs,* but they do remember that there are six *ons* in the combination. How many possible sequences are there for them to try in order to access the lab?

CHAPTER 5

PROBABILITY

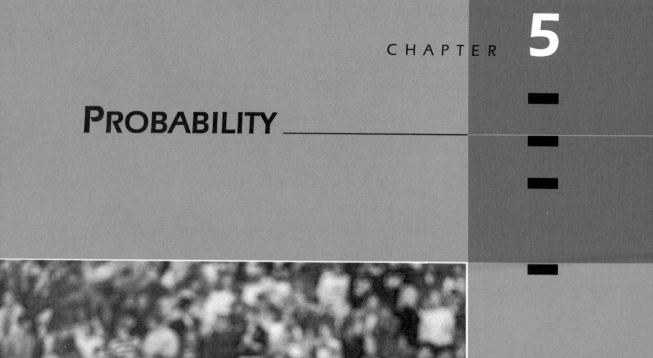

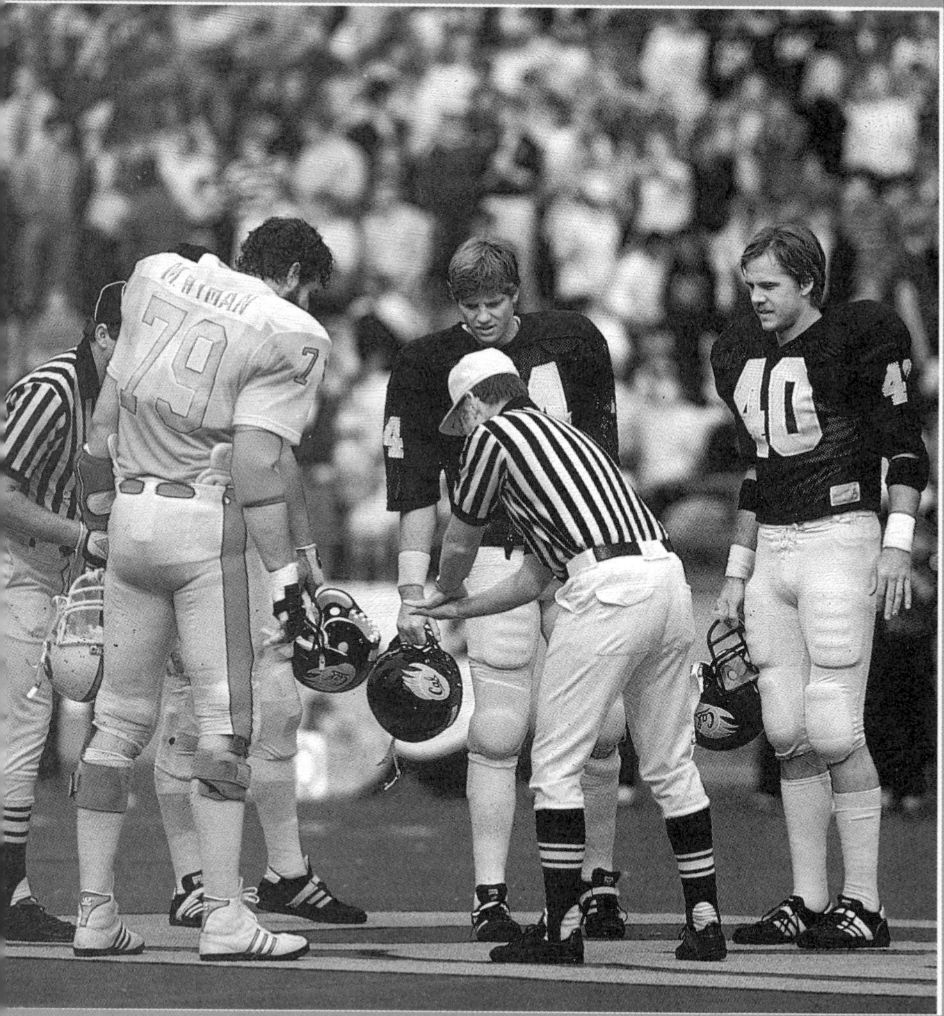

When an official at a football game flips a coin into the air, we know for certain that the coin will fall to the ground. The outcome of this action can be predicted with certainty because of gravitational attraction, which is known to hold for all objects in the universe. On the other hand, we cannot state with certainty that the coin will land heads-up. We know that landing heads-up or tails-up is a matter of chance and cannot be predicted with certainty by any law. Thus, when we flip a coin, we can be certain only that it will fall and that it will land either heads-up or tails-up. The study of chance events, such as the toss of a coin, is called **probability.** Our purpose in this chapter is to explain some of the basic ideas in this branch of mathematics.

The study of probability is not limited to the tossing of coins. Most occurrences in life can be thought of as sequences of chance events. For instance:

- Weather forecasting. In fact, the TV weather forecast usually includes the "probability" of rain or of snow.
- The dimensions of mass-produced parts.
- The chance meeting of your parents, and your subsequent birth.
- The results of public opinion polls. Polls are usually accompanied by a statement of the probable error of the results.

It is the aim of this chapter to give you some understanding of the language used to discuss chance events. Some of the words that occur early in any discussion of probability are *experiment, trial,* and *outcome.* While you may have an understanding of these words as they are used in everyday language, you must now learn their meaning when they are applied to the study of probability.

5.1

An **experiment** is the observation of any physical phenomena. This includes, but is not restricted to, what goes on in laboratories. In probability, the word *experiment* applies to such observable actions as tossing a coin, measuring the diameter of ball bearings, and taking a public opinion poll. The repetition of an experiment is called a **trial** and the possible results of a trial are called **outcomes.** If an experiment consists of tossing a coin ten times and recording whether the coin lands heads-up or tails-up, then each toss is a trial and the possible outcomes are *heads* and *tails.*

The coin referred to in the previous discussion has two sides. Our intuition tells us that if we were to perform a large number of trials of tossing the coin, we should expect heads to come up half the time and tails to come up half the time. We describe this by saying that the outcomes of the experiment are **equally likely** to occur. When outcomes of each trial are equally likely, an experiment is described as being *fair*. Unless stated otherwise, the experiments we describe are fair.

Example 1

(a) A die is a cube with its sides, or faces, marked with one to six dots, as shown here.

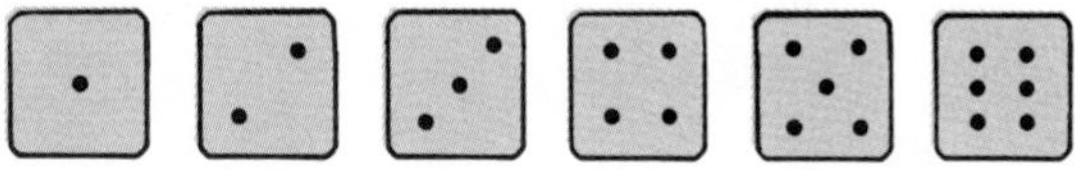

If the die is fair, each of the sides is equally likely to turn up when the die is rolled.

(b) Each of the 52 cards in an ordinary deck of playing cards is equally likely to be dealt from the deck if the deck is thoroughly shuffled before the card is dealt. ■

If an experiment is repeated a large number of times, the ratio of the number of times a certain outcome occurs to the number of trials performed is called the **relative frequency** of that outcome. Thus, if we toss a coin 1000 times and *heads* is recorded on 491 of the 1000 trials, the relative frequency of *heads* is

$$\frac{491}{1000} = 0.491$$

DEFINITION Relative Frequency of an Outcome

If an experiment is repeated n times and a certain outcome occurs k times, then the **relative frequency** of the outcome is

$$\frac{k}{n}$$

In the study of probability, we consider both the relative frequencies of outcomes and the generation of models that will predict the relative frequency of an outcome if an experiment is repeated *a large number of times.* A model that predicts relative frequency is called a **probability model.** In constructing probability models for a given outcome we want to find a number p, called the **probability** of the outcome, that will predict the relative frequency of the outcome. Thus, if the relative frequency of an outcome is $\frac{k}{n}$, then

$$p \approx \frac{k}{n}$$

when the number of trials n is large. The larger n is, the more accurately p should predict the value of $\frac{k}{n}$. This is consistent with our intuitive understanding of the meaning of probability. Referring again to the experiment of tossing a coin, we assign the probability of $\frac{1}{2}$ to the outcome *heads,* since we anticipate that about half the outcomes should be *heads* when the experiment is repeated *a large number of times.*

Example 2 Assign probabilities to each of the following outcomes:

(a) A 2 is obtained on the roll of a single die.
(b) A club is drawn from a deck of 52 playing cards.
(c) A red ball is drawn from a box that contains three red balls and four green balls.
(d) A given type of weather pattern is followed by rain about once every ten times it occurs.

Solution

(a) A die is a cube with six sides, or faces, numbered 1, 2, 3, 4, 5, and 6. Assume we perform the experiment of rolling the die a large number of times, and for each trial we record the number showing on the top face of the die. Since each side is equally likely to turn up when the die is rolled, we expect about $\frac{1}{6}$ of the recorded numbers to be 2's. Therefore, we assign a probability of $\frac{1}{6}$ to the outcome of rolling a 2. By a similar argument, we would assign a $\frac{1}{6}$ probability to each of the six possible outcomes.
(b) A standard deck of 52 playing cards is divided equally into four suits: clubs, diamonds, hearts, and spades. Since one-quarter of the cards are clubs, we expect that after a large number of repetitions of drawing a card, replacing it, drawing a new card, and so on, about $\frac{1}{4}$ of the cards drawn should be clubs. Therefore, we assign a $\frac{1}{4}$ probability to the outcome of drawing a club from the deck.
(c) The box contains seven balls: three red and four green. The experiment consists of drawing a ball from the box, recording its color, and replacing it before drawing the next ball. Since $\frac{3}{7}$ of the balls are red, we expect that after a large number of repetitions of this experiment, about $\frac{3}{7}$ of the balls drawn from the box should be red. Therefore, we assign a probability of $\frac{3}{7}$ to drawing a red ball from the box.
(d) If it rains once every ten times a certain weather pattern occurs, the weather forecaster would assign a probability of 0.1 to the possibility of rain when this pattern is present. ▪

Sample Spaces

When a die is rolled, we know that the outcomes of the trial will be one of the six numbers 1, 2, 3, 4, 5, and 6, because these are the only possible outcomes of the experiment. The set of all possible outcomes of an experiment is called the **sample space** of the experiment. The sample space for the die-rolling experiment is the set

$$S = \{1, 2, 3, 4, 5, 6\}$$

Since S has a finite number of elements, it is called a **finite** sample space. In this book we consider only probability models that have finite sample spaces.

When discussing sample spaces, the outcomes of an experiment, which make up the sample space, are frequently called **sample points.** For instance, 1, 2, 3, 4, 5, and 6 are the sample points in the sample space for the die-rolling experiment. We use both terms, *outcome* and *sample point,* in this chapter.

Example 3

(a) An experiment consists of drawing one ball from a jar that contains three red balls, two blue balls, and seven green balls. The only possible outcomes are R (a red ball), B (a blue ball), or G (a green ball). Therefore, the sample space for this experiment is the set

$$S = \{R, B, G\}$$

(b) A penny and a dime are tossed onto a table and the outcome recorded. In each case, the outcome of the penny is recorded first and that of the dime second. Thus, HT denotes a head for the penny and a tail for the dime. The sample space for this experiment is the set

$$S = \{HH, HT, TH, TT\}$$

(c) The flavors of ice cream ordered more than 50 times a day at the Nickelodeon Ice Cream Shoppe are recorded. The sample space of this experiment is $S = \{$Chocolate, French Vanilla, Mint Chocolate Chip, Berry Supreme, Sherbet Surprise$\}$. ■

The outcomes, or sample points, of a given experiment can be recorded in various ways, depending to a large extent upon the experimenter. As a result, the sample space for an experiment is not necessarily fixed. The next example shows how two different sample spaces can arise for the same experiment.

Example 4

(a) A researcher studying families with three children decides to record the gender of each child by listing, in order, the gender of the first child, then that of the second, and finally that of the third. For instance, a family whose first two children are girls and whose third child is a boy is recorded as GGB. Write the sample space for this experiment.

(b) A second researcher studying the same data decides to record the number of girls in the family. For instance, a family with two girls and one boy is recorded as a 2. Write the sample space for this experiment.

Solution

(a) The sample space for this experiment consists of the following set of eight sample points:

$$S = \{GGG, GGB, GBG, BGG, BBG, BGB, GBB, BBB\}$$

(b) The sample space for this experiment consists of the following set of four sample points:

$$S = \{3, 2, 1, 0\}$$

■

EVENTS

An **event** is a subset of a sample space S. For instance, in Example 4(a), the fact that the family has two girls is an event E, represented by

$$E = \{GGB, GBG, BGG\}$$

which is a subset of S. Since S and $\varnothing$ are subsets of S, they are events. We interpret the event S to be an event that must occur and the event $\varnothing$ to be an event that cannot occur. If the event consists of a single outcome, it is called a **simple event.** In Example 4(a), the event $\{GGB\}$ is a simple event.

Example 5

(a) The sample space for the roll of a die is $S = \{1, 2, 3, 4, 5, 6\}$. The event E_1, rolling an even number, is the set $E_1 = \{2, 4, 6\}$. The event E_2, rolling a number less than 5, is the set $E_2 = \{1, 2, 3, 4\}$. The event E_3, rolling a number greater than 6, is $E_3 = \varnothing$, since a number greater than 6 is impossible.

(b) The sample space for tossing a penny and a dime is $S = \{HH, HT, TH, TT\}$. The event E_1, at least one of the coins is heads, is the set $E_1 = \{HH, HT, TH\}$. The event E_2, two heads are showing, is the set $E_2 = \{HH\}$. Notice that E_2 is a simple event. ■

COMMENT: *The fact that events are sets means that new events can be constructed from other events by using the set operations of union, intersection, and complement.*

Example 6 On a roll of a die the event E_1, rolling an even number, is the set $E_1 = \{2, 4, 6\}$; the event E_2, rolling a number less than 5, is the set $E_2 = \{1, 2, 3, 4\}$.

(a) The event E_3, rolling an even number that is less than 5, is the set $E_3 = \{2, 4\}$. Notice that event E_3 is also the intersection of E_1 and E_2. Thus,

$$E_3 = E_1 \cap E_2 = \{2, 4, 6\} \cap \{1, 2, 3, 4\} = \{2, 4\}$$

(b) The event E_4, rolling an even number or a number less than 5, is the set $E_4 = \{1, 2, 3, 4, 6\}$. This is the union of E_1 and E_2. Thus,

$$E_4 = E_1 \cup E_2 = \{2, 4, 6\} \cup \{1, 2, 3, 4\} = \{1, 2, 3, 4, 6\}$$

(c) The event E_5, rolling a number greater than 4, is the set $E_5 = \{5, 6\}$. Another way of expressing this event is as the complement of E_2. The sample points in E_2' are the sample points in $S = \{1, 2, 3, 4, 5, 6\}$ that are not in E_2. Thus,

$$E_5 = E_2' = \{1, 2, 3, 4\}' = \{5, 6\}$$ ■

Mutually Exclusive Events

When we toss a coin, we understand that only one of the two outcomes in the sample space can occur; that is, if H (heads) occurs when the coin is tossed, then T (tails) cannot occur. In the study of probability, we describe this by saying that the two events are **mutually exclusive.** Events that are mutually exclusive have no elements in common, so the intersection of such events is the empty set. We use this as a definition of mutually exclusive events.

DEFINITION Mutually Exclusive Events

The events A and B are **mutually exclusive** if

$$A \cap B = \varnothing$$

COMMENT: *In set theory, we use the word* disjoint *to describe two sets that have no points in common. In probability, we use* mutually exclusive *to describe the same concept.*

Example 7 Consider the sample space

$$S = \{GGG, GGB, GBG, BGG, BGB, GBB, BBG, BBB\}$$

where the sample point GGB indicates that the first child is a girl, the second is a girl, and the third is a boy. If we define events A and B as follows—

$$A = \{GGB, GBG, BGG\} \quad \text{(two girls and one boy)}$$
$$B = \{BBG, BGB, GBB\} \quad \text{(two boys and one girl)}$$

then events A and B are mutually exclusive events, because they have no sample points in common; that is, $A \cap B = \varnothing$. ■

EXERCISES SECTION 5.1

An experiment consists of dealing a card from a well-shuffled deck of 52 playing cards. Assign a probability to each event in Exercises 1–6.

1. A heart is dealt.

2. A spade is dealt.

3. A king is dealt.

4. A three is dealt.

5. A jack of diamonds is dealt.

6. A four of clubs is dealt.

An experiment consists of drawing a ball from a jar that contains three white balls, four red balls, and six blue balls. In Exercises 7–10, assign a probability to each event.

7. A white ball is drawn.

8. A blue ball is drawn.

9. A brown ball is drawn.

10. A ball of any color is drawn.

11. An experiment consists of rolling a die 500 times and recording the number of times a 5 turns up. If a 5 turns up 82 times, what is the relative frequency of rolling a 5?

12. An experiment consists of tossing a penny 2000 times and recording the number of tails. If tails occurs 1005 times, what is the relative frequency of tails?

13. An experiment consists of spinning the needle shown in the figure 1000 times and recording the number of times it stops in section 1. If it stops in this section 422 times, what is the relative frequency of the needle's stopping in section 1?

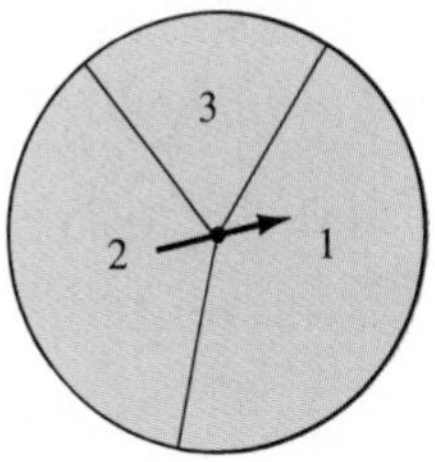

14. If the needle in the figure is spun 150 times and 73 of those times it stops in section 2, what is the relative frequency of the needle's stopping in section 2?

15. Give the sample space of an experiment that consists of rolling a die and recording whether the number that turns up is odd or even.

16. Give the sample space of an experiment that consists of drawing a card from a deck of 52 playing cards and recording its suit (clubs, diamonds, hearts, spades).

Exercises 17 and 18 refer to an experiment that consists of tossing a penny, a nickel, and a dime.

17. (a) Give the sample space, if the sample points are of the form HHT (which indicates heads for the penny, heads for the nickel, and tails for the dime).
(b) Give the sample points in the event E: at least one head turns up.
(c) Give the sample points in the event F: two heads turn up.

18. (a) Give the sample space, if the sample points indicate the number of heads that turn up.
(b) Give the sample points in the event E: at least one head turns up.
(c) Give the sample points in the event F: two heads turn up.

Exercises 19 and 20 refer to an experiment that consists of rolling a pair of dice: one red die and one green die.

19. (a) Give the sample space, if the sample points indicate the sum of the numbers on the two dice.
(b) Give the sample points in the event A: the sum is an even number.
(c) Give the sample points in the event B: the sum is a number greater than 7.
(d) Are events A and B mutually exclusive events?
(e) Give the sample points in the event $C = A \cap B$.

20. (a) Give the sample space, if the sample points indicate the number that turns up on the red die, followed by the number that turns up on the green die; that is, a 2 on the red die and 5 on the green die are recorded as 25.
(b) Give the sample points in the event A: a 3 on the red die.
(c) Give the sample points in the event B: the same number showing on both dice.
(d) Are events A and B mutually exclusive events?

(e) Give the sample points in the event $C = A \cap B$.

In Exercises 21–30, give the sample points in each of the events:

An experiment consists of drawing a ball from a jar that contains ten balls numbered 1–10. The events A, B, and C are defined as

$$A = \{2, 4, 6, 8, 10\}$$
$$B = \{1, 2, 3, 4, 5\}$$
$$C = \{5, 7, 9\}$$

21. $A \cup B$
22. $B \cup C$
23. $A \cup C$
24. $B \cap C$
25. $A \cap C$
26. $A \cap B$
27. A'
28. C'
29. $A \cap B'$
30. $B \cap C'$

In Exercises 31–38, give the sample points in each of the events:

A personnel manager plans to interview eight candidates for a new job. The candidates are John, Jerry, Kristin, Fred, Julie, Hylda, Pete, and Ben. Three of the candidates, John, Fred, and Julie, are company employees and the rest are from an employment agency. Events A, B, and C are defined as

A: The candidate is male.
B: The candidate is female.
C: The candidate is a company employee.

31. $A \cup C$
32. $A \cup B$
33. $A \cap B$
34. $B \cap C$
35. A'
36. C'
37. $A \cap C'$
38. $B \cap C'$

5.2 CALCULATING PROBABILITY

In Section 5.1 we defined the probability of the outcome of an experiment as a number that will predict the relative frequency of the outcome when the experiment is repeated a large number of times. In this section we discuss the process of assigning probabilities to equally likely outcomes of an experiment. Recall that if each outcome of an experiment is just as likely to occur as any other, the outcomes are said to be **equally likely outcomes.** If an experiment is performed a large number of times, our intuition tells us that each of the equally likely outcomes should occur about the same number of times. For instance, when we roll a die a large number of times, each of its six sides should turn up about the same number of times. Therefore, each of the six numbers would have the same relative frequency, so we should assign the same probability to each of the six equally likely outcomes. Since we expect each side to turn up about $\frac{1}{6}$ of the time, we assign the probability $\frac{1}{6}$ to each of the six. We use this example to motivate the following probability model for experiments in which there are n mutually exclusive and equally likely outcomes:

Probability Model for Equally Likely Outcomes

If the sample space S of an experiment contains n mutually exclusive and equally likely sample points, then a probability of $\frac{1}{n}$ is assigned to each of the n sample points.

COMMENT: *A probability model that assigns the probability $1/n$ to each of the n sample points in the sample space is called a* ***uniform probability model.*** *Some examples of uniform probability models are given in Example 1.*

Example 1 Each of the following experiments is described by a uniform probability model:

(a) The experiment of tossing a coin is described by a uniform probability model, because we assign a probability of $\frac{1}{2}$ to each of the two mutually exclusive and equally likely simple events $\{H\}$ and $\{T\}$.

(b) The experiment of rolling a die is described by a uniform probability model, because we assign a probability of $\frac{1}{6}$ to each of the six mutually exclusive and equally likely simple events $\{1\}$, $\{2\}$, $\{3\}$, $\{4\}$, $\{5\}$, and $\{6\}$.

(c) The experiment in which the girl/boy sequence for a family with three children is recorded is described by a uniform probability model, because we assign a probability of $\frac{1}{8}$ to each of the eight mutually exclusive and equally likely simple events $\{GGG\}$, $\{GGB\}$, $\{GBG\}$, $\{BGG\}$, $\{GBB\}$, $\{BGB\}$, $\{BBG\}$, and $\{BBB\}$. ■

If event E is a subset of a sample space that is described by a uniform probability model, then the following principle holds. This principle is basic to the process of calculating probabilities of events.

Probability Principle for Uniform Probability Models

If S is a sample space with n mutually exclusive and equally likely sample points and if E is an event that contains k of these sample points, then the probability of event E is

$$P(E) = \frac{k}{n}$$

This principle is also stated in the form of the following addition principle: If event $E = \{s_1, s_2, \ldots, s_k\}$, where $\{s_1\}, \{s_2\}, \ldots, \{s_k\}$ are k of the n mutually exclusive and equally likely simple events from a sample space S, and if $P(\{s_1\}) = P(\{s_2\}) = \cdots = P(\{s_k\}) = \frac{1}{n}$, then

$$P(E) = P(\{s_1\}) + P(\{s_2\}) + \cdots + P(\{s_k\}) = \underbrace{\frac{1}{n} + \frac{1}{n} + \cdots + \frac{1}{n}}_{k \text{ fractions}} = \frac{k}{n}$$

Example 2 If a die is rolled, what is the probability of (a) rolling a number greater than 4 and of (b) rolling an even number?

Solution

(a) Let E_1 be the event "rolling a number greater than 4"; then $E_1 = \{5, 6\}$. Since E_1 consists of two of the possible six sample points, the probability of rolling a number greater than 4 is

$$P(E_1) = \frac{2}{6} = \frac{1}{3}$$

(b) Let E_2 be the event "rolling an even number"; then $E_2 = \{2, 4, 6\}$. Since E_2 contains three of the possible six sample points, the probability of rolling an even number is

$$P(E_2) = \frac{3}{6} = \frac{1}{2}$$

Example 3 A card is drawn from a deck of 52 playing cards. What is the probability of each of the following events?

(a) K: a king is drawn.
(b) H: a heart is drawn.
(c) F: a face card is drawn.

Solution

(a) There are four kings in the deck, so the probability of drawing a king is

$$P(K) = \frac{4}{52} = \frac{1}{13}$$

(b) There are 13 hearts in the deck, so the probability of drawing a heart is

$$P(H) = \frac{13}{52} = \frac{1}{4}$$

(c) There are 12 face cards in the deck, so the probability of drawing a face card is

$$P(F) = \frac{12}{52} = \frac{3}{13}$$

Example 4 A box contains 50 light bulbs, 3 of which are defective. If a bulb is selected at random from the box, what is the probability that it is a defective bulb?

Solution The expression *selected at random* means that each bulb has an equally likely chance of being selected. Let D be the event "a defective bulb is selected." Since 3 of the 50 bulbs are defective, it follows from the probability principle that the probability of selecting one of them is

$$P(D) = \frac{3}{50}$$

■

If we look closely at the probability assignments that we have been making, two important characteristics emerge:

- First, we note that the relative frequency of an outcome of an experiment is always a positive number that is less than or equal to 1. Consequently, this must also be true for the probability of each outcome of an experiment.
- Second, for a given experiment, the sum of the relative frequencies of the equally likely outcomes is 1, so the sum of the probabilities assigned to these outcomes must also be 1. For instance, when a die is rolled, each of the six outcomes is assigned a probability of $\frac{1}{6}$, so $\frac{1}{6} + \frac{1}{6} + \frac{1}{6} + \frac{1}{6} + \frac{1}{6} + \frac{1}{6} = 1$. This property holds for all probability models, not just for uniform probability models.

Properties of Probability Models

Let $S = \{s_1, s_2, \ldots, s_n\}$ be a sample space of an experiment with n mutually exclusive sample points. With each sample point s_k, we associate a probability p_k, which has the following properties:

$$0 \leq p_k \leq 1$$

and

$$\sum_{k=1}^{n} p_k = p_1 + p_2 + \cdots + p_n = 1$$

The first property means that the probability of an event must be greater than or equal to zero and less than or equal to 1. The second means that the sum of the probabilities of the sample points of the sample space must be 1. As a result of the second property, the probability that the sample space S will occur is

$$P(S) = 1$$

Finally, the probability of an event that is not in the sample space is zero; that is,

$$P(\varnothing) = 0$$

COMMENT: *Given the above properties, we can easily compute the probability that event A will* not *occur if we know the probability that it will occur. The sum of the probabilities of the sample points in the sample space is 1. Therefore, if $P(A)$ is the probability that A will occur, and the probability that A will not occur is $P(A')$, then $P(A) + P(A') = 1$. Solving for $P(A')$, we get*

$$P(A') = 1 - P(A)$$

Example 5

(a) When a die is rolled, the probability of getting a 5 is $\frac{1}{6}$. Therefore, the probability of *not* getting a 5 is

$$P(\text{not a } 5) = 1 - P(5) = 1 - \frac{1}{6} = \frac{5}{6}$$

(b) When one card is drawn from a deck of 52 playing cards, the probability of its being an ace is $\frac{1}{13}$, so the probability of *not* drawing an ace is

$$P(\text{not an ace}) = 1 - P(\text{ace}) = 1 - \frac{1}{13} = \frac{12}{13}$$

Using the probability principle to compute probabilities requires that we be able to determine both the number of desirable outcomes and the total number of possible outcomes. We frequently use the counting techniques discussed in Chapter 4 to determine these numbers.

Example 6 Calculate the probability of getting three heads in five tosses of a coin.

Solution To calculate the number of ways the coin can land in five tosses, we note that on each toss, the coin can land in two ways; so there are $2 \cdot 2 \cdot 2 \cdot 2 \cdot 2 = 2^5$ ways the coin can land in five tosses. The number of ways we can get three heads in five tosses is the combination of five objects taken three at a time. Recall (Sections 4.4 and 4.5) that this combination is given by $C_{5,3} = \frac{5!}{3!2!} = 10$. Therefore, the probability of getting three heads in five tosses of a coin is

$$P(3 \text{ heads}) = \frac{C_{5,3}}{2^5} = \frac{10}{32} = \frac{5}{16} \approx 0.31$$

Example 7 A box containing five cards numbered 1 through 5 is used to generate a five-digit number for a lottery. The cards are drawn one at a time and are not used again, and each number is recorded in one of the five spaces shown, starting on the left.

What is the probability of guessing the lottery number?

Solution The first space can be filled by any one of the five numbers, the second by any one of the remaining four numbers, and so on. Therefore, there are $5 \cdot 4 \cdot 3 \cdot 2 \cdot 1 = 5! = 120$ ways of ordering the five numbers. If we assume that each of the five-digit numbers is equally likely, the probability of guessing the number is

$$\frac{1}{5!} = \frac{1}{120} \approx 0.0083$$

■

Example 8 Three balls are to be drawn at random from a box containing five red balls and seven black balls. Calculate the probability that all three of the balls will be red.

Solution The number of ways three red balls can be drawn from the five red balls in the box is

$$C_{5,3} = \frac{5!}{3!2!} = \frac{5 \cdot 4}{2 \cdot 1} = 10$$

and the total number of ways any three balls can be drawn from the twelve balls in the box is

$$C_{12,3} = \frac{12!}{3!9!} = \frac{12 \cdot 11 \cdot 10}{3 \cdot 2 \cdot 1} = 220$$

Therefore, the probability of drawing three red balls is

$$P(3 \text{ red balls}) = \frac{C_{5,3}}{C_{12,3}} = \frac{10}{220} = \frac{1}{22} \approx 0.045$$

■

Example 9 What is the probability that, of the five balls drawn at random from the box in Example 8, two will be red and three will be black?

Solution The number of ways two red balls can be drawn from the five red balls in the box is

$$C_{5,2} = \frac{5!}{2!3!} = \frac{5 \cdot 4}{2 \cdot 1} = 10$$

The number of ways three black balls can be drawn from the seven black balls in the box is

$$C_{7,3} = \frac{7!}{3!4!} = \frac{7 \cdot 6 \cdot 5}{3 \cdot 2 \cdot 1} = 35$$

Therefore, by the counting principle, the number of ways two red balls and three black balls can be drawn from the box is $C_{5,2} \cdot C_{7,3} = 10 \cdot 35 = 350$. In addition, the number of ways five balls can be drawn from the twelve in the box is

$$C_{12,5} = \frac{12!}{5!7!} = 792$$

Finally, the probability of drawing two red balls and three black balls is

$$P(2 \text{ red and } 3 \text{ black}) = \frac{C_{5,2} \cdot C_{7,3}}{C_{12,5}} = \frac{350}{792} = \frac{175}{396} \approx 0.44$$

■

Example 10 In the game of poker, a full house is a five-card hand that consists of three cards of one kind and two of another. For instance, a full house of kings and tens has three kings and two tens.

(a) Calculate the probability of a full house of kings and tens.
(b) Calculate the probability of a full house.

Solution

(a) There are four kings and four tens, so there are $C_{4,3} = \frac{4!}{3!1!} = 4$ ways of getting three kings and $C_{4,2} = \frac{4!}{2!2!} = \frac{4 \cdot 3}{2 \cdot 1} = 6$ ways of getting two tens. It follows from the counting principle that there are $C_{4,3} \cdot C_{4,2} = 4 \cdot 6 = 24$ ways of getting three kings and two tens. Since there are $C_{52,5} = \frac{52!}{5!47!} = \frac{52 \cdot 51 \cdot 50 \cdot 49 \cdot 48}{5 \cdot 4 \cdot 3 \cdot 2 \cdot 1} = 2{,}598{,}960$ ways of dealing five cards, the probability of a full house of three kings and two tens is

$$P(3 \text{ kings and } 2 \text{ tens}) = \frac{C_{4,3} \cdot C_{4,2}}{C_{52,5}} = \frac{24}{2{,}598{,}960} \approx 0.0000092$$

(b) To find the probability of a full house of any kind, we note that there are 13 card denominations: A, 2, 3, 4, 5, 6, 7, 8, 9, 10, J, Q, K. Since there are $C_{4,3}$ ways of getting three of a kind and there are 13 kinds, by the counting principle there are $13 \cdot C_{4,3}$ ways of getting three of a kind. Once a three-of-a-kind has been accounted for, there are 12 denominations from which to draw the two-of-a-kind. Two of a kind can be obtained from four possibilities in $C_{4,2}$ ways, so there are $12 \cdot C_{4,2}$ ways of getting two of a kind to go with the three of a kind. It follows by the counting principle that there are $(13 \cdot C_{4,3})(12 \cdot C_{4,2})$ ways of getting a full house. As in (a), there are $C_{52,5}$ ways of dealing five cards, so the probability of a full house is

$$P(\text{full house}) = \frac{(13 \cdot C_{4,3})(12 \cdot C_{4,2})}{C_{52,5}} = \frac{13 \cdot 4 \cdot 12 \cdot 6}{2{,}598{,}960} \approx 0.00144$$

■

Example 11 A couple wants to determine the probability of having two boys in a family of three children.

Solution Since there are $2^3 = 8$ ways they can have three children, and of these, $C_{3,2}$ ways of having two boys and a girl, the probability is

$$P(2 \text{ boys}) = \frac{C_{3,2}}{8} = \frac{3}{8}$$

■

Example 12 Your stockbroker supplies you with a list of ten stocks, from which you choose three. You later find out that seven of the ten are growth stocks and the others are income-producing stocks. What is the probability that you inadvertently chose three growth stocks?

Solution We assume that your choices of stocks were equally likely. The number of ways that you could have chosen three growth stocks from the seven is given by $C_{7,3}$ and the total number of ways of choosing the three stocks from the list of ten is given by $C_{10,3}$. Hence

$$P(3 \text{ growth stocks}) = \frac{C_{7,3}}{C_{10,3}} = \frac{\dfrac{7 \cdot 6 \cdot 5}{3 \cdot 2 \cdot 1}}{\dfrac{10 \cdot 9 \cdot 8}{3 \cdot 2 \cdot 1}} = \frac{7}{24}$$

or about 30%. ▪

Odds

The probability of an event A is often given indirectly by two numbers called **the odds in favor of A.** The odds in favor of an event are defined as the ratio of the probability that the event will occur to the probability that it will not occur. Thus, if $P(A)$ is the probability that A will occur and $1 - P(A)$ is the probability that it will not occur, then the odds in favor of A are given by

$$O(A) = \frac{P(A)}{1 - P(A)}$$

The odds in favor of A are expressed as the ratio of two integers. Thus, if $O(A) = \frac{3}{2}$, we say, "The odds in favor of A are 3 to 2."

Example 13 What are the odds of getting either a 5 or a 6 on a single roll of a die?

Solution The probability of getting a 5 or a 6 is $\frac{1}{3}$. Therefore, the odds in favor of a 5 or a 6 are

$$O(5 \text{ or } 6) = \frac{\frac{1}{3}}{1 - \frac{1}{3}} = \frac{\frac{1}{3}}{\frac{2}{3}} = \frac{1}{2}$$

Thus, the odds in favor of a 5 or a 6 are 1 to 2. ▪

Example 14 If the odds in favor of an event A are 5 to 2, what is the probability of A?

Solution By the definition of odds, we have

$$\frac{P(A)}{1 - P(A)} = \frac{5}{2}$$

Solving for $P(A)$ yields

$$2P(A) = 5 - 5P(A)$$
$$7P(A) = 5$$
$$P(A) = \frac{5}{7}$$

Thus, the probability of the event A is $\frac{5}{7}$. ■

EXERCISES SECTION 5.2

For Exercises 1–4, a card is drawn at random from a deck of 52 playing cards. Calculate the probability that:

1. The card is a heart.

2. The card is red.

3. The card is the king of clubs.

4. The card is the ten of hearts.

5. A box contains three bolts and five nuts. If an item is drawn at random from the box, what is the probability that it is a bolt? What is the probability that it is a nut?

6. A piggy bank contains 20 dimes and 15 nickels. Assuming each coin has an equally likely chance of falling out when the bank is turned over and shaken, what is the probability that the first coin to fall out will be a dime?

For Exercises 7–10, a die is rolled and the number that turns up is recorded. Calculate the probability that:

7. A 3 or a 4 will turn up.

8. An odd number will turn up.

9. An even number greater than 3 will turn up.

10. A number less than 5 will turn up.

11. Three pennies are tossed onto a tabletop. Calculate the probability that exactly two of the pennies will show heads.

12. Assuming that each child born has an equally likely chance of being a boy or a girl, what is the probability that all three children in a family will be girls?

For Exercises 13–16, a box contains three red balls, four white balls, and six yellow balls.

3 red 4 white 6 yellow

A ball is drawn at random from the box. Calculate the probability that:

13. The ball is yellow.

14. The ball is red.

15. The ball is not red.

16. The ball is not yellow.

17. Six coins are tossed onto the floor. What is the probability that exactly three will turn up tails?

18. What is the probability of getting five heads if seven coins are tossed onto a tabletop?

19. Five boys and eight girls place their names in a hat. If seven names are drawn at random from the hat, what is the probability that three will be boys and four will be girls?

For Exercises 20–22, a box contains six quarters, five dimes, and fifteen nickels. Two coins are drawn at random from the box. Calculate the probability that:

20. Both coins have the same value.

21. Both coins are nickels.

22. One coin is a dime and the other is a nickel.

For Exercises 23–28, three cards are drawn at random from a deck of 52 playing cards. Calculate the probability that:

23. The three cards are kings.

24. The three cards are aces.

25. The three cards are hearts.

26. The three cards are red.

27. There are two hearts and one spade.

28. There are two face cards and one 10.

For Exercises 29–32, a five-card hand is dealt from a deck of 52 playing cards. Calculate the probability that the hand contains:

29. Exactly two aces.

30. All clubs (called a flush).

31. Two aces, two kings, and a 3.

32. Three hearts and two spades.

33. What are the odds in favor of a 6 on a single roll of a die?

34. What are the odds in favor of exactly two aces in a five-card poker hand?

35. Suppose the probability of your passing a mathematics test is 0.3. What are the odds in favor of your passing?

36. If the probability of the Forty-Niners winning the Super Bowl is $\frac{2}{9}$, what are the odds in favor of their winning the Super Bowl?

37. What is the probability of Son of Secretariat's winning a horse race if the odds in favor of his winning are 8 to 3?

38. What is the probability of a stock's being successful and returning good quarterly dividends if the odds in favor of this are 1 to 3?

39. If the probability of success of a new sporting-goods store is 0.35 percent, what are the odds in favor of its success?

40. You select a car from a lot of 200 cars. You are told that ten of the cars are defective. What is the probability that you selected a defective car?

41. People who claim to have ESP are challenged to participate in an experiment in which they are asked to match a sequence of cards with a sequence being arranged elsewhere. By chance alone, what is the probability that a subject would match a sequence of four cards colored red, white, blue, and orange?

42. The odds *against* an event E's occurring are defined as $\frac{P(E')}{P(E)}$. If E is your selecting a defective car in Exercise 40, what are the odds against E?

5.3 THE ADDITION RULE

This section is concerned with the probability of an event that arises from the union of two events. Consider a game in which you can advance ten spaces if you get either a 2 or a 5 on a single roll of the die; or consider a power plant manager's attempt to estimate the probability that the plant will be down for repairs due to the failure of either a diesel engine or a generator. In both of these examples, the probability of the union of two or more events is considered.

THE ADDITION RULE FOR MUTUALLY EXCLUSIVE EVENTS

Recall that two events are *mutually exclusive* if the occurrence of one of them excludes the occurrence of the other. For instance, in the game mentioned above, on a single roll of the die, the event E_1 (a 2 is showing) and the event E_2 (a 5 is showing) are mutually exclusive events, since rolling a 2 excludes the possibility of rolling a 5.

Suppose we want to compute the probability of getting either a 2 or a 5 on a single roll of the die. The event E (a 2 is showing or a 5 is showing) is the union of the mutually exclusive events E_1 (a 2 is showing) and E_2 (a 5 is showing); that is,

$$E = E_1 \cup E_2$$

Since event E includes two of the possible six outcomes, the probability of getting a 2 or a 5 on a single roll of the die is

$$P(E) = P(E_1 \cup E_2) = \frac{2}{6} = \frac{1+1}{6} = \frac{1}{6} + \frac{1}{6} = P(E_1) + P(E_2)$$

The fact that the probability of the union of these two mutually exclusive events turns out to be the sum of the probabilities of the events making up the union is true for mutually exclusive events in general. The generalization for more than two mutually exclusive events is stated below.

The Addition Rule for Mutually Exclusive Events

If $E_1, E_2, \ldots, E_n$ are mutually exclusive events, the probability of the union $E_1 \cup E_2 \cup \cdots \cup E_n$ of these events is the sum of the probabilities of the individual events; that is,

$$P(E_1 \cup E_2 \cup \cdots \cup E_n) = P(E_1) + P(E_2) + \cdots + P(E_n)$$

COMMENT: *Notice that this rule applies* only *to mutually exclusive events. For instance, it cannot be used to compute the probability of getting an ace or a heart on a single draw from a deck of playing cards, since drawing an ace and drawing a heart are not mutually exclusive events. (You could draw the ace of hearts!)*

Example 1 A box contains five red balls, four white balls, and seven black balls. A ball is drawn at random from the box. Calculate the probability that it is a red ball (R) or a white ball (W).

Solution The events R and W are mutually exclusive; that is, if you draw a red ball you cannot draw a white ball. The probability of drawing a red ball is $P(R) = \frac{5}{16}$ and the probability of drawing a white ball is $P(W) = \frac{4}{16}$. Therefore, by the addition principle, the probability of getting a red ball or a white ball on a single draw is

$$P(R \cup W) = P(R) + P(W) = \frac{5}{16} + \frac{4}{16} = \frac{9}{16}$$

■

Example 2 A card is drawn from a deck of 52 playing cards. Calculate the probability that the card is an ace or a two or a face card.

Solution Drawing an ace, drawing a two, and drawing a face card are mutually exclusive events. Let A be the event "an ace is drawn," T be the event "a two is drawn," and F be the event "a face card is drawn." There are four aces, four twos, and twelve face cards in the deck, so the probabilities of these events are

$$P(A) = \frac{4}{52}, \qquad P(T) = \frac{4}{52}, \qquad P(F) = \frac{12}{52}$$

Therefore, by the addition principle, the probability that an ace or a two or a face card is drawn is

$$P(A \cup T \cup F) = P(A) + P(T) + P(F) = \frac{4}{52} + \frac{4}{52} + \frac{12}{52} = \frac{20}{52} = \frac{5}{13}$$ ■

THE GENERAL ADDITION RULE

As we noted earlier, the addition rule for the union of mutually exclusive events cannot be used to calculate the probability of drawing an ace or a heart from a deck of cards, because drawing an ace and drawing a heart are not mutually exclusive events. To see how the probability of getting an ace or a heart on a single draw is computed, let A be the event "an ace is drawn" and H be the event "a heart is drawn." The event "drawing an ace or drawing a heart" is then represented by $A \cup H$. Since 4 of the 52 cards are aces, the probability of drawing an ace is

$$P(A) = \frac{4}{52}$$

Similarly, since 13 of the cards are hearts, the probability of drawing a heart is

$$P(H) = \frac{13}{52}$$

However, the ace of hearts is in both A and H, so $P(A \cup H) \neq P(A) + P(H)$. To get around this problem, let H_1 be the event "a heart, other than the ace of hearts, is drawn." Since we have excluded the ace of hearts from H_1, there are only 12 elements in this event. Therefore, the probability of drawing a heart, other than the ace of hearts, is

$$P(H_1) = \frac{12}{52}$$

Since the ace of hearts is excluded from the hearts in H_1, it follows that A and H_1 are mutually exclusive events. Thus,

$$P(A \cup H_1) = P(A) + P(H_1) = \frac{4}{52} + \frac{12}{52} = \frac{16}{52} \tag{5.1}$$

Rather than go through the process of modifying the events to make them mutually exclusive, as we did above, we observe that

$$P(H_1) = P(H) - P(A \cap H) = \frac{13}{52} - \frac{1}{52} = \frac{12}{52}$$

where $P(A \cap H) = \frac{1}{52}$ is the probability of drawing the ace of hearts. If we make the substitution $P(H_1) = P(H) - P(A \cap H)$ in Equation 5.1, the probability of drawing an ace or a heart is given by

$$P(A \cup H) = P(A) + P(H_1) = P(A) + P(H) - P(A \cap H) = \frac{4}{52} + \frac{13}{52} - \frac{1}{52} = \frac{16}{52}$$

which agrees with the previous result. This procedure can be used to find the probability of the union of any two events, whether they are mutually exclusive or not. The result is stated as a general addition rule.

The General Addition Rule

If E_1 and E_2 are events from the sample space S, then the probability of their union $E_1 \cup E_2$ is given by

$$P(E_1 \cup E_2) = P(E_1) + P(E_2) - P(E_1 \cap E_2)$$

The validity of this rule can be established by using Venn diagrams to show that $P(E_1 \cap E_2)$ must be subtracted from $P(E_1) + P(E_2)$ when E_1 and E_2 are not mutually exclusive events. Figure 5.1 shows E_1 and E_2 as intersecting circles to indicate that they are not mutually exclusive events. Assign probabilities p_1, p_2, and p_3 to each of the separate regions shown. Then we see that $P(E_1 \cup E_2) = p_1 + p_2 + p_3$, $P(E_1 \cap E_2) = p_3$, $P(E_1) = p_1 + p_3$, and $P(E_2) = p_2 + p_3$. We proceed as follows:

$$\begin{aligned} P(E_1 \cup E_2) &= p_1 + p_2 + p_3 \\ &= p_1 + p_3 + p_2 + p_3 - p_3 && \text{Adding and subtracting } p_3 \\ &= (p_1 + p_3) + (p_2 + p_3) - p_3 && \text{Grouping terms} \\ &= P(E_1) + P(E_2) - P(E_1 \cap E_2) \end{aligned}$$

Figure 5.1

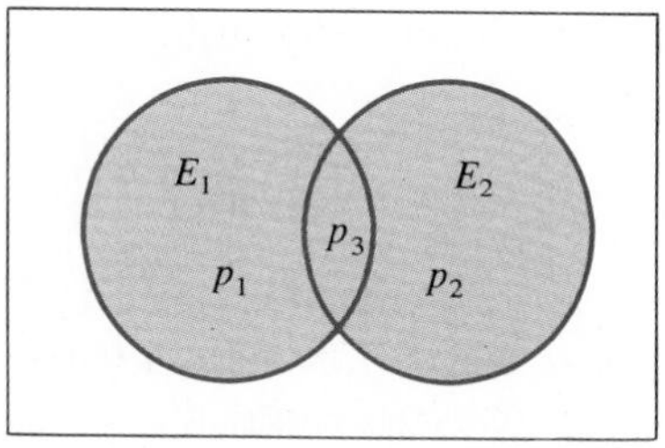

Example 3 A number is to be chosen from

$$S = \{1, 2, 3, 4, 5, 6, 7, 8, 9, 10, 11, 12\}$$

Let event A be "an odd number is chosen" and event B be "a number divisible by 3 is chosen." Find the probability that the number chosen is an odd number or a number divisible by 3.

Solution The odd numbers in S are

$$A = \{1, 3, 5, 7, 9, 11\}$$

Since six of the twelve numbers in S are odd numbers, the probability of an odd number is

$$P(A) = \frac{6}{12}$$

The numbers in S that are divisible by 3 are

$$B = \{3, 6, 9, 12\}$$

Since there are four numbers in B, the probability that the chosen number is divisible by 3 is

$$P(B) = \frac{4}{12}$$

Events A and B are not mutually exclusive because

$$A \cap B = \{3, 9\}$$

This means that two of the twelve numbers in S are odd numbers that are divisible by 3, so the probability that the number is odd and divisible by 3 is

$$P(A \cap B) = \frac{2}{12}$$

Finally, the probability that the number drawn is odd or divisible by 3 is

$$P(A \cup B) = P(A) + P(B) - P(A \cap B) = \frac{6}{12} + \frac{4}{12} - \frac{2}{12} = \frac{8}{12} = \frac{2}{3}$$

Example 4 A company has 75 employees, of whom 23 speak German, 15 speak Spanish, and 7 speak both. What is the probability that an employee chosen at random speaks German or Spanish?

Solution The events are:

G: person speaks German
S: person speaks Spanish
$G \cap S$: person speaks German and Spanish
$G \cup S$: person speaks German or Spanish

The probabilities of the given events are

$$P(G) = \frac{23}{75} \qquad P(S) = \frac{15}{75} \qquad P(G \cap S) = \frac{7}{75}$$

Since G and S are not mutually exclusive events, the probability that the person speaks German or Spanish is

$$P(G \cup S) = P(G) + P(S) - P(G \cap S) = \frac{23}{75} + \frac{15}{75} - \frac{7}{75} = \frac{31}{75}$$

■

COMMENT: *Percentages are not probabilities, but the two numbers are related. Since percentage is (relative frequency) × 100, it follows that we can obtain probability from percentage by*

$$\textbf{probability} = \frac{\textbf{percentage}}{100}$$

For example, if 5% of the bolts in a box are defective, the probability of drawing a defective bolt is $\frac{5}{100} = 0.05$.

Example 5 The power plant mentioned at the beginning of this section consists of a diesel engine and a generator. Records of the plant reveal that the diesel engine is down for repair on 2% of the days, the generator is down for repair on 1% of the days, and both are down at the same time on 0.3% of the days. If a day of the year is picked at random, what is the probability that the diesel engine or the generator will be down for repair on this day?

Solution The events are:

D: the diesel engine is down for repair

G: the generator is down for repair

We interpret the fact that the diesel engine is down 2% of the time to mean that the probability of event D is 0.02; that is,

$$P(D) = 0.02$$

Similarly, the probability that the generator is down is

$$P(G) = 0.01$$

Notice that D and G are not mutually exclusive, because the diesel engine and the generator can be down at the same time, an event that is denoted by $D \cap G$. The probability of this event is

$$P(D \cap G) = 0.003$$

If either the diesel engine or the generator is down, we say the system is down. Denoting the event "the system is down" by $D \cup G$, the probability of this event is

$$P(D \cup G) = P(D) + P(G) - P(D \cap G) = 0.02 + 0.01 - 0.003 = 0.027$$

Thus, the probability that the system is down on a given day is 0.027. Or, in terms of percentages, the system is down on 2.7% of the days.

■

EXERCISES SECTION 5.3

In Exercises 1–10, use the addition rule for mutually exclusive events to calculate each probability.

1. What is the probability of rolling a 5 or a 6 on a single roll of a die?
2. What is the probability of rolling a number less than 3 on a single roll of a die?
3. A box contains three red balls, four black balls, and two white balls. A ball is drawn from the box. Calculate the probability that it is a red ball or a black ball.
4. Calculate the probability that the ball in Exercise 3 is black or white.
5. One card is drawn from a deck of 52 playing cards. Calculate the probability that it is an ace or a face card.
6. One card is drawn from a deck of 52 playing cards. Calculate the probability that it is a ten or a jack or a king.
7. If A, B, and C are mutually exclusive events and $P(A) = 0.4$, $P(B) = 0.3$, and $P(C) = 0.1$, calculate the probability that:
   (a) A or B occurs.
   (b) Neither A nor B occurs.
   (c) A or B or C occurs.
8. If E, F, and G are mutually exclusive events and $P(E) = 0.35$, $P(F) = 0.28$, and $P(G) = 0.16$, calculate the probability that:
   (a) E or G occurs.
   (b) G or F occurs.
   (c) Neither E nor F nor G occurs.
9. A psychology researcher conducts a color preference experiment in which white mice have a choice of pushing a red button, a green button, or a yellow button to get food. Experience shows that the mice choose the red button 23% of the time, the green button 32% of the time, and the yellow button 39% of the time. If a mouse is chosen at random, calculate the probability that:
   (a) The mouse will push the red button or the green button.
   (b) The mouse will push neither the red button nor the green button.
   (c) The mouse will not push any button.
10. An automatic lathe produces parts that fall into one of three categories: "acceptable," "need to be reworked," and "must be scrapped." The engineer in charge of the process finds that 87% of the parts are acceptable, 9% need to be reworked, and 4% must be scrapped. If a part is chosen at random, calculate the probability that:
    (a) The part is acceptable or needs to be reworked.
    (b) The part is not acceptable.

Use the following information to answer Exercises 11–16:

Two dice are rolled and the sum of the two numbers is recorded. There are 11 possible outcomes in this sample space, namely

$$S = \{2, 3, 4, 5, 6, 7, 8, 9, 10, 11, 12\}$$

Each sum is not equally likely, as you can see from the following table:

Sum	Outcomes	Number of Ways to Get Sum
2	(1, 1)	1
3	(1, 2), (2, 1)	2
4	(1, 3), (3, 1), (2, 2)	3
5	(1, 4), (4, 1), (2, 3), (3, 2)	4
6	(1, 5), (5, 1), (2, 4), (4, 2), (3, 3)	5
7	(1, 6), (6, 1), (2, 5), (5, 2) (3, 4), (4, 3)	6
8	(2, 6), (6, 2), (3, 5), (5, 3), (4, 4)	5
9	(3, 6), (6, 3), (4, 5), (5, 4)	4
10	(4, 6), (6, 4), (5, 5)	3
11	(5, 6), (6, 5)	2
12	(6, 6)	1

In Exercises 11–16, calculate the probability of rolling a pair of dice to obtain:

11. A 3 or a 7.
12. A 2 or a 10.
13. A 2 or an 11 or a 12.
14. Not an 11.
15. Neither a 5 nor a 7.
16. A number greater than 9.

In Exercises 17–30, use the general addition rule to calculate the indicated probability.

A card is drawn from a deck of 52 playing cards. Calculate the probability that the card drawn is:

17. A king or a club.

18. A 10 or a spade.

19. An ace or a red card.

20. A 3 or a black card.

21. A newspaper survey of readers finds that 45% of the population buy the *Daily Times*, 27% buy the *Herald*, and 11% buy both. If a person is picked at random, calculate the probability that:
(a) The person buys the *Daily Times* or the *Herald*.
(b) The person does not buy either of these papers.

22. The weather forecaster at WEEX is correct 78% of the time and the forecaster at WWWA is correct 81% of the time. If 62% of the time they are both correct, what is the probability that one of the two forecasters is correct?

23. In a class of 50 students, 14 scored above 90 on a math exam, 10 scored above 90 on an English exam, and 3 scored above 90 on both exams. A student is chosen at random from the class. What is the probability that the student scored above 90 on the math exam or the English exam?

24. A number is drawn at random from $S = \{4, 5, 6, 7, 10, 11, 12, 14, 15\}$. Calculate the probability that:
(a) An even number or a number divisible by 3 is drawn.
(b) An odd number or a number divisible by 5 is drawn.
(c) A number less than 12 or greater than 10 is drawn.

For Exercises 25–28, a red die and a blue die are rolled and the sum of the two dice is recorded. The sample space for this is shown in the table preceding Exercise 11, in which the first number is the number on the red die and the second is the number on the blue die. For a given roll of the dice, calculate the probability that:

25. The red die shows a 5 or the sum of the two dice is 11.

26. The blue die shows a 3 or the sum of the two dice is 7.

27. The sum of the two dice is 5 or 6 or the red die shows a 2.

28. The sum of the two dice is 11 or 12 or the red die shows a 6.

For Exercises 29 and 30, a psychology researcher has three white mice, five white rabbits, four brown guinea pigs, and seven spotted guinea pigs in a holding pen. If the door is opened, any one of the animals in the pen is equally likely to come through the door. Calculate the probability that the first to come through the door is:

29. A rabbit or a white animal.

30. A brown animal or a guinea pig.

5.4 THE MULTIPLICATION RULE

Consider a radar detector manufacturer who is concerned about the probability that an item selected at random from the production line will have no defective components. This manufacturer is interested in the event that neither the switch nor the warning beeper nor the warning light is defective. This is the *intersection* of three events. In order to determine the probability that an item has no defective components, we need to develop a rule for the probability of an event formed by the intersection of other events. Note that in the previous section, when considering the probability of an event formed by the *union* of other events, we used the addition rule to evaluate these probabilities. However, for us to use this rule, the probability of the intersection of the two events had to be known. In this section we wish to develop ways to determine the probability of an intersection of events.

INDEPENDENT EVENTS

A box contains three red balls, five black balls, and two green balls. Two balls are drawn in succession from the box. What is the probability that both balls are red? The answer to this question depends upon how the balls are drawn from the box. If the first ball is replaced before the second is drawn, the answer is $\frac{9}{100}$; if the first ball is not replaced before the second is drawn, the answer is $\frac{6}{90}$. The explanation of the difference between these two answers lies in the nature of the two events. The events in which the first ball is replaced before the second is drawn are typical independent events. Two events E and F are said to be **independent** if the probability that event F will occur is not influenced by event E and the probability that event E will occur is not influenced by event F. If, in our example, the first ball is replaced in the box before the second ball is drawn, then the probability of drawing a red ball on the second draw is not influenced by what happened on the first draw. We describe this by saying that the events "a red ball on the first draw" and "a red ball on the second draw" are independent events. Events that are not independent are called **dependent.** For the moment, we will consider only independent events.

■ **Example 1** Two cards are drawn in succession from a deck of 52 playing cards. Let event D_1 be "the ace of spades is the first card drawn" and event D_2 be "the two of clubs is the second card drawn." If the first card is replaced in the deck before the second is drawn, then we conclude that D_1 and D_2 are independent events. If the first card is not replaced before the second is drawn, then we conclude that D_1 and D_2 are dependent events. ■

■ **Example 2** Consider the two boxes shown in the figure. Suppose one ball is to be drawn from each box. Let event R_1 be "a red ball from Box 1" and event R_2 be "a red ball from Box 2." Then we conclude that R_1 and R_2 are independent events, because what happens in drawing a ball from Box 1 does not influence what happens in drawing a ball from Box 2, and vice versa.

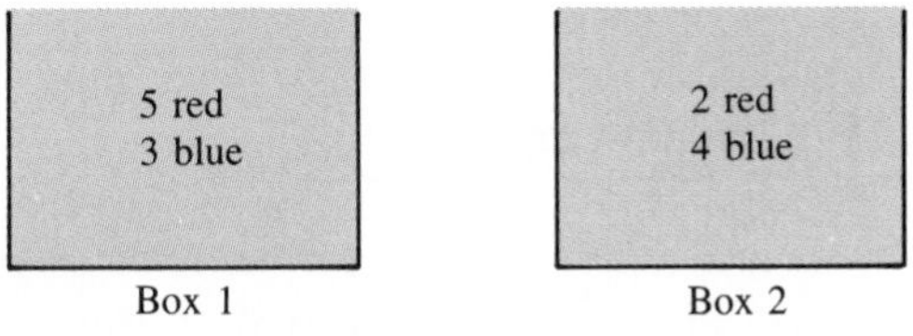

■

COMMENT: *The concept of independent events should not be confused with that of mutually exclusive events. The two concepts are quite different. To illustrate the difference, suppose an experiment consists of drawing two cards in succession from a deck of playing cards. Further, suppose that the first card is replaced before the second is drawn. Then the events A (the first card is an ace) and B (the second card is a heart) are independent, but they are not mutually exclusive. They are not mutually exclusive because the card drawn could be the ace of hearts.*

Next, we want to consider the probability that two successive independent events will occur. Consider an experiment that consists of drawing one ball from each of the boxes shown in Example 2. What is the probability that two red balls will be drawn? As noted in Example 2, the events R_1 (a red ball from Box 1) and R_2 (a red ball from Box 2) are independent events. Since a red ball can be drawn from Box 1 in five ways out of the eight balls in the box, the probability of R_1 is $P(R_1) = \frac{5}{8}$; a red ball can be drawn from Box 2 in two ways out of the six balls in the box, so the probability of R_2 is $P(R_2) = \frac{2}{6}$ or $\frac{1}{3}$. Further, it follows from the counting principle that two red balls can be drawn from the two boxes in $5 \cdot 2$ ways out of $8 \cdot 6$ possible outcomes. Denoting the probability that both balls drawn from the boxes will be red by $P(R_1 \cap R_2)$, we have

$$P(R_1 \cap R_2) = \frac{5 \cdot 2}{8 \cdot 6} = \frac{5}{8} \cdot \frac{2}{6} = P(R_1) \cdot P(R_2)$$

That is, the probability that both balls will be red is the product of the probabilities of the independent events. This multiplication rule holds for independent events in general.

The Multiplication Rule for Independent Events

If $E_1, E_2, \ldots, E_n$ are independent events of a sample space S, then the probability that E_1 and E_2 and ... and E_n will occur is the product of the probabilities of the independent events. Denoting E_1 and E_2 and ... and E_n by $E_1 \cap E_2 \cap \cdots \cap E_n$, we have

$$P(E_1 \cap E_2 \cap \cdots \cap E_n) = P(E_1) \cdot P(E_2) \cdot \cdots \cdot P(E_n)$$

Example 3 Consider a jar containing five red balls and three yellow balls. A ball is drawn from the jar and replaced, and then a second ball is drawn. Calculate the probability that the first ball will be red and the second will be yellow.

Solution Let event R_1 be "a red ball on the first draw" and event Y_2 be "a yellow ball on the second draw." We note that R_1 and Y_2 are independent events, because the first ball is replaced before the second is drawn. The probability of a red ball on the first draw is $P(R_1) = \frac{5}{8}$; the probability of a yellow ball on the second draw is $P(Y_2) = \frac{3}{8}$. The probability of a red ball on the first and a yellow ball on the second is denoted by $P(R_1 \cap Y_2)$ and given by the multiplication rule to be

$$P(R_1 \cap Y_2) = P(R_1)P(Y_2) = \frac{5}{8} \cdot \frac{3}{8} = \frac{15}{64}$$

■

Example 4 Four coins are tossed onto a tabletop. What is the probability of getting four heads?

Solution The probability of any coin's coming up heads is $\frac{1}{2}$. Since we assume that each coin is independent of the others, the probability of four heads is

$$P(\text{four heads}) = P(H_1 \cap H_2 \cap H_3 \cap H_4) = P(H_1) \cdot P(H_2) \cdot P(H_3) \cdot P(H_4)$$
$$= \frac{1}{2} \cdot \frac{1}{2} \cdot \frac{1}{2} \cdot \frac{1}{2} = \frac{1}{16}$$

Example 5 At the beginning of this section, we mentioned a manufacturer of a radar detector that has a switch, a warning beeper, and a warning light. Company records show that 1.5% of the switches are defective, 1.0% of the beepers are defective, and 2.1% of the lights are defective. If a unit is taken at random from the production line and checked for defects, what is the probability that it has none of these defects?

Solution Let S, B, and L be the events

S: the switch is not defective
B: the beeper is not defective
L: the light is not defective

We assume that these events are independent because they involve separate components. The probabilities of the complements of these events are given, so the probabilities of the events are

$$P(S) = 1 - P(S') = 1 - 0.015 = 0.985$$
$$P(B) = 1 - P(B') = 1 - 0.010 = 0.990$$
$$P(L) = 1 - P(L') = 1 - 0.021 = 0.979$$

If we denote the event "no defects" by $S \cap B \cap L$, then the probability of this event is

$$P(\text{no defects}) = P(S \cap B \cap L) = P(S) \cdot P(B) \cdot P(L)$$
$$= (0.985)(0.990)(0.979) \approx 0.955$$

Or we could say that 95.5% of the units have none of the indicated defects.

Example 6 A game wheel with equally spaced numbers from 1 to 100 is turned and allowed to come to rest. The number on which it stops is recorded. Calculate the probability that no number will be repeated in ten trials.

Solution Consider the probability of not getting a duplicate number. After the first number is recorded, there are 99 numbers that could turn up that would not duplicate the first number, so the probability that the second number will not duplicate the first number is $\frac{99}{100}$. After the second trial, there are 98 possible numbers that will not duplicate either of the first two numbers, so the probability that the third number will not duplicate either of the first two numbers is $\frac{98}{100}$. Continuing in this manner, the probability of not getting a duplicate in ten trials is the product of the probabilities of the ten independent outcomes:

$$P(\text{no duplicate in 10 trials}) = \frac{100}{100} \cdot \frac{99}{100} \cdot \frac{98}{100} \cdot \cdots \cdot \frac{92}{100} \cdot \frac{91}{100} \approx 0.628$$

Thus, the probability of a duplicate number in ten trials is

$$\begin{aligned} P(\text{a duplicate in 10 trials}) &= 1 - P(\text{no duplicate in 10 trials}) \\ &\approx 1 - 0.628 = 0.372 \end{aligned}$$

■

Conditional Probability

The probability of an event may depend upon the information that we have available. To illustrate this point, consider the probability that a card, drawn from a deck of 52 playing cards, is a king. We know the probability of drawing a king is $\frac{4}{52}$, because there are four kings in the deck. Now suppose that someone tells us that the card that was drawn is a face card. This additional information allows us to revise the probability of its being a king to $\frac{4}{12}$, because there are only 12 face cards in the deck. The probability that the card is a king, given that it is a face card, is called **conditional probability.** If K is the event "king" and F is the event "face card," then the event "a king, given that it is a face card" is denoted $K|F$. The conditional probability that the card is a king, given that the card is a face card, is then written as

$$P(K|F) = \frac{4}{12}$$

With this understanding of the conditional probability $P(K|F)$, suppose we now divide the numerator and the denominator by 52, the number of cards in the deck. Thus,

$$P(K|F) = \frac{\frac{4}{52}}{\frac{12}{52}}$$

The $\frac{4}{52}$ in the numerator can be interpreted as $P(K \cap F)$, and the $\frac{12}{52}$ in the denominator as $P(F)$. Using these substitutions, we have

$$P(K|F) = \frac{P(K \cap F)}{P(F)}$$

This form of $P(K|F)$ is used as the definition of conditional probability.

DEFINITION Conditional Probability

If A and B are events of a sample space S, and $P(B) \neq 0$, then the **conditional probability** of A given that B has occurred is

$$P(A|B) = \frac{P(A \cap B)}{P(B)}$$

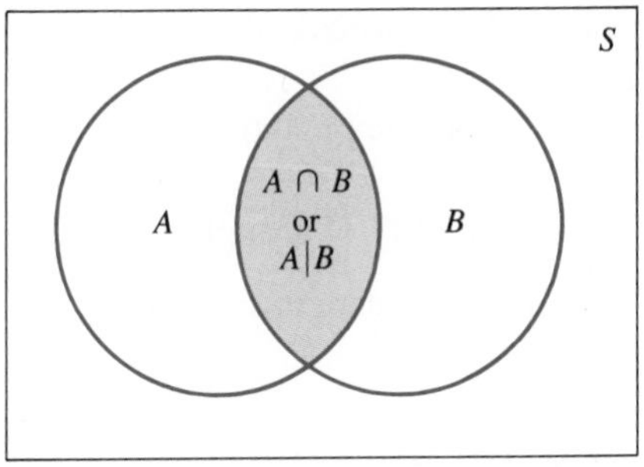

Figure 5.2

This definition can be justified by using the Venn diagram in Figure 5.2. Notice that events $A|B$ and $A \cap B$ are both represented by the dark shaded region in the figure. Assume there are n elements in the sample space S. Further, assume there are k elements in B, p of which are in the dark shaded region. Then the probability of B is $P(B) = \frac{k}{n}$, and since there are p elements in $A|B$, the probability of this event is $P(A|B) = \frac{p}{k}$. Similarly, there are p elements in $A \cap B$, so $P(A \cap B) = \frac{p}{n}$. Thus,

$$P(A|B) = \frac{p}{k} = \frac{\frac{p}{n}}{\frac{k}{n}} = \frac{P(A \cap B)}{P(B)}$$

Although we define conditional probability for dependent events as given above, most of the applications that we will see are related to the multiplication rule that arises when we solve this expression for $P(A \cap B)$.

Multiplication Rule for Dependent Events

If A and B are events and if A depends upon B, the probability of the intersection of A and B is

$$P(A \cap B) = P(B) \cdot P(A|B)$$

Similarly, if B depends upon A, the probability of the intersection of A and B is

$$P(A \cap B) = P(A) \cdot P(B|A)$$

COMMENT: *It is important to note that for independent events, $P(A|B) = P(A)$. As a matter of fact, this formula is used as a definition of an independent event, and is the* *only way we have to verify that two events are independent. Clearly, the multiplication rule given for independent events is a special case of the multiplication rule given here.*

Example 7 Calculate the probability of successively drawing two spades from a deck of playing cards if the first card is not replaced.

Solution The events are:

S_1: a spade on the first draw

S_2: a spade on the second draw

The probability of a spade on the first draw is $P(S_1) = \frac{13}{52}$. Since the first card is not returned to the deck, there are only 51 cards in the deck when the second card is drawn; further, if the first card was a spade, there are only 12 spades in the deck when the second card is drawn. The probability of drawing a spade on the second draw is the conditional probability $P(S_2|S_1) = \frac{12}{51}$. Therefore, by the multiplication rule, the probability of drawing two spades in succession is

$$P(S_1 \cap S_2) = P(S_1) \cdot P(S_2|S_1) = \frac{13}{52} \cdot \frac{12}{51} = \frac{1}{17}$$

■

Example 8 A box contains seven red balls and six white balls. What is the probability of drawing a red ball first and a white ball second, if the first ball is not replaced?

Solution The probability of a red ball on the first draw is $P(R_1) = \frac{7}{13}$. Since the first ball is not replaced before the second is drawn, there are only 12 balls (6 of which are white) left in the box. The conditional probability of a white ball on the second draw is then $P(W_2|R_1) = \frac{6}{12}$. Finally, the desired probability is

$$P(R_1 \cap W_2) = P(R_1) \cdot P(W_2|R_1) = \frac{7}{13} \cdot \frac{6}{12} = \frac{7}{26}$$

■

Example 9 A sports car club has 50 members: 33 men and 17 women. Of the men, 37% own Corvettes; of the women, 59% own Triumphs. If a person is chosen at random from the club roster, what is the probability that the person is a man who owns a Corvette?

Solution Let event A be "person is a man" and event B be "person owns a Corvette." Then $A \cap B$ represents a man who owns a Corvette and $B|A$ represents the conditional event "the person owns a Corvette given that the person is a man." The probability that the person is a man is $P(A) = \frac{33}{50}$. The probability that the person owns a Corvette is not given in the statement of the problem, but the conditional probability that the person owns a Corvette, given that the person is a man, is given as $P(B|A) = 0.37$. Thus, by the product rule, the probability that the person selected is a man who owns a Corvette is

$$P(A \cap B) = P(A) \cdot P(B|A) = \frac{33}{50} \cdot (0.37) \approx 0.24$$

■

In Section 5.3 we used the addition rule to calculate the probabilities of certain events, and in this section we introduced the multiplication rule. The next example shows a problem in which both the addition rule and the multiplication rule are used in calculating the probability of an event.

Example 10 Two boxes of red and green balls are shown below. A box is selected at random, and a ball is drawn at random from this box. Calculate the probability that:

(a) Box 1 is selected and a green ball is selected.
(b) Box 2 is selected and a green ball is selected.
(c) A green ball is selected.

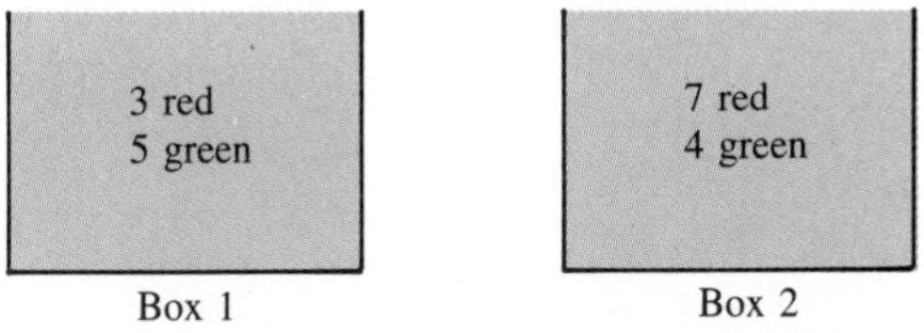

Solution

(a) The probability that Box 1 is selected is $P(\text{Box } 1) = \frac{1}{2}$, because we assume each box is equally likely to be selected. The probability that a green ball is selected, given that Box 1 was selected, is $P(G|\text{Box } 1) = \frac{5}{8}$. Thus, by the product rule, the probability that Box 1 and a green ball are selected is

$$P(\text{Box } 1 \cap G) = P(\text{Box } 1) \cdot P(G|\text{Box } 1) = \frac{1}{2} \cdot \frac{5}{8} = \frac{5}{16}$$

(b) The probability that Box 2 and a green ball are selected is, by a similar analysis, given by

$$P(\text{Box } 2 \cap G) = P(\text{Box } 2) \cdot P(G|\text{Box2}) = \frac{1}{2} \cdot \frac{4}{11} = \frac{2}{11}$$

(c) Selecting a green ball means that either "Box 1 and G" or "Box 2 and G" has occurred. Since only one of these two events can occur, it follows that the addition rule for mutually exclusive events should be used to find the probability of a green ball. Thus, the probability of a green ball is the sum of the probabilities of these two mutually exclusive events, and is given by

$$P(G) = P(\text{Box } 1 \cap G) + P(\text{Box } 2 \cap G) = \frac{5}{16} + \frac{2}{11} = \frac{87}{176} \approx 0.49$$

■

COMMENT: Suppose the balls in Box 2 of Example 10 are dumped into Box 1, so that it contains 19 balls, 10 red and 9 green. The probability of drawing a green ball from this box is then $\frac{9}{19} \simeq 0.47$, which is not the result obtained in Example 10 when the 19 balls were in two separate boxes. Can you explain why these two results differ? For an explanation see the April 1989 issue of ***Mathematics Teacher****, p. 250.*

Exercises Section 5.4

For Exercises 1–4, two cards are drawn in succession from a deck of 52 playing cards. Calculate the probability that:

1. Two hearts are drawn, if the first card is replaced before the second is drawn.

2. The first card is an ace and the second card is a face card, if the first card is replaced before the second is drawn.

3. Two hearts are drawn, if the first card is not replaced.

4. The first card is an ace and the second card is a face card, if the first card is not replaced.

5. Three cards are drawn (without replacement) from a deck of 52 playing cards. Calculate the probability that the first card is a heart, the second is a diamond, and the third is a club.

6. One card is drawn from each of three separate decks of playing cards. Calculate the probability that all three cards are hearts.

7. Box 1 contains seven red balls and four black balls; Box 2 contains four red balls and nine black balls. A ball is drawn at random from each box. Calculate the probability that (a) both balls are red; (b) a red ball is drawn from Box 1 and a black ball from Box 2.

8. Two cards are drawn (without replacement) from a deck of 52 playing cards. What is the probability that the first card is red and the second is a black face card?

9. Two people are chosen at random from a group of ten people, which includes only one married couple. Calculate the probability that the married couple will be chosen.

10. The names Tom, Dick, Ann, Clyde, Jane, and Sam are placed in a hat. What is the probability that the first three names drawn will be, in order, Ann, Dick, and Clyde?

For Exercises 11–14, a box contains three red balls, seven black balls, and five white balls. Two balls are drawn at random (without replacement) from the box. Calculate the probability that:

11. A red ball or a black ball is drawn first and a white ball second.

12. A red ball or a white ball is drawn first and a black ball second.

13. One red ball and one black ball are drawn.

14. One red ball and one white ball are drawn.

For Exercises 15 and 16, a company buys three boxes of 100 machined parts, one from each of three different companies. There are three defective parts in the box from Company A, two defective parts in the box from Company B, and two defective parts in the box from Company C. One part is selected at random from each box. Calculate the probability that:

15. All three parts are defective.

16. At least two of the three are defective.

For Exercises 17 and 18, five cards are drawn without replacement at random from a deck of 52 playing cards. Calculate the probability that:

17. Exactly three of the cards are clubs.

18. Exactly two of the cards are kings.

For Exercises 19–23, consider three boxes of balls:

3 red 2 black	5 red 4 black	2 red 7 black
Box 1	Box 2	Box 3

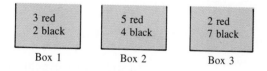

A ball is drawn at random from Box 1. If it is red, the second ball is drawn from Box 2; if it is black, the second ball is drawn from Box 3. Calculate the

probability that:

19. Both balls are red.

20. Both balls are black.

21. The first ball is red and the second ball is black.

22. The first ball is black and the second ball is red.

23. One ball is red and one ball is black.

24. Box 1 contains two green balls and one orange ball; Box 2 contains one green ball and two orange balls. One ball is selected at random from Box 1 and transferred to Box 2, and a ball is then drawn at random from Box 2. What is the probability that the ball drawn from Box 2 is green?

25. Fifty tickets, numbered from 1 to 50, are placed in a box. A ticket is drawn at random from the box. Its number is recorded and it is replaced before another ticket is drawn.
(a) If five tickets are drawn in succession, what is the probability that the same ticket number will be drawn twice?
(b) How many times must a drawing be made in order that the probability that the same ticket number is drawn twice will be 0.5 or greater?

26. A calculator consists of a display, a processor, and a keyboard. Records show that 98.5% of the displays are good, 97.1% of the processors are good, and 95.3% of the keyboards are good. If we assume that the quality of each type of component is independent, what is the probability that a calculator chosen at random is good?

27. Fifty students take a math test and a physical fitness test. Forty-two of the students pass the math test; forty-five pass the physical fitness test. Under the assumption that these are independent events, calculate the probability that a student selected at random passed both tests.

28. Forty solid-fuel rocket motors were test-fired. Thirty-one ignited correctly and nine failed to ignite. Of those that failed to ignite, 42% had moisture in the fuel core. Calculate the probability that a rocket motor tested in the future will not ignite and will have moisture in the fuel core.

29. A hardware store has eight bins of nails. One of the bins contains finishing nails, of which 25% are aluminum. If a nail is selected at random from one of the bins, what is the probability that it is an aluminum finishing nail?

30. Bob has a drawer full of socks that are not paired. He takes two socks from the drawer without looking. What is the probability that he will get 2 blue socks in a drawer full of 5 blue, 9 white, and 6 black socks?

31. Two boxes of red and green balls are shown in the figure.

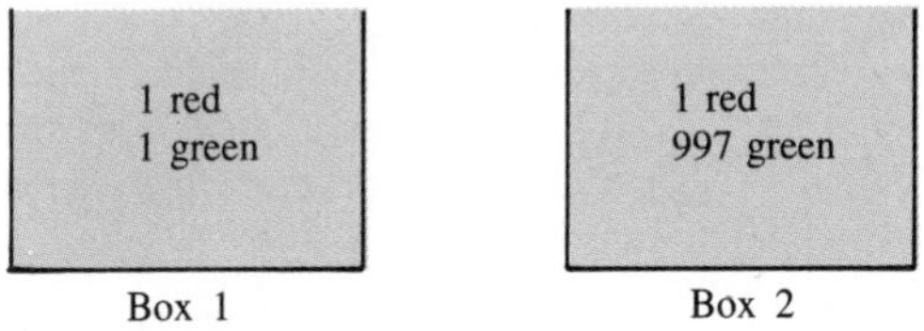

A box is selected at random, and a ball is drawn at random from this box. Calculate the probability that:
(a) Box 1 is selected and a green ball is selected.
(b) Box 2 is selected and a green ball is selected.
(c) A green ball is selected.
Now all 1000 balls are put into the same box. Calculate the probability that:
(d) A green ball is selected.

5.5 STOCHASTIC PROCESSES

If a sequence of experiments is performed in which the outcome of one experiment depends upon the outcome of another experiment, the sequence of experiments is called a **stochastic process.** Many (in fact most) of our daily tasks are stochastic in the sense that one follows from the other. Hence, we are interested in determining the probabilities associated with such dependent events. To illustrate a stochastic process, consider the two boxes of balls shown in Figure 5.3 and the sequence of experiments associated with choosing a box at random and then choosing a ball at random from this box. This is a stochastic process, consisting of the experiment "select a box at random" followed by the experiment "select a ball at random."

Figure 5.3

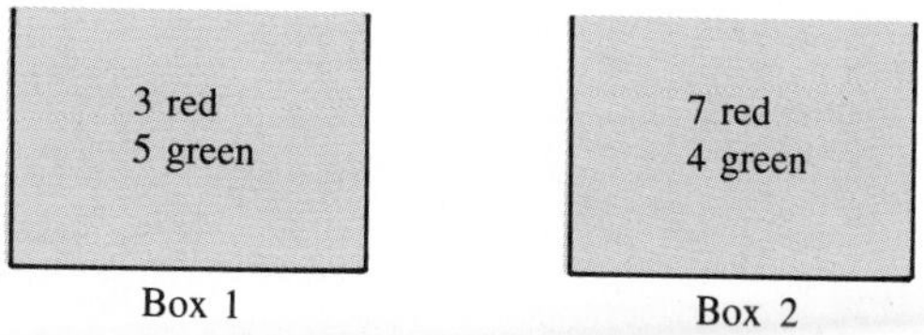

The outcomes of the first experiment are *Box 1* and *Box 2*; the outcomes of the second experiment are *red* and *green.* Clearly, the outcome of the second experiment depends upon the outcome of the first experiment. Because stochastic processes involve branching sequences of experiments, they are often represented by tree diagrams, as illustrated in the next example.

Example 1 Consider the two boxes of balls in Figure 5.3. If we choose a box at random and then choose a ball at random from this box, what is the probability that the ball is green? Notice that this problem was solved in Example 10 of Section 5.4. Here we use a tree diagram to represent the solution. The tree diagram is not the solution; it is just an aid to understanding the stochastic process.

Solution As noted above, this sequence of experiments is a stochastic process. We use a tree diagram to represent it. The various outcomes of each experiment are represented as branches emanating from a point, with the probability of the outcome indicated along the branch. For example, the experiment "select a box at random," which has the outcomes *Box 1* and *Box 2,* is shown in Figure 5.4. Since each box has an equal chance of being selected, each is assigned a probability of $\frac{1}{2}$, as shown in Figure 5.4. The experiment "select a ball at random" has outcomes of *red* and *green,* whose probabilities depend upon the outcome of the

Figure 5.4

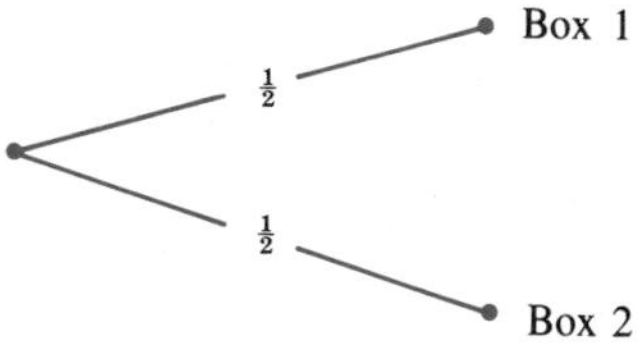

first experiment. We represent a sequence of experiments by stringing together the tree diagrams of the individual experiments. Figure 5.5 shows the tree diagram for the two experiments in this example. An outcome is indicated at the end of each branch, and the associated probability is indicated on the branch.

Figure 5.5

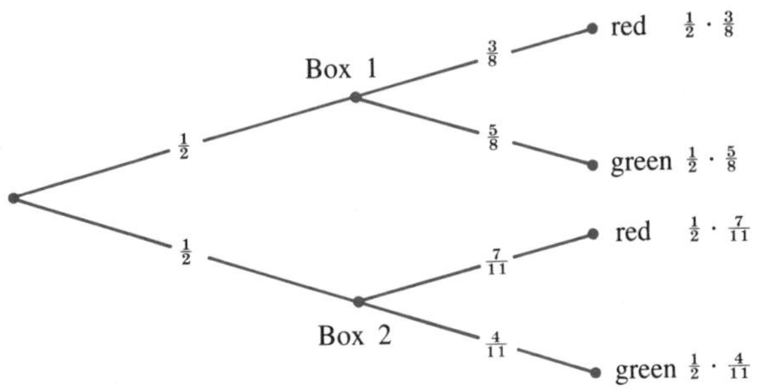

The probability of a sequence of outcomes is readily found by multiplying the probabilities along the appropriate path of the tree diagram. For instance, the probability of selecting a green ball from Box 1 is the product of the probability that Box 1 is selected and the probability that a green ball is selected from Box 1. This is written

$$P(\text{Box } 1 \cap G) = P(\text{Box } 1) \cdot P(G|\text{Box } 1) = \frac{1}{2} \cdot \frac{5}{8} = \frac{5}{16}$$

Similarly, the probability that a green ball is selected from Box 2 is

$$P(\text{Box } 2 \cap G) = P(\text{Box } 2) \cdot P(G|\text{Box } 2) = \frac{1}{2} \cdot \frac{4}{11} = \frac{2}{11}$$

To find the probability that a green ball is selected, we note that there are two ways to get a green ball: we can select either the path *Box 1 and green* or the path *Box 2 and green.* Since these two outcomes are represented by different paths in the tree diagram, we conclude that they are mutually exclusive. The probability of a green ball can then be found by adding the probabilities of these two branches. Thus,

$$P(G) = P(\text{Box } 1 \cap G) + P(\text{Box } 2 \cap G) = \frac{5}{16} + \frac{2}{11} = \frac{87}{176} \approx 0.49$$

■

Example 2 If a coin is tossed three times, what is the probability of getting two heads and one tail?

Solution We treat this problem as a sequence of three experiments in which the outcomes are H and T. This stochastic process is represented in the following tree diagram in Figure 5.6.

Figure 5.6

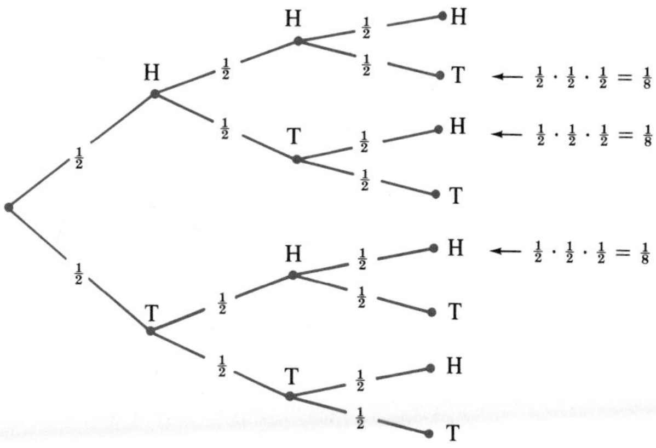

The figure shows that there are three mutually exclusive events that result in two heads and one tail—namely, HHT, HTH, and THH. Since all of the probabilities in the figure are $\frac{1}{2}$, the probability of each path is $\frac{1}{2} \cdot \frac{1}{2} \cdot \frac{1}{2} = \frac{1}{8}$. The probability of two heads and one tail is then the sum of the probabilities of the three mutually exclusive events shown in the figure. Thus,

$$P(2 \text{ heads and } 1 \text{ tail}) = P(HHT) + P(HTH) + P(THH) = \frac{1}{8} + \frac{1}{8} + \frac{1}{8} = \frac{3}{8}$$ ■

Example 3 Two cards are drawn without replacement from a deck of 52 playing cards. Draw a tree diagram for this stochastic process, assuming we are interested in the probability that the first card is a face card and the second one is a two.

Solution We can think of this in terms of a sequence of two experiments in which the outcome of the first experiment is *face card* and that of the second is *two*. The tree diagram is then given in Figure 5.7.

Figure 5.7

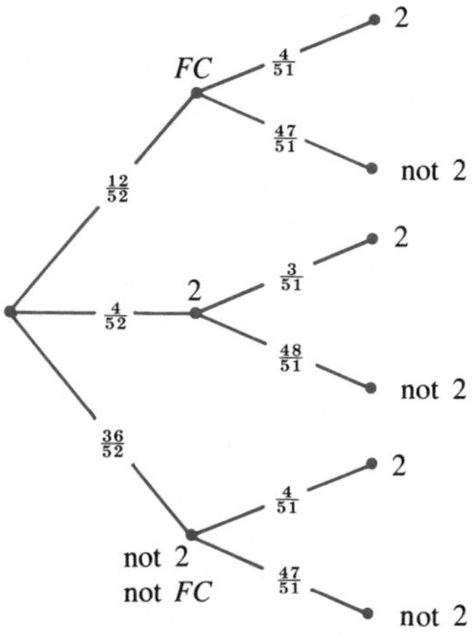

■

Example 4 A restaurateur owns three restaurants: one in Tampa, one in Orlando, and one in Daytona. The company that supplies coffee wants to do some market research on the coffee preferences of the patrons of these three restaurants. The waitresses in Tampa report that 45% of the customers who drink coffee prefer regular coffee and 55% prefer decaf; in Orlando, 58% prefer regular coffee and 42% prefer decaf; and in Daytona, 33% prefer regular and 67% prefer decaf. If the company selects a coffee drinker at random, calculate the probability that:

(a) The coffee drinker is from Orlando and prefers decaf.
(b) The coffee drinker prefers regular coffee.
(c) The restaurant is in Daytona if the person prefers regular coffee.

Solution Assuming that each restaurant has an equal chance of being selected, the stochastic process is represented in the tree diagram in Figure 5.8.

Figure 5.8

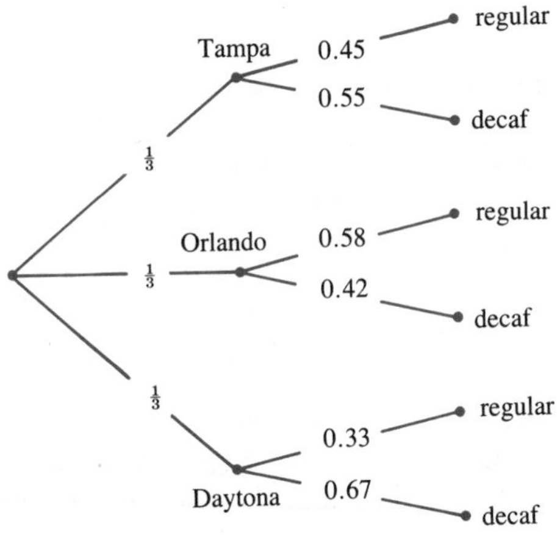

(a) From the figure, the probability of a decaf drinker from Orlando is

$$P(\text{Orlando} \cap D) = P(\text{Orlando}) \cdot P(D|\text{Orlando}) = \frac{1}{3}(0.42) = 0.14$$

(b) From the figure, the probability of a regular coffee drinker is

$$P(R) = P(\text{Tampa} \cap R) + P(\text{Orlando} \cap R) + P(\text{Daytona} \cap R)$$
$$= \frac{1}{3}(0.45) + \frac{1}{3}(0.58) + \frac{1}{3}(0.33) = 0.15 + 0.19 + 0.11 = 0.45$$

(c) Here, we want to find the conditional probability that the person is from Daytona, if we know the person prefers regular coffee. Using the definition of the conditional probability of a person's being from Daytona, given that the person prefers regular coffee, we have

$$P(\text{Daytona}|R) = \frac{P(\text{Daytona} \cap R)}{P(R)} = \frac{0.11}{0.45} \approx 0.24$$

Thus, there is about a 24% chance that the person is from Daytona if it is known that the person prefers regular coffee. ▪

EXERCISES SECTION 5.5

Draw a tree diagram to represent each of the stochastic processes in the exercise set, then use the tree diagram to assist you in calculating the required probabilities.

For Exercises 1–6, two boxes of marbles are represented in the figure below.

7 red 4 blue	4 red 9 blue
Box 1	Box 2

A box is selected at random and a marble is then drawn at random from this box. Draw a tree diagram for this stochastic process. Calculate the probability that:

1. A red marble is selected from Box 1.
2. A blue marble is selected from Box 2.
3. A red marble is selected.
4. A blue marble is selected.
5. The marble came from Box 1, given that it is red.
6. The marble came from Box 2, given that it is blue.

For Exercises 7–12, two cards are drawn from a deck of 52 playing cards. Calculate the probability that

7. An ace is drawn first and a king second, if the first card is replaced before the second is drawn.
8. A ten is drawn first and a jack second, if the first card is replaced before the second is drawn.
9. An ace is drawn first and a king second, if the first card is not replaced.
10. A ten is drawn first and a jack second, if the first card is not replaced.
11. A heart is drawn first and a black card second, if the first card is not replaced.
12. A queen is drawn first and a face card second, if the first card is not replaced.

For Exercises 13–15, three boxes of red and black balls are represented in the figure.

3 red 2 black	5 red 4 black	2 red 7 black
Box 1	Box 2	Box 3

A ball is drawn from Box 1; if it is red, the second ball is drawn from Box 2, and if it is black, the second ball is drawn from Box 3. Draw a tree diagram for this stochastic process. Calculate the probability that:

13. Both balls are red.
14. The first ball is red and the second is black.
15. One ball is red and one ball is black.
16. Box A contains two black balls and one white ball; Box B contains one black ball and two white balls. One ball is transferred from Box A to Box B and then a ball is drawn at random from Box B. Draw a tree diagram for this stochastic process and calculate the probability that the ball drawn is black.
17. Box A contains two red balls and two white balls; Box B contains three red balls and four white balls; Box C contains two red balls and five white balls. If a box is selected at random and a ball drawn at random from this box, what is the probability that the ball is red? Draw a tree diagram to represent this stochastic process.
18. A company receives two shipments of motors and generators. One shipment contains 50 motors and 25 generators; the other, 40 motors and 20 generators. If one item is selected from each shipment, what is the probability of selecting one motor and one generator? Draw a tree diagram of the stochastic process.
19. Three ordinary coins and one modified coin, which is weighted to turn up heads 60% of the time, are tossed onto a table and two of the coins turn up heads. If one of the coins showing heads is picked at random, what is the probability that it is the modified coin? Draw a tree diagram for this stochastic process.
20. Solve the problem in Exercise 19 if three of the four coins turn up heads.

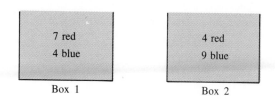

For Exercises 21–24, a politician hires a marketing firm to conduct a public opinion poll of voters in Iowa to see if there is enough public recognition to pursue the governorship of the state. The table shows the results of the poll.

	Political Preference	Recognized Politician's Name	Did Not Recognize Politician's Name
Democrat	985	345	640
Republican	544	402	142
Independent	471	250	221

If a voter is selected at random, what is the probability that

21. The voter is a Democrat who recognized the politician's name.

22. The voter is a Republican who did not recognize the politician's name.

23. The voter is an independent if it is known that the voter recognized the politician's name.

24. The voter is a Democrat if it is known that the voter did not recognize the politician's name.

5.6 BAYES' FORMULA

Suppose NASA scientists can calculate the probability that a space probe will fail to reach a certain trajectory if it is known that three of the probe's thrusters have failed to work. A question that might be of interest to the scientists is "If the probe fails to reach the desired trajectory, what is the probability that three of the thrusters failed to work?" This problem is typical of problems in which an outcome is known and the experimenter wants to know the probability that a certain event "caused" the outcome. An approach to solving a similar problem is discussed in the next example, culminating in a formula for systematically treating problems in which we want to know the probability of the "cause" of an outcome.

Example 1 Consider the two boxes of red and green balls shown in the figure. An experiment consists of selecting a box at random and then selecting a ball at random from this box. If a green ball is selected, what is the probability that it came from Box 1?

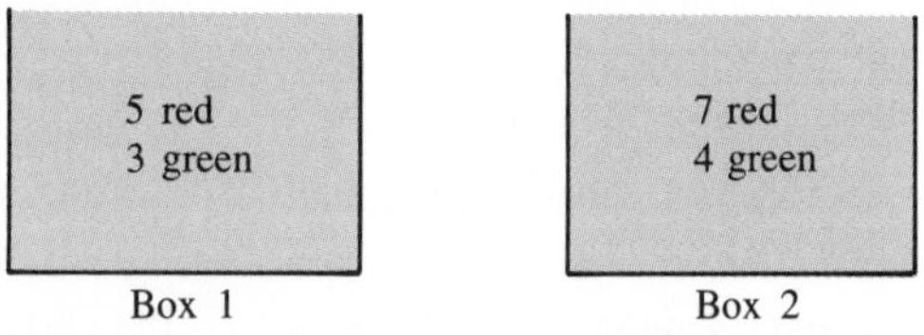

Solution The probability that the ball came from Box 1, given that a green ball was selected, is the conditional probability $P(\text{Box } 1 | G)$, which is given by

$$P(\text{Box } 1 | G) = \frac{P(\text{Box } 1 \cap G)}{P(G)} \tag{1}$$

The numerator is given by

$$P(\text{Box } 1 \cap G) = P(\text{Box } 1) \cdot P(G | \text{Box } 1) = \frac{1}{2} \cdot \frac{3}{8} = \frac{3}{16} \tag{2}$$

The denominator is given by

$$P(G) = P(\text{Box } 1 \cap G) + P(\text{Box } 2 \cap G)$$

$$P(G) = P(\text{Box } 1) \cdot P(G|\text{Box } 1) + P(\text{Box } 2) \cdot P(G|\text{Box } 2) = \frac{1}{2} \cdot \frac{3}{8} + \frac{1}{2} \cdot \frac{4}{11} \quad \textbf{(3)}$$

Thus, the conditional probability that the ball was selected from Box 1 if it is green is

$$P(\text{Box } 1|G) = \frac{P(\text{Box } 1 \cap G)}{P(G)} = \frac{\frac{3}{16}}{\frac{1}{2} \cdot \frac{3}{8} + \frac{1}{2} \cdot \frac{4}{11}} \approx 0.51$$

■

If in Example 1, instead of using the numbers from (2) and (3) in (1), we use the formulas for $P(\text{Box } 1 \cap G)$ and $P(G)$, we get the formula

$$P(\text{Box } 1|G) = \frac{P(\text{Box } 1) \cdot P(G|\text{Box } 1)}{P(\text{Box } 1) \cdot P(G|\text{Box } 1) + P(\text{Box } 2) \cdot P(G|\text{Box } 2)}$$

This formula is a special case of Bayes' formula, which we now state in general terms for n mutually exclusive events as "causes" of some possible event.

Bayes' Formula

Let $A_1, A_2, \ldots, A_n$ be n mutually exclusive events such that the union of these events is the sample space S, and let B be any possible event. Then the probability that A_i was the "cause" of B is

$$P(A_i|B) = \frac{P(A_i) \cdot P(B|A_i)}{P(A_1) \cdot P(B|A_1) + P(A_2) \cdot P(B|A_2) + \cdots + P(A_n) \cdot P(B|A_n)}$$

COMMENT: *The important special case for $n = 2$ is the case that we will use most often. Bayes' formula for A_1 given B is*

$$P(A_1|B) = \frac{P(A_1) \cdot P(B|A_1)}{P(A_1) \cdot P(B|A_1) + P(A_2) \cdot P(B|A_2)}$$

where A_1 and A_2 are complementary sets. A similar formula holds for A_2 given B.

HISTORICAL COMMENT: *Bayes' formula is named for Thomas Bayes (1702–1763), who wrote a noteworthy paper on probability.*

Bayes' formula can be interpreted in terms of a tree diagram as follows: If we let B be an outcome of an experiment and $A_1, A_2, \ldots, A_k$ be the mutually exclusive events whose union is the sample space S, then the branches of the tree that lead to B, as shown in Figure 5.9, give the probabilities that path i leads to B. The probability of $A_i|B$ is then the probability that path i leads to B, divided by the sum of the probabilities of all paths that lead to B.

Figure 5.9

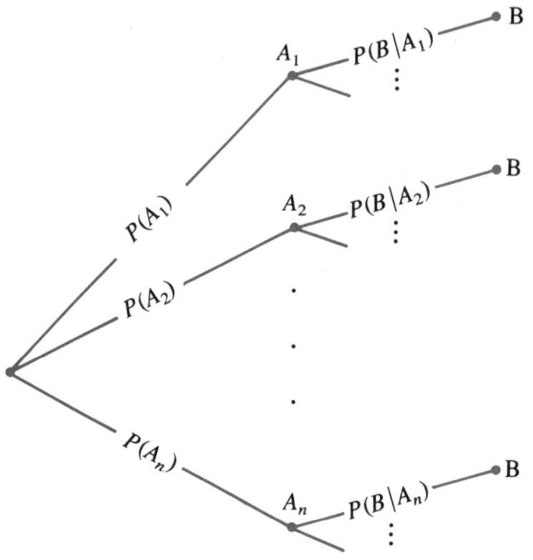

Example 2 An industrial engineer finds that 97% of the parts that are acceptable are correctly passed by the quality control inspector as acceptable and 3% of the acceptable parts are mistakenly rejected as defective. Further, the study shows that 5% of the defective parts that are inspected are passed as acceptable. A part is taken at random from a box of 1000 parts, 80 of which are defective. Use Bayes' formula to calculate the probability that the part is acceptable, given that the inspector passes it as acceptable.

Solution The events of interest are:

A_1: the part is acceptable
A_2: the part is defective
B: the part is passed as acceptable

Figure 5.10

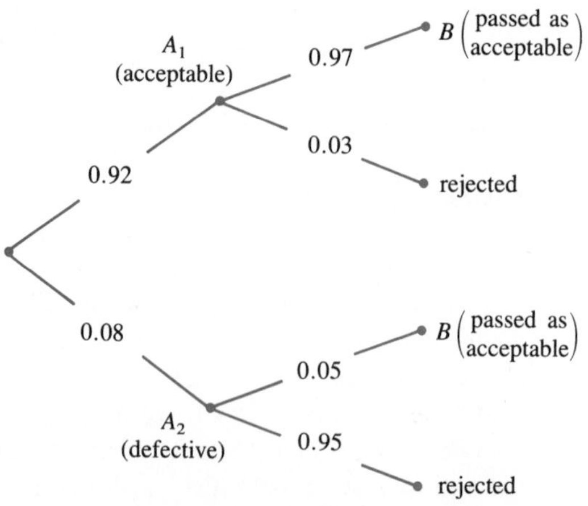

We want $P(A_1|B)$, the probability that the part is acceptable if it is passed as acceptable. We note that the appropriate probabilities are

$P(A_1) = \frac{920}{1000} = 0.92$ *The probability of selecting an acceptable part*

$P(A_2) = \frac{80}{1000} = 0.08$ *The probability of selecting a defective part*

$P(B|A_1) = 0.97$ *The probability that an acceptable part is passed as acceptable*

$P(B|A_2) = 0.05$ *The probability that a defective part is passed as acceptable*

The tree diagram for this stochastic process is shown in Figure 5.10. Thus, the probability that a part is acceptable when it is passed as acceptable by the quality inspector is given by Bayes' formula to be

$$P(A_1|B) = \frac{P(A_1) \cdot P(B|A_1)}{P(A_1) \cdot P(B|A_1) + P(A_2) \cdot P(B|A_2)} = \frac{(0.92)(0.97)}{(0.92)(0.97) + (0.08)(0.05)}$$
$$\approx 0.996$$

Note that this means that about 4 parts in 1000 will be passed as acceptable when they are really unacceptable. ■

Example 3 A marketing class is studying buying trends among American consumers. A survey of car buyers in the U. S. shows that 35% of car buyers buy cars built in the Orient, 45% buy cars built in the U.S., and 20% buy cars built in Europe. The table below shows, for each of the three areas, the proportion of car buyers and the proportion of repeat buyers—that is, consumers who buy the same brand they previously bought.

	Proportion of Buyers	Proportion of Repeat Buyers
Car built in Orient	0.35	0.71
Car built in U. S.	0.45	0.68
Car built in Europe	0.20	0.82

If a person selected at random is a repeat buyer, what is the probability that the person's car was built in the U. S.?

Solution Let A_1 be "car built in Orient," A_2 be "car built in U. S.," and A_3 be "car built in Europe," and let B be "person is a repeat buyer." Then the probability that the car was built in the U. S., given that the person is a repeat buyer, is denoted by $P(A_2|B)$ and given by Bayes' formula to be

$$P(A_2|B) = \frac{P(A_2) \cdot P(B|A_2)}{P(A_1) \cdot P(B|A_1) + P(A_2) \cdot P(B|A_2) + P(A_3) \cdot P(B|A_3)}$$
$$= \frac{(0.45)(0.68)}{(0.35)(0.71) + (0.45)(0.68) + (0.20)(0.82)} \approx 0.43$$

Thus, if we know that the consumer is a repeat buyer, the probability that the person buys U. S. cars is 0.43. ■

EXERCISES SECTION 5.6

For Exercises 1 and 2, consider two boxes of blue and white balls as shown.

Box 1	Box 2
8 blue 3 white	5 blue 9 white

A box is selected at random and a ball is selected at random from this box.

1. If a white ball is selected, what is the probability that it came from Box 1?
2. If a blue ball is selected, what is the probability that it came from Box 2?

For Exercises 3 and 4, consider three boxes of red and green balls as shown.

Box 1	Box 2	Box 3
4 red 3 green	5 red 7 green	2 red 4 green

A ball is drawn at random from Box 1. If it is red, a ball is drawn from Box 2; if it is green, a ball is drawn from Box 3.

3. If the second ball drawn is red, what is the probability that a red ball was drawn from Box 1?
4. If the second ball drawn is green, what is the probability that a red ball was drawn from Box 1?

A survey of registered voters was conducted to find out what percentage of voters favor a national health program. The results of the survey are shown in the table.

	Political Party	Favor Nat'l Health
Democrat	48%	89%
Republican	35%	78%
Independent	17%	52%

Assume that these data are representative of the U.S. population and that a registered voter is selected at random.

5. What is the probability that the person is a Republican who favors a national health program?
6. What is the probability that the person is a Democrat who favors a national health program?
7. If the person favors a national health program, what is the probability that the person is a Democrat?
8. If the person favors a national health program, what is the probability that the person is an independent?

A new test for diabetes will detect the disease 98% of the time in a person who has the disease and will fail to detect it 2% of the time. In addition, the test will falsely detect the disease 7% of the time in a healthy person. If the test is given to a person selected at random from a group of subjects, 95 of whom are healthy and 5 of whom have diabetes, what is the probability that:

9. Diabetes will be detected if the person has the disease?
10. Diabetes will be falsely detected if the person does not have the disease?
11. The person has diabetes if the test detects the disease?
12. The person does not have diabetes if the test fails to detect the disease?

A geologist claims to have developed a test that can be used to identify oil deposits by analyzing satellite photographs. The geologist estimates that there is a 0.8 probability that the test will detect an oil field, and that there is a 0.3 probability that it will falsely detect an oil field where no field exists. Data indicate that 10% of the exploratory drillings in suspected oil fields will yield oil. Calculate the probability that:

13. An oil field exists if the test indicates that an oil field exists.
14. An oil field does not exist if the test indicates that an oil field exists.

A television sports show conducts a survey of its viewers to see whether they think salaries are too high in professional sports. The station records the gender of each respondent and the number of respondents who think that salaries are too high in baseball, basketball, and football. The responses are shown in the table.

Gender	Number Surveyed	Baseball	Basketball	Football
Male	150	123	140	105
Female	39	30	25	11

Thus, 150 males were surveyed, of whom 123 thought that baseball salaries are too high, and so on. Based on this survey, calculate the probability that if a viewer is chosen at random:

15. The viewer is female if the viewer thinks basketball salaries are too high.

16. The viewer is male if the viewer thinks football salaries are too high.

17. The viewer is a male if the viewer thinks that baseball salaries are too high.

18. The viewer is a female if the viewer thinks that football salaries are too high.

IMPORTANT WORDS AND PHRASES CHAPTER 5

In this chapter you have been exposed to the fundamental language of probability theory. Here are some of the important words and phrases for you to recall:

addition rule
Bayes' formula
conditional probability
dependent events
equally likely outcomes
event
experiment
independent events
multiplication rule
  for dependent events
  for independent events
mutually exclusive events
odds
outcomes
probability model
random selection
relative frequency
sample space
stochastic process
trial
uniform probability model

REVIEW EXERCISES CHAPTER 5

1. An experiment consists of rolling a die 1000 times and recording the number of times a 6 comes up. If a 6 comes up 165 times, what is the relative frequency of rolling a 6?

2. Give the sample space of an experiment that consists of drawing a card from a deck of 52 playing cards and recording its suit.

3. A red die and a blue die are rolled.
(a) Give the sample space if the sample points indicate that the sum of the numbers on the two dice is odd.
(b) Let A be the event "the sum of the two dice is odd." Let B be the event "the sum is less than 5." Are A and B mutually exclusive?
(c) Give the sample points in event $A \cap B$.

4. (a) Give the sample space for the experiment of having four children.
(b) Give the sample space for the event "the oldest child is a boy."

5. Juanita draws a card from a deck of 52 playing cards. Calculate the probability that:
(a) The card is a spade.
(b) The card is black.
(c) The card is the queen of spades.

6. Jim has a drawer containing eight blue, five black, and six white socks. If two socks are drawn, what is the probability that Jim will draw a pair of blue socks?

7. In a family with two children, what is the probability that both are boys?

8. A bowl contains four red, five white, and eight yellow balls. A ball is drawn.
   (a) What is the probability that the ball is red?

   Assuming that the first ball is red and is not replaced:

   (b) What is the probability that the second ball is red?
   (c) What is the probability that a red ball is not drawn in either draw?
9. Five coins are tossed. What is the probability that exactly two of them will be heads?
10. Calculate the probability of drawing a flush when a five-card hand is dealt from a well-shuffled deck of 52 cards. (Recall that a flush consists of all cards of the same suit.)
11. An economist says that the odds in favor of a recession are 6 to 5. What is the probability of a recession?
12. An assembly line produces mufflers for automobiles. The inspector finds that 78% of the mufflers are acceptable, 10% need reworking, and the rest need to be scrapped. If a part is chosen at random, what is the probability that:
    (a) It is acceptable?
    (b) It needs to be scrapped?
13. The weather forecaster at station WIBV is correct 82% of the time; the forecaster at neighboring station WILA, 65% of the time. What is the probability that on a given occasion, one of the two will be correct?
14. If Tanya selects a security from a list of three growth stocks, seven income stocks, and five bonds, what is the probability that she will select a bond or a growth stock?
15. Two cards are drawn from a deck of 52 playing cards. Compute the probability of drawing two aces in succession if:
    (a) The first card is an ace, and it is returned to the deck before the second draw is made.
    (b) The first card is an ace, and it is not replaced in the deck.
16. A manufacturer of automobiles receives 1000 car radios from each of three different suppliers. Unknown to the manufacturer, there are five defective radios from supplier A, seven from B, and only two from C. As a means of quality checking, one radio is selected at random from each of the shipments. What is the probability that:
    (a) All the radios selected are in working order?
    (b) At least one of the selected radios is defective?
17. A computer from the Electronic Computer company consists of a keyboard, a monitor, a CPU, and a printer. The suppliers of these four units claim to have a percentage-of-defects record of 2.5%, 1%, 4%, and 6%, respectively. What claim of reliability can the Electronic Computer company make, in terms of the percentage of its components that are probably defective?
18. A retailer receives two shipments of TV sets. The first shipment, from company A, is known historically to be 5% defective. The second, from company B, is known to be 3% defective. If one item is selected from each shipment, what is the probability of selecting one good TV and one defective TV? Draw a tree diagram of the stochastic process.
19. Box A contains three red balls and two white balls. Box B contains four red balls and six white balls. Box C contains three red balls and seven white balls. If a box is selected at random and a ball drawn at random from this box, what is the probability that the ball is red? Draw a tree diagram to represent the stochastic process.
20. A new test for Alzheimer's Disease will detect the disease 95% of the time in a person who has Alzheimer's and will fail to detect it 5% of the time. In addition, the test will falsely detect the disease 15% of the time in a healthy person. If the test is given to a person selected at random from a group of subjects, 90 of whom are healthy and 10 of whom have Alzheimer's, what is the probability that:
    (a) Alzheimer's will be detected if the person has the disease?
    (b) Alzheimer's will be falsely detected if the person does not have the disease?
    (c) The person has Alzheimer's if the test detects the disease?

CHAPTER 6

PROBABILITY AND STATISTICS

Probability and the judgments made on the basis of probability affect us daily. A morning weather prediction of a 70% chance of rain makes us pick up an umbrella on the way out. The frequency of repairs on particular car models, as listed in *Consumer Reports*, guides our selection of a used car. The published admission rate for a certain college may encourage or discourage our application. These probabilities for weather, car repair frequency, and admission rate are all based on information that was gathered and studied. Statistics is the branch of mathematics that deals with the collection, analysis, interpretation, and presentation of numerical data. The interpretation of, and the inferences made from, the numerical data depend on the concepts of probability we studied in Chapter 5. In this chapter, we will look at the relationship between probability and statistics.

6.1 Random Variables and Probability Distributions

Recall that a sample space consists of the possible outcomes of an experiment. The probability of each of these outcomes is determined or assigned. However, the outcomes in these sample spaces are generally not numerical. Here are a few we have looked at:

$$A = \{AH, 2H, 3H, \ldots, AD, 2D, 3D, \ldots, AS, 2S, 3S, \ldots, AC, 2C, 3C, \ldots, KC\}$$
$$= \{x \mid x \text{ is a card drawn from a deck of 52 playing cards}\}$$

$$B = \{GG, RG, GR, RR\}$$
$$= \{xy \mid x \text{ is the color of the first ball drawn and } y \text{ is the color of the second ball drawn from a box containing two red and three green balls}\}$$

$$C = \{1, 2, 3, 4, 5, 6\} = \{x \mid x \text{ is the number on the face of a die tossed once}\}$$

$$D = \{HH, HT, TH, TT\}$$
$$= \{xy \mid x \text{ is the face of a coin on the first toss and } y \text{ is the face of a coin on the second toss}\}$$

Notice that only sample space C is made up of numbers. Since statistics deals with numbers, we will look at a way to pair each element of a sample space with a number and then with its probability.

Function

The concept of a function, which was discussed in Section 1.3, will help us to pair each element in a sample space with a number. Recall that a function is any pairing of numbers in which no two pairs with the same first element have different second elements. The pairings can be formed by a formula, such as $y = x^2$, or by simply listing a set of ordered pairs of numbers. Sometimes it is convenient to use a table to indicate a functional pairing. For instance, the following table helps us see that $y = x^2$ assigns to y a number that is the square of x.

x	$y = x^2$
1	1
2	4
3	9
4	16

Recall that a function assigns a unique value of y to each x. This means that the numbers in the next table do not represent a functional pairing, because $x = 2$ is paired with $y = 3$ in one case and with $y = -1$ in another.

x	y
3	4
2	3
0	0
2	-1

The fact that a function assigns a unique value of y to each x in the domain does not mean that it assigns a *different* value of y to each x, only that it does not assign a different second element to the same first element. The next table defines a function in which 3 is assigned to each value of x.

x	y
1	3
2	3
0	3
5	3
2	3

In this functional pairing, a number has been assigned to each outcome of an experiment according to a stated rule.

A Random Variable

The idea of functional pairing can be extended to probability experiments by assigning a number to each outcome of an experiment. To illustrate what we mean, consider an experiment in which the genders of the children in families with three children are recorded. The eight possible outcomes of this experiment are *GGG*, *GGB*, *GBG*, *BGG*, *BBG*, *BGB*, *GBB*, and *BBB*. If we are only interested in how many girls are in the family, then we can assign a number to each outcome of the experiment that corresponds to the number of girls. This is shown in Figure 6.1.

Figure 6.1

GGG	*GGB*	*GBG*	*BGG*	*BBG*	*BGB*	*GBB*	*BBB*
↓	↓	↓	↓	↓	↓	↓	↓
3	2	2	2	1	1	1	0

The diagram in Figure 6.1, showing the number paired with each outcome of the experiment, defines a functional pairing in which the first element in the pair is the outcome, listing the sequence of girls and boys, and the second element in the pair is the number of girls. This kind of functional pairing occurs so frequently in probability that we give it a special name.

DEFINITION Random Variable

A rule X that assigns a unique number to each outcome e_i of an experiment is called a **random variable.** We denote random variables by uppercase letters such as X, Y, and Z.

COMMENT: *The term* random variable *is chosen to reflect the fact that numerical values are being assigned to the random outcomes of an experiment.*

Example 1 Consider an experiment that consists of tossing a coin twice and recording the sequence of heads and tails that occurs. Let X represent the random variable that assigns to each outcome the number of heads that occur. Table 6.1 lists the outcomes of the experiment and the corresponding values of the random variable X. The fact that X assigns the number 2 to the outcome $e_1 = HH$ can be denoted $X(e_1) = X(HH) = 2$.

TABLE 6.1

Outcome, e_i	Value of $X(e_i)$
$e_1 = HH$	2
$e_2 = HT$	1
$e_3 = TH$	1
$e_4 = TT$	0

■

Example 2 Two students are chosen at random from a classroom and asked to indicate whether they are Democrats, Republicans, or independents. What are the values of the random variable Y, if Y denotes the number who answer "Democrat" to the question?

Solution Let d denote Democrat, r denote Republican, and i denote independent. The following tree diagram indicates the possible outcomes for the two students.

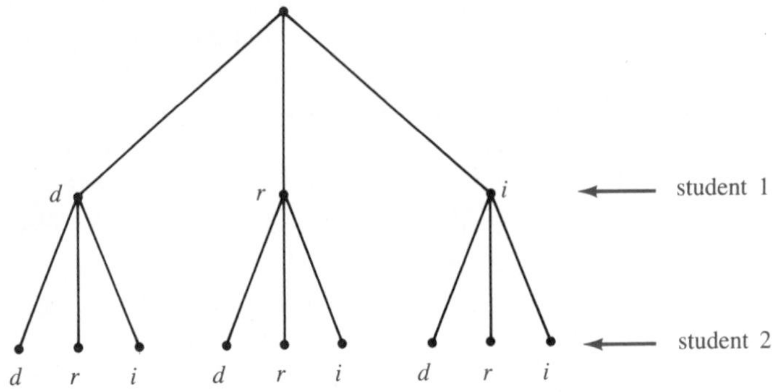

The possible outcomes are *dd, dr, di, rd, rr, ri, id, ir, ii*. The outcome *dr* means that the first student chosen is a Democrat and the second is a Republican. The outcomes and the values of the random variable Y are shown in Table 6.2.

TABLE 6.2

Outcome, e_i	Value of $Y(e_i)$
dd	2
dr	1
di	1
rd	1
rr	0
ri	0
id	1
ir	0
ii	0

■

PROBABILITY DISTRIBUTIONS

Associated with each outcome of an experiment is a probability. Since a random variable is a rule that assigns a number to each of the outcomes of an experiment, it follows that the values of a random variable also occur with certain probabilities. The values of a random variable and their corresponding probabilities, considered as a whole, are called a **probability distribution.** Thus, a probability distribution is a functional pairing of each value of a random variable with its probability of occurrence. In Chapter 5 we used the term *uniform probability distribution* to describe a sample space in which each of the sample points has the same probability of occurrence. The next two examples show probability distributions that are not uniform.

Example 3 Consider the experiment of tossing a coin twice and recording the sequence of heads and tails. Determine the probability distribution of the random variable Y that assigns the number of heads to each outcome.

Solution This is the experiment described in Example 1. If we assume that each of the four outcomes in Table 6.1 is equally likely, then each outcome has a probability of $\frac{1}{4}$ of occurring. From the table, we see that X takes on the value 2 for one of the sample points, the value 1 for two of the sample points, and the value 0 for one of the sample points. Thus, the probability that $X = 2$ is $\frac{1}{4}$, that $X = 1$ is $\frac{1}{4} + \frac{1}{4} = \frac{1}{2}$, and that $X = 0$ is $\frac{1}{4}$. The probability distribution for this random variable is shown in Table 6.3.

TABLE 6.3

Value of X	$P(X)$
0	$\frac{1}{4}$
1	$\frac{1}{2}$
2	$\frac{1}{4}$

■

COMMENT: *Notice that in Example 3 the random variable X assigns 1 to both HT and TH. Since these are mutually exclusive outcomes, the probability assigned to X = 1 is computed using the addition rule; that is,*

$$P(1) = P(HT \cup TH) = P(HT) + P(TH) = \frac{1}{4} + \frac{1}{4} = \frac{1}{2}$$

Observe in Table 6.3 that the sum of the probabilities is one.

Example 4 A box contains three red balls and seven green balls. Two balls are drawn in sequence from the box; the first ball is replaced before the second is drawn, and the color of each ball is recorded. Let the random variable X be the number of red balls drawn from the box. Determine the probability distribution for this random variable.

Solution The possible outcomes of this experiment are RR, RG, GR, GG, where RG means that the first ball is red and the second is green. To determine the probability of the outcome RG, we note that the probability of drawing a red ball is $\frac{3}{10}$ and the probability of drawing a green ball is $\frac{7}{10}$. Since we assume the two balls are selected independently from the box, the probability of a red ball and then a green ball is given by the product $\frac{3}{10} \cdot \frac{7}{10} = \frac{21}{100}$. The other probabilities are computed, and the results shown in Table 6.4 with the values of the random variable X.

TABLE 6.4

Outcome, e_i	Probability	Value of $X(e_i)$
RR	$\frac{3}{10} \cdot \frac{3}{10} = \frac{9}{100}$	2
RG	$\frac{3}{10} \cdot \frac{7}{10} = \frac{21}{100}$	1
GR	$\frac{7}{10} \cdot \frac{3}{10} = \frac{21}{100}$	1
GG	$\frac{7}{10} \cdot \frac{7}{10} = \frac{49}{100}$	0

Notice that the random variable X has the value 1 for both RG and GR. Since these are mutually exclusive outcomes, the probability of $X = 1$ is given by the sum $\frac{21}{100} + \frac{21}{100} = \frac{42}{100}$. The probability distribution for X is then given in Table 6.5.

TABLE 6.5

Value of $X(e_i)$	Probability
0	$\frac{49}{100}$
1	$\frac{42}{100}$
2	$\frac{9}{100}$

HISTOGRAMS

A probability distribution such as the one in the preceding example assigns a probability to each value of the random variable; that is, $(0, \frac{49}{100})$, $(1, \frac{42}{100})$, $(2, \frac{9}{100})$. Consequently, we can represent the pairings graphically by using the x-axis for the value of the random variable X and the y-axis for the probability of X. The graph is shown in Figure 6.2.

Figure 6.2

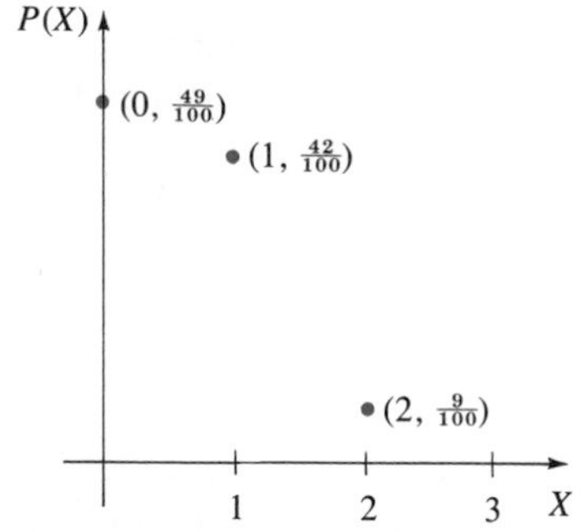

Another form of the graph of a probability distribution is the histogram, or bar graph. In a **histogram,** each value of the random variable is represented by a rectangle whose height corresponds to its probability. In most cases, we make all the bars the same width, and draw them so their sides touch each other, as shown in Figure 6.3. This is the same probability distribution represented by the points in Figure 6.2. Histograms help to display important characteristics of a probability distribution, such as symmetry.

Figure 6.3

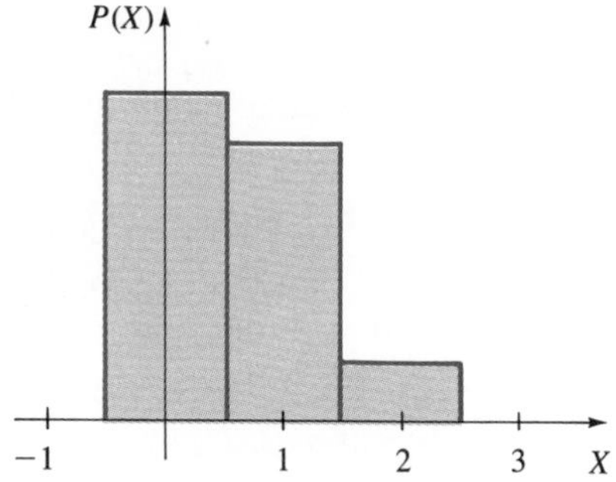

COMMENT: *The width of the bars in a histogram is somewhat arbitrary, but rectangles whose width is* 1 *are typical because in this case* the area of the rectangle corresponds to the probability of the value of the random variable. *This means that the total area of the histogram is* 1. *Why? We will say more about this aspect of histograms in Section 6.5.*

The values x of a random variable X represent mutually exclusive events. Consequently, the probability of $X < x$ is the sum of the probabilities of those values of X that are less than x. This is illustrated in the next example.

Example 5 The manager of a quick-lube service station records, to the nearest minute, the time required to change the oil and lubricate each car. The probability distribution of this random variable is represented by the histogram in

Figure 6.4(a). Based on this probability distribution, what is the probability that the next car to come through the station will take:

(a) Less than ten minutes?
(b) More than fifteen minutes?
(c) More than eight minutes and less than twelve minutes?

Figure 6.4(a)

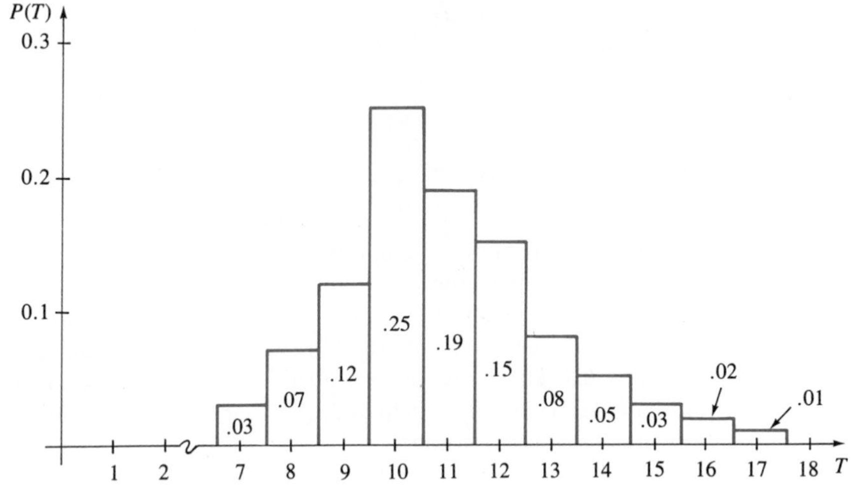

Solution

(a) Representing the probability that it will take less than ten minutes by $P(T < 10)$, we have

$$P(T < 10) = P(T = 7) + P(T = 8) + P(T = 9)$$
$$= 0.03 + 0.07 + 0.12 = 0.22$$

The shaded portion of the histogram in Figure 6.4(b), which represents 22% of the area, corresponds to this probability.

Figure 6.4(b)

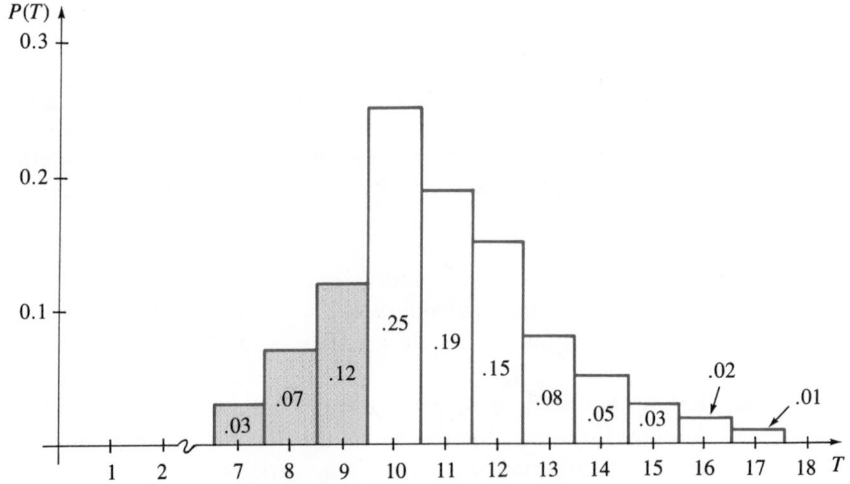

(b) The probability that it will take more than fifteen minutes is

$$P(T > 15) = P(T = 16) + P(T = 17) = 0.02 + 0.01 = 0.03$$

This probability corresponds to 3% of the area of the histogram, as shown by the shaded region in Figure 6.4(c).

Figure 6.4(c)

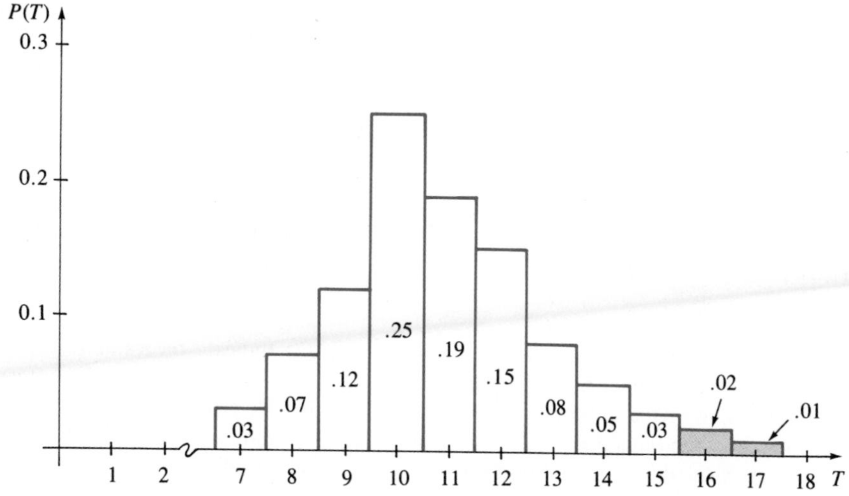

(c) The probability that it will take more than eight minutes and less than twelve minutes is

$$P(8 < T < 12) = P(T = 9) + P(T = 10) + P(T = 11)$$
$$= 0.12 + 0.25 + 0.19 = 0.56$$

The shaded region in Figure 6.4(d) shows the area corresponding to this probability.

Figure 6.4(d)

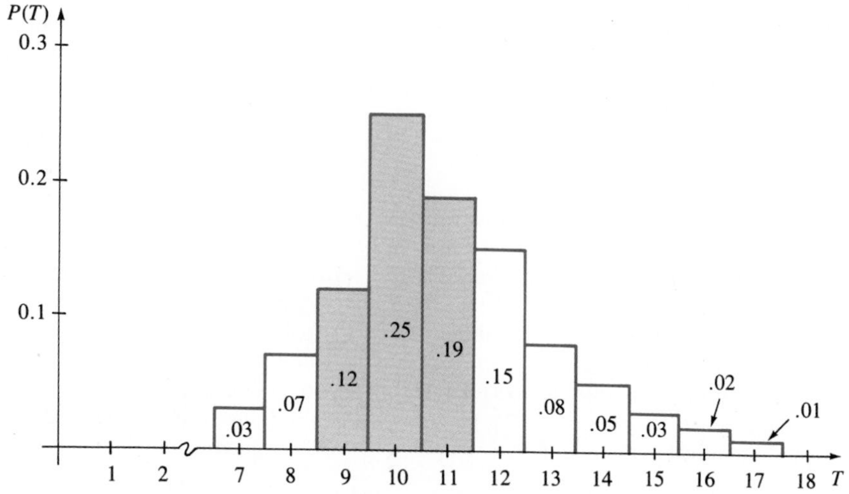

■

EXERCISES SECTION 6.1

In Exercises 1–8, indicate which of the sets of ordered pairs define functions.

1. $\{(-1, 1), (0, 2), (1, 5)\}$
2. $\{(0, 0), (-1, 2), (3, 3)\}$
3. $\{(2, 3), (4, -1), (5, 2), (0, 3)\}$
4. $\{(-1, 0), (0, 3), (1, 0)\}$
5. $\{(3, 0), (1, 2), (5, 0), (1, 3)\}$
6. $\{(2, 1), (-1, 1), (0, 1), (7, 0)\}$
7. $\{(0, 2), (1, 2), (2, 2), (3, 2), (4, 2)\}$
8. $\{(5, 0), (-2, 0), (5, -1)\}$
9. Three coins are tossed on a table. Let the random variable be the number of times that tails appears.
   (a) Indicate the value of the random variable assigned to each of the outcomes.
   (b) Determine the probability distribution for this random variable.
   (c) Draw the histogram.
10. Four coins are tossed on a table. Let the random variable be the number of times that heads appears.
    (a) Indicate the value of the random variable assigned to each of the outcomes.
    (b) Determine the probability distribution for this random variable.
    (c) Draw the histogram.
11. Two dice are rolled. Let the random variable be the sum of the dots on the two dice.
    (a) Indicate the value of the random variable assigned to each of the outcomes.
    (b) Determine the probability distribution for this random variable.
    (c) Draw the histogram.
12. Two balls are drawn at random (with replacement) from a box that contains three red balls, five black balls, and two white balls. Let the random variable be the number of black balls.
    (a) Indicate the value of the random variable assigned to each of the outcomes.
    (b) Determine the probability distribution for this random variable.
    (c) Draw the histogram.
13. A roulette wheel has 38 pockets into which the ball may fall. Eighteen are colored red, eighteen are colored black, and two are not colored. If the ball lands on red you win \$2; otherwise you lose \$1. Let the random variable be the amount you win.
    (a) Indicate the value of the random variable assigned to each of the outcomes.
    (b) Determine the probability distribution for this random variable.
    (c) Draw the histogram.
14. A card is dealt from a well-shuffled deck of playing cards. If the card is a face card you win \$5; otherwise you lose \$1. Let the random variable be the amount you win.
    (a) Indicate the value of the random variable assigned to each of the outcomes.
    (b) Determine the probability distribution for this random variable.
    (c) Draw the histogram.
15. Each day at coffee break Carroll and Ben flip a coin to see who will buy the coffee. If the coin comes up heads, Carroll buys; otherwise Ben buys. Assume coffee costs \$0.50. Let the random variable be the cost to Carroll.
    (a) Indicate the value of the random variable assigned to each of the outcomes.
    (b) Determine the probability distribution for this random variable.
    (c) Draw the histogram.
16. A psychologist studies families with four children. Let the random variable be the number of girls in a group of four children.
    (a) Indicate the value of the random variable assigned to each of the outcomes.
    (b) Determine the probability distribution for this random variable.
    (c) Draw the histogram.
17. A machine produces electrical contacts to be used in computers. Past studies show that 3% of the parts are defective. Two parts are selected at random from the machine. Let the random variable be the number of defects.
    (a) Indicate the value of the random variable assigned to each of the outcomes.

(b) Determine the probability distribution for this random variable.
(c) Draw the histogram.

18. A coin is weighted so that the probability of heads is $\frac{2}{5}$ and that of tails is $\frac{3}{5}$. The coin is tossed three times and the number of heads recorded.
(a) Indicate the value of the random variable assigned to each of the outcomes.
(b) Determine the probability distribution for this random variable.
(c) Draw the histogram.

19. A histogram of a probability distribution is shown in the figure below. What is the probability that the value of the random variable X is:
(a) Equal to 2?
(b) Less than 1?
(c) Greater than 0 and less than 3?

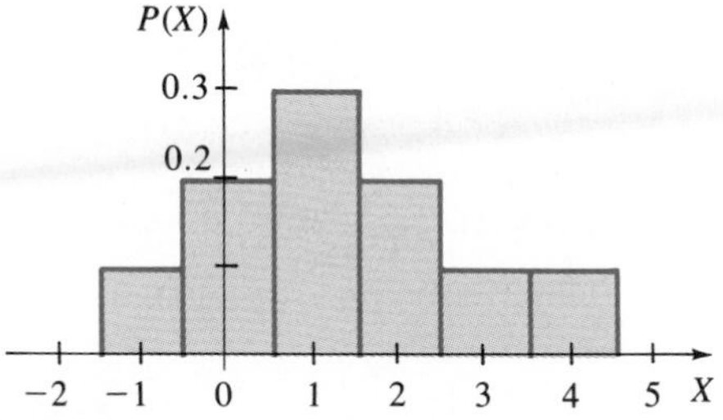

20. A histogram of a probability distribution is shown in the figure below. What is the probability that the value of the random variable X is:
(a) Equal to 3?
(b) Less than 1?
(c) Greater than 0 and less than 3?

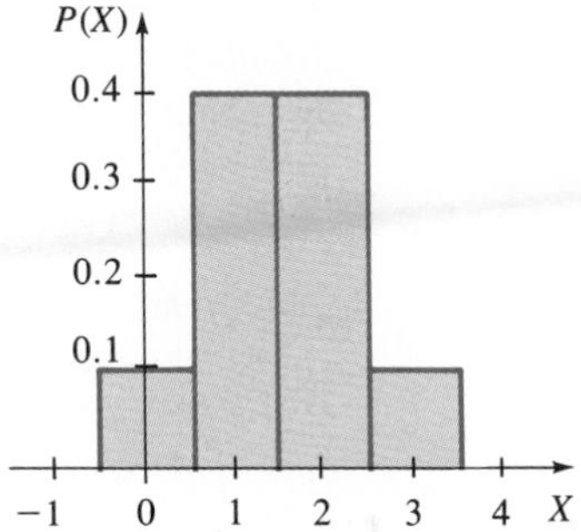

6.2 EXPECTED VALUE

Probability distributions play a central role in the study of probability; later in this chapter we will discuss two important probability distributions. To prepare for this discussion, consider the probability distributions represented by the histograms shown in Figure 6.5.

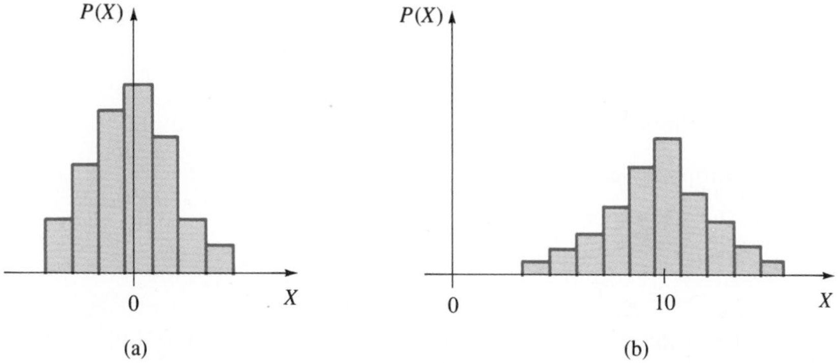

Figure 6.5

If you were asked to describe the differences between the two histograms in Figure 6.5, you might say that histogram (a) seems to have its values centered about

$X = 0$ and histogram (b) seems to have its values centered about $X = 10$. Another observation you might make is that histogram (a) seems to be more concentrated about its central value than does histogram (b). Implicit in this discussion of the two histograms is the attempt to describe the probability distributions by giving some measure of what might be called the *center* of the distribution and some measure of the way in which the values of the random variable are *distributed* about this central value.

The quantity we use to measure the central tendency of a probability distribution of a random variable X is called the **expected value of X** and is closely related to the idea of an arithmetic average. You are already familiar with the idea of using an average value to represent an entire set of numerical data. We talk about the average height or the average weight of a group of people; students are interested in the class average on the exam; the government publishes the average income of the citizens; and so on.

COMMENT: *Before discussing the concept of arithmetic average it is convenient to introduce a compact notation for writing the sum of the n numbers $x_1, x_2, \ldots, x_n$. The Greek letter Σ ("sigma") is used to indicate the summation of terms. Thus, $x_1 + x_2 + \cdots + x_n$ can be represented by*

$$\sum_{i=1}^{n} x_i$$

For example, the sum of the first four positive integers can be written as

$$\sum_{i=1}^{4} i = 1 + 2 + 3 + 4$$

We will use "sigma notation" for sums throughout this chapter.

The **arithmetic mean,** or **average,** of a set of numbers is the sum of the numbers in the set divided by the total number of numbers in the set. The average of the k numbers $\{x_1, x_2, \ldots, x_k\}$ is denoted by $\bar{x}$ and is given by

$$\bar{x} = \frac{x_1 + x_2 + \cdots + x_k}{k}$$

Recalling that $x_1 + x_2 + \cdots + x_k = \sum_{i=1}^{k} x_i$, the formula for $\bar{x}$ can be written more compactly as

$$\bar{x} = \frac{\sum_{i=1}^{k} x_i}{k}$$

For example, the average of 3, 1, 2, 3, 2, 5, 2, 5, and 2 is

$$\bar{x} = \frac{\sum_{i=1}^{9} x_i}{9} = \frac{3 + 1 + 2 + 3 + 2 + 5 + 2 + 5 + 2}{9} \approx 2.8$$

Notice that the value of the average is not one of the numbers in the set. In general, there is no requirement that the average be a number in the set. For instance, the average number of children in a family is 1.8. Clearly, there cannot be a family with 1.8 children. More important at this point, we note that in this example the sum for $\bar{x}$ can be written

$$\bar{x} = \frac{1+2+2+2+2+3+3+5+5}{9} = \frac{(1)+4(2)+2(3)+2(5)}{9}$$

$$= \frac{1}{9}(1) + \frac{4}{9}(2) + \frac{2}{9}(3) + \frac{2}{9}(5) \approx 2.8$$

Our reason for writing $\bar{x}$ in this form is to show that **$\bar{x}$ can be thought of as the sum of the products of the numbers times their relative frequencies**. Thus, the relative frequency of the number 1 is $\frac{1}{9}$, that of the number 2 is $\frac{4}{9}$, and so on. If 1, 2, 3, and 5 are the values of a random variable X and the relative frequency of each number corresponds to its probability, then the probability distribution for this random variable is given in the following table:

Value of X	Probability
1	$\frac{1}{9}$
2	$\frac{4}{9}$
3	$\frac{2}{9}$
5	$\frac{2}{9}$

The mean of a random variable X is often represented by the Greek letter μ_x, or simply μ if the random variable is understood. The mean of this random variable is then given by $\mu \approx 2.8$. The mean of a random variable X is also called the expected value of X and is denoted by $E(X)$. In this section, we will use this terminology, and we define $E(X)$ as follows:

DEFINITION Expected Value of a Random Variable

If $x_1, x_2, \ldots, x_n$ are values of a random variable X and $p_1, p_2, \ldots, p_n$ are the corresponding probabilities, then the **expected value of X** is denoted by $E(X)$ and defined by

$$E(X) = x_1p_1 + x_2p_2 + \cdots + x_np_n = \sum_{i=1}^{n} x_ip_i$$

COMMENT: *The expected value $E(X)$ is the* center *of the probability distribution of the random variable X in the sense in which it is the balance point of its histogram; that is, the area of the histogram to the left of $E(X)$ is the same as that to the right of $E(X)$.*

Example 1 Four coins are tossed on a table and the number of heads is recorded. Find the expected number of heads for this experiment.

Solution The probability distribution for this random variable is given in the table and illustrated by the histogram in Figure 6.6.

x_i No. of Heads	p_i Probability
0	$\frac{1}{16}$
1	$\frac{4}{16}$
2	$\frac{6}{16}$
3	$\frac{4}{16}$
4	$\frac{1}{16}$

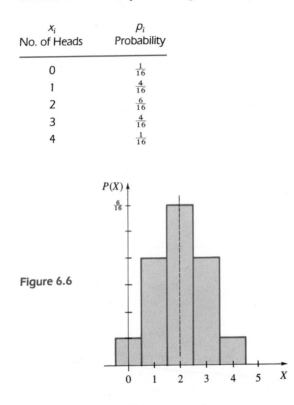

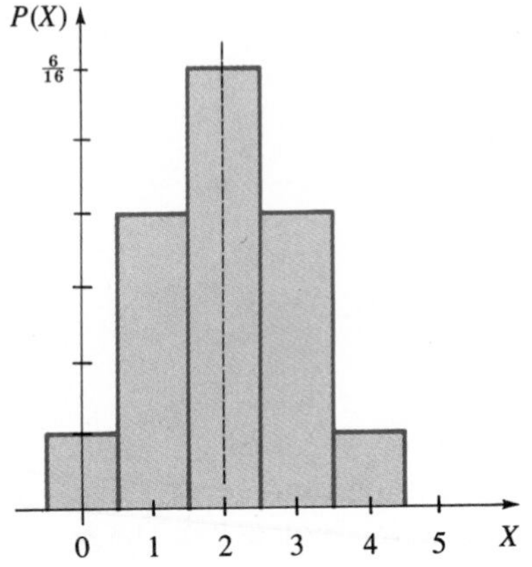

Figure 6.6

The expected value of X is then given by

$$E(X) = 0 \cdot \frac{1}{16} + 1 \cdot \frac{4}{16} + 2 \cdot \frac{6}{16} + 3 \cdot \frac{4}{16} + 4 \cdot \frac{1}{16} = \frac{32}{16} = 2$$

The expected value of X is shown in Figure 6.6 as a dashed line. Notice that $E(X)$ marks the center of the histogram. ■

COMMENT: *We interpret the expected value $E(X)$ of a random variable X as the long-term average value of X when the experiment is repeated a large number of times. Thus, in Example 1, $E(X) = 2$ indicates that if the four coins are tossed a large number of times, the average number of heads obtained is expected to be about 2.*

Example 2 Two cards are dealt from a standard deck of 52 playing cards. Let X represent the random variable that assigns the number of hearts dealt. Calculate the expected number of hearts.

Solution The values of X are 0, 1, and 2. The probability of no hearts is $\frac{C_{39,2}}{C_{52,2}} \approx$ 0.559, the probability of one heart is $\frac{C_{13,1} \cdot C_{39,1}}{C_{52,2}} \approx 0.382$, and the probability of two hearts is $\frac{C_{13,2}}{C_{52,2}} \approx 0.059$. The corresponding probability distribution is shown in the table.

No. of Hearts	Probability
0	0.559
1	0.382
2	0.059

The expected number of hearts is then the sum of the values of X times their probabilities of occurrence. Thus,

$$E(X) = 0(0.559) + 1(0.382) + 2(0.059) = 0.5$$

Notice that the expected value 0.5 is not one of the values of the random variable. Keep in mind that the expected value of a random variable X is interpreted as the long-term average value of X when the experiment is repeated a large number of times.

■

Expected value is frequently used to analyze games of chance. In such cases, the random variable assigns dollar amounts to the outcomes of winning and losing. The expected value of the random variable then tells us whether we can expect to win or lose if we continue to play a game of chance for a long period of time. In these problems, the random variable assigns the *net* gain or *net* loss. That is, the amount of the bet, or the investment, is subtracted from the winnings, or return on the investment, to determine the values of the random variable. The next three examples illustrate this procedure.

Example 3 The parents of marching band members at West Carrollton High School are holding a raffle to raise money for the band. The first prize is \$500; the second prize is \$200. They plan to sell 2000 raffle tickets for \$1 each. If you buy a raffle ticket, what is your expected value for this raffle?

Solution From your perspective as a ticket holder, there are three outcomes of interest—namely, winning first prize, winning second prize, and not winning either of the prizes. Let the random variable X assign the net gain of $\$500 - \$1 = \$499$ to winning the first prize, $\$200 - \$1 = \$199$ to winning the second prize, and the net loss of $-\$1$ to not winning either prize. Since there are 2000 tickets sold, your chance of winning the first prize is $\frac{1}{2000}$, your chance of winning the second prize is $\frac{1}{2000}$, and your chance of losing your dollar is $\frac{1998}{2000}$. The expected value of X is then

$$E(X) = 499 \cdot \frac{1}{2000} + 199 \cdot \frac{1}{2000} + (-1) \cdot \frac{1998}{2000}$$
$$= -\frac{1300}{2000} = -0.65$$

We interpret this result to mean that if you continue to play this raffle over a long period of time, your average loss will be \$0.65 per raffle. You can expect to win the raffle once in a while, but in the long run you will lose \$0.65 for each time you participate. ▪

Example 4 A game consists of rolling a die and betting \$2 on the number that will turn up. If a 1 or a 2 turns up you get back \$5 and if a 6 turns up you get back \$2; otherwise, you lose \$2. Calculate the expected value of this game.

Solution Let the random variable X assign the net gain or loss to each outcome. Thus, if a 1 or a 2 turns up you win \$5 − \$2 = \$3 and if a 6 turns up you win \$2 − \$2 = 0; otherwise, you win −\$2 (that is, you lose \$2). The probability distribution for this random variable is shown in the table.

Value of X	Probability
3	$\frac{2}{6}$
0	$\frac{1}{6}$
−2	$\frac{3}{6}$

The expected value is then given by

$$E(X) = 3 \cdot \frac{2}{6} + 0 \cdot \frac{1}{6} + (-2) \cdot \frac{3}{6} = 0$$

The expected value of zero means that if you continue to play this game, in the long term you can expect to break even. ▪

COMMENT: *If the expected value of a game is zero—that is, $E(X) = 0$—we consider the game to be* **fair,** *in the sense that in the long term we can expect to break even. Clearly, if you are running a gambling casino you do not want "fair" games. Gambling casinos make money because all of the games have negative expected values for the players. The casino managers know that even though an occasional player will be a big winner, in the long term the expected value of the random variable is in their favor.*

The concept of expected value can also be used to help us make decisions concerning events in our lives, such as investment decisions.

Example 5 Assume that you are considering two investments. In one option, you must invest \$2000, with a possible return of \$12,000. Assume you know that there is a probability of 0.8 that you will get a return of \$12,000 and a probability of 0.2 that you will get back \$1000. In the other option, you must invest \$5000, with a possible return of \$20,000. Assume there is a probability of 0.9 that you will get a return of \$20,000 and a probability of 0.1 that you will get back nothing. Based on this information and on the concept of expected value, which is the best investment for you?

Solution In the first case, let the values of the random variable be the net returns of \$12,000 − \$2000 = \$10,000 and \$1000 − \$2000 = −\$1000. In the second, let the values of the random variable be \$20,000 − \$5000 = \$15,000 and \$0 − \$5000 = −\$5000. The expected value in the first case is

$$E(X) = 10{,}000(0.8) + (-1000)(0.2) = 7800$$

The expected value in the second case is

$$E(X) = 15{,}000(0.9) + (-5000)(0.1) = 13{,}000$$

Based on the computed expected values, you should choose the second option, since it gives the greatest expected value. Of course, there are other factors that must be considered in making your decision, such as whether you can raise the money required and how much you can afford to lose. ■

EXERCISES SECTION 6.2

In Exercises 1–4, draw a histogram of the probability distribution and calculate the expected value of the random variable.

1.

Value of X	Probability
1	$\frac{1}{8}$
2	$\frac{1}{4}$
3	$\frac{1}{2}$
4	$\frac{1}{8}$

2.

Value of X	Probability
−10	$\frac{1}{7}$
0	$\frac{2}{7}$
10	$\frac{4}{7}$

3.

Value of X	Probability
−10	0.1
0	0.2
10	0.2
20	0.3
30	0.2

4.

Value of X	Probability
2	0.2
5	0.3
8	0.4
10	0.1

5. A box contains ten red balls and seven white balls. Two balls are drawn without replacement. Calculate the expected number of red balls drawn.

6. An experiment consists of recording the number of boys born in families with four children. Calculate the expected number of boys.

7. Two dice are rolled and the sum of the spots recorded. Calculate the expected number for this random variable.

8. Two dice are rolled and the larger of the two numbers recorded. If both show the same number, then that number is recorded. Calculate the expected value for this random variable.

9. A box contains five quarters, fifteen dimes, and seven nickels. A coin is drawn at random from the box. Calculate the expected value of the draw.

10. A box contains seven quarters, three dimes, and twelve nickels. A coin is drawn at random from the box. Calculate the expected value of the draw.

11. A box contains 40 light bulbs, 15 of which are defective. Three bulbs are selected at random from the box without replacement. Calculate the expected number of good bulbs in the selection.

12. Two students are selected at random from a class of 30 students in a calculus class. If six of the students are engineering majors, what is the expected number of engineering majors selected?

13. Two cards are dealt from a deck of 52 playing cards. Calculate the expected number of face cards dealt.

14. Two cards are dealt from a deck of 52 playing cards. Calculate the expected number of red cards dealt.

15. A roulette wheel has 38 pockets into which a ball may fall: 18 red pockets, 18 black pockets, and 2 green pockets. If you bet \$1 on red and the ball lands in red, you win \$1; otherwise, you lose \$1. Calculate the expected value for this game.

16. Suppose that the roulette game played in Exercise 15 is as follows: If you bet on red and the ball lands in red, you win \$1; if it lands in green, you get your \$1 back; if it lands in black, you lose \$1. Calculate the expected value of this game.

17. A raffle has prizes of \$100, \$50, and \$25. The sponsors of the raffle sell 500 tickets at \$1 each, and each ticket is eligible for all three prizes. Calculate your expected winnings if:
(a) You buy one ticket.
(b) You buy five tickets.

18. A raffle in which the prize is \$500 sells 1200 tickets at \$1 each. Calculate the expected value for this raffle.

19. A contractor bids on the job of constructing a new motel. There is a probability of 0.8 that the profit on the job will be \$25,000, a probability of 0.1 that there will be no profit, and a probability of 0.1 that there will be a loss of \$10,000 (due to bad weather, strikes, and so on). Calculate the expected profit to the contractor.

20. A psychology student records the number of errors made by paid subjects in a learning experiment. The following table shows the probability distribution for the experiment.

Number of Errors	Probability
0	0.2
1	0.3
2	0.3
3	0.1
4	0.1

Draw a histogram for the probability distribution and calculate the expected number of errors.

21. You have the opportunity to choose between two sales jobs. If you take the job in Chicago, there is a 50% chance that you will earn \$45,000 and a 50% chance that you will earn \$30,000. If you take the job in Los Angeles, there is a 70% chance that you will earn \$20,000 and a 30% chance that you will earn \$60,000. If you base your decision on expected value, which job should you take?

6.3 VARIANCE AND STANDARD DEVIATION

The discussion of the expected value and the mean in the previous section emphasized the central tendency of a random variable. The mean and the expected value focused our attention on the point of highest concentration, or the balance point. Now we want to consider how concentrated—how tightly grouped—the sample points are about the mean. Another way of looking at this concentration is to consider how widespread the sample space is, or how far the points in the sample space deviate from the mean.

VARIANCE

The histograms shown in Figure 6.7 represent probability distributions that have the same expected value or mean. It is clear from the figure that the variability of the values of the random variable in Figure 6.7(a) is greater than that pictured in Figure 6.7(b). Even with an untrained eye, you can tell that the values in (b) are more concentrated about the mean value μ than are those of (a). In this section, we will develop a measure of the variability of a random variable.

Figure 6.7

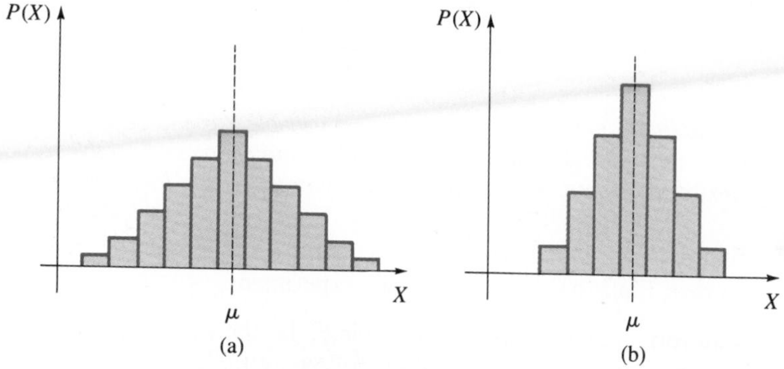

Let X be a random variable whose mean is μ. The deviation of the value x from its mean is then represented by

$$x - \mu$$

This difference tells us how far x is from μ. For example, if $x - \mu = 2$, x is 2 units to the right of μ, and if $x - \mu = -3$, x is 3 units to the left of μ. To express the variability of numbers about their mean, we might assume that a good measure of this variability would be the sum of the deviations of the numbers from the mean. To see that this is not a good measure of variability of numbers about their mean, consider the set of numbers $\{5, 6, 8, 9\}$. The mean of these four numbers is 7, and the sum of the deviations of the numbers from their mean is

$$(5 - 7) + (6 - 7) + (8 - 7) + (9 - 7) = (-2) + (-1) + 1 + 2 = 0$$

The sum of the deviations is actually zero. Recalling that we interpret the mean as the balance point of a probability distribution, it follows that the sum of the differences of the values of a random variable from its mean is *always* zero. Consequently, such a sum is an unacceptable measure of the variability of the values of a random variable. It turns out that whereas the sum of the deviations from the mean is not an acceptable measure of variability of a random variable, the sum of the squares of these deviations is very useful. Actually, we use the expected value of the squared deviations as a measure of the variability of X. To this end, we define the variance of a random variable X.

DEFINITION The Variance of a Random Variable

Let $x_1, x_2, \ldots, x_n$ be the values of a random variable X with respective probabilities $p_1, p_2, \ldots, p_n$. If the expected value of X is μ, then the **variance** of X is denoted by var(X) and is defined by

$$\text{var}(X) = p_1(x_1 - \mu)^2 + p_2(x_2 - \mu)^2 + \cdots + p_n(x_n - \mu)^2$$
$$= \sum_{i=1}^{n} p_i(x_i - \mu)^2$$

COMMENT: *The variance of a random variable X is also denoted by σ_x^2, read as "sigma sub x squared." We will use var(X) and σ_x^2 interchangeably in this section. If the random variable is obvious, the subscript is usually dropped and the expression becomes σ^2.*

Example 1 Three coins are tossed on a table and the number of heads is recorded. Find the variance for this experiment.

Solution Let the random variable X be the number of heads. The probability distribution for X is given in the following table:

x_i No. of Heads	p_i Probability
0	$\frac{1}{8}$
1	$\frac{3}{8}$
2	$\frac{3}{8}$
3	$\frac{1}{8}$

To find the variance, we first need to calculate the expected value μ. Thus,

$$\mu = \sum_{i=1}^{4} p_i x_i = \frac{1}{8}(0) + \frac{3}{8}(1) + \frac{3}{8}(2) + \frac{1}{8}(3) = \frac{12}{8} = 1.5$$

The variance is then given by

$$\sigma^2 = \text{var}(X) = \sum_{i=1}^{4} p_i(x_i - \mu)^2$$
$$= \frac{1}{8}(0 - 1.5)^2 + \frac{3}{8}(1 - 1.5)^2 + \frac{3}{8}(2 - 1.5)^2 + \frac{1}{8}(3 - 1.5)^2 = \frac{6}{8} = 0.75$$

■

Standard Deviation

Although the variance is a useful measure of the variability of X, it has two serious drawbacks. First, since we use squared deviations to get it, the size of the variance does not indicate the actual variation. Second, any units that are associated with X become squared units in the variance. For instance, if the units of X were pounds, the units of variance would be square pounds. To avoid these two problems, we use the square root of the variance to measure the variability of X. The square root of the variance, which is called the **standard deviation** of X, is denoted by σ_x or, if the random variable is understood, by σ.

DEFINITION The Standard Deviation of a Random Variable

If the variance of a random variable X is var(X), then the standard deviation of X is denoted by σ_x and defined by

$$\sigma_x = \sqrt{\text{var}(X)}$$

Example 2 In Example 1 we computed the variance of the number of heads when three coins are tossed. Calculate the standard deviation for this experiment.

Solution In Example 1, we found var(X) = 0.75. Since the standard deviation is the square root of the variance, we have

$$\sigma = \sqrt{0.75} \approx 0.866$$

■

Example 3 A box contains 1000 microchips. Ninety-five percent of the chips have no defects, 4% have one defect, and 1% have two defects. Calculate the expected number of defects and the standard deviation of the defects.

Solution The values of the random variable X are 0, 1, and 2. The probability of no defects is 0.95, of one defect is 0.04, and of two defects is 0.01. Therefore the expected number of defects is

$$\mu = \sum_{i=1}^{3} p_i x_i = 0.95(0) + 0.04(1) + 0.01(2) = 0.06$$

The standard deviation of the defects is then

$$\sigma = \sqrt{\sum_{i=1}^{3} p_i(x_i - \mu)^2} = \sqrt{0.95(0 - 0.06)^2 + 0.04(1 - 0.06)^2 + 0.01(2 - 0.06)^2}$$
$$= \sqrt{0.00342 + 0.035344 + 0.037636} = \sqrt{0.0764} \approx 0.276 \text{ defect}$$

■

Sometimes, to facilitate the calculation of expected value and standard deviation, we arrange the terms in tabular form. This approach is particularly useful when there are numerous terms to be summed. The next example illustrates this process.

Example 4 The probability distribution of a random variable Y is shown in the histogram below. Calculate the expected value and standard deviation of Y.

Figure 6.8

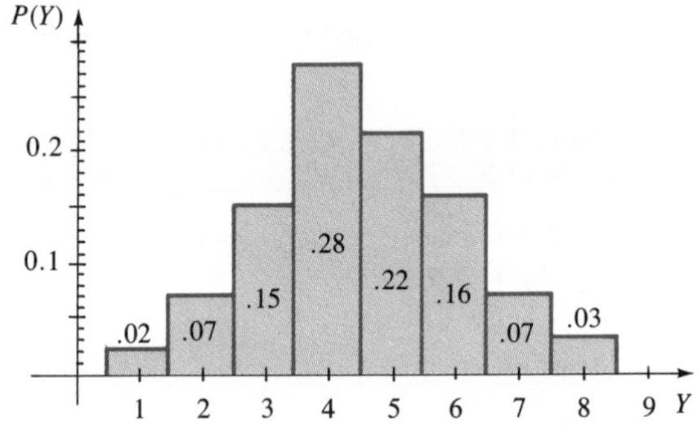

Solution The probability distribution of Y is given in the table below, along with appropriately labeled columns that are used in calculating the expected value and the variance of Y. The column format organizes the computations in easy-to-follow steps.

y_i	p_i	$p_i y_i$	$(y_i - \mu)^2$	$p_i(y_i - \mu)^2$	
1	0.02	0.02	12.3904	0.2478	
2	0.07	0.14	6.3504	0.4445	
3	0.15	0.45	2.3104	0.3466	
4	0.28	1.12	0.2704	0.0757	
5	0.22	1.10	0.2304	0.0507	
6	0.16	0.96	2.1904	0.3505	
7	0.07	0.49	6.1504	0.4305	
8	0.03	0.24	12.1104	0.3633	
		$\mu = 4.52$		$\text{var}(Y) = 2.3096$	← SUMS

Notice that the sum of the entries in the column labeled $p_i y_i$ gives the expected value of Y, and the sum of the entries in the column labeled $p_i(y_i - \mu)^2$ gives the variance of Y. The desired mean and standard deviation are then

$$\mu = 4.52$$

and

$$\sigma = \sqrt{2.3096} \approx 1.52$$ ■

Chebyshev's Inequality

We have defined the variance and standard deviation of a random variable as measures of the variability of the random variable. We understand that a random variable with $\sigma = 1.77$ has more variability than, say, one with $\sigma = 0.85$, but beyond that we have assigned no meaning to a standard deviation of 1.77. It turns out that if we know the standard deviation, we can specify the percentage of values of the random variable that are within k units of the mean. For example, we can say that at least 75% of the values of a random variable will lie within 2σ units of the mean or, equivalently, that the probability that a value lies within 2σ of the mean is at least 0.75. The probability that the value of a random variable will lie within k standard deviations of the mean is given in the following theorem.

THEOREM Chebyshev's Inequality

If X is a random variable with mean μ and standard deviation σ, the probability that a value of X will lie within k standard deviations of the mean is at least

$$1 - \frac{1}{k^2}$$

That is, the probability that a value of X lies between $\mu - k\sigma$ and $\mu + k\sigma$ is

$$P(\mu - k\sigma \leq X \leq \mu + k\sigma) \geq 1 - \frac{1}{k^2}$$

COMMENTS:

1. *Chebyshev's theorem assures us that the probability that a value of a random variable X is within k standard deviations of the mean is* at least $1 - \frac{1}{k^2}$. *It may be greater than this, but it cannot be less.*
2. *The result in Chebyshev's theorem is independent of the probability distribution of the random variable, so* the theorem holds for any probability distribution.
3. *Recall that a percentage can be obtained from probability by multiplying probability by 100. Thus, another way to interpret Chebyshev's theorem is that at least* $\left(1 - \frac{1}{k^2}\right) \times 100$ *percent of the values of X lie within $k\sigma$ units of the mean.*

Example 5 Let X be a random variable with $\mu = 25$ and $\sigma = 3$. What is the probability that a value of X lies within two standard deviations of the mean? What is the interval defined by $\mu - 2\sigma \leq X \leq \mu + 2\sigma$?

Solution According to Chebyshev's theorem, the probability that a value of X will be within two standard deviations of the mean is at least

$$1 - \frac{1}{2^2} = 1 - \frac{1}{4} = \frac{3}{4}$$

Since $\mu - 2\sigma = 25 - 2(3) = 19$ and $\mu + 2\sigma = 25 + 2(3) = 31$, it follows that

$$P(19 \leq X \leq 31) \geq \frac{3}{4}$$

Notice that we can also say that at least 75% of the values of X lie in the interval $19 \leq X \leq 31$. ■

Example 6 The mean number of work-related accidents per year for a given manufacturing company is 60, with a standard deviation of 4. Use the Chebyshev inequality to estimate the probability that the number of accidents in a given year will be between 50 and 70.

Solution We want to estimate $P(50 \leq X \leq 70)$ for a random variable with $\mu = 60$ and $\sigma = 4$. Since this probability is greater than or equal to $1 - \frac{1}{k^2}$, we need to find k for the conditions given in this problem. Observing that

$$\mu - k\sigma = 50 \qquad \text{and} \qquad \mu + k\sigma = 70$$

and substituting $\mu = 60$ and $\sigma = 4$ into these equations, we have

$$60 - 4k = 50 \qquad \text{and} \qquad 60 + 4k = 70$$

Both of these equations yield the value $k = \frac{5}{2}$. Thus, the probability that the number of accidents in a given year will be between 50 and 70 is

$$P(50 \leq X \leq 70) \geq 1 - \frac{1}{\left(\frac{5}{2}\right)^2} = \frac{21}{25}$$ ■

EXERCISES SECTION 6.3

In Exercises 1–4, find the expected value and the standard deviation of the random variable.

1.

Value of X	Probability
1	$\frac{1}{8}$
2	$\frac{1}{4}$
3	$\frac{1}{2}$
4	$\frac{1}{8}$

2.

Value of X	Probability
−10	$\frac{1}{7}$
0	$\frac{2}{7}$
10	$\frac{4}{7}$

3.

Value of X	Probability
−10	0.1
0	0.2
10	0.2
20	0.3
30	0.2

4.

Value of X	Probability
2	0.2
5	0.3
8	0.4
10	0.1

5. A box contains ten red balls and seven white balls. Two balls are drawn without replacement. Let X be the number of red balls drawn. Calculate the mean and standard deviation of X.

6. An experiment consists of recording the number of boys born in families with four children. Calculate the mean and standard deviation of the number of boys.

7. Two dice are rolled and their sum recorded. Calculate the mean and standard deviation for this random variable.

8. Two dice are rolled and the larger of the two numbers recorded. If both show the same number, then that number is recorded. Calculate the mean and standard deviation for this random variable.

9. A box contains five quarters, fifteen dimes, and seven nickels. A coin is drawn at random from the box. Calculate the mean and standard deviation of the value of the draw.

10. A box contains seven quarters, three dimes, and twelve nickels. A coin is drawn at random from the box. Calculate the mean and standard deviation of the draw.

11. A box contains 40 light bulbs, 15 of which are defective. Three bulbs are selected at random from the box. Calculate the mean and standard deviation of the number of good bulbs in the sample of three bulbs.

12. A roulette wheel has 38 pockets into which a ball may fall: 18 red pockets, 18 black pockets, and 2 green pockets. If you bet \$1 on red and the ball lands in red, you win \$1; otherwise, you lose \$1. Calculate the mean and standard deviation of your winnings for this game.

13. A psychology student records the number of errors made by paid subjects in a learning experiment. The following table shows the probability distribution for the experiment:

Number of Errors	Probability
0	0.2
1	0.3
2	0.3
3	0.1
4	0.1

Calculate the mean and standard deviation for the number of errors.

14. A raffle has prizes of \$100, \$50, and \$25. The sponsors of the raffle sell 500 tickets at \$1 each. Calculate the mean and standard deviation for the winnings on the raffle. Assume each ticket is eligible to win each prize.

15. The probability distribution of a random variable X has a mean μ and standard deviation σ. Use Chebyshev's inequality to calculate the probability that a value of X will lie within k standard deviations of the mean for:
(a) $k = 2$ (b) $k = 3$ (c) $k = 1.8$

16. The probability distribution of a random variable X has a mean of 25 and a standard deviation of 2. Indicate the interval that is within 3σ units of the mean. Use Chebyshev's inequality to find the probability that a value of X lies within this interval.

17. The probability distribution of a random variable Z has a mean of 110 and a standard deviation of 5. Indicate the interval that is within 2.5σ units of the mean. Use Chebyshev's inequality to find the probability that a value of X lies within this interval.

18. The probability distribution of a random variable X has a mean of 18 and a standard deviation of 2. Use Chebyshev's inequality to find the probability that a value of X lies between 15 and 21.

19. The probability distribution of a random variable Y has a mean of 40 and a standard deviation of 3. Use Chebyshev's inequality to find the probability that a value of Y lies between 30 and 50.

20. The probability distribution of a random variable X has a mean of 75 and a standard deviation of 4. Calculate the endpoints of the interval that will contain 90% of the values of X.

21. The headlights manufactured by a certain company have a mean life of 500 hours and a standard deviation of 20 hr. Calculate the probability that a headlight chosen at random will last between 450 and 550 hr.

22. A preservative added to bread dough keeps the bread from spoiling for 4 days, with a standard deviation of 0.2 day. Calculate the probability that a loaf of bread chosen at random will last between 3.5 and 4.5 days.

6.4 THE BINOMIAL PROBABILITY DISTRIBUTION

A marketing research study found that of those customers who bought Rise soap during a one-week study, 58% chose a package with *SPECIAL VALUE* printed across the label over an identical product without this statement. The researchers now want to know what the probability is that 150 of the next 200 customers will select the package with *SPECIAL VALUE* printed on the label.

The quality-assurance group of an electronic equipment manufacturer estimates that 2% of the company's products are defective. The quality-control manager is interested in the probability that there are fewer than three defective items in a sample of ten pieces of equipment.

The solution to problems like these requires that we pull together several ideas we have previously developed—namely, the counting principle, the binomial expansion, and the mean and standard deviation. We begin by looking at some familiar problems from a slightly different point of view.

BERNOULLI TRIALS

In many cases, an experiment consists of a sequence of independent trials, each of which has two possible outcomes. For example, the experiment of tossing a penny five times consists of five independent trials, each with an outcome of either heads or tails. Experiments of this type are called **Bernoulli experiments** or **processes,** and it is common practice to refer to the two outcomes as *success* and *failure*. We are interested in Bernoulli experiments because they are used extensively in applications. Part of the reason for this is that the outcomes of many experiments can be classified as successes or failures even though they have nothing to do with success or failure in the usual sense. For instance, in dealing cards from a deck of playing cards there are 52 outcomes, but if we win $100 if a face card is dealt, then we think of a face card as a success and the other cards in the deck as failures. Formally, we define a Bernoulli experiment as follows:

DEFINITION Bernoulli Experiment

A Bernoulli experiment consists of a sequence of trials that have the following properties:

1. There are a finite number of trials.
2. The trials are independent of each other.
3. There are two outcomes to each trial: "success" and "failure."
4. The probability of "success" is constant for each trial.

HISTORICAL COMMENT: *Jacques Bernoulli (1654–1705) was a Swiss mathematician who contributed to probability theory and to differential and integral calculus. The Bernoullis are the most famous family in the history of mathematics, because several generations of Bernoullis achieved fame in mathematics and science.*

Example 1 Suppose we roll a die ten times. If a 3 turns up, we win \$5; otherwise, we lose \$1. To show that this experiment can be thought of as a Bernoulli experiment, we must show that the four properties given above are satisfied. We note that there are a finite number of trials and that the outcome of each roll is independent of the others. Further, there are only two outcomes: "success" if a 3 turns up, and "failure" otherwise. Finally, the probability of success, $p = \frac{1}{6}$, is constant for each trial. Since each of the four properties is satisfied, we conclude that this is a Bernoulli experiment. ▪

COMMENT: *The terms "success" and "failure" do not always reflect our everyday understanding of these words:*

(a) *Suppose we are inspecting parts coming off an assembly line for defects. The inspector might assume that finding a defect constitutes a "success" in the sense of a Bernoulli experiment.*

(b) *In testing a virus to see whether it causes cancer in white mice, a researcher might consider it a "success" if it caused cancer and a "failure" if it did not.*

Binomial Probabilities

In studying Bernoulli experiments, the probability of a "success" is usually represented by p, and the probability of a "failure" by q. Since there are only two possible outcomes, it follows that for Bernoulli experiments

$$p + q = 1$$

or, solving for q,

$$q = 1 - p$$

Example 2 If the probability of "success" is $p = 0.3$, then the probability of "failure" is $q = 1 - 0.3 = 0.7$. ▪

Consider a Bernoulli experiment in which a die is rolled five times and the number of 1s recorded. In this case, a 1 corresponds to "success" and the probability of a success is $p = \frac{1}{6}$. Suppose we wish to calculate the probability of exactly one 1 in the five trials; then we are interested in the set of possible outcomes

$$\begin{matrix} s & f & f & f & f \\ f & s & f & f & f \\ f & f & s & f & f \\ f & f & f & s & f \\ f & f & f & f & s \end{matrix}$$

We see that there are five possible events that have one success and four failures. While we obtained this result by enumerating the five outcomes, we could also have observed that the number of ways of getting one success in five trials is equivalent to the number of combinations of five objects taken one at a time—that is, $C_{5,1} = \dfrac{5!}{1!4!} = 5$.

Since the trials within each event are independent and the probability of failure is $q = 1 - \frac{1}{6} = \frac{5}{6}$, the probability of any one of these events is the same as the first, which we will calculate. The probability of a success followed by four failures is

$$\frac{1}{6} \cdot \frac{5}{6} \cdot \frac{5}{6} \cdot \frac{5}{6} \cdot \frac{5}{6} = \left(\frac{1}{6}\right)\left(\frac{5}{6}\right)^4$$

Now, the events *sffff*, *fsfff*, *ffsff*, *fffsf*, and *ffffs* are mutually exclusive, so the probability of P(*sffff* or *fsfff* or *ffsff* or *fffsf* or *ffffs*) is just the sum of the individual probabilities. Thus,

$$P(\text{one success}) = C_{5,1}\left(\frac{1}{6}\right)\left(\frac{5}{6}\right)^4$$

To calculate the probability of two 1s in five rolls of the die, we note that there are $C_{5,2} = \dfrac{5!}{2!3!} = 10$ ways for the two 1s to occur. One such sequence is

$$s \quad s \quad f \quad f \quad f$$

The probability of this event is

$$\frac{1}{6} \cdot \frac{1}{6} \cdot \frac{5}{6} \cdot \frac{5}{6} \cdot \frac{5}{6} = \left(\frac{1}{6}\right)^2\left(\frac{5}{6}\right)^3$$

The probability of one of the ten mutually exclusive events "two 1s" is then

$$P(\text{two successes}) = C_{5,2}\left(\frac{1}{6}\right)^2\left(\frac{5}{6}\right)^3$$

Although we have only calculated the probabilities for one success in five trials and two successes in five trials, we can see a pattern in the results. We infer from these results that the probability of three successes in five trials should be

$$P(\text{three successes}) = C_{5,3}\left(\frac{1}{6}\right)^3\left(\frac{5}{6}\right)^2$$

Notice that in each instance the exponent of the probability of success is equal to the number of successes, and the exponent of the probability of failure is equal to the number of trials minus the number of successes. We could continue by writing the probability of four successes, but we will omit this step and go directly to the general formula for r successes in n trials. Using the pattern established in the previous example of rolling a die as our model, with p as the probability of a success and q as the probability of failure, we infer the following formula for r successes in n trials of a Bernoulli experiment:

DEFINITION Binomial Probability Function

The probability of exactly r successes in n trials of a Bernoulli experiment is given by

$$P(r \text{ successes}) = C_{n,r}p^r q^{n-r}$$

where p is the probability of a success and $q = 1 - p$ is the probability of a failure.

COMMENT: *The binomial probability function gets its name from the fact that it is the same function used to express the terms in the expanded form of the binomial* $(p + q)^n$. *(See Section 4.5.) The formula* $C_{n,r}p^r q^{n-r}$ *gives the* r*th term in this expansion.*

CALCULATOR COMMENT: *Although there are tables of binomial probabilities, you should be able to use your calculator to make the necessary calculations.*

Example 3 A basketball player makes 85% of his foul shots. Assuming that his foul-shooting attempts are independent trials, calculate:

(a) The probability that he will make exactly three foul shots in five attempts.
(b) The probability that he will make at least three foul shots in five attempts.

Solution We assume that each attempt is an independent trial and that "making the foul shot" is a "success." Since he makes 85% of his foul shots, the probability of a success is $p = 0.85$, and $q = 1 - 0.85 = 0.15$. Let X be the number of foul shots made.

(a) Using the binomial probability function, the probability of three successes in five attempts is

$$P(X = 3) = C_{5,3}(0.85)^3(0.15)^2 \approx 0.138$$

(b) The probability of at least three foul shots means $X \geq 3$; that is, 3 or 4 or 5 foul shots. Since these events are mutually exclusive, the addition rule for probabilities yields

$$\begin{aligned} P(X \geq 3) &= P(3 \text{ successes}) + P(4 \text{ successes}) + P(5 \text{ successes}) \\ &= C_{5,3}(0.85)^3(0.15)^2 + C_{5,4}(0.85)^4(0.15)^1 + C_{5,5}(0.85)^5(0.15)^0 \\ &\approx 0.138 + 0.392 + 0.444 = 0.974 \end{aligned}$$

■

Example 4 A biologist finds that 70% of the subjects vaccinated with a new serum do not develop flu symptoms when exposed to a flu virus. If 20 vaccinated subjects are exposed to a flu virus, what is the probability that exactly 15 of them will not develop flu symptoms?

Solution Let a success be "the subject does not develop flu symptoms" and let X be the number of successes. The probability of a success is $p = 0.7$; the probability of a failure is $q = 1 - 0.7 = 0.3$. Then the probability of 15 successes in 20 trials is

$$P(X = 15) = C_{20,15}(0.7)^{15}(0.3)^5 \approx 0.179$$

■

Example 5 Consider a Bernoulli experiment in which the probability of a success is $p = 0.4$. Draw a histogram for the probability distribution for the binomial random variable X, which is the number of successes in five trials of the experiment.

Solution The probabilities of the possible values of X are

$$\begin{aligned} P(X = 0) &= C_{5,0}(0.4)^0(0.6)^5 \approx 0.078 \\ P(X = 1) &= C_{5,1}(0.4)^1(0.6)^4 \approx 0.259 \\ P(X = 2) &= C_{5,2}(0.4)^2(0.6)^3 \approx 0.346 \\ P(X = 3) &= C_{5,3}(0.4)^3(0.6)^2 \approx 0.230 \\ P(X = 4) &= C_{5,4}(0.4)^4(0.6)^1 \approx 0.077 \\ P(X = 5) &= C_{5,5}(0.4)^5(0.6)^0 \approx 0.010 \end{aligned}$$

The desired probability distribution is then as shown in the table below, and the histogram is shown in Figure 6.9.

X	Probability
0	0.078
1	0.259
2	0.346
3	0.230
4	0.077
5	0.010

■

We discussed the mean and the standard deviation of the probability distribution of a random variable in Sections 6.2 and 6.3. The mean and standard deviation of a binomial probability distribution, which we give without proof, are given by the following formulas:

Mean and Standard Deviation of a Binomial Random Variable

If X is a binomial random variable representing the number of successes in n trials of a Bernoulli experiment in which the probability of a success is p and the probability of failure is q, then the mean of X is

$$\mu = np$$

and the standard deviation of X is

$$\sigma = \sqrt{npq}$$

Figure 6.9

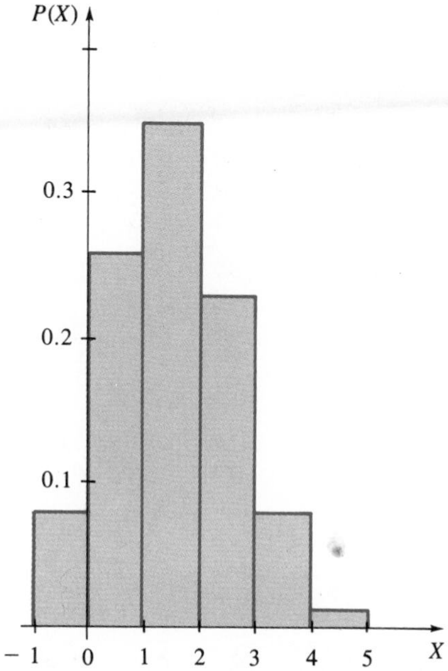

Example 6 Calculate the mean and standard deviation for the binomial random variable in Example 5.

Solution From Example 5, $n = 5$, $p = 0.4$, and $q = 0.6$. Thus,

$$\mu = np = 5(0.4) = 2$$
$$\sigma = \sqrt{npq} = \sqrt{5(0.4)(0.6)} \approx 1.1$$

COMMENT: *If the method of Section 6.2 is used to compute μ for the binomial random variable in Example 5, the value is $\mu = 1.999$, to three decimal places. The difference between 1.999 and 2 can be attributed to round-off error.*

To conclude this section, we note that the probability of r successes in n trials for a given p is always the same, and so the binomial probabilities can be organized into binomial probability tables. A portion of the table of binomial probabilities given in the Appendix is duplicated here for easy reference. The number of trials

n and the number of successes r are given in the far left columns and the probability of a success p is given across the top of the table. Suppose we want to use the table to find the probability of $r = 4$ successes in $n = 7$ trials, given that the probability of a success is $p = 0.2$. Enter the left column at $n = 7$ and $r = 4$, then move across the table to the column labeled 0.20. The correct probability is 0.029. Note that the number of successes, r, is repeated in the far right column for convenience.

Table of Binomial Probabilities

		p													
n	r	.01	.05	.10	.20	.30	.40	.50	.60	.70	.80	.90	.95	.99	r
2	0	.980	.902	.810	.640	.490	.360	.250	.160	.090	.040	.010	.002	.000	0
	1	.020	.095	.180	.320	.420	.480	.500	.480	.420	.320	.180	.095	.020	1
	2	.000	.002	.010	.040	.090	.160	.250	.360	.490	.640	.810	.902	.980	2
3	0	.970	.857	.729	.512	.343	.216	.125	.064	.027	.008	.001	.000	.000	0
	1	.029	.135	.243	.384	.441	.432	.375	.288	.189	.096	.027	.007	.000	1
	2	.000	.007	.027	.096	.189	.288	.375	.432	.441	.384	.243	.135	.029	2
	3	.000	.000	.001	.008	.027	.064	.125	.216	.343	.512	.729	.857	.970	3
4	0	.961	.815	.656	.410	.240	.130	.062	.026	.008	.002	.000	.000	.000	0
	1	.039	.171	.292	.410	.412	.346	.250	.154	.076	.026	.004	.000	.000	1
	2	.001	.014	.049	.154	.265	.346	.375	.346	.265	.154	.049	.014	.001	2
	3	.000	.000	.004	.026	.076	.154	.250	.346	.412	.410	.292	.171	.039	3
	4	.000	.000	.000	.002	.008	.026	.062	.130	.240	.410	.656	.815	.961	4
5	0	.951	.774	.590	.328	.168	.078	.031	.010	.002	.000	.000	.000	.000	0
	1	.048	.204	.328	.410	.360	.259	.156	.077	.028	.006	.000	.000	.000	1
	2	.001	.021	.073	.205	.309	.346	.312	.230	.132	.051	.008	.001	.000	2
	3	.000	.001	.008	.051	.132	.230	.312	.346	.309	.205	.073	.021	.001	3
	4	.000	.000	.000	.006	.028	.077	.156	.259	.360	.410	.328	.204	.048	4
	5	.000	.000	.000	.000	.002	.010	.031	.078	.168	.328	.590	.774	.951	5
6	0	.941	.735	.531	.262	.118	.047	.016	.004	.001	.000	.000	.000	.000	0
	1	.057	.232	.354	.393	.303	.187	.094	.037	.010	.002	.000	.000	.000	1
	2	.001	.031	.098	.246	.324	.311	.234	.138	.060	.015	.001	.000	.000	2
	3	.000	.002	.015	.082	.185	.276	.312	.276	.185	.082	.015	.002	.000	3
	4	.000	.000	.001	.015	.060	.138	.234	.311	.324	.246	.098	.031	.001	4
	5	.000	.000	.000	.002	.010	.037	.094	.187	.303	.393	.354	.232	.057	5
	6	.000	.000	.000	.000	.001	.004	.016	.047	.118	.262	.531	.735	.941	6
7	0	.932	.698	.478	.210	.082	.028	.008	.002	.000	.000	.000	.000	.000	0
	1	.066	.257	.372	.367	.247	.131	.055	.017	.004	.000	.000	.000	.000	1
	2	.002	.041	.124	.275	.318	.261	.164	.077	.025	.004	.000	.000	.000	2
	3	.000	.004	.023	.115	.227	.290	.273	.194	.097	.029	.003	.000	.000	3
	4	.000	.000	.003	.029	.097	.194	.273	.290	.227	.115	.023	.004	.000	4
	5	.000	.000	.000	.004	.025	.077	.164	.261	.318	.275	.124	.041	.002	5
	6	.000	.000	.000	.000	.004	.017	.055	.131	.247	.367	.372	.257	.066	6
	7	.000	.000	.000	.000	.000	.002	.008	.028	.082	.210	.478	.698	.932	7

EXERCISES SECTION 6.4

In each of the following exercises evaluate the probability of the given Bernoulli trials.

A coin is tossed six times. Calculate the probability of:

1. Exactly four heads.
2. Exactly five heads.
3. (a) Exactly zero heads.
   (b) Exactly one head.
   (c) At least one head.
4. More than one head.

A die is rolled seven times. Calculate the probability of:

5. Exactly three 2s.
6. Exactly five 3s.
7. (a) Exactly six 3s.
   (b) Exactly seven 3s.
   (c) More than five 3s.
8. Fewer than six 3s.

The probability of contracting a certain contagious disease is 0.75. If ten people are exposed to the disease, calculate the probability that:

9. Exactly 8 people contract the disease.
10. Exactly 5 people contract the disease.
11. More than 9 people contract the disease.
12. Fewer than 10 people contract the disease.

A company that builds digital radios buys transistors from an electronics warehouse. The company estimates that 2% of the transistors are defective. If a radio has eight transistors, what is the probability that a radio chosen at random from the assembly line will have:

13. No defects?
14. One defect?
15. At most two defects?
16. More than two defects?

Three cards are dealt from a deck of 52 playing cards. What is the probability of being dealt:

17. Exactly one face card?
18. Exactly two face cards?
19. At least two face cards?
20. More than one face card?

Suppose that 12% of all graduates in this year's class are marketing majors. If the alumni office randomly picks a group of 15 graduates for a survey, what is the probability that the group contains:

21. Exactly 2 marketing majors?
22. Fewer than 2 marketing majors?
23. A die is rolled five times. A number greater than 4 is considered a success.
    (a) Calculate the probability of rolling no successes, one success, two, three, four, and five successes in five trials.
    (b) Draw a histogram of this binomial probability distribution.
    (c) Calculate the mean and standard deviation of the random variable associated with this probability distribution.
24. In a certain voting district 45% of the registered voters are Democrats, 40% are Republicans, and 15% are independent.
    (a) Calculate the probability of picking no Democrats, one Democrat, two, three, four, and five Democrats in a randomly selected group of five voters.
    (b) Draw a histogram of this binomial probability distribution.
    (c) Calculate the mean and standard deviation of the random variable associated with this probability distribution.
25. A baseball player has a batting average of 0.270.
    (a) Calculate the probability that the player will have no hits, one hit, two, three, and four hits in four times at bat.
    (b) Draw a histogram of this binomial probability distribution.
    (c) Calculate the mean and standard deviation of the random variable associated with this probability distribution.
26. A maintenance contractor estimates that 3% of the new light bulbs installed to replace burned-out bulbs will burn out within one week of installation.

Three new bulbs are installed by the maintenance crew.

(a) Calculate the probability that no bulbs, one, two, and three light bulbs will fail in the first week.
(b) Draw the histogram of this probability distribution.
(c) Calculate the mean and standard deviation of the random variable associated with this probability distribution.

27. Consider a Bernoulli experiment in which the probability of a success is 0.4. Use the table of binomial probabilities in the appendix to find the indicated probability distributions:

(a) The probability distribution associated with no successes, one, two, and three successes in three trials. Draw the histogram.
(b) The probability distribution associated with no successes, one, two, three, four, five, and six successes in six trials. Draw the histogram.
(c) The probability distribution associated with no successes, one, two, three, four, five, six, seven, eight, nine, ten, eleven, and twelve successes in twelve trials. Draw the histogram.
(d) Based on the histograms in (a), (b), and (c), what would you say seems to be happening to the probability distributions as the number of trials increases?

6.5 THE NORMAL PROBABILITY DISTRIBUTION

In Section 6.1 we mentioned that it is convenient to draw histograms in which the width of the rectangles is equal to one unit because the area of the rectangle is then numerically equal to the probability of the outcome represented by the rectangle. This also means that the area encompassed by the histogram is equal to 1, because the sum of the probabilities is 1.

Recall from the previous section that the binomial probability distribution is given by

$$P(r \text{ successes}) = C_{n,r}p^r q^{n-r}$$

for $r = 0, 1, 2, \ldots, n$. If we draw the histogram for the binomial probability distribution with $p = 0.5$ and $n = 12$, as shown in Figure 6.10, the area of each rectangle corresponds to the probability of the corresponding outcome, and the total area of the histogram is 1. Also of interest is the shape of the histogram, which appears to have the largest concentration of values in the vicinity of the mean, $\mu = 6$, and seems to be symmetric with respect to the mean. We note that this is the typical shape of histograms of binomial probability distributions in which p is close to 0.5 and n is greater than 12.

THE NORMAL PROBABILITY DISTRIBUTION

Consider an experiment in which the values of the random variable X are the diameters of 1000 washers. If we were to use a meter stick to make the measurements, we could expect to measure the diameters to an accuracy of about 0.5 mm. If instead of a meter stick we were given a micrometer, we could expect to measure the diameter to an accuracy of about 0.005 mm. There are other instruments that would give us even more accurate measurements, but ultimately we would have to admit that any measurement we make is only an approximation of the actual diameter, which may be any real number within a certain interval. In this sense,

Figure 6.10

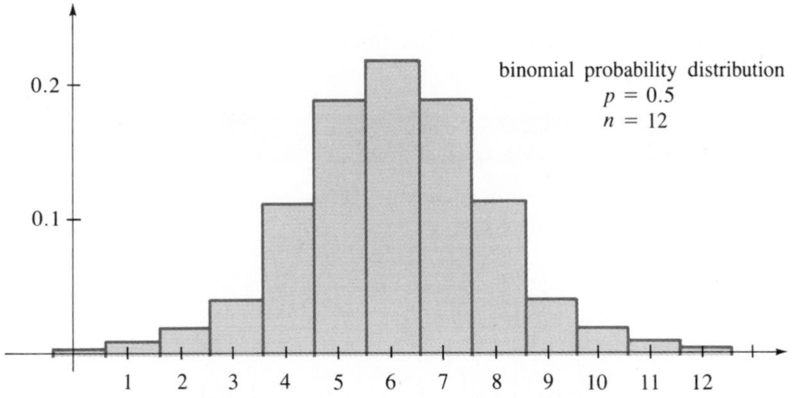

the values of this random variable are not a finite set of discrete numbers, as they have been in the previous sections, but an infinite set of real numbers on some interval. A random variable that can take on any real number within a given interval is called a *continuous* random variable. In this section we discuss probability distributions associated with continuous random variables. Such probability distributions are very naturally called *continuous probability distributions.* Although there are many different continuous probability distributions, the **normal probability distribution** is used in probability theory more than any of the others. One reason for this is that in most experiments that involve the measurement of length, weight, or time, the measurements are described by normal probability distributions.

Just as the probability distribution of a finite random variable is graphically represented by a histogram, the probability distribution of a continuous random variable is graphically represented by a smooth curve. Unlike a finite probability distribution, which may be given by a table of values, a continuous probability distribution is defined by a function, called a *probability density function,* whose domain coincides with the interval of values taken on by the random variable. A probability density function f for a continuous random variable X has the following properties:

- $f(x) \geq 0$ for all x.
- The area of the region bounded by the graph of f and the x-axis is equal to 1.

The graph of a typical probability density function is shown in Figure 6.11.

Figure 6.11

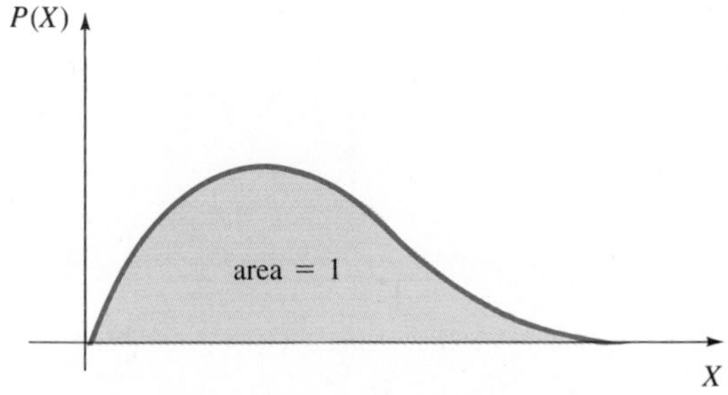

As noted above, the most important continuous probability distribution is called the normal probability distribution.

DEFINITION Equation for a Normal Curve

If X is a normal random variable with mean μ and standard deviation σ, then the probability of x is

$$P(x) = \frac{1}{\sqrt{2\pi}\sigma} e^{-\frac{1}{2}\left(\frac{x-\mu}{\sigma}\right)^2}$$

The mean μ and standard deviation σ of a continuous random variable have the same meaning as they do for finite random variables; that is, the mean μ is a measure of the central tendency of the probability distribution, and the standard deviation σ is a measure of its dispersion about the mean. The graph of a normal probability density function (a *normal curve*) is shown in Figure 6.12. A typical normal curve has a bell shape similar to that of the graph shown below.

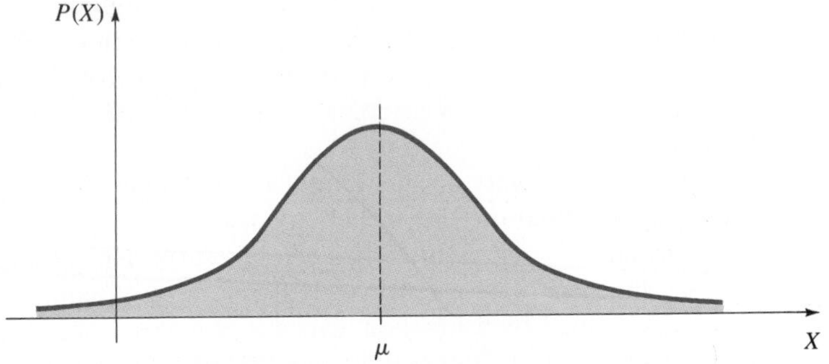

Figure 6.12

The normal curve is completely determined by its mean μ and the standard deviation σ. If μ is changed, the normal curve is translated either left or right along the x-axis (see Figure 6.13). If σ is changed, the spread about the mean is either increased or decreased (see Figure 6.14). In either case, the area bounded by the normal curve and the x-axis remains constant at 1.

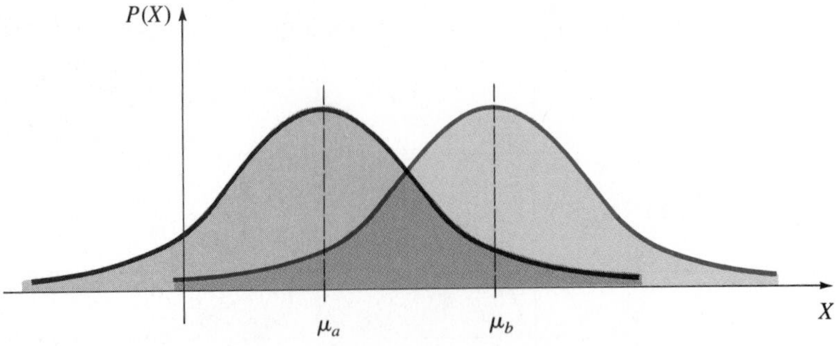

Figure 6.13

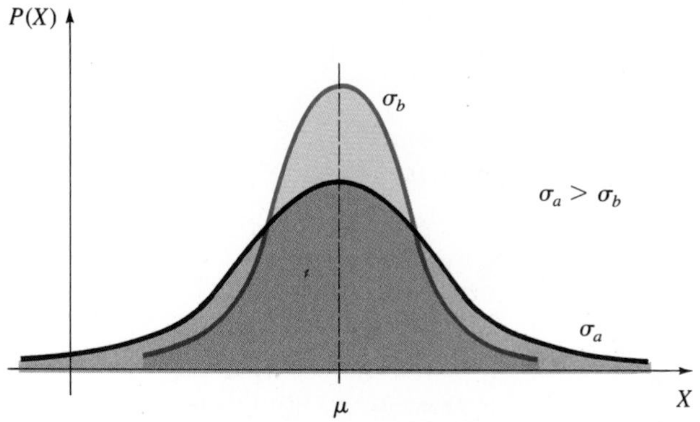

Figure 6.14

HISTORICAL COMMENT: *The normal curve is also called the* Gaussian curve, *in honor of Carl Friedrich Gauss (1777–1855), a German mathematician.*

As noted previously, the normal curve is completely determined by the mean μ and the standard deviation σ of the random variable X. However, the following properties are shared by all normal curves:

- The normal curve is symmetric with respect to its mean μ.
- The normal curve has a bell shape that extends indefinitely to the left and right of the mean.
- The normal curve is concave-down on the interval $\mu - \sigma \leq x \leq \mu + \sigma$ and concave-up outside this interval. (See Figure 6.15.)

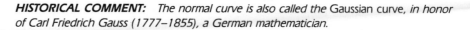

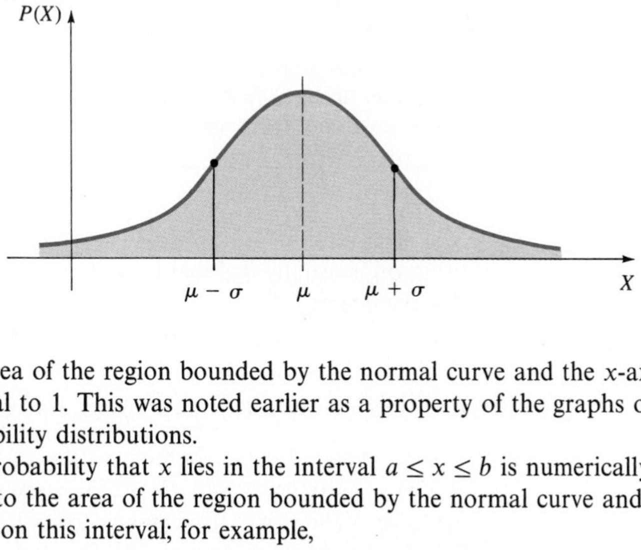

Figure 6.15

- The area of the region bounded by the normal curve and the x-axis is equal to 1. This was noted earlier as a property of the graphs of all probability distributions.
- The probability that x lies in the interval $a \leq x \leq b$ is numerically equal to the area of the region bounded by the normal curve and the x-axis on this interval; for example,

$$P(a \leq x \leq b) = \text{area of shaded region in Figure 6.16.}$$

Figure 6.16

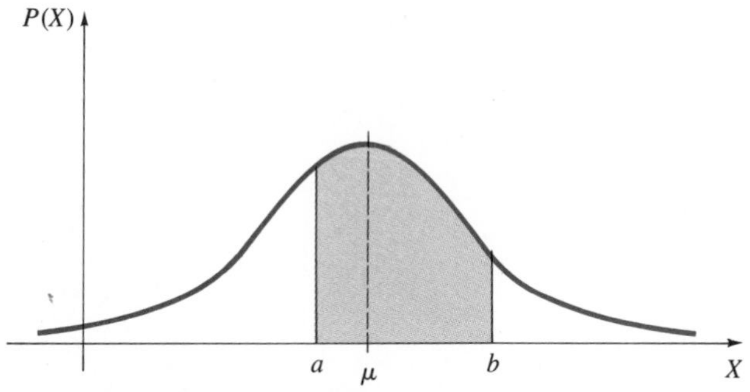

If we are to use this last property, we must be able to calculate the area of the region bounded by a normal curve and the x-axis on some interval $a \leq x \leq b$, a process that requires a knowledge of integral calculus. Further, we must be able to calculate this area for *any* normal curve. Fortunately, there is another property of normal curves that makes this task relatively easy. This property is that *any normal curve can be transformed into any other normal curve. Transformed* means that the mean, the standard deviation, and the area bounded by the curve on a given interval can be related algebraically to the mean, the standard deviation, and the area under another normal curve. Because of this property, we need only make the area calculations for a single normal curve, a curve that will act as a standard. Then we can transform any other normal curve into this standard curve. The normal curve we choose for this standard curve is called the *standard normal curve,* and has a mean of zero and a standard deviation of 1. We use this curve to make the area calculations because it is the easiest normal curve to work with.

DEFINITION The Standard Normal Curve

The normal probability distribution with $\mu = 0$ and $\sigma = 1$ is called the **standard normal probability distribution** and the corresponding normal curve is called the **standard normal curve.**

The random variable associated with the standard normal probability distribution is usually denoted by Z. Probabilities for values of the random variable Z in the interval $Z \leq z_1$ are calculated for the standard normal curve and arranged in a table of normal probabilities, as shown in Table 2 of the Appendix.

Using the Table of Standard Normal Probabilities

Table 2, Appendix A gives areas to the left of z_1—that is, probabilities for values of $Z \leq z_1$. The value of z_1 is given in the far left column and across the top of the table. Values of Z from -3.49 to 3.49, in increments of 0.01, can be found in Table 2, Appendix A. The probability that the value of Z is in the interval $Z \leq z_1$ is the number at the intersection of the appropriate row and column representing

z_1. The following example shows how to use Table 2, Appendix to calculate probabilities for the standard normal variable Z. Later, we will show how the standard normal curve is used for other normal distributions of a random variable X.

Example 1 Given a standard normal random variable Z, use Table 2, Appendix to find:

(a) $P(Z \leq 1.37)$.
(b) $P(Z \geq 2.40)$.
(c) $P(-1.55 \leq Z \leq 0)$.
(d) $P(0.58 \leq Z \leq 0.90)$.

Solution

(a) The region under the standard normal curve that corresponds to $P(Z \leq 1.37)$ is shown in Figure 6.17. The probability is obtained from Table 2, Appendix by locating 1.3 in the left column and 0.07 along the top of the table. The probability 0.9147 is the number located at the intersection of the row labeled 1.3 and the column labeled 0.07. Thus,

$$P(Z \leq 1.37) = 0.9147$$

Figure 6.17

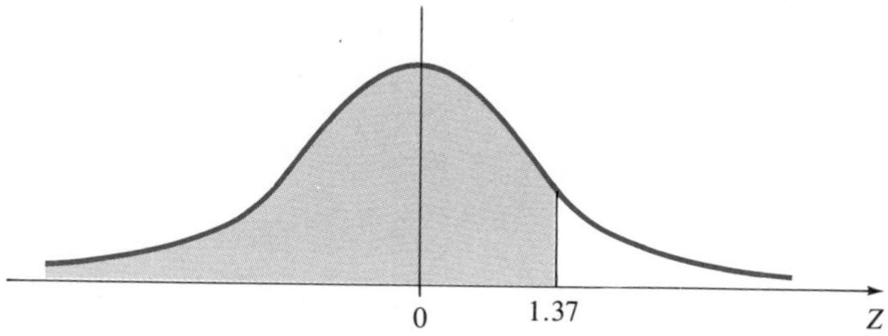

(b) The desired probability corresponds to that portion of the region under the normal curve to the right of $Z = 2.40$. (See Figure 6.18.) To find this area, we subtract the area to the left of 2.40 from 1.0. We subtract from 1.0 since this is the total area under the normal curve.

$$P(Z \geq 2.40) = 1.0 - P(Z \leq 2.40) = 1.0 - 0.9918 = 0.0082$$

Figure 6.18

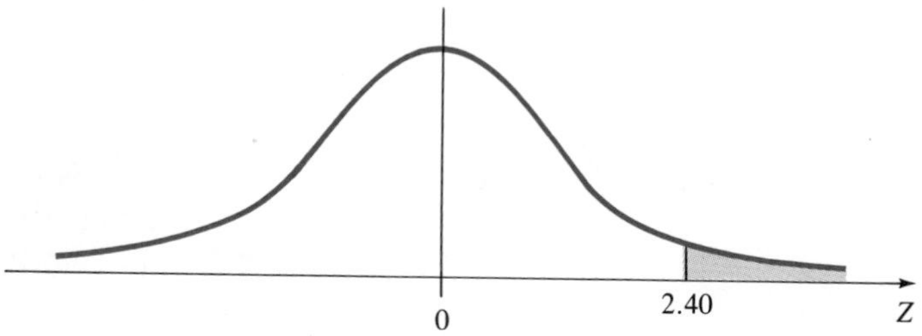

(c) Figure 6.19 shows the desired region. Using Table 2, Appendix, we find that the area to the left of $z_1 = -1.55$ is 0.0606 and the area to the left of $z_2 = 0$ is 0.5000. Therefore, the desired probability corresponds to the difference between these two areas:

$$\begin{aligned} P(-1.55 \le Z \le 0) &= P(Z \le 0) - P(Z \le -1.55) \\ &= 0.5000 - 0.0606 = 0.4394 \end{aligned}$$

Figure 6.19

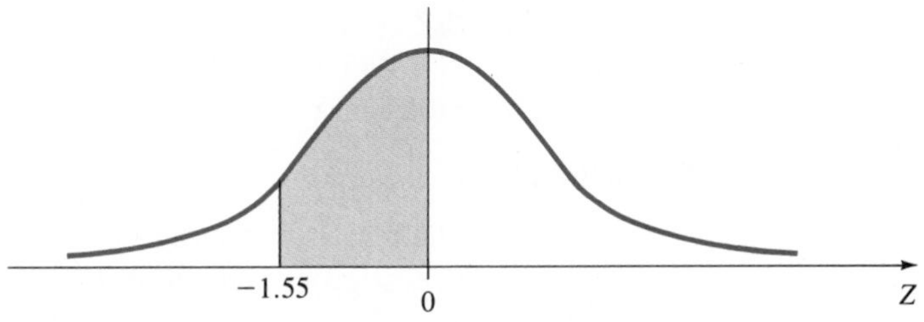

(d) The area of this region, which is shown in Figure 6.20, can be found by subtracting the area to the left of 0.58 from the area to the left of 0.90. Thus, the desired probability is

$$\begin{aligned} P(0.58 \le Z \le 0.90) &= P(Z \le 0.90) - P(Z \le 0.58) \\ &= 0.8159 - 0.7190 = 0.0969 \end{aligned}$$

Figure 6.20

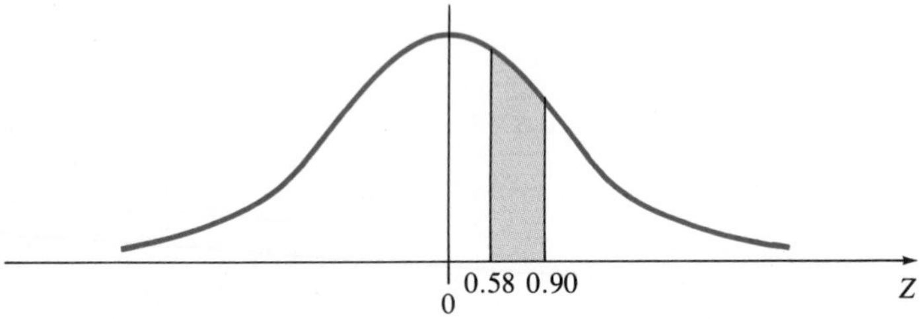

Example 2 If Z is a standard normal random variable, find the value z_1 for which $P(Z \le z_1) = 0.9850$.

Solution In this case, we know the probability that the value of Z is in some interval to the left of z_1 and we want to use Table 2, Appendix to find z_1. The situation is illustrated in Figure 6.21. To find the desired value for z_1, we scan the

Figure 6.21

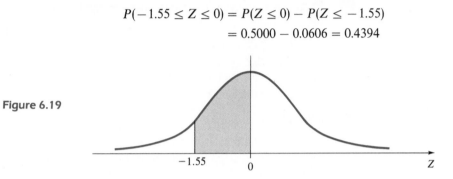

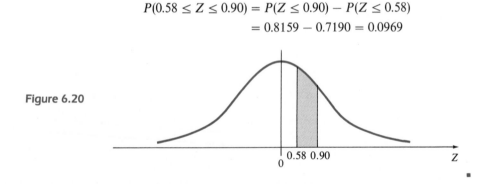

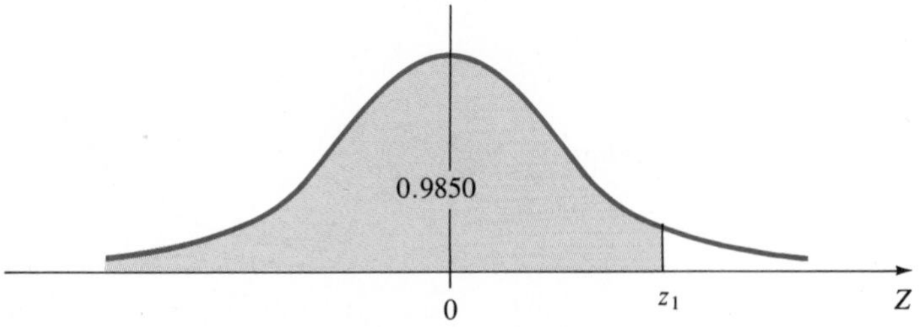

probabilities in Table 2, Appendix until we find 0.9850, or the entry closest to this number. In this instance, we locate 0.9850 in the row labeled 2.1 and the column labeled 0.07, so $z_1 = 2.17$ is the correct number. ■

Next, we will show how the table of standard normal probabilities is used to calculate probabilities for any normal probability distribution. The question that we need to answer is "How is a normal distribution with a mean of μ and a standard deviation of σ transformed into a normal distribution with a mean of zero and a standard deviation of 1?" To answer this question, we note that if μ is subtracted from each value of X, the normal curve is moved, or translated, left or right by μ units. Thus, by making the calculation $x - \mu$ for each value of X, the mean of the distribution of X is aligned with zero. Next, we note that if we divide each value of X by σ, the spread of the distribution is expanded or contracted by σ units. This means that if we divide a normal distribution, whose standard deviation is σ, by σ, we will end up with a normal distribution with a standard deviation of 1. Combining these two operations into a single formula for a standard normal variable, we have the following transformation equation:

The Standard Normal Variable

To transform a normal random variable X with a mean of μ and a standard deviation of σ into a standard normal variable Z with a mean of 0 and a standard deviation of 1, use the equation

$$z = \frac{x - \mu}{\sigma} \tag{6.1}$$

where μ is the mean and σ is the standard deviation of the given normal random variable.

The relation between the standard normal curve with a mean of 0 and a standard deviation of 1 and a normal curve with a mean of μ and a standard deviation of σ is shown below. The shaded areas are equal, because z in (a) is the transform of x in (b).

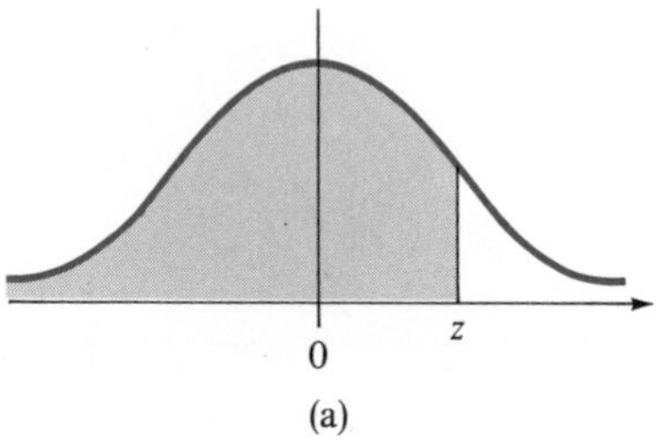

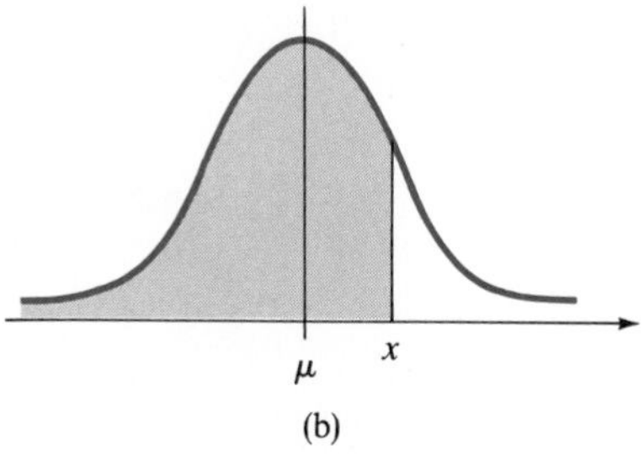

Example 3 Let X be a normal random variable with $\mu = 50$ and $\sigma = 4$. Use Equation 6.1 and Table 2, Appendix to find the following probabilities:

(a) $P(50 \leq X \leq 60)$
(b) $P(X \leq 60)$
(c) $P(X \geq 55)$
(d) $P(42 \leq X \leq 55)$

Solution

(a) The probability $P(50 \leq X \leq 60)$ is shown in Figure 6.22. Using Equation 6.1 to transform $x = 50$ and $x = 60$, we have

$$z_1 = \frac{50 - 50}{4} = 0 \qquad \text{and} \qquad z_2 = \frac{60 - 50}{4} = 2.5$$

Therefore, $P(50 \leq X \leq 60) = P(0 \leq Z \leq 2.5)$. This relationship is shown in Figure 6.22. From Table 2, Appendix, we get

$$\begin{aligned} P(50 \leq X \leq 60) &= P(0 \leq Z \leq 2.5) \\ &= P(Z \leq 2.5) - P(Z \leq 0) = 0.9938 - 0.5000 = 0.4938 \end{aligned}$$

Figure 6.22

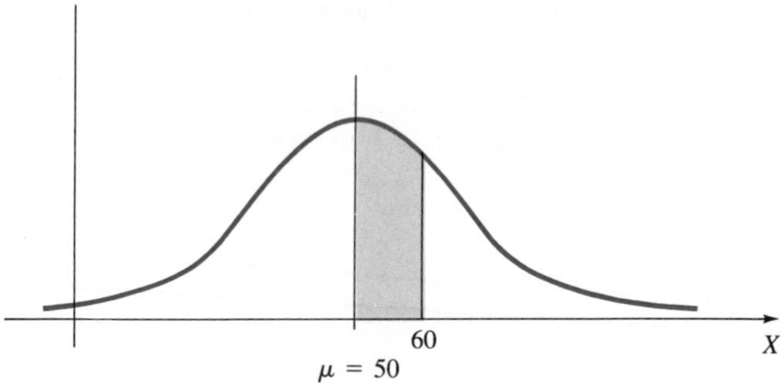

(b) $P(X \leq 60)$ corresponds to the area to the left of $x = 60$ or $z = 2.5$. (See Figure 6.23.) The required calculation is

$$P(X \leq 60) = P(Z \leq 2.5) = 0.9938$$

Figure 6.23

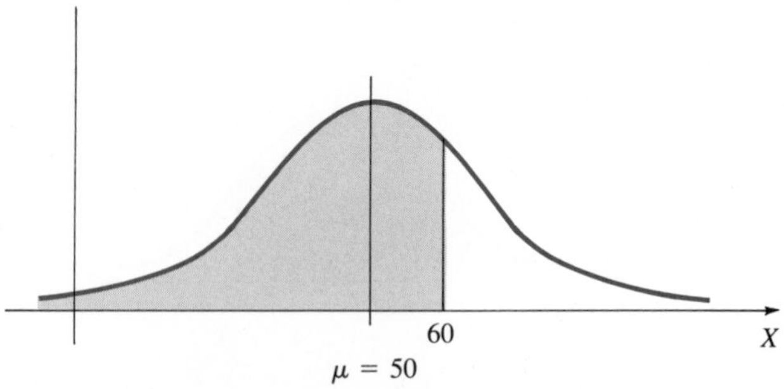

(c) The area corresponding to $P(X \geq 55)$ is shown in Figure 6.24. We use Equation 6.1 to transform $x = 55$ into the standard normal variable as follows:

$$z = \frac{55 - 50}{4} = 1.25$$

Since the desired area is to the right of $x = 55$ or $z = 1.25$, we make the following calculation to find $P(X \geq 55)$:

$$P(X \geq 55) = P(Z \geq 1.25) = 1.0 - P(Z \leq 1.25) = 1.0 - 0.8944 = 0.1056$$

Figure 6.24

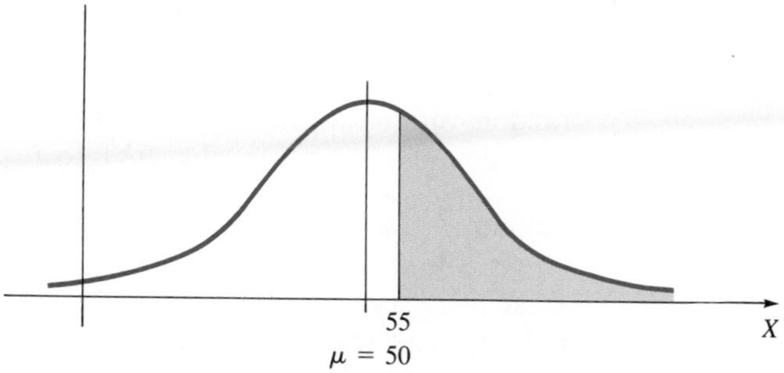

(d) To find $P(42 \leq X \leq 55)$ we use Equation 6.1 to transform the endpoints of the interval into values of the standard normal variable. Thus,

$$z_1 = \frac{42 - 50}{4} = -2.00 \qquad \text{and} \qquad z_2 = 1.25 \text{ (from part (c))}$$

The required calculation is

$$\begin{aligned} P(42 \leq X \leq 55) &= P(-2.00 \leq Z \leq 1.25) \\ &= P(Z \leq 1.25) - P(Z \leq -2.00) \\ &= 0.8944 - 0.0228 = 0.8716 \end{aligned}$$

The normal curve is shown in Figure 6.25.

Figure 6.25

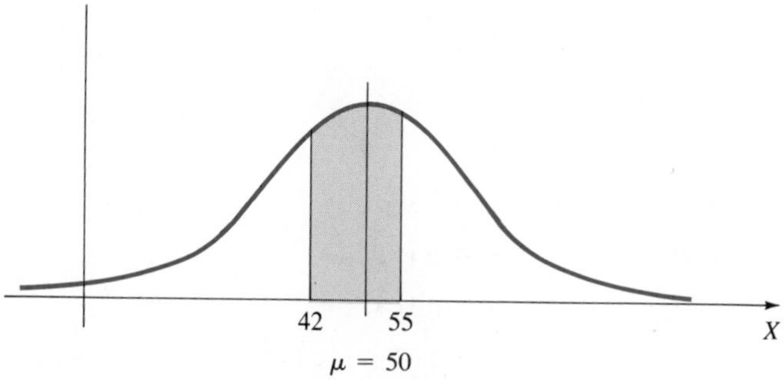

■

COMMENT: *Chebyshev's theorem, which was discussed in Section 6.3, assures us that at least* $1 - \frac{1}{k^2}$ *percent of the values of a random variable lie within k standard deviations of the mean. If we know that the random variable has a normal distribution, we can improve the Chebyshev estimate. For example, consider a normal distribution with* $\mu = 500$ *and* $\sigma = 30$. *Chebyshev's theorem assures us that at least* $\left(1 - \frac{1}{2^2}\right) \cdot 100 = 75\%$ *of the values lie within* 2σ *of the mean; that is, at least 75% of the values lie within the interval 440 to 560. The fact that we know that the variable is normally distributed allows us to obtain more accurate results. Specifically, a normal distribution has 95.44% of its values within this interval. (Calculate the area under the standard normal curve from* $z_1 = -2$ *to* $z_2 = 2$.)

Example 4 The SAT scores for entering students at Middle University are a random variable X that is normally distributed with a mean of 1000 and a standard deviation of 40. Between what SAT scores do 90% of the student scores lie?

Solution Using the symmetry of the normal curve, we note that 45% of the SAT scores will lie above 1000 and 45% below 1000. Therefore, we wish to find the value of z_1 such that the area to the left of z_1 is equal to $0.5 + 0.45 = 0.95$. We note that 0.95 is midway between the table values of 0.9495 ($z = 1.64$) and 0.9505 ($z = 1.65$). We conclude that $z = 1.645$ is the appropriate value. This means that the boundaries of the desired interval are ± 1.645 standard deviations from the mean. Since the mean is 1000 and the standard deviation is 40, we conclude that 90% of the SAT scores lie within the interval

$$1000 - 1.645(40) \leq X \leq 1000 + 1.645(40)$$
$$934 \leq X \leq 1066$$

EXERCISES SECTION 6.5

In Exercises 1–14, if the random variable Z has a standard normal distribution with a mean of 0 and a standard deviation of 1, find the probability that z takes on a value in the indicated interval:

1. $z \leq 1.07$
2. $z \leq 0.56$
3. $z \leq -0.5$
4. $z \leq -1.7$
5. $0 \leq z \leq 1.5$
6. $0 \leq z \leq 2.2$
7. $0 \leq z \leq 0.67$
8. $0 \leq z \leq 1.14$
9. $-1.12 \leq z \leq 0$
10. $-0.31 \leq z \leq 0$
11. $-0.34 \leq z \leq 1.07$
12. $-2.02 \leq z \leq 1.55$
13. $z \geq 1.19$
14. $z \geq 2.31$

In Exercises 15–20, find the value of z_1 *corresponding to the given conditions.*

15. 96% of the total area is to the left of z_1.
16. 75% of the total area is to the left of z_1.
17. 5% of the total area is to the right of z_1.
18. 1% of the total area is to the right of z_1.
19. 10% of the total area is between $z = 0$ and $z = z_1$, and $z_1 > 0$.

20. 35% of the total area is between $z = 0$ and $z = z_1$, and $z_1 > 0$.

In Exercises 21–30, X is a normally distributed random variable with the indicated mean and standard deviation. Find the probability that X takes on a value in the given interval.

21. Given $\mu = 7.3$ and $\sigma = 0.4$, find $P(X \leq 7.5)$.

22. Given $\mu = 44.1$ and $\sigma = 1.8$, find $P(X \leq 40)$.

23. Given $\mu = 125$ and $\sigma = 6.5$, find $P(X \geq 120)$.

24. Given $\mu = 78.3$ and $\sigma = 5.1$, find $P(X \geq 75)$.

25. Given $\mu = 200$ and $\sigma = 15$, find $P(205 \leq X \leq 210)$.

26. Given $\mu = 1.75$ and $\sigma = 0.2$, find $P(1.8 \leq X \leq 2.0)$.

27. Given $\mu = 10$ and $\sigma = 3$, find $P(10 \leq X \leq 12)$.

28. Given $\mu = 25$ and $\sigma = 2$, find $P(25 \leq X \leq 28)$.

29. Given $\mu = 0.95$ and $\sigma = 0.08$, find $P(0 \leq X \leq 0.97)$.

30. Given $\mu = 330$ and $\sigma = 12.5$, find $P(0 \leq X \leq 350)$.

31. A runner finds that her times for 10-km races are normally distributed with a mean of 42.5 min and a standard deviation of 1.3 min. What is the probability that she will run the next race in less than 42.0 min?

32. A quality-control inspector finds that the cylinders of a new gasoline engine have a mean diameter of 3.15 in. and a standard deviation of 0.02 in. Assuming that the measurements are normally distributed, find the probability that the diameter of a cylinder will measure between 3.12 and 3.18 in.

33. The government requires the batteries that it purchases for its vehicles to have a mean life of 40 months with a standard deviation of 6 months. If battery life is normally distributed, what percentage of these batteries will fail in less than 36 months?

34. A manufacturer of bicycle wheels requires that the axles used in the wheel have a diameter between 0.55 cm and 0.58 cm. If the axles being supplied are normally distributed with a mean diameter of 0.56 cm and a standard deviation of 0.02 cm, what percentage of the axles fall within the acceptable range?

35. The mean annual rainfall in Ripley, Ohio is 52.7 in. with a standard deviation of 5.3 in. If annual rainfall is normally distributed, what is the probability that the annual rainfall will exceed 60 in.?

36. The thickness of sheet steel produced by a rolling mill is normally distributed with a mean of 0.25 in. and standard deviation of 0.005 in. If only sheets whose thickness is between 0.249 in. and 0.251 in. are acceptable, what percentage of sheets are rejected?

37. A company that manufactures water heaters provides a ten-year limited warranty on its heaters. Company records show that the mean life of the water heaters is 12 years with a standard deviation of 1.1 years. If heater life is normally distributed, what percentage of the water heaters will fail in less than ten years?

38. If the water heater company in Exercise 37 wants the probability that a water heater will fail before the end of the warranty period to be less than 0.005, what warranty period should it allow for the heaters?

6.6 Normal Approximation of the Binomial Distribution

Recall from Section 6.4 that a Bernoulli process is one with a finite number of independent trials, in which the outcome of each trial is classified as "success" or "failure." To calculate the probability of r successes in n trials of a Bernoulli process we use the formula

$$P(r \text{ successes}) = \frac{n!}{r!(n-r)!} p^r q^{n-r}$$

If $n < 10$, this formula is fairly easy to use, but it gets more difficult as n gets larger. For instance, calculating the probability of getting fewer than 40 heads in 100 tosses of a coin is a formidable task. One of the major reasons we study the normal distribution is that it can be used to approximate the binomial distribution. As an example of how this is done, suppose a coin is tossed 50 times and we want to compute the probability that the number of heads will be between 27 and 30, inclusive. Since this is a Bernoulli experiment, we can calculate this probability by calculating the probabilities of 27, 28, 29, and 30 heads and adding those probabilities; that is,

$$P(27 \le H \le 30) = P(27) + P(28) + P(29) + P(30)$$

where

$$P(27) = \frac{50!}{27!23!}(0.5)^{27}(0.5)^{23}$$

$$P(28) = \frac{50!}{28!22!}(0.5)^{28}(0.5)^{22}$$

$$P(29) = \frac{50!}{29!21!}(0.5)^{29}(0.5)^{21}$$

$$P(30) = \frac{50!}{30!20!}(0.5)^{30}(0.5)^{20}$$

The evaluation of these probabilities is a nasty job without a calculator or a computer. Fortunately, we had access to a calculator, and found that

$$P(27 \le H \le 30) = 0.2764$$

The histogram for this binomial experiment is shown in Figure 6.26. The mean of the distribution is $\mu = np = 50(0.5) = 25$ and the standard deviation is $\sigma = \sqrt{npq} = \sqrt{(50)(0.5)(0.5)} \approx 3.54$.

Figure 6.26

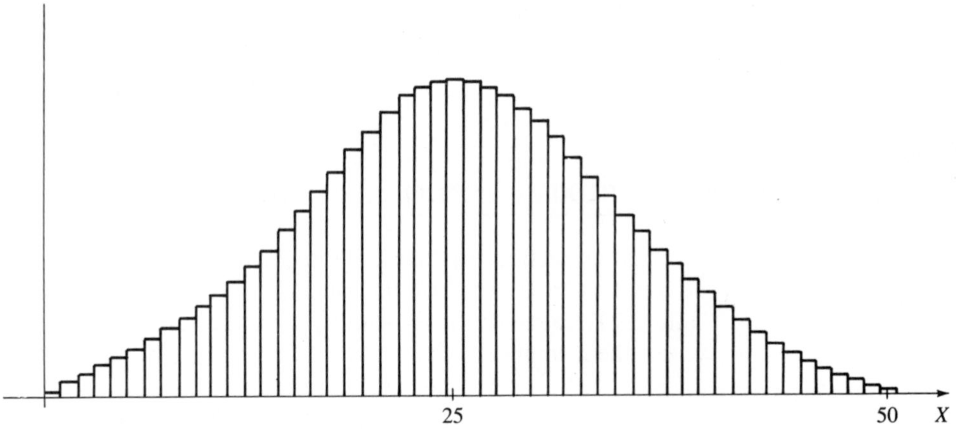

Clearly, the contour of the binomial distribution in Figure 6.26 is similar to the bell shape of the normal probability distribution. To emphasize the similarity between the two distributions, we have redrawn the histogram for the binomial distribution in Figure 6.27 and superimposed on it a normal curve, whose mean is 25 and whose standard deviation is 3.54. Of course, the binomial distribution is discrete and the normal distribution is continuous, but you can see that the area under the normal curve would be very close to that of the binomial histogram.

Figure 6.27

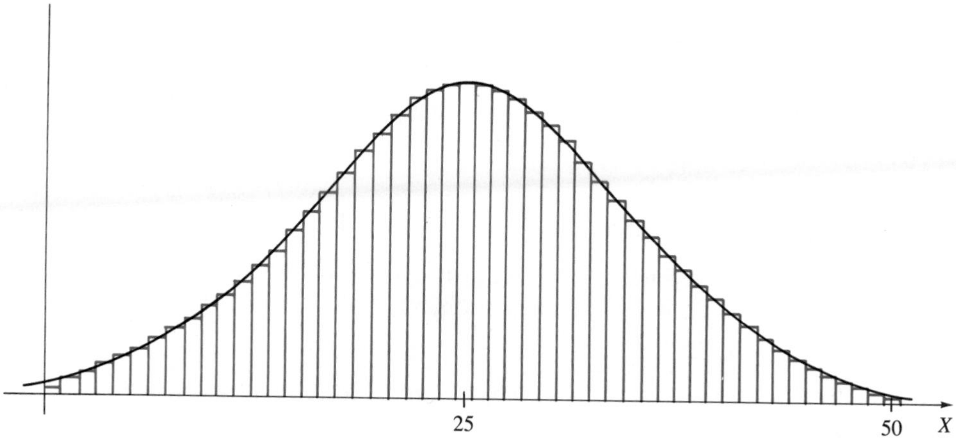

COMMENT: *In general, normal probabilities can be used to approximate binomial probabilities when the number of trials is greater than 12 and the probability of a success is close to 0.5. A useful rule of thumb is that the normal distribution can be used to approximate the binomial probabilities when* $np > 5$.

HISTORICAL COMMENT: *Abraham DeMoivre (1667–1754) was the first person to recognize that the normal curve can be used to approximate binomial probabilities. His work was later generalized by P. S. Laplace (1749–1827).*

When using the normal distribution to approximate binomial probabilities, we expand the interval by one-half unit at each endpoint in order to make the continuous distribution match the discrete distribution. For example, to use the normal distribution to approximate the binomial probability of getting 27, 28, 29, or 30 heads on 50 tosses of a coin, we compute the normal probability for the interval $26.5 \leq H \leq 30.5$. Figure 6.28 shows that the area under the normal curve from $x = 26.5$ to $x = 30.5$ is approximately equal to that of the shaded portion of the binomial histogram.

To find the area under the normal curve from $x_1 = 26.5$ to $x_2 = 30.5$ we first compute the corresponding z-values. For $x_1 = 26.5$, $z_1 = \dfrac{26.5 - 25}{3.54} \approx 0.42$ and for

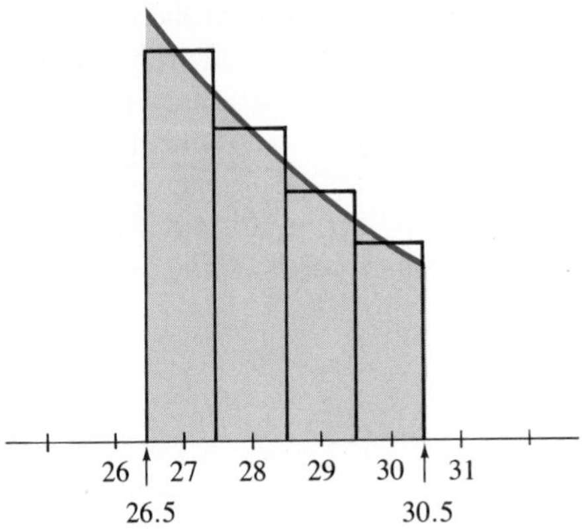

Figure 6.28

$x_2 = 30.5$, $z_2 = \dfrac{30.5 - 25}{3.54} \approx 1.55$. From the table of normal probabilities (Table 2, Appendix), $z_1 = 0.42$ gives an area of 0.6628 and $z_2 = 1.55$ gives an area of 0.9394. To find the desired probability, we subtract the first of these two numbers from the second. Thus,

$$P(26.5 \leq H \leq 30.5) = 0.9394 - 0.6628 = 0.2766$$

Notice that this is very close to the binomial probability 0.2764 that we computed at the beginning of this section.

Example 1 Use the normal distribution to estimate the probability that exactly 30 heads will occur when a coin is tossed 50 times.

Solution Since we cannot calculate the normal probability of $x = 30$, we use the interval $x_1 = 29.5$ to $x_2 = 30.5$. The corresponding z-scores for these endpoints are $z_1 = \dfrac{29.5 - 25}{3.54} \approx 1.27$ and $z_2 = \dfrac{30.5 - 25}{3.54} \approx 1.55$. Using Table 2, Appendix, the area to the left of $z_1 = 1.27$ is 0.8980 and the area to the left of $z_2 = 1.55$ is 0.9394. Therefore,

$$P(30 \text{ heads}) \approx P(29.5 \leq \text{heads} \leq 30.5) = 0.9394 - 0.8980 = 0.0414.$$

We know that the probability of exactly 30 heads is $P(30 \text{ heads}) = C_{50,30}(0.5)^{30}(0.5)^{20} \approx 0.0419$. This agrees well with the result obtained using the normal distribution. ■

Example 2 A basketball player shoots 72% from the foul line. Use the normal probability distribution to approximate the probability that the player will make fewer than 20 of the next 25 attempts.

Solution This is a Bernoulli experiment in which the probability of a success (making a foul shot) is 0.72 and the number of trials is 25. We will use the normal distribution with $\mu = np = 25(0.72) = 18$, which is greater than 5, and $\sigma = \sqrt{npq} = \sqrt{(25)(0.72)(0.28)} \approx 2.24$ to approximate the desired binomial probability. Notice that to make fewer than 20 shots means to make between no shots and 19 shots. Therefore, we use $x_1 = -0.5$ to $x_2 = 19.5$. The corresponding z-scores are

$$z_1 = \frac{-0.5 - 18}{2.24} \approx -8.26$$

and

$$z_2 = \frac{19.5 - 18}{2.24} \approx 0.67$$

The desired probability corresponds to the area under the normal curve from $z_1 = -8.26$ to $z_2 = 0.67$. Table 2, Appendix does not include $z_1 = -8.26$ because the area to the left of this z-score is so small that we may replace it with zero. The area to the left of $z_2 = 0.67$ is 0.7486. Therefore,

$$\begin{aligned} P(X < 20) &\approx P(-0.5 \le X \le 19.5) \\ &= P(-8.26 \le Z \le 0.67) = 0.7486 - 0 = 0.7486 \end{aligned}$$

We conclude that there is a probability of about 0.75 that the player will make fewer than 20 of the next 25 attempts. ■

Example 3 A health researcher finds that there is a probability of 0.4 that people exposed to a cold virus will develop the symptoms of a cold. Calculate the probability that if 1000 people are exposed to the virus, between 375 and 400 of them will develop the symptoms of a cold.

Solution This is a Bernoulli experiment with a probability of success (symptoms of a cold develop) of 0.4. We will use a normal distribution with a mean of $\mu = np = 1000(0.4) = 400$ and a standard deviation of $\sigma = \sqrt{npq} = \sqrt{(1000)(0.4)(0.6)} \approx 15.49$ to approximate the desired binomial probability. The z-scores for $x_1 = 374.5$ and $x_2 = 400.5$ are $z_1 = \dfrac{374.5 - 400}{15.49} \approx -1.65$ and $z_2 = \dfrac{400.5 - 400}{15.49} \approx 0.03$, respectively. Using Table 2, Appendix, we find that the area to the left of $z_1 = -1.65$ is 0.0495 and the area to the left of $z_2 = 0.03$ is 0.5120. Therefore, the probability that between 375 and 400 people will develop cold symptoms is

$$\begin{aligned} P(375 \le \text{successes} \le 400) &\approx P(374.5 \le X \le 400.5) \\ &= P(-1.65 \le Z \le 0.03) \\ &= 0.5120 - 0.0495 = 0.4625 \end{aligned}$$

■

EXERCISES SECTION 6.6

In Exercises 1–8, assume that n represents the number of trials and p represents the probability of a success in a Bernoulli experiment. A rule of thumb states that the normal distribution can be used to approximate the binomial distribution if $np > 5$. Using this rule, indicate whether the normal distribution is a suitable approximation of each binomial distribution.

1. $n = 100$, $p = 0.007$

2. $n = 50$, $p = 0.04$

3. $n = 15$, $p = 0.3$

4. $n = 13$, $p = 0.45$

5. $n = 750$, $p = 0.008$

6. $n = 200$, $p = 0.035$

7. $n = 250$, $p = 0.012$

8. $n = 1500$, $p = 0.002$

In Exercises 9–14, use the normal distribution to approximate each binomial probability.

9. $n = 15$, $p = 0.4$; use the normal distribution to approximate $P(7 \leq S \leq 10)$.

10. $n = 20$, $p = 0.3$; use the normal distribution to approximate $P(5 \leq S \leq 8)$.

11. $n = 350$, $p = 0.5$; use the normal distribution to approximate $P(S \leq 170)$.

12. $n = 200$, $p = 0.4$; use the normal distribution to approximate $P(S \leq 100)$.

13. $n = 100$, $p = 0.1$; use the normal distribution to approximate $P(S \geq 12)$.

14. $n = 100$, $p = 0.2$; use the normal distribution to approximate $P(S \geq 23)$.

15. The probability that a subject in a psychology experiment will respond correctly to a stimulus is 0.4. The experiment is to be repeated 15 times.
(a) Calculate the probability of exactly eight successes.
(b) Use the normal distribution to approximate the probability of eight successes.

16. A coin is tossed 20 times and the number of heads is recorded.
(a) Calculate the probability of exactly 12 heads.
(b) Use the normal distribution to approximate the probability of exactly 12 heads.

17. A stockbroker believes that 32% of the stocks traded on the New York Stock Exchange will go up in value each month. The broker selects 20 stocks at random. Use the normal distribution to approximate the probability that fewer than four of the stocks will show an increase in value.

18. The probability that a number greater than 4 will turn up on the face of a die is $\frac{1}{3}$. A die is rolled 100 times. Use the normal distribution to approximate the probability that the number showing will be greater than 4 between 30 and 40 times, inclusive.

19. A purchasing agent for a company that builds snow-blowers buys bolts from the Ajax Company. Past records show that 6% of the bolts produced by the Ajax Company are defective. If a sample of 100 bolts is selected at random from a shipment of 5000 bolts, what is the probability that the sample will contain fewer than five defective bolts?

20. Fifty-two percent of the undergraduates at the University of Dayton are male. Two hundred students, picked at random, were surveyed regarding the quality of food provided in the cafeterias. What is the probability that fewer than 50% of the students surveyed were male?

6.7 STATISTICS

Consider the following situation: A quality engineer has a box that contains 10,000 computer chips and wishes to know something about the mean and standard deviation of the life of the chips, but no such information is available. One approach to the problem would be to test each chip until it failed. However, there are two obvious shortcomings to this approach: (1) it would take a long time to test each of the 10,000 chips, and (2) since life testing is destructive (the test is performed until the chip fails), there would be 10,000 defective chips when the test was completed. Clearly, neither approach is acceptable. A better way to find the mean and standard deviation of the life of the computer chips in the box is to use **statistics,** a branch of mathematics that is concerned with the collection, analysis, interpretation, and presentation of numerical data. Statistical methods can be used to estimate the mean and standard deviation of the life of the computer chips without testing every chip.

In statistics, the set of objects of interest, such as the box of 10,000 computer chips, is called the **population.** Statisticians are concerned with methods of estimating properties of a population based on the properties obtained from a small *random* sample taken from the population. A property of a population (such as a mean) that is used to describe the population is called a **parameter** of the population. The values of parameters are constants and are assumed to be unknown; otherwise, it would not be necessary to use the methods of statistics to estimate them. If the mean of a sample is used to estimate the value of the mean of the population, then the mean of the sample is called a **statistic.** Thus, a statistic is an estimate of the value of a parameter. In this section, we look at statistics as they are used to measure the central tendency and dispersion of quantitative data.

COMMENT: *It is useful to note that most popular brands of scientific and business calculators are programmed to calculate means and standard deviations.*

UNGROUPED DATA

The quantitative data analyzed in statistics are usually classified as being grouped or ungrouped. The data listed in Table 6.6, showing the time in seconds required for an individual to complete a specified task, are **ungrouped,** in the sense that the table contains the individual data points in the order in which they were recorded. The same data are shown in Table 6.7, in which the data are **grouped** into the three intervals labeled 0–4, 4–8, and 8–12 seconds. We note that there are three measurements, 1.3, 2.1, and 3.2, in the interval 0–4 seconds; five measurements, 4.0, 5.7, 6.9, 7.2, and 7.4, in the interval 4–8 seconds; and two measurements, 10.4 and 11.5, in the interval 8–12 seconds.

TABLE 6.6

3.2, 11.5, 1.3, 5.7, 6.9,
7.4, 2.1, 10.4, 7.2, 4.0

TABLE 6.7

Time (seconds)	Frequency
0–4	3
4–8	5
8–12	2

Notice that the upper limit of the first interval is the lower limit of the second interval. In order to avoid confusion when entering data into such a table, we use the convention that the interval 4–8 includes numbers from 4 up to but not including 8. Thus, 4 goes into the interval 4–8 and not into the interval 0–4.

First we will work with problems involving ungrouped data and later with grouped data. We begin with a discussion of several measures of the central tendency of the data—namely, the mean, the median, and the mode. The mean was introduced in Section 6.2 in connection with the expected value of a random variable, so we will treat it first. Measures of central tendency are often referred to as *averages*.

Measures of Central Tendency

Suppose life tests are performed on 50 computer chips selected at random from the box, and suppose further that each entry below represents a chip life in hundreds of hours. Thus, the entry 85 means that the life of that computer chip was 8500 hours.

TABLE 6.8

85	90	88	83	93	99	87	90	78	89
81	91	79	98	95	86	80	92	102	93
91	100	77	84	101	97	85	94	100	82
76	87	106	107	96	85	93	94	91	97
88	92	96	98	84	104	95	103	80	93

The mean of these data is

$$\bar{x} = \frac{\sum x_i}{n} = \frac{85 + 90 + 88 + \cdots + 93}{50} = 91.1$$

Thus, the mean life of the 50 chips is 9110 hours. It can be shown that the mean of a sample of a population can be used to estimate the mean of the population. Accepting this fact without proof, we conclude that the mean of the population is about 9110 hours.

Another measure of central tendency is the median. The **median** of the set of numbers $\{x_1, x_2, \ldots, x_n\}$ is the middle number of the set when the numbers are arranged in increasing order of magnitude. If there are an even number of data points, then the median is the average of the two middle values.

Example 1

(a) Find the median of the five numbers 7, 3, 8, 4, and 11.
(b) Find the median of the six numbers 12, 15, 17, 12, 14, and 19.
(c) Find the median life of the 50 computer chips listed in Table 6.8.

Solution

(a) Arranging the five numbers in increasing order yields

3, 4, 7, 8, 11

The median is the middle number, 7.

(b) Arranging the six numbers in increasing order yields

12, 12, 14, 15, 17, 19

Since there are an even number of entries, the median is the mean of the middle two numbers, 14 and 15. Thus,

$$\text{median} = \frac{14 + 15}{2} = 14.5$$

(c) To find the median life of the computer chips, we first arrange the numbers from the smallest to the largest, including any repetitions. Actually, we only need to order the first 26 numbers, since the median of the 50 numbers will be the mean of the 25th and 26th numbers.

76	77	78	79	80	80	81	82	83	84
84	85	85	85	86	87	87	88	88	89
90	90	91	91	91	92	...			

Since there are 50 sample points, we get the median by averaging the 25th and 26th values, 91 and 92. Thus,

$$\text{median} = \frac{91 + 92}{2} = 91.5$$

In other words, the median life of the sample of 50 computer chips is 9150 hours. Notice that the median is the middle value of the data, so that half of the values are less than the median and half are greater than the median. ■

COMMENT: *If the mean and the median are both given, a mean that is less than the median suggests the presence of values that are considerably smaller than the others. Similarly, if the mean is larger than the median, it suggests the presence of extremely large values relative to the others.*

Example 2 Consider a company with five employees, whose salaries are given below.

Jones	\$50,000
Tsui	\$33,000
Smith	\$32,000
Gomez	\$32,000
Frank	\$30,000

Compare the mean and the median salaries.

Solution The mean of these five numbers is \$35,400 and the median is \$32,000. Clearly the mean is not representative of the five numbers. The mean reflects the fact that one person earns considerably more than the others. In this situation, the median is a better indicator of the central tendency of the salaries. Which measure you choose often depends upon what you want the data to say. For example, the management would want to use the mean value of \$35,400, because it makes it look as if they pay their employees more than they do. The union, on the other hand, would choose to use the median value of \$32,000, so that they could argue that the employees are underpaid. ■

A third measure of central tendency is called the mode. The **mode** of a set of numbers is the number that occurs with the greatest frequency. Since there may be more than one number with the same frequency of occurrence, there can be more than one mode. If there are two modal values, the distribution is said to be **bimodal.** A set of data without any repeated numbers has no mode.

Example 3

(a) Find the mode of the set of numbers 2, 4, 2, 5, 3, 2, 7.
(b) Find the mode of the set of numbers 2, 5, 1, 1, 5, 6, 1, 5, 3.
(c) Find the mode of the computer chip data in Table 6.8.

Solution

(a) The mode of the set of numbers 2, 4, 2, 5, 3, 2, 7 is 2, because it occurs three times.
(b) The set of numbers 2, 5, 1, 1, 5, 6, 1, 5, 3 is bimodal, because 5 and 1 both occur three times.
(c) A study of the computer chip data in Table 6.8 reveals that 93 has the highest frequency (it occurs four times), so the mode is 9300 hours. Because of the nature of the mode, it is most commonly used in connection with grouped data. ▪

MEASURES OF DISPERSION

To obtain the standard deviation of the sample, we recall from Section 6.3 that the standard deviation of a random variable X is defined by

$$\sigma_x = \sqrt{\sum p_i(x_i - \mu)^2}$$

where μ is the mean of the random variable and p_i is the probability associated with x_i. If each x_i is equally likely, then $p_i = \frac{1}{n}$, where n is the number of values. If we make the assumption that each chip life is equally likely, then the standard deviation of a population with mean μ is

$$\sigma_x = \sqrt{\frac{\sum(x_i - \mu)^2}{n}}$$

It is usually impossible, or impractical, to determine the mean and standard deviation of a population. Thus, we use statistics to estimate these values. Earlier we stated that the sample mean $\bar{x}$ is used to estimate the population mean μ. Similarly, we use the sample standard deviation s_x to estimate the population standard deviation σ_x. However, to get good agreement between s_x and σ_x, it can be shown that the sum of the squared differences of the sample should be divided by $n - 1$ instead of n. Thus, the formula for the standard deviation of a sample is

$$s_x = \sqrt{\frac{\sum(x_i - \bar{x})^2}{n - 1}}$$

where $\bar{x}$ is the sample mean. We will use this formula without proof.

Example 4 The following sample of four ACT scores is selected at random from the class of first-year students:

18, 23, 21, 28

Use this sample to estimate the mean and standard deviation of the ACT scores of the first-year students.

Solution The population mean μ is estimated by $\bar{x}$, where

$$\bar{x} = \frac{18 + 23 + 21 + 28}{4} = 22.5$$

The population standard deviation is estimated by s_x, where

$$s_x = \sqrt{\frac{(18 - 22.5)^2 + (23 - 22.5)^2 + (21 - 22.5)^2 + (28 - 22.5)^2}{4 - 1}} \approx 4.2$$

Based on this random sample, we estimate the population mean to be 22.5 and the standard deviation to be 4.2. ■

Example 5 Find s_x for the sample of computer chip data in Table 6.8.

Solution We previously found that $\bar{x} = 91.1$, so

$$s_x = \sqrt{\frac{(85 - 91.1)^2 + (90 - 91.1)^2 + \cdots + (93 - 91.1)^2}{49}} \approx 7.872$$

Thus, we estimate that the standard deviation of the life of the computer chips is 787.2 hours. ■

Needless to say, the computation shown above for s_x is, at best, tedious. This is where the *Stat Pack* on your calculator comes to the rescue. The *Stat Pack* takes much of the drudgery out of the process of calculating means and standard deviations.

CALCULATOR COMMENT: *The following is an explanation of how a* Stat Pack *on a generic calculator works:*

1. *Put the calculator in STAT mode. This usually involves pressing a key labeled STAT.*
2. *Once the calculator is in STAT mode, entries are made by pressing a key labeled DATA after each number is in the display register. Most calculators show how many entries have been made each time the* [DATA] *key is pressed. This feature can be used to see whether you have made the correct number of entries.*
3. *After all the data have been entered, the mean is obtained by pressing a key labeled* $\bar{x}$, μ, *or* μ_x. *The standard deviation is then obtained by pressing a key labeled* σ_x *or* s_x.

A note is in order here concerning standard deviation. Most calculators have both a [σ_x] *key and a* [s_x] *key, and you will note that the two keys give slightly different answers. The reason for this has been mentioned above. Recall that the* [σ_x] *key gives the standard deviation of a population according to the formula*

$$\sigma_x = \sqrt{\frac{\sum(x_i - \mu)^2}{n}}$$

whereas the $\boxed{s_x}$ *key gives the sample estimate of the standard deviation of the population according to the formula*

$$s_x = \sqrt{\frac{\sum(x_i - \bar{x})^2}{n-1}}$$

On some calculators the keys for the standard deviation for a population and for a sample are labeled σ_n *and* σ_{n-1}. *For large values of* n *(such as* $n > 12$*) there is very little difference between the values given by the two formulas so, for simplicity, we will always use* s_x.

Another measure of dispersion is the range. The **range** is the difference between the smallest number and the largest number. For example, the range for the computer chip data in Table 6.8 is

$$\text{range} = 107 - 76 = 31$$

The range indicates a spread of 3100 hours.

Grouped Data

Histograms One of the most basic methods used to describe numerical data is the histogram, which was introduced in Section 6.1 to represent probability distributions. This same device is used in the study of descriptive statistics to help us visualize the distribution of the collected data. For instance, a histogram of the data in Table 6.9 is shown in Figure 6.29. Table 6.9 is often referred to as a *frequency table,* since it gives the frequency of occurrence in each interval.

TABLE 6.9

Chip Life (hr)	Frequency
75–80	4
80–85	7
85–90	9
90–95	13
95–100	9
100–105	6
105–110	2

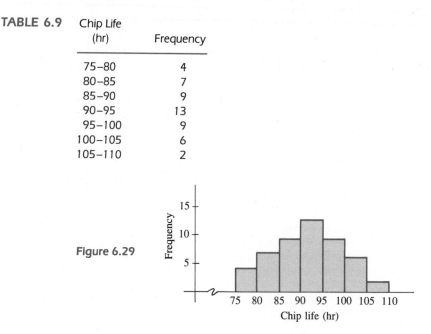

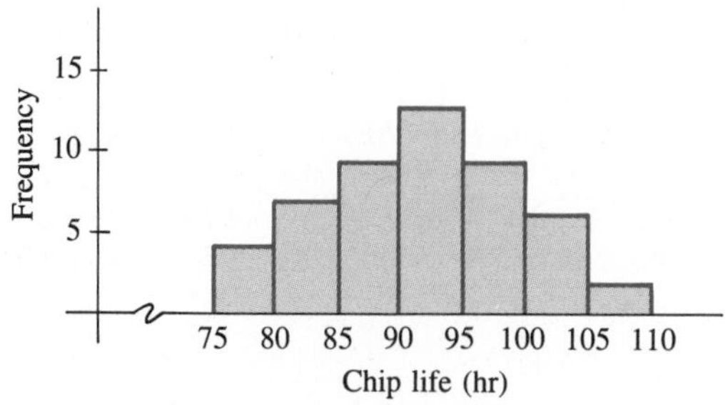

Figure 6.29

The numerical intervals (75–80, 80–85, 85–90, and so on) into which the data in Table 6.9 are grouped are called **class intervals.** The number of class intervals is arbitrary, but a rule of thumb is to choose somewhere between five and fifteen intervals of convenient length. Notice that the upper limit of the first interval is the lower limit of the second interval. Recall that in order to avoid confusion when entering data into such a table, we use the convention that the interval 75–80 includes numbers from 75 up to *but not including* 80. Thus, the number 80 goes into the interval 80–85.

To see how to compute the mean and standard deviation of grouped data, consider the following problem.

Example 6 Compute the mean of the set of numbers 2, 4, 2, 1, 2, 4, and 5.

Solution As ungrouped data, we have

$$\bar{x} = \frac{\sum x_i}{n} = \frac{2 + 4 + 2 + 1 + 2 + 4 + 5}{7} \approx 2.86$$

Notice that there are three 2s and two 4s, so $\bar{x}$ can also be written

$$\bar{x} = \frac{3(2) + 2(4) + 1 + 5}{7} \approx 2.86$$

where we interpret 3(2) to mean that the value $x = 2$ is multiplied by its frequency, $f = 3$. ■

Example 6 illustrates an approach that we use to calculate the mean and standard deviation of grouped data. The formulas for $\bar{x}$ and s_x are as follows:

Mean and Standard Deviation Formulas for Grouped Data

For grouped data in which x_i is the midpoint of the ith interval and f_i is the frequency of the ith interval, the sample mean is given by

$$\bar{x} = \frac{\sum f_i x_i}{n}$$

Notice that this is the formula in Section 6.2 for the expected value $E(X)$, where

$$p_i = \frac{f_i}{n}$$

The sample standard deviation is given by

$$s_x = \sqrt{\frac{\sum f_i (x_i - \bar{x})^2}{n - 1}}$$

Example 7 Use the formulas for the mean and standard deviation of grouped data to find $\bar{x}$ and s_x for the data in Table 6.9. Compare these results with those obtained for ungrouped data in Example 5.

Solution Table 6.9 is repeated here with the addition of the midpoint—the mean value of each interval—added to the table.

TABLE 6.9

Chip Life (hr)	Midpoint x_i	Frequency f_i
75–80	77.5	4
80–85	82.5	7
85–90	87.5	9
90–95	92.5	13
95–100	97.5	9
100–105	102.5	6
105–110	107.5	2

The sample mean is

$$\bar{x} = \frac{4(77.5) + 7(82.5) + \cdots + 2(107.5)}{50} = 91.7$$

and the standard deviation is

$$s_x = \sqrt{\frac{4(77.5 - 91.7)^2 + 7(82.5 - 91.7)^2 + \cdots + (107.5 - 91.7)^2}{50 - 1}} \approx 7.913$$

Notice that $\bar{x}$ and s_x are slightly different from the values $\bar{x} = 91.1$ and $s_x = 7.872$ computed in Example 5. The difference is caused by the definition of the interval width and the assumption that the values are equally spaced over each interval, which is not the case. We are usually willing to accept these slight differences because of the convenience of working with grouped data. ■

CALCULATOR COMMENT: *Again we note that many calculators are programmed to calculate the mean and standard deviation of grouped data. The format for a generic calculator is the same as for ungrouped data, except that the frequency is multiplied by the midpoint of the interval before the* DATA *key is pressed. The order in which x_i and f_i are entered is important. Usually the midpoint value x_i is entered first.*

EXERCISES SECTION 6.7

In Exercises 1–6, find and compare the mean, median, and mode of each set of numbers.

1. 2, 5, 1, 3, 4, 5, 5, 2, 1.

2. 10, 10, 9, 11, 8, 12, 10, 8, 13.

3. 20, 18, 24, 23, 22, 17.

4. 1.5, 1.7, 1.9, 1.6, 2.2, 2.0.

5. 3, 2, 2, 3, 2, 4, 1, 3, 5.

6. 5, 8, 6, 8, 5, 8, 5, 7, 5, 8.

For Exercises 7–12, assuming the sets of numbers in Exercises 1–6 represent random samples of populations, use $\bar{x}$ and s_x to estimate the mean and standard deviation of each population. Do not use the Stat Pack *on your calculator.*

For Exercises 13–18, use the Stat Pack *on your calculator to compute $\bar{x}$ and s_x for the sets of numbers in Exercises 1–6.*

In Exercises 19–22, calculate $\bar{x}$ and s_x for the grouped data. Do not use your Stat Pack.

19.

Speed (mph)	No. of Cars
0–10	3
10–20	7
20–30	12
30–40	9
40–50	5
50–60	1

20.

Time (hr)	No. of Customers
1–3	5
3–5	7
5–7	11
7–9	8
9–11	3
11–13	2

21.

Weight (lb)	No. of Babies
1–4	5
4–7	3
7–10	6
10–13	4
13–16	1

22.

Cost ($)	No. of Books
10–15	6
15–20	15
20–25	107
25–30	200
30–35	93
35–40	25

For Exercises 23–26, use the Stat Pack *on your calculator to find $\bar{x}$ and s_x for the grouped data in Exercises 19–22.*

27. The tires on a fleet of 20 company cars were replaced when the tread depth was reduced to a certain level. The following table indicates the miles (in thousands) on each odometer when the tires were replaced:

41.2	40.5	39.2	43.2	42.9	37.5
41.6	41.7	43.8	40.7	35.6	38.5
43.3	39.6	40.0	41.4	39.9	42.6
45.9	37.1				

(a) Calculate the mean and standard deviation of the ungrouped data. Use the *Stat Pack.*

(b) Make a frequency table using intervals of 2000 miles, starting with 34,000 miles.

(c) Draw a histogram of the frequency table in (b).

(d) Calculate the mean and standard deviation of the grouped data. Use the *Stat Pack.*

(e) Compare the values of $\bar{x}$ and s_x obtained in (a) and (d).

28. A sports historian takes a sample of the total points scored by the Cleveland Browns from a listing of their scores from 1950 to 1990. The 20 scores are shown in the following table:

13	28	21	21	35	12	7	20	0	21
14	14	21	38	30	3	12	7	49	20

(a) Calculate the mean and standard deviation of the ungrouped scores. Use the *Stat Pack* on your calculator.

(b) Make a frequency table using intervals of six points, starting with zero.

(c) Draw a histogram of the frequency table in (b).

(d) Calculate the mean and standard deviation of the grouped scores in (b). Use the *Stat Pack* on your calculator.

(e) Compare the values of $\bar{x}$ and s_x obtained in (a) and (d).

29. An environmentalist studying mercury levels in the fish in Lake Ontario takes a sample of fish each week and measures the mercury content in parts per million (ppm). The results of the study are listed in the following table:

3.2	2.1	2.4	2.2	1.6	2.5	2.3	3.1	1.9
1.9	1.4	3.2	2.6	2.3	2.4	1.8	2.2	2.1
2.8	3.0	1.8	2.5	2.1	2.0	3.1	1.3	2.8

(a) Calculate the mean and standard deviation of the ungrouped mercury levels. Use the *Stat Pack* on your calculator.

(b) Make a frequency table using intervals of 0.3 ppm, starting with 1.2 ppm.

(c) Draw a histogram of the frequency table in (b).

(d) Calculate the mean and standard deviation of the grouped mercury levels in (b). Use the *Stat Pack* on your calculator.

(e) Compare the values of $\bar{x}$ and s_x obtained in (a) and (d).

30. A market researcher conducts a survey of 25 households selected at random from the telephone book to find out how many TV sets they own. The following table gives the number of TV sets in each household.

3	2	2	1	2	3	3	4	2	0
3	2	1	5	4	1	2	5	3	4
2	1	4	3	2					

(a) Calculate the mean and standard deviation of the ungrouped numbers. Use the *Stat Pack* on your calculator.
(b) Make a frequency table using intervals of 1, starting with zero.
(c) Draw a histogram of the frequency table in (b).
(d) Calculate the mean and standard deviation of the grouped numbers in (b). Use the *Stat Pack* on your calculator.
(e) Compare the values of $\bar{x}$ and s_x obtained in (a) and (d).

IMPORTANT WORDS AND PHRASES CHAPTER 6

The phrases *random variable, distribution, probability, binomial,* and *normal* dominate the vocabulary of this chapter. Here is a list of the new words and phrases you should have learned:

arithmetic mean
Bernoulli experiment
bimodal distribution
binomial probability
Chebyshev's inequality and theorem
expected value of a random variable
histogram
mean (of a population)
measure of central tendency
measure of dispersion
median of a population
mode of a population
normal approximation of the binomial
normal density function
parameter of a population
population
probability of the value of a random variable
probability distribution
  binomial probability distribution
  normal probability distribution
random sample
random variable
range of a set of data
sample mean and standard deviation
sigma notation
standard deviation
standard normal variable/ normal random variable
statistic
variance

REVIEW EXERCISES CHAPTER 6

1. Two balls are drawn without replacement at random from a box that contains four orange balls and seven blue balls. Let the random variable be the number of orange balls.
(a) Indicate the values of the random variable.
(b) Determine the probability distribution for this random variable.
(c) Draw the histogram for this random variable.
(d) Calculate the expected value and variance for this random variable.

2. A card is dealt from a deck of 52 playing cards. If a spade is dealt, you win \$2; otherwise, you lose \$1.
(a) Indicate the values of the random variable.
(b) Determine the probability distribution for this random variable.
(c) Draw the histogram for this random variable.

(d) Calculate the expected value and variance for this random variable.

3. Six out of the ten subjects in a medical experiment are male. Two of the subjects are selected at random from the test group. Let the random variable be the number of males.
   (a) Indicate the values of the random variable.
   (b) Determine the probability distribution for this random variable.
   (c) Draw the histogram for this random variable.
   (d) Calculate the expected value and variance for this random variable.

4. A die is weighted so that the probability of an even number is 0.6 and that of an odd number is 0.4. The die is rolled three times and the number of odd numbers is recorded.
   (a) Indicate the values of the random variable.
   (b) Determine the probability distribution for this random variable.
   (c) Draw the histogram for this random variable.
   (d) Calculate the expected value and variance for this random variable.

5. A buyer for the luggage department of a department store can buy either luggage made in the U. S. or luggage made in Korea. If the Korean-made luggage is purchased, there is a 70% chance that the department will make a profit of \$100,000, a 20% chance that it will break even, and a 10% chance that it will lose \$15,000. If the domestic luggage is purchased, there is a 50% chance that the department will make a profit of \$200,000, a 30% chance that it will break even, and a 20% chance that it will lose \$50,000. Based on expected value, what should the buyer do?

6. A game that consists of dealing two cards from a well-shuffled deck of 52 playing cards costs \$2 to play. If both cards are hearts, your net win is \$8; if both cards are red, your net win is \$3; otherwise, you lose your \$2. Find the expected value of this game. Does the game favor the player?

7. A random variable X has a mean of 25 and a standard deviation of 4. Indicate the interval that is within 1.5σ units of the mean. Use Chebyshev's inequality to find the probability that a value of X lies within 1.5σ units of the mean.

8. A random variable X has a mean of 100 and a standard deviation of 5. Indicate the interval that is within 1.2σ units of the mean. Use Chebyshev's inequality to find the probability that a value of X lies within 1.2σ units of the mean.

9. A random variable X has a mean of 72 and a standard deviation of 3.2. Use Chebyshev's inequality to find the probability that a value of X lies between 60 and 84.

10. A random variable X has a mean of 110 and a standard deviation of 3. Use Chebyshev's inequality to find the probability that a value of X lies between 105 and 115.

11. Amy Smith is an 85% foul-shooter. Calculate the probability that:
    (a) She will make exactly eight out of ten foul shots.
    (b) She will make eight or more foul shots.
    (c) She will make all ten shots.

12. Market research shows that 35% of the population smoke cigarettes. If eight people are chosen at random from the population, calculate the probability that:
    (a) All eight are cigarette smokers.
    (b) Seven of the eight do not smoke cigarettes.
    (c) Fewer than two are cigarette smokers.

13. Draw a histogram for a binomial experiment with four trials, with $p = 0.4$.

14. Draw a histogram for a binomial experiment with two trials, with $p = 0.2$.

For Exercises 15–22, if a random variable Z has a standard normal distribution with a mean of zero and a standard deviation of 1, find the probability that the random variable takes on a value in each indicated interval.

15. $Z \leq 2.35$

16. $Z \leq 0.76$

17. $0 \leq Z \leq 1.07$

18. $0 \leq Z \leq 2.22$

19. $-0.23 \leq Z \leq 0.5$

20. $-1.1 \leq Z \leq -0.39$

21. $Z \geq -1.03$

22. $Z \geq 1.54$

23. Given that Z is a standard normal random variable, calculate the value of z_1 such that there is a probability of 0.3 that $0 \leq Z \leq z_1$.

24. Given that Z is a standard normal random variable, calculate the value of z_1 such that there is a probability of 0.4 that $0 \leq Z \leq z_1$.

In Exercises 25–30, X is a normally distributed random variable with the indicated mean and standard deviation. Find the probability that X takes on a value in each given interval.

25. Given $\mu = 15$ and $\sigma = 2$, find $P(X \le 18)$.

26. Given $\mu = 37$ and $\sigma = 3$, find $P(X \le 35)$.

27. Given $\mu = 7.5$ and $\sigma = 1.2$, find $P(7.0 \le X \le 8.5)$.

28. Given $\mu = 0.1$ and $\sigma = 0.02$, find $P(0.09 \le X \le 0.12)$.

29. Given $\mu = 33$ and $\sigma = 2.5$, find $P(X \ge 30)$.

30. Given $\mu = 5.3$ and $\sigma = 0.3$, find $P(X \ge 5.1)$.

In Exercises 31–36, use the normal distribution to approximate each indicated binomial probability.

31. Given $n = 20$, $p = 0.45$, approximate $P(S \le 11)$.

32. Given $n = 50$, $p = 0.3$, approximate $P(S \le 20)$.

33. Given $n = 50$, $p = 0.52$, approximate $P(24 \le S \le 30)$.

34. Given $n = 100$, $p = 0.6$, approximate $P(50 \le S \le 65)$.

35. Given $n = 2500$, $p = 0.35$, approximate $P(S \ge 1000)$.

36. Given $n = 150$, $p = 0.84$, approximate $P(S \ge 130)$.

37. An industrial engineer studying the production of new cars at an assembly plant counts the number of cars coming off the assembly line in a one-hour period. The following table indicates the number of cars coming off the line in 20 different one-hour periods during the study:

30	27	28	28	26	19	23	25	27	28
24	29	25	20	31	22	28	26	25	25

(a) Calculate the mean, median, mode, and standard deviation of the ungrouped data.
(b) Make a frequency table, using intervals of 3, starting at 18.
(c) Draw a histogram for the frequency table in (b).
(d) Calculate the mean and standard deviation of the grouped data.
(e) Compare the values of $\bar{x}$ and s_x in (a) and (d).

38. An investment broker follows ten stocks on the New York Stock Exchange for 30 days. The following table indicates how many of the stocks went up in value on each of the 30 days:

3	5	1	7	2	5	4	5	3	0
2	4	5	5	3	2	6	3	4	5
1	2	2	4	2	6	5	3	6	4

(a) Calculate the mean, median, mode, and standard deviation of the ungrouped data.
(b) Make a frequency table, using intervals of 2, starting at zero.
(c) Draw a histogram for the frequency table in (b).
(d) Calculate the mean and standard deviation of the grouped data.
(e) Compare the values of $\bar{x}$ and s_x in (a) and (d).

CHAPTER 7

Markov Chains and Game Theory

Basketball players play a game called Twenty-one, in which a player shoots until he or she misses or scores 21 points. When the first shooter misses, the second shooter takes his or her turn until a miss occurs, and so on. Consider the following situation: Sam makes 60% of his shots and Leon makes 70% of his shots. If Sam starts the game as the shooter, what is the probability that he will take the second shot in the game? Given that Sam shoots first, there are two possibilities: "Sam makes his first shot, so he takes the second shot" or "Sam misses his first shot, so Leon takes the second shot." This is an example of a stochastic process, called a **Markov chain,** in which the probability of the next outcome of an experiment depends only upon the present state of the system. Our game of Twenty-one has two states, namely:

State 1: Sam takes the next shot.

State 2: Leon takes the next shot.

Since Sam makes 60% of his shots, there is a probability of 0.6 that he will take the next shot (State 1) and a probability of 0.4 that Leon will take the next shot (State 2). Similarly, if Leon is the shooter, there is a probability of 0.7 that he will take the next shot (State 2) and a probability of 0.3 that Sam will take the next shot (State 1). In either case, the next state depends only upon the present state and the probabilities of remaining in that state or changing to another state.

COMMENT: *Informally, a Markov chain is a sequence of events in which the probability of the outcome of each event depends only on the previous event.*

The study of Markov chains permits the determination of probabilities of sequential and long-term behavior of a system when the probabilities of the identifiable states are known. The following examples will give you some idea of the kinds of problems that might be analyzed using Markov chains:

- The long-term ridership trends of commuters using public transportation.
- The voting trends of a segment of the registered voters in the nation.
- The long-term housing preferences of the adult population.
- The percentage of people on public assistance from one year to the next.

HISTORICAL COMMENT: *Markov chains are named for the Russian mathematician A. A. Markov (1856–1922), who developed much of the original theory of stochastic processes.*

7.1 THE TRANSITION MATRIX

To continue our discussion of Sam's and Leon's game of Twenty-one, we observe that the two states of the process and their corresponding probabilities can be organized into the following 2×2 matrix:

$$\begin{array}{ccc} & & \text{Next Shooter} \\ & & \begin{array}{cc}\text{Sam} & \text{Leon}\end{array} \\ \text{Present Shooter} & \begin{array}{c}\text{Sam} \\ \text{Leon}\end{array} & \begin{bmatrix} 0.6 & 0.4 \\ 0.3 & 0.7 \end{bmatrix} \end{array}$$

This matrix shows the probabilities of moving from one state to another. The first row gives the probabilities associated with Sam; the second row, the probabilities associated with Leon. The first column gives the probabilities that Sam will take the next shot; the second column, the probabilities that Leon will take the next shot. For instance, we interpret the probabilities in column 1 as follows: If Sam is shooting, there is a probability of 0.6 that he will take the second shot (that is, the system remains in State 1); if Leon is shooting, there is a probability of 0.3 that Sam will take the next shot (that is, a change from State 2 to State 1). In this sense, the numbers in the matrix represent probabilities of transition from one state to another.

A matrix of transition probabilities is called a **transition matrix.** We identify the following properties of transition matrices:

DEFINITION Properties of Transition Matrices

- *The number of rows and the number of columns are equal.*
- *Each entry in the matrix is a probability.*
- *The sum of the probabilities in any row is equal to 1.*

Example 1 A real-estate broker conducts a study of the housing preferences of those adults who rent apartments, buy condominiums, or buy houses. The broker believes that a person's next choice of housing depends only upon his or her present housing state. Interpret the following transition matrix in the context of this study:

$$\begin{array}{ccc} & & \text{Next Move} \\ & & \begin{array}{ccc}\text{Rent} & \text{Condo} & \text{House}\end{array} \\ \text{Present Housing} & \begin{array}{c}\text{Rent} \\ \text{Condo} \\ \text{House}\end{array} & \begin{bmatrix} 0.2 & 0.4 & 0.4 \\ 0.3 & 0.5 & 0.2 \\ 0.1 & 0.2 & 0.7 \end{bmatrix} \end{array}$$

Solution Three states are represented by this transition matrix; namely:

State 1: Rents

State 2: Buys condo

State 3: Buys house

The first row of the transition matrix indicates that there is a probability of 0.2 that a person who rents will rent the next time he or she changes housing accommodations, a probability of 0.4 that he or she will buy a condominium, and a probability of 0.4 that he or she will buy a house. A renter who buys a condo or a house is said to "change state." The second row, which represents those who presently own condominiums, shows that there is a probability of 0.3 that they will rent the next time they move, a probability of 0.5 that they will buy another condo, and a probability of 0.2 that they will buy a house. Finally, the third row shows that there is a probability of 0.1 that those who own houses will rent their next place, a probability of 0.2 that they will buy a condo, and a probability of 0.7 that they will buy another house. Notice that the sum of the entries in any row is 1. ∎

Example 2 A marketing analyst studying consumer buying habits for Big Value Food Stores finds that 60% of the people who try Ballpark brand hot dogs will buy them the next time they buy hot dogs, while 25% will switch to All-American brand hot dogs, and the remaining 15% will switch to Coney Island brand hot dogs. Of those who try All-American hot dogs, 50% will buy them the next time they buy hot dogs, 40% will switch to Ballpark hot dogs, and 10% to Coney Island hot dogs. Finally, 55% of those who try Coney Island hot dogs will buy them again, 30% will switch to Ballpark brand, and 15% will switch to All-American brand. We assign the following three states:

State 1: Consumer buys Ballpark hot dogs
State 2: Consumer buys All-American hot dogs
State 3: Consumer buys Coney Island hot dogs

The transition matrix for these three states is

$$\begin{array}{cc} & \text{Next Purchase} \\ & \begin{array}{ccc}\text{Ballpark} & \text{All-American} & \text{Coney Island}\end{array} \\ \text{Purchase}\ \begin{array}{r}\text{Ballpark}\\ \text{All-American}\\ \text{Coney Island}\end{array} & \begin{bmatrix} 0.60 & 0.25 & 0.15 \\ 0.40 & 0.50 & 0.10 \\ 0.30 & 0.15 & 0.55 \end{bmatrix}\end{array}$$

∎

Example 3 Metro Bus Service conducts a study of those people who ride the bus to work. The study shows that there is a probability of 0.88 that a person who rides a bus one day will ride it the next day, and a probability of 0.12 that he or she will not ride the bus the next day. It also shows that there is a probability of 0.25 that a person who did not ride the bus on a given day will ride it the next day, and a probability of 0.75 that the person will not ride it the next day. Construct a transition matrix for bus ridership.

Solution The first row of the transition matrix consists of the two probabilities associated with State 1: "Rides the bus." The second row consists of the two probabilities associated with State 2: "Does not ride the bus." Thus, the desired transition matrix is

$$\begin{array}{cc} & \text{Next day} \\ & \begin{array}{cc}\text{Rider} & \text{Nonrider}\end{array} \\ \text{Current Day}\ \begin{array}{r}\text{Rider}\\ \text{Nonrider}\end{array} & \begin{bmatrix} 0.88 & 0.12 \\ 0.25 & 0.75 \end{bmatrix}\end{array}$$

∎

TREE DIAGRAMS AND CHANGE OF STATE

The matrix in Example 3 shows the transition matrix for bus ridership from one day to the next. Under the assumption that these transition probabilities are constant from day to day, what can we say about bus ridership over the next two-day period? For instance, if a person rides the bus on Monday, what is the probability that that person will ride it on Wednesday? The answer to this question can be found in the tree diagram below, which shows all of the possibilities of ridership over a two-day period.

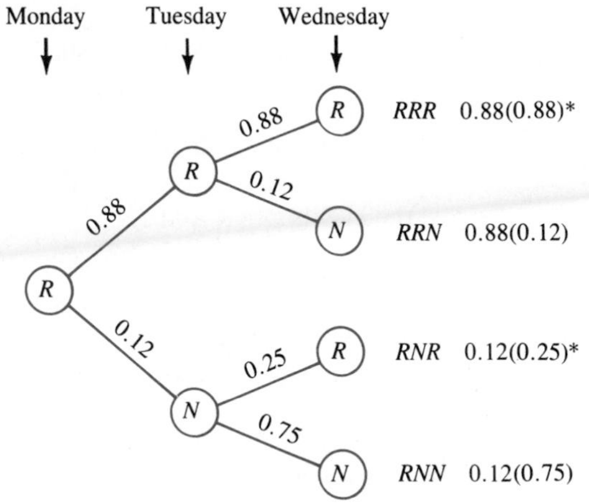

Figure 7.1

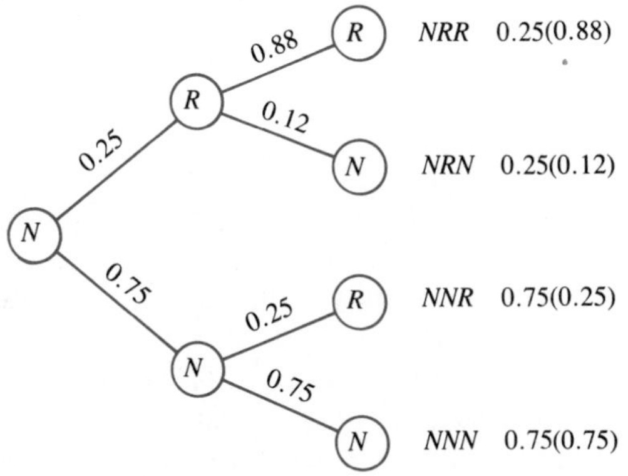

If we assume that a person is a rider on Monday and will ride on Wednesday, there are two possible paths to consider. One is "Rides the bus Tuesday and Wednesday." We diagram this as

$$RRR = \text{Rider} \rightarrow \text{Rider} \rightarrow \text{Rider}$$

The other possible path is "Does not ride on Tuesday but rides on Wednesday."

$$RNR = \text{Rider} \rightarrow \text{Nonrider} \rightarrow \text{Rider}$$

These two paths are marked with asterisks in Figure 7.1.

- We see from Figure 7.1 that $P(RRR) = 0.88(0.88)$ and $P(RNR) = 0.12(0.25)$. Thus, the probability that a person who rides on Monday will ride on Wednesday is the sum of the probabilities associated with the events RRR and RNR; that is,

$$0.88(0.88) + 0.12(0.25) = 0.8044$$

Similarly, Figure 7.1 can be used to compute the other state probabilities for our ridership problem.

- The probability that a person who rides on Monday will not ride on Wednesday is given by the sum of the probabilities associated with the events RRN and RNN. Following these paths in Figure 7.1, we get

$$0.88(0.12) + 0.12(0.75) = 0.1956$$

- The probability that a person who does not ride on Monday will ride on Wednesday is given by the sum of the probabilities associated with the events NRR and NNR. Figure 7.1 yields

$$0.25(0.88) + 0.75(0.25) = 0.4075$$

- The probability that a person who does not ride on Monday will not ride on Wednesday is given by the sum of the probabilities associated with the events NRN and NNN. Thus,

$$0.25(0.12) + 0.75(0.75) = 0.5925$$

This "tree-diagram method" of computing the probabilities for ridership becomes unmanageable if we want to find the probability of a person's remaining in, or changing, his or her state over three, four, or more days. These computations can be made more efficiently by using matrix algebra, but before we show how, we will introduce the idea of a state vector.

State Vectors

A $1 \times n$ matrix has 1 row and n columns, and is called a **row vector.** For instance, $[0.88 \;\; 0.12]$ is a 1×2 row vector of probabilities corresponding to the first row of the transition matrix for our bus ridership problem. The first entry in $[0.88 \;\; 0.12]$ is the probability that a person who rides the bus today will ride it tomorrow, and the second entry is the probability that a person who does not ride today will ride tomorrow. We call $[0.88 \;\; 0.12]$ a **state vector** because it gives the probabilities that the system is in a particular state. Notice that the entries in our sample state vector are nonnegative and their sum is 1. In general, we have the following definition of a state vector:

DEFINITION State Vector

Suppose a system has k states. A **state vector** is a $1 \times k$ row vector $[p_1 \;\; p_2 \;\; \cdots \;\; p_k]$, where $p_1 + p_2 + \cdots + p_k = 1$ and each p_i represents the probability of a system's being in the state i at a particular time.

A person who rides the bus on Monday may be considered to be a rider (State 1) with probability 1 and to be a nonrider (State 2) with probability zero. We represent this by the 1×2 row vector $R_0 = [1 \ \ 0]$, and call it the **initial-state vector** of the system. The state vector $R_0 = [1 \ \ 0]$ is an initial-state vector in the sense that it represents the state of the system at our first observation. Similarly, $N_0 = [0 \ \ 1]$ is the initial-state vector for the system if a person is considered to be a nonrider on Monday. There are other possible initial-state vectors that could be used; for example, if the person flips a coin to decide whether to ride the bus or not, the initial-state vector will be $[0.5 \ \ 0.5]$.

COMMENT: *We often think of a system as progressing from its initial state through a sequence of different states, and in this way generating a chain of events. Keep in mind, however, that the probabilities for the next state vector depend only upon the present state vector and the transition matrix, and not on the sequence of past state vectors that preceded the present state vector.*

Transition Matrices and State Vectors

If we know that a person rode the bus on Monday, we cannot tell with certainty whether he or she will ride the bus on Tuesday. The best we can do is give the probabilities of ridership (0.88) and nonridership (0.12), which we write as the state vector $R_1 = [0.88 \ \ 0.12]$. Previously, we used a tree diagram to compute state vectors for our bus ridership problem, although we didn't use the term *state vector* at that time. Now notice what happens in the bus ridership problem when we multiply the initial-state vector R_0 and the transition matrix T.

$$R_0T = \underbrace{[1 \quad 0]}_{\substack{\text{Initial-state vector}\\ \text{for ridership}}} \underbrace{\begin{bmatrix} 0.88 & 0.12 \\ 0.25 & 0.75 \end{bmatrix}}_{\substack{\text{Transition}\\ \text{matrix}}} = \underbrace{[0.88 \quad 0.12]}_{\substack{\text{State vector for}\\ \text{ridership for Tuesday}}} = R_1$$

We see that the product of the initial-state vector and the transition matrix is the state vector R_1 for Tuesday. Similarly, the state vector R_2 for Wednesday is obtained by multiplying the state vector R_1 by the transition matrix T. Thus,

$$R_1T = R_0T^2 = \underbrace{[0.88 \quad 0.12]}_{\substack{\text{State vector for}\\ \text{ridership for Tuesday}}} \underbrace{\begin{bmatrix} 0.88 & 0.12 \\ 0.25 & 0.75 \end{bmatrix}}_{\substack{\text{Transition}\\ \text{matrix}}} = \underbrace{[0.8044 \quad 0.1956]}_{\substack{\text{State vector for}\\ \text{ridership for Wednesday}}} = R_2$$

Thus, we see that the matrix computations give the same results as the tree diagram in Figure 7.1.

The preceding discussion is intended to motivate a theorem that tells us how to obtain the state vector for the next trial in a Markov-chain process if the state vector of the present trial is known. Notice that the state vector for Wednesday (the second day from the initial-state vector) can be obtained by multiplying the initial-state vector R_0 by the square of the transition matrix T; that is, $R_2 = R_0T^2$.

We anticipate that the state vector for the third day from the initial-state vector should be given by $R_3 = R_0T^3$, which, as Theorem 7.1 verifies, is true. Under the assumption that the transition probabilities remain constant from day to day, this process may be carried on for as many days as we wish.

THEOREM 7.1

If a state vector after the nth trial of a Markov process is X_n and T is the transition matrix, then the state vector for the $(n + 1)$st trial is

$$X_{n+1} = X_nT = X_0T^{n+1}$$

where X_0 is the initial-state vector.

Example 4 Find the state vector for Thursday for the bus ridership problem, if the initial-state vector on Monday is [1 0].

$$R_3 = R_0T^3 = [1 \quad 0]\begin{bmatrix} 0.88 & 0.12 \\ 0.25 & 0.75 \end{bmatrix}^3$$

We know the state vector for Wednesday is $R_2 = [0.8044 \quad 0.1956]$. Therefore, the state vector R_3 for Thursday is

$$R_3 = R_0T^3 = R_2T = \underbrace{[0.8044 \quad 0.1956]}_{\substack{\text{State vector for} \\ \text{ridership for Wednesday}}}\underbrace{\begin{bmatrix} 0.88 & 0.12 \\ 0.25 & 0.75 \end{bmatrix}}_{\substack{\text{Transition} \\ \text{vector}}} = \underbrace{[0.7568 \quad 0.2432]}_{\substack{\text{State vector for} \\ \text{ridership for Thursday}}}$$

Example 5 Suppose an executive of the bus company assumes that only 50% of possible riders ride the bus on a given day. Using the same transition matrix as before, what percentages of this population would ride on the next day?

Solution The initial-state vector for this problem is [0.5 0.5]. Hence, the state vector for the next day is

$$[0.5 \quad 0.5]\begin{bmatrix} 0.88 & 0.12 \\ 0.25 & 0.75 \end{bmatrix} = [0.565 \quad 0.435]$$

This means that 56.5% of the population would ride on the day after 50% rode. ■

Example 6 Assume that the realtor in Example 1 wants to use the transition matrix there to predict the states of the present renters after two moves. Use a Markov chain to make the prediction.

Solution The initial-state vector for renting is $[1 \ 0 \ 0]$. The state vector for renters after one move is obtained by multiplying the initial-state vector by the transition matrix in Example 1. Thus,

$$[1 \quad 0 \quad 0]\begin{bmatrix} 0.2 & 0.4 & 0.4 \\ 0.3 & 0.5 & 0.2 \\ 0.1 & 0.2 & 0.7 \end{bmatrix} = [0.2 \quad 0.4 \quad 0.4]$$

To determine the state vector for renters after two moves, we multiply the state vector $[0.2 \ 0.4 \ 0.4]$ by the transition matrix:

$$[0.2 \quad 0.4 \quad 0.4]\begin{bmatrix} 0.2 & 0.4 & 0.4 \\ 0.3 & 0.5 & 0.2 \\ 0.1 & 0.2 & 0.7 \end{bmatrix} = [0.20 \quad 0.36 \quad 0.44]$$

We interpret the state vector $[0.20 \ 0.36 \ 0.44]$ to mean that after two moves 20% of the renters will be still renting, 36% of them will have purchased condos, and 44% of them will have purchased a home. ▪

Example 7 Assume that the realtor in Example 1 finds that at present 30% of the adults are renting, 20% own condos, and 50% own houses.

(a) What is the initial-state vector?
(b) What is the state vector after their next move?

Solution

(a) The initial-state vector is obtained from the given percentages and written as

$$[0.30 \quad 0.20 \quad 0.50]$$

(b) The state vector after their next move is obtained by multiplying the initial-state vector by the transition matrix in Example 1. Thus,

$$[0.30 \quad 0.20 \quad 0.50]\begin{bmatrix} 0.2 & 0.4 & 0.4 \\ 0.3 & 0.5 & 0.2 \\ 0.1 & 0.2 & 0.7 \end{bmatrix} = [0.17 \quad 0.32 \quad 0.51]$$

We interpret the state vector $[0.17 \ 0.32 \ 0.51]$ to mean that after their next move 17% of the adults will rent, 32% will own condos, and 51% will own houses. ▪

EXERCISES SECTION 7.1

In Exercises 1–8, identify those matrices that qualify as transition matrices. If a matrix does not qualify, indicate why.

1. $\begin{bmatrix} \frac{1}{2} & \frac{1}{2} \\ 0 & 1 \end{bmatrix}$

2. $\begin{bmatrix} 1 & 0 \\ \frac{1}{4} & \frac{3}{4} \end{bmatrix}$

3. $\begin{bmatrix} \frac{1}{3} & \frac{1}{3} & \frac{1}{3} \\ \frac{1}{2} & \frac{1}{2} & 0 \end{bmatrix}$

4. $\begin{bmatrix} 1 & 0 \\ 0 & 1 \\ \frac{1}{2} & \frac{1}{2} \end{bmatrix}$

5. $\begin{bmatrix} \frac{1}{3} & \frac{2}{3} & 0 \\ 0 & \frac{1}{2} & \frac{1}{2} \\ \frac{2}{3} & 0 & \frac{2}{3} \end{bmatrix}$

6. $\begin{bmatrix} \frac{1}{2} & \frac{1}{2} & 0 \\ 0 & \frac{1}{3} & \frac{2}{3} \\ \frac{1}{6} & \frac{1}{3} & \frac{1}{2} \end{bmatrix}$

7. $\begin{bmatrix} 1 & 0 & 0 \\ 0.5 & 0.5 & 0 \\ 0.1 & 0.8 & 0.1 \end{bmatrix}$

8. $\begin{bmatrix} 0.1 & 0.2 & 0.7 \\ 0.6 & 0 & 0.4 \\ 0 & 0.9 & 0.1 \end{bmatrix}$

9. A video game shows a frog sitting on one of two lily pads in a pond. If the frog is sitting on Pad 1 when the game starts, there is a probability of 0.4 that it will remain on Pad 1 and a probability of 0.6 that it will jump to Pad 2. If the frog is on Pad 2 when the game starts, there is a probability of 0.3 that it will remain on Pad 2 and a probability of 0.7 that it will jump to Pad 1. Write a transition matrix for the state of the frog and draw a tree diagram for the first two jumps.

10. In the game of straight pool, the first player shoots until he or she misses, at which time the second player shoots until he or she misses, and so on, until the game is completed. Ann makes 72% of her shots and Joe makes 65% of his shots. Write a transition matrix for this game.

11. A video game will show two different monsters on the screen, but only one at a time. If Godzilla is on the screen when a game ends, there is a probability of 0.2 that Godzilla will start the next game and a probability of 0.8 that King Kong will start the next game. If King Kong is on the screen when a game ends, there is a probability of 0.4 that he will remain on the screen and a probability of 0.6 that Godzilla will be the next threat. Write the transition matrix for this game.

12. Sometimes Professor Martinez collects homework from her students and sometimes she doesn't. The probability that she will collect it two days in a row is 0.3 and the probability that she will collect it today if she didn't collect it yesterday is 0.8. Write the transition matrix for her system.

13. A reporter for *Car News* magazine conducts an annual survey of car buyers on July 1, 1992, and finds that 70% of those who owned Japanese cars on July 1, 1991 still own Japanese cars, 20% say they switched to an American car, and 10% say they now own a European car. Of the American car owners from 1991, 73% say they still own an American car, 15% say they now own a Japanese car, and 12% say they now own a European car. Finally, 65% of the European car owners from 1991 say they have not switched, 20% say they now own an American car, and 15% now own a Japanese car. Write the transition matrix for car-buyer preferences.

14. A sociologist finds that there is a constant movement of people between the poor class, the middle class, and the wealthy class. Census data show that 80% of the poor in 1980 remained poor in 1990, 19% moved up to the middle class, and 1% moved to the wealthy class. Of those classified as middle class in 1980, 70% remained middle class in 1990, 10% dropped into the poor class, and 20% moved up to the wealthy class. Finally, 92% of the wealthy remained in that class, 7% dropped to the middle class, and 1% dropped to the poor class. Write the transition matrix for this process, assuming the data are the same from decade to decade.

15. Draw a tree diagram for the movement of people between classes, as described in Exercise 14, from 1980 to 2000 beginning with a poor person in 1980.

16. Draw a tree diagram for the car buyers in Exercise 13, beginning with an American-car owner and continuing for two years.

In Exercises 17–20, multiply each state vector by the transition matrix

$$\begin{bmatrix} 0.3 & 0.7 \\ 0.8 & 0.2 \end{bmatrix}$$

17. $[0.6 \quad 0.4]$

18. $[0.2 \quad 0.8]$

19. $[0.1 \quad 0.9]$

20. $[0.5 \quad 0.5]$

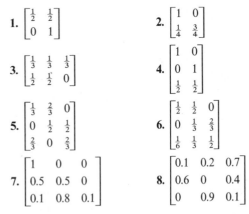

In Exercises 21–24, multiply each state vector by the transition matrix

$$\begin{bmatrix} 0.1 & 0.1 & 0.8 \\ 0.7 & 0.3 & 0 \\ 0.5 & 0.2 & 0.3 \end{bmatrix}$$

21. $[0.2 \quad 0.3 \quad 0.5]$

22. $[0.1 \quad 0.7 \quad 0.2]$

23. $[0 \quad 0.2 \quad 0.8]$

24. $[0.3 \quad 0 \quad 0.7]$

25. The transition matrix for the game of Twenty-one described at the beginning of this section is

		Next Shooter	
		Sam	Leon
Present Shooter	Sam	0.6	0.4
	Leon	0.3	0.7

(a) Find the probability that Sam will take the fourth shot if Leon starts the game.
(b) Find the probability that Sam will take the fourth shot if he starts the game.

26. The bus ridership transition matrix from Example 3 is

		Next Day	
		Rider	Nonrider
Current Day	Rider	0.88	0.12
	Nonrider	0.25	0.75

(a) Find the probability that a person who rides the bus on Monday will ride it on Friday.
(b) Find the probability that a person who does not ride the bus on Monday will ride it on Thursday.

27. The transition matrix for the real-estate buyers in Example 1 is

		Next Move		
		Rent	Condo	House
Present Housing	Rent	0.2	0.4	0.4
	Condo	0.3	0.5	0.2
	House	0.1	0.2	0.7

Assume that at present 35% of these adults rent, 20% own condos, and 45% own houses.
(a) Write an initial-state vector representing the present living conditions of adults.
(b) Find the percentage in each category after two moves.

28. The transition matrix for the hot dog brands in Example 2 is

		Next Purchase		
		Ballpark	All-American	Coney Island
Purchase	Ballpark	0.60	0.25	0.15
	All-American	0.40	0.50	0.10
	Coney Island	0.30	0.15	0.55

Assume that Ballpark has 40% of the market, All-American has 35%, and Coney Island has 25%.
(a) Write an initial-state vector reflecting the market share of each brand.
(b) Find the percentage of the market held by each brand after the next purchase.

29. A study is made of females whose IQ is greater than 120 and their female children. Eighty-two percent of the female children born to mothers with IQs greater than 120 have IQs greater than 120; 45% of the female children born to mothers with IQs less than 120 have IQs greater than 120. Set up a transition matrix for this study and predict the relative proportions of IQ among the female grandchildren of a grandmother.

7.2 Regular Markov Chains

To continue our discussion of the bus ridership problem in Section 7.1, suppose that you are the president of Metro Bus Service and that you are looking at the transition matrix for bus ridership.

	Rider	Nonrider
Rider	0.88	0.12
Nonrider	0.25	0.75

One thing you would probably want to know is "Are there any long-term trends in ridership that can be predicted by this model?" It turns out in this instance that we can answer your question by using Theorem 7.1 to compute successive ridership state vectors. The results of this computation are shown in Example 1.

Example 1 In Example 4 of Section 7.1, we calculated the first two state vectors obtained from the initial-state vector $[1 \quad 0]$. Seven additional state vectors are computed in this same manner and the results listed in the table below. The initial-state vector is labeled 0 in the table. The probabilities in the table are rounded off to two decimal places.

Days from Initial Day		Rides Bus	Does Not Ride Bus
0	$[1 \quad 0]T^0 =$	[1	0]
1	$[1 \quad 0]T^1 =$	[0.88	0.12]
2	$[1 \quad 0]T^2 =$	[0.80	0.20]
3	$[1 \quad 0]T^3 =$	[0.76	0.24]
4	$[1 \quad 0]T^4 =$	[0.73	0.27]
5	$[1 \quad 0]T^5 =$	[0.71	0.29]
6	$[1 \quad 0]T^6 =$	[0.70	0.30]
7	$[1 \quad 0]T^7 =$	[0.69	0.31]
8	$[1 \quad 0]T^8 =$	[0.68	0.32]
9	$[1 \quad 0]T^9 =$	[0.68	0.32]

Since the ninth state vector is the same as the eighth, we can conclude that all subsequent state vectors after the eighth will be the same. In particular, the table shows that in the long run, the state vectors stabilize at $[0.68 \quad 0.32]$. This state vector is called the **equilibrium-state vector** of the system. The implication of the equilibrium-state vector in the bus ridership problem is that after a long period of time, 68% of those who initially rode the bus will continue to ride and 32% of those who initially rode will no longer ride. ▪

The next example is intended to show that the equilibrium-state vector for bus ridership is independent of the initial-state vector.

Example 2 Find the equilibrium-state vector for the bus ridership problem if we start with a nonrider on the first day.

Solution Since we begin on the first day with a nonrider, the initial-state vector is $[0 \quad 1]$. Thus, the state vector for the second day is

$$[0 \quad 1]\begin{bmatrix} 0.88 & 0.12 \\ 0.25 & 0.75 \end{bmatrix} = [0.25 \quad 0.75]$$

For the third day,

$$[0.25 \quad 0.75]\begin{bmatrix} 0.88 & 0.12 \\ 0.25 & 0.75 \end{bmatrix} = [0.41 \quad 0.59]$$

Continuing in this manner, we arrive at the successive state vectors listed in the table below. The entries have been rounded off to two decimal places.

Days from Initial Day		Rides Bus	Does Not Ride Bus
0	$[0\ \ 1]T^0$ =	[0	1]
1	$[0\ \ 1]T^1$ =	[0.25	0.75]
2	$[0\ \ 1]T^2$ =	[0.41	0.59]
3	$[0\ \ 1]T^3$ =	[0.51	0.49]
4	$[0\ \ 1]T^4$ =	[0.57	0.43]
5	$[0\ \ 1]T^5$ =	[0.61	0.39]
6	$[0\ \ 1]T^6$ =	[0.63	0.37]
7	$[0\ \ 1]T^7$ =	[0.65	0.35]
8	$[0\ \ 1]T^8$ =	[0.66	0.34]
9	$[0\ \ 1]T^9$ =	[0.67	0.33]
10	$[0\ \ 1]T^{10}$ =	[0.68	0.32]
11	$[0\ \ 1]T^{11}$ =	[0.68	0.32]

Since the last two state vectors are the same, we conclude that $[0.68\ \ 0.32]$ is the equilibrium-state vector. We note that this is the same result obtained in the previous example using the initial-state vector $[1\ \ 0]$—that is, starting with a person who rides on the first day. The fact that it took more days to reach the equilibrium-state vector is unimportant, because we are looking for long-term trends. ▪

The results of Examples 1 and 2 demonstrate the fact that if a transition matrix has an equilibrium-state vector, it is independent of the initial-state vector. We state this fact, without proof, in the form of a theorem.

THEOREM 7.2

If A is a transition matrix and if A has an equilibrium-state vector V, then the equilibrium-state vector is independent of the initial-state vector.

It can be shown that not all transition matrices have equilibrium-state vectors. A transition matrix that has an equilibrium-state vector is said to be *regular* and is defined as follows.

DEFINITION Regular Transition Matrix

A transition matrix A is **regular** if there exists a power of A whose elements are only positive numbers. That is, there must exist a positive integer k such that A^k contains only positive elements.

COMMENT: *Of the following three matrices, A has a zero entry, B has a negative entry, and C has only positive entries.*

$$A = \begin{bmatrix} 1 & 2 \\ 0 & 3 \end{bmatrix} \quad B = \begin{bmatrix} -1 & 2 \\ 3 & 4 \end{bmatrix} \quad C = \begin{bmatrix} 2 & 1 \\ 3 & 7 \end{bmatrix}$$

The definition of a regular transition matrix A does not require that A contain only positive entries; it requires that there exist a power of A that contains only positive entries. The next example illustrates a regular transition matrix that has a zero entry.

Example 3 The transition matrix

$$A = \begin{bmatrix} 0 & 1 \\ \frac{3}{4} & \frac{1}{4} \end{bmatrix}$$

is regular, because

$$A^2 = \begin{bmatrix} 0 & 1 \\ \frac{3}{4} & \frac{1}{4} \end{bmatrix} \begin{bmatrix} 0 & 1 \\ \frac{3}{4} & \frac{1}{4} \end{bmatrix} = \begin{bmatrix} \frac{3}{4} & \frac{1}{4} \\ \frac{3}{16} & \frac{13}{16} \end{bmatrix}$$

has only positive elements. ■

Example 4 The transition matrix

$$B = \begin{bmatrix} 0 & 1 \\ 1 & 0 \end{bmatrix}$$

is not regular. To see this, we note that

$$B^2 = \begin{bmatrix} 0 & 1 \\ 1 & 0 \end{bmatrix} \begin{bmatrix} 0 & 1 \\ 1 & 0 \end{bmatrix} = \begin{bmatrix} 1 & 0 \\ 0 & 1 \end{bmatrix}$$

$$B^3 = B \cdot B^2 = \begin{bmatrix} 0 & 1 \\ 1 & 0 \end{bmatrix} \begin{bmatrix} 1 & 0 \\ 0 & 1 \end{bmatrix} = \begin{bmatrix} 0 & 1 \\ 1 & 0 \end{bmatrix}$$

$$B^4 = B \cdot B^3 = \begin{bmatrix} 0 & 1 \\ 1 & 0 \end{bmatrix} \begin{bmatrix} 0 & 1 \\ 1 & 0 \end{bmatrix} = \begin{bmatrix} 1 & 0 \\ 0 & 1 \end{bmatrix}$$

If we continue this process, it will eventually become clear that even powers of B yield the identity matrix and odd powers of B yield the given matrix B. In either case, the power of B will have two zero elements. Since no power of B has only positive entries, B is not regular. ■

COMMENT: *If B is the transition matrix in Example 4, then* $[1 \;\; 0]B^0 = [1 \;\; 0]$, $[1 \;\; 0]B^1 = [0 \;\; 1]$, $[1 \;\; 0]B^2 = [1 \;\; 0]$, $[1 \;\; 0]B^3 = [0 \;\; 1]$, $[1 \;\; 0]B^4 = [1 \;\; 0]$, *and so on. Clearly, the products* $[1 \;\; 0]B^n$ *alternate between* $[1 \;\; 0]$ *and* $[0 \;\; 1]$. *Hence, there is no equilibrium-state vector for this matrix.*

The next theorem, which we give without proof, tells us that every regular transition matrix has an equilibrium-state vector.

THEOREM 7.3

Let T be an $n \times n$ regular transition matrix for a Markov chain. Then T has a $1 \times n$ equilibrium-state vector $V = [v_1 \ v_2 \ \cdots \ v_n]$ that satisfies the matrix equation

$$VT = V$$

Theorem 7.3 not only assures us that every regular transition matrix T has an equilibrium-state vector V; it also indicates that we can find V by solving the matrix equation $VT = V$. Example 5 shows that to find V we must solve a system of equations.

COMMENT: *Equilibrium-state vectors for a regular transition matrix T may also be found by repeatedly forming the matrix product $((((AT)T)T)T) \cdots$ until a constant vector is obtained, where A is an initial-state vector. This is the approach we took in Examples 1 and 2 and is particularly easy on a calculator with matrix functions.*

Example 5 Find the equilibrium-state vector V for the transition matrix

$$T = \begin{bmatrix} 0 & 1 \\ \frac{3}{4} & \frac{1}{4} \end{bmatrix}$$

Solution We showed in Example 3 that this is a regular transition matrix. Let $V = [v_1 \ v_2]$. Then we want V to satisfy the matrix equation $VT = V$; that is,

$$[v_1 \quad v_2]\begin{bmatrix} 0 & 1 \\ \frac{3}{4} & \frac{1}{4} \end{bmatrix} = [v_1 \quad v_2]$$

Expanding the left side, we have

$$[\tfrac{3}{4}v_2 \quad v_1 + \tfrac{1}{4}v_2] = [v_1 \quad v_2]$$

Equating the corresponding elements in the two row vectors, we get the following system of equations:

$$\frac{3}{4}v_2 = v_1 \qquad \text{and} \qquad v_1 + \frac{1}{4}v_2 = v_2$$

Rewriting both equations so that the variables v_1 and v_2 are to the left of the equal sign, we have

$$-v_1 + \frac{3}{4}v_2 = 0 \qquad \text{and} \qquad v_1 - \frac{3}{4}v_2 = 0$$

If the first equation is multiplied by -1, we see that both equations are really the same equation, which means that the system is dependent. Recall that a dependent system does not have a unique solution. Since dependent systems do not have a unique solution, we need to identify another equation that relates v_1 and v_2. To this end, we use the fact that V is a probability state vector; that is, v_1 and v_2

must satisfy the equation $v_1 + v_2 = 1$. Using this equation with either of the equations in the dependent system will allow us to solve for v_1 and v_2. We choose to solve the system

$$-v_1 + \frac{3}{4}v_2 = 0$$

$$v_1 + v_2 = 1$$

Writing the system in augmented matrix form, we have

$$\left[\begin{array}{cc|c} -1 & \frac{3}{4} & 0 \\ 1 & 1 & 1 \end{array}\right]$$

By performing the appropriate row operations, we get

$$\left[\begin{array}{cc|c} 1 & 0 & \frac{3}{7} \\ 0 & 1 & \frac{4}{7} \end{array}\right]$$

The solution to the system is then $v_1 = \frac{3}{7}$ and $v_2 = \frac{4}{7}$. We conclude that the equilibrium-state vector for T is $V = [\frac{3}{7}\ \frac{4}{7}]$. ■

COMMENT: *The system of equations obtained from $VT = V$ is always a dependent system. To get a unique solution, we augment the system of equations with the equation that expresses the fact that the sum of the probabilities is equal to 1.*

AN ALTERNATE PROCEDURE FOR SOLVING $VT = V$

The system $VT = V$ can also be written as $VT - V = 0$ or $V(T - I) = 0$, where V is the $1 \times n$ row vector to be determined, T is the $n \times n$ transition matrix, and I is the $n \times n$ identity matrix. However, the system $V(T - I) = 0$ is not in the form that we need to perform Gauss-Jordan elimination. To solve the system $V(T - I) = 0$, we first take its transpose. Recall that the transpose of the matrix A is the matrix A^t, obtained by interchanging the rows and columns of A. It can then be shown that the system $V(T - I) = 0$ is equivalent to the system $(T - I)^t V^t = 0$, where $(T - I)^t$ and V^t are the transposes of $T - I$ and V, respectively. The system $(T - I)^t V^t = 0$ may be solved using Gauss-Jordan elimination.

Example 6 Find the equilibrium-state vector for the bus ridership with transition matrix

	Rider	Nonrider
Rider	0.88	0.12
Nonrider	0.25	0.75

Solution The transition matrix is regular, because it has only positive entries. If we let T represent the transition matrix and $V = [v_1\ \ v_2]$ represent the equilibrium-state vector, then we wish to solve

$$[v_1\ \ v_2]\begin{bmatrix} 0.88 & 0.12 \\ 0.25 & 0.75 \end{bmatrix} = [v_1\ \ v_2]$$

Following the suggested alternate procedure, we write the system $VT = V$ in the form $V(T - I) = 0$:

$$[v_1 \quad v_2]\left(\begin{bmatrix} 0.88 & 0.12 \\ 0.25 & 0.75 \end{bmatrix} - \begin{bmatrix} 1 & 0 \\ 0 & 1 \end{bmatrix}\right) = [0 \quad 0]$$

Taking the transpose of this system, we solve the equivalent system $(T - I)^t V^t = 0$:

$$\left(\begin{bmatrix} 0.88 & 0.12 \\ 0.25 & 0.75 \end{bmatrix} - \begin{bmatrix} 1 & 0 \\ 0 & 1 \end{bmatrix}\right)^t [v_1 \quad v_2]^t = \begin{bmatrix} 0 \\ 0 \end{bmatrix}$$

Simplifying the matrix equation (don't forget to take the transpose!), we get

$$\begin{bmatrix} -0.12 & 0.25 \\ 0.12 & -0.25 \end{bmatrix}\begin{bmatrix} v_1 \\ v_2 \end{bmatrix} = \begin{bmatrix} 0 \\ 0 \end{bmatrix}$$

This dependent system yields the solution $v_1 = \frac{25}{12}v_2$. Using this result along with the condition $v_1 + v_2 = 1$, we obtain $v_1 = \frac{25}{37} \approx 0.68$ and $v_2 = \frac{12}{37} \approx 0.32$. Thus, the equilibrium-state vector is $V = [0.68 \quad 0.32]$, which agrees with our observations in Examples 1 and 2. ■

Example 7 The transition matrix below was presented in Example 1 of Section 7.1 to indicate the housing preferences of adults who rent, own condos, and own houses.

		Next Move		
		Rent	Condo	House
	Rent	0.2	0.4	0.4
Present Housing	Condo	0.3	0.5	0.2
	House	0.1	0.2	0.7

Find the equilibrium-state vector for this regular transition matrix.

Solution Since all the entries in this matrix are positive numbers, it is a regular transition matrix; therefore, it is possible to find an equilibrium-state vector. If we represent the equilibrium-state vector by $V = [v_1 \ v_2 \ v_3]$, then we wish to solve the matrix equation

$$[v_1 \quad v_2 \quad v_3]\begin{bmatrix} 0.2 & 0.4 & 0.4 \\ 0.3 & 0.5 & 0.2 \\ 0.1 & 0.2 & 0.7 \end{bmatrix} = [v_1 \quad v_2 \quad v_3]$$

along with the condition that $v_1 + v_2 + v_3 = 1$. The matrix equation can be put into the form

$$\left(\begin{bmatrix} 0.2 & 0.4 & 0.4 \\ 0.3 & 0.5 & 0.2 \\ 0.1 & 0.2 & 0.7 \end{bmatrix} - \begin{bmatrix} 1 & 0 & 0 \\ 0 & 1 & 0 \\ 0 & 0 & 1 \end{bmatrix}\right)^t [v_1 \quad v_2 \quad v_3]^t = \begin{bmatrix} 0 \\ 0 \\ 0 \end{bmatrix}$$

After simplifying and taking the indicated transposes, this becomes

$$\begin{bmatrix} -0.8 & 0.3 & 0.1 \\ 0.4 & -0.5 & 0.2 \\ 0.4 & 0.2 & -0.3 \end{bmatrix} \begin{bmatrix} v_1 \\ v_2 \\ v_3 \end{bmatrix} = \begin{bmatrix} 0 \\ 0 \\ 0 \end{bmatrix}$$

This is a dependent system, which can be reduced using Gauss-Jordan elimination on the coefficient matrix. One possible reduction is

$$\begin{bmatrix} -0.8 & 0.3 & 0.1 \\ 0.4 & -0.5 & 0.2 \\ 0.4 & 0.2 & -0.3 \end{bmatrix} \xrightarrow[10③ \to ③]{10① \to ①,\ 10② \to ②} \begin{bmatrix} -8 & 3 & 1 \\ 4 & -5 & 2 \\ 4 & 2 & -3 \end{bmatrix} \xrightarrow[3① + ③ \to 3]{-2① + ② \to ②}$$

$$\begin{bmatrix} -8 & 3 & 1 \\ 20 & -11 & 0 \\ -20 & 11 & 0 \end{bmatrix} \xrightarrow{② + ③ \to ③} \begin{bmatrix} -8 & 3 & 1 \\ 20 & -11 & 0 \\ 0 & 0 & 0 \end{bmatrix}$$

(Note: There are other equivalent reductions.) From the equation represented by the second row of this last matrix, we obtain $v_1 = \frac{11}{20}v_2$. Using this in the equation represented by the first row, we get $v_3 = \frac{28}{20}v_2$. Finally, using these expressions for v_1 and v_3 in the equation $v_1 + v_2 + v_3 = 1$, we obtain

$$V = \begin{bmatrix} \frac{11}{59} & \frac{20}{59} & \frac{28}{59} \end{bmatrix} \approx [0.19 \quad 0.34 \quad 0.47]$$

We interpret this to mean that in the long run, 19% of the adults will rent, 34% will own condos, and 47% will own houses. ■

EXERCISES SECTION 7.2

In Exercises 1–8, explain why each transition matrix is regular.

1. $\begin{bmatrix} 0.4 & 0.6 \\ 0.5 & 0.5 \end{bmatrix}$

2. $\begin{bmatrix} 0.1 & 0.9 \\ 0.7 & 0.3 \end{bmatrix}$

3. $\begin{bmatrix} 0 & 1 \\ 0.2 & 0.8 \end{bmatrix}$

4. $\begin{bmatrix} 0.5 & 0.5 \\ 1 & 0 \end{bmatrix}$

5. $\begin{bmatrix} 0.2 & 0.2 & 0.6 \\ 0.3 & 0.4 & 0.3 \\ 0.5 & 0.4 & 0.1 \end{bmatrix}$

6. $\begin{bmatrix} 0.3 & 0.7 & 0 \\ 0.2 & 0.1 & 0.7 \\ 0.5 & 0 & 0.5 \end{bmatrix}$

7. $\begin{bmatrix} 0.5 & 0.5 & 0 \\ 0.4 & 0.4 & 0.2 \\ 0 & 0.9 & 0.1 \end{bmatrix}$

8. $\begin{bmatrix} 0.1 & 0.3 & 0.6 \\ 0.2 & 0.3 & 0.5 \\ 0.8 & 0.1 & 0.1 \end{bmatrix}$

9. Use the method of Example 1 to find the equilibrium-state vector for the regular transition matrix

$$\begin{bmatrix} 0.3 & 0.7 \\ 0.8 & 0.2 \end{bmatrix}$$

if the initial-state vector is (a) [1 0], (b) [0 1], (c) [0.4 0.6].

10. Use Theorem 7.3 to find the equilibrium-state vector for the transition matrix in Exercise 9.

11. Use the method of Example 1 to find the equilibrium-state vector for the regular transition matrix

$$\begin{bmatrix} 0.1 & 0.9 \\ 0.5 & 0.5 \end{bmatrix}$$

if the initial-state vector is (a) [1 0], (b) [0 1], (c) [0.3 0.7]

12. Use Theorem 7.3 to find the equilibrium-state vector for the transition matrix in Exercise 11.

In Exercises 13–20, use Theorem 7.3 to find the equilibrium-state vector for each regular transition matrix.

13. $\begin{bmatrix} 0.4 & 0.6 \\ 0.5 & 0.5 \end{bmatrix}$

14. $\begin{bmatrix} 0.1 & 0.9 \\ 0.7 & 0.3 \end{bmatrix}$

15. $\begin{bmatrix} 0 & 1 \\ 0.2 & 0.8 \end{bmatrix}$

16. $\begin{bmatrix} 0.5 & 0.5 \\ 1 & 0 \end{bmatrix}$

17. $\begin{bmatrix} 0.2 & 0.2 & 0.6 \\ 0.3 & 0.4 & 0.3 \\ 0.5 & 0.4 & 0.1 \end{bmatrix}$

18. $\begin{bmatrix} 0.3 & 0.7 & 0 \\ 0.2 & 0.1 & 0.7 \\ 0.5 & 0 & 0.5 \end{bmatrix}$

19. $\begin{bmatrix} 0.5 & 0.5 & 0 \\ 0.4 & 0.4 & 0.2 \\ 0 & 0.9 & 0.1 \end{bmatrix}$

20. $\begin{bmatrix} 0.1 & 0.3 & 0.6 \\ 0.2 & 0.3 & 0.5 \\ 0.8 & 0.1 & 0.1 \end{bmatrix}$

21. The transition matrix for the game of Twenty-one between Sam and Leon (see Section 7.1) is

		Next Shooter: Sam	Next Shooter: Leon
Present Shooter	Sam	0.6	0.4
	Leon	0.3	0.7

Use the method of Example 1 to find the equilibrium-state vector for this transition matrix if:
(a) Sam starts the game.
(b) Leon starts the game.
(c) The person who starts is decided by the flip of a coin.

22. Use Theorem 7.3 to find the equilibrium-state vector for the game described in Exercise 21.

23. Sometimes Professor Martinez collects homework and sometimes she doesn't. The probability that she will collect it two days in a row is 0.3 and the probability that she will collect it today if she didn't collect it yesterday is 0.8. Use the method of Example 1 to find the equilibrium-state vector if:
(a) Homework was collected the first day.
(b) Homework was not collected the first day.
(c) There is a 50-50 chance that homework was collected the first day.

24. Use Theorem 7.3 to find the equilibrium-state vector for the homework-collection process in Exercise 23.

25. A video game shows two different monsters on the screen, but only one at a time. If Godzilla is on the screen when a game ends, there is a probability of 0.2 that Godzilla will be on the screen when the next game starts. If King Kong is on the screen when the game ends, there is a probability of 0.4 that King Kong will be on the screen when the next game starts. Use Theorem 7.3 to find the equilibrium-state vector for this process. In the long run, what percentage of the games start with Godzilla and what percentage start with King Kong?

26. Exercise 14 of Section 7.1 describes the dynamics of people moving among the poor, middle, and wealthy classes. Use Theorem 7.3 to find the equilibrium-state vector for this process.

27. A car rental company has offices in Cincinnati, Dayton, and Columbus. A person who rents a car in one of these three cities can return it to an office in any one of the three. An accountant in the controller's office of the company reports that 85% of the rentals from Cincinnati are returned in Cincinnati, 10% are returned in Dayton, and 5% are returned in Columbus. From Dayton, 75% of the rentals are returned in Dayton, 12% in Cincinnati, and 13% in Columbus. Of those cars rented in Columbus, 90% are returned in Columbus, 5% in Dayton, and 5% in Cincinnati. Set up a transition matrix for this process. Use Theorem 7.3 to find the percentage of cars that will end up in each city if the company does not transfer cars between cities.

28. A psychologist studies the traffic pattern of people as they enter a department store by one of three doors and then leave the store by one of the same three doors. As a result of the study, the psychologist sets up the following transition matrix to represent the process:

		Exit: Door 1	Exit: Door 2	Exit: Door 3
Enter	Door 1	0.7	0.2	0.1
	Door 2	0.4	0.3	0.3
	Door 3	0.2	0.6	0.2

Use Theorem 7.3 to find the percentage of customers using each door in the long run.

7.3 DECISION-MAKING: GAME THEORY

Suppose you are on a television game show and you have already won \$5000. At the end of the show you have the option of keeping the \$5000 or trading it for an unknown prize, which may be either a new Lincoln Town Car or an empty box. What decision would you make? Would you take the \$5000 and be safe, or would you take a chance and pick the unknown prize? Mathematics cannot be used to explain why you make the decisions that you do, but it can sometimes be used to assist you in making a rational decision. In this section, we describe how expected value and game theory can be used to help make decisions.

EXPECTED VALUE

Suppose it costs you \$5 to play a game in which two coins are tossed—if two heads show you get back \$12, if one head shows you get back \$3, and if no heads show you lose your \$5. Is it wise to play this game? To help us make this decision, we use the concept of the expected value of a random variable, which we discussed in Section 6.2. The definition of expected value is given here for convenience.

DEFINITION Expected Value of a Random Variable

If $x_1, x_2, \ldots, x_k$ are values of the random variable X and if $p_1, p_2, \ldots, p_k$ are probabilities associated with the values of the random variable, then the expected value of X is denoted by $E(X)$ and defined by

$$E(X) = x_1p_1 + x_2p_2 + \cdots + x_kp_k$$

Example 1 In the game described above, we interpret the values of the random variable to be $x_1 = \$7$ (you win $\$12 - \$5 = \$7$ if two heads show), $x_2 = -\$2$ (you lose $\$5 - \$3 = \$2$ if one head shows), and $x_3 = -\$5$ (you lose \$5 if no heads show). The probability that two heads will show is $p_1 = \frac{1}{4}$, that one head will show is $p_2 = \frac{1}{2}$, and that no heads will show is $p_3 = \frac{1}{4}$. Therefore, the expected value of this game is

$$E(X) = x_1p_1 + x_2p_2 + x_3p_3 = \$7\left(\frac{1}{4}\right) + (-\$2)\left(\frac{1}{2}\right) + (-\$5)\left(\frac{1}{4}\right) = -\$0.50$$

Recall from Section 6.2 that we interpret $E(X) = 0$ to be a fair game, $E(X) < 0$ to favor the house, and $E(X) > 0$ to favor the player. In this case $E(X) = -\$0.50$, which means that in the long run you can expect to lose 50 cents per game. Obviously, you would not be wise to play this game, at least not over the long run. Of course, there may be other factors that influence your decision, such as the knowledge that the proceeds of the game are for charity, in which case you may elect to play even though you know the game favors the house. ▪

Sometimes we use expected value to make choices between two or more options. We assume that the option with the largest expected value is the best choice. The next example shows this decision process.

Example 2 Suppose that you have \$10,000 to invest in the stock market, and that you can invest either in blue-chip stocks (low-risk) or in speculative stocks (high-risk). If you invest in blue-chip stocks and the market goes up 100 points, your investment will be worth \$12,000; if the market goes down 100 points, it will be worth \$9000. If you invest in speculative stocks and the market goes up 100 points, your investment will be worth \$15,000; it will be worth \$7000 if the market goes down 100 points. If the probability that the market will go up 100 points is 0.4 and the probability that it will go down 100 points is 0.3, what is your best investment strategy? Assume the best investment strategy is the one with the largest expected value.

Solution The expected value for the blue-chip stocks is

$$\$12{,}000(0.4) + \$9000(0.3) = \$7500$$

and the expected value for the speculative stocks is

$$\$15{,}000(0.4) + \$7000(0.3) = \$8100$$

Based on the expected value for the given conditions, the speculative stocks are your best investment. We note that the expected-value approach to decision-making is only as good as the assumptions we make about the various options. For instance, if the probability that the market will go up 100 points changes, then the expected values computed above no longer hold. ■

GAME THEORY

The word *game* is used here to mean any competitive situation existing between two or more opponents; it could be two people playing checkers, or it could be two or more businesses competing for customers. **Game theory** deals with the process of making decisions when you are in competition with an opponent and can control your own strategy but *not* that of your opponent.

HISTORICAL COMMENT: *The theory of games originated in the seventeenth century, but it was not until the early part of this century that it gained acceptance as a decision-making tool. The principles of game theory have been used for decision-making in business and government since the 1940s. Most of the credit for developing game theory is given to the Hungarian-born mathematician John von Neumann (1903–1957).*

Consider two players A and B who are playing a form of the matching pennies game. Each player selects H (heads) or T (tails) without knowing what choice the other player has made, and then the choices are compared. The choice that a player makes is called a **strategy.** The rules of the game are as follows:

- If A matches B, then B pays A a penny.
- If A does not match B, then A pays B a penny.

A game of this kind is called a *two-person game*. The payoff of this game can be represented by the following *payoff matrix*:

$$\begin{array}{cc} & \text{Player B} \\ & \begin{array}{cc} H & T \end{array} \\ \text{Player A} \begin{array}{c} H \\ T \end{array} & \begin{bmatrix} 1 & -1 \\ -1 & 1 \end{bmatrix} \end{array}$$

We have used the convention that payoffs to A are positive and payoffs to B are negative. Therefore, in the payoff matrix a 1 indicates that B pays A a penny, and a -1 indicates that A pays B a penny. For instance, if A chooses tails and B chooses heads, the payoff is -1, which means A pays B a penny. Games that may be represented in the form of a matrix are called **matrix games.** In two-person matrix games, the number of rows in the matrix corresponds to the number of strategies available to one player, and the number of columns corresponds to the number of strategies available to the other player.

Notice that in the matching pennies game the money won by one player is lost by the other player, so if we add the money won and the money lost, the sum is zero. For this reason, games of this type are called **zero-sum games.** Players in the stock market are not playing a zero-sum game, since the money lost does not necessarily equal the money gained.

The matrix games considered in this section are two-person zero-sum games. The purpose of game theory is to find strategies that are most profitable to the players. A strategy that maximizes winnings and minimizes losses is called an **optimum strategy.** We are interested in how players in a zero-sum matrix game can optimize their rewards. Remember, *each player must choose a strategy without knowing what strategy their opponent has chosen.* Otherwise, there would be no game theory.

Example 3 Consider a two-person matrix game in which the payoff matrix is

$$\begin{array}{cc} & \text{Player B} \\ \text{Player A} & \begin{bmatrix} 1 & -2 & -4 \\ 2 & 0 & 1 \\ -2 & -3 & 5 \end{bmatrix} \end{array}$$

The game is played as follows:

- Player A chooses a row of the matrix and Player B chooses a column of the matrix.
- If the intersection of the row and column chosen is a positive number, Player B pays Player A that many dollars.
- If the indicated intersection is a negative number, A pays B that many dollars.
- If the intersection is zero, there is no payoff.

For example, if A chooses row 1 and B chooses column 2, then the intersection is -2, which means that A pays B \$2. The objective of each player is to choose a strategy that will optimize his or her winnings. ■

Strictly Determined Games: Pure Strategies

A game is called a **strictly determined game** if the best strategy for the players is strictly determined by the payoff matrix. The matrix game defined in Example 3 is a strictly determined matrix game.

To see why the 3×3 matrix game of Example 3 is strictly determined, imagine that you are Player A, also called the "row player." What strategy would you choose? Since you are choosing rows, it might seem that your best strategy is to pick row 3, because you could win \$5 if Player B chose column 3, but you could also lose \$3 if B chose column 2. Or you might decide that row 2 is a better strategy, because you will win \$2 if B chooses column 1 and the worst you can do is break even. Likewise, Player B will see that if he or she chooses column 2, there is a possible payoff of \$3 if A chooses row 3, and the worst that can happen is a zero payoff if A chooses row 2. We want to know if there is a best strategy for each player.

We see that in this game, a knowledge of our opponent's strategy would not change our best strategy, unless, of course, our opponent decided not to adopt his or her best strategy. This is a *strictly determined game.* Recall that a positive entry means the row player receives the payoff and a negative entry means the column player receives the payoff. It is important to notice that *in a strictly determined game the smallest entry in the best row strategy is the same as the largest entry in the best column strategy.* This entry represents the "value" of the game and is called a **saddle point.** The saddle point in the game of Example 3 is the zero entry, which means that the value of the game is zero. We say that the game is *fair* if its value is zero; otherwise, the game is said to be *unfair*.

A player who always chooses the same strategy is said to be playing a **pure strategy.** In a strictly determined game, both players should choose a pure strategy that includes the saddle point. Thus, for strictly determined games the row player should use the following strategy:

Optimum Pure Strategy for the Row Player in a Strictly Determined Game

1. Determine the least element in each row.
2. Choose the row that contains the largest of the least row elements.

Example 4 The least element in each row of the following matrix game is circled. The largest of these elements is zero, so the best pure strategy for Player A is to play row 2.

$$\text{Player A} \overset{\text{Player B}}{\begin{bmatrix} 1 & -2 & \textcircled{-4} \\ 2 & \textcircled{0} & 1 \\ -2 & \textcircled{-3} & 5 \end{bmatrix}}$$

■

For strictly determined games the column player should choose the following pure strategy:

Optimum Pure Strategy for the Column Player in a Strictly Determined Game

1. Determine the largest element in each column.
2. Choose the column which contains the least of the largest column elements.

Example 5 The largest element in each column of the following matrix game is circled. The smallest of these elements is zero, so the best pure strategy for Player B is column 2.

$$\text{Player A} \overset{\text{Player B}}{\begin{bmatrix} 1 & -2 & -4 \\ \textcircled{2} & \textcircled{0} & 1 \\ -2 & -3 & \textcircled{5} \end{bmatrix}}$$

■

COMMENT: *From Examples 4 and 5 we see that the saddle point for this game is zero. All strictly determined games have at least one saddle point. It can be shown that if a strictly determined game has more than one saddle point, all of the saddle points must have the same value.*

IDENTIFYING STRICTLY DETERMINED GAMES

We can identify a strictly determined matrix game by showing that it has at least one saddle point. To do this we proceed as follows:

1. List the minimum entry in each row.
2. List the maximum entry in each column.
3. Determine the largest of the minimum row entries.
4. Determine the smallest of the maximum column entries.
5. If the largest of the minimum row entries is equal to the smallest of the maximum column entries, then the game has a saddle point and is therefore a strictly determined game.

Example 6 Show that the following matrix defines a strictly determined game and determine the optimum strategy:

$$\begin{array}{r c c} & \begin{bmatrix} 1 & -2 & -1 \\ 3 & 1 & -3 \\ 2 & 4 & 1 \end{bmatrix} & \begin{array}{c} \text{Minimum row entry} \\ -2 \\ -3 \\ 1 \end{array} \\ \text{Maximum column entry} & \begin{array}{ccc} 3 & 4 & 1 \end{array} & \end{array}$$

Solution We have indicated the row minima and the column maxima outside the matrix. Since the largest row minimum and the smallest column maximum are both the same entry, 1, we conclude that this entry is a saddle point. Therefore, this is a strictly determined game with value 1. The optimum strategy dictates that the row player choose row 3 and that the column player choose column 3. ▪

EXERCISES SECTION 7.3

In Exercises 1–4, find the expected value of the random variable.

1.

Value of X	*Probability*
1	$\frac{1}{8}$
2	$\frac{1}{4}$
3	$\frac{1}{2}$
4	$\frac{1}{8}$

2.

Value of X	*Probability*
−10	$\frac{1}{7}$
0	$\frac{2}{7}$
10	$\frac{4}{7}$

3.

Value of X	*Probability*
−10	0.1
0	0.2
10	0.2
20	0.3
30	0.2

4.

Value of X	*Probability*
2	0.2
5	0.3
8	0.4
10	0.1

5. A box contains five quarters, fifteen dimes, and seven nickels. A coin is drawn at random from the box. Calculate the expected value of the draw.

6. A roulette wheel has 38 pockets into which a ball may fall: 18 red pockets, 18 black pockets, and 2 green pockets. If you bet $1 on red and the ball lands in red, you win $1; otherwise, you lose $1. Calculate the expected value for this game.

7. A raffle in which the prize is $500 sells 1200 tickets at $1 each. Calculate the expected value for this raffle.

8. You have the opportunity to choose between two sales jobs. If you take the job in Chicago, there is a 50% chance that you will earn $45,000 and a 50% chance that you will earn $30,000. If you take the job in Los Angeles, there is a 70% chance that you will earn $20,000 and a 30% chance that you will earn $60,000. If you base your decision on expected value, which job should you take?

9. Suppose that you have won $5,000 on a game show and you have a chance for the Big Jackpot at the end of the show. The rule is that you have the option of keeping your $5,000 or trading it in for one of three hidden prizes. One of the prizes is worth $25,000, one is worth $2,500, and one is worth $500.

Based on expected value, should you keep your $5,000 or trade it?

10. A contractor bids on the job of constructing a new motel. There is a probability of 0.8 that the profit on the job will be $25,000, a probability of 0.1 that there will be no profit, and a probability of 0.1 that there will be a loss of $10,000 (due to bad weather, strikes, and so on). Calculate the expected profit to the contractor.

In Exercises 11–20, decide whether each matrix game is strictly determined. If the game is strictly determined, indicate the saddle point(s) and whether the game is fair or unfair. Give the best pure strategy for each player.

11. $\begin{bmatrix} 1 & -2 \\ -2 & 1 \end{bmatrix}$ **12.** $\begin{bmatrix} -1 & 2 \\ -2 & 1 \end{bmatrix}$

13. $\begin{bmatrix} 1 & -2 \\ 0 & 3 \end{bmatrix}$ **14.** $\begin{bmatrix} -1 & 2 \\ 3 & 1 \end{bmatrix}$

15. $\begin{bmatrix} -2 & -3 \\ 4 & 3 \end{bmatrix}$ **16.** $\begin{bmatrix} -1 & 0 \\ -2 & -1 \end{bmatrix}$

17. $\begin{bmatrix} \frac{1}{2} & \frac{1}{2} \\ \frac{2}{3} & \frac{1}{3} \end{bmatrix}$ **18.** $\begin{bmatrix} -\frac{1}{2} & \frac{1}{2} \\ \frac{1}{2} & -\frac{1}{2} \end{bmatrix}$

19. $\begin{bmatrix} 2 & -3 \\ 4 & -1 \end{bmatrix}$ **20.** $\begin{bmatrix} 2 & -2 \\ -5 & 5 \end{bmatrix}$

In Exercises 21–24, find the saddle point(s) for each strictly determined matrix game.

21. $\begin{bmatrix} 1 & 1 & 0 \\ 5 & 6 & 2 \\ -1 & 3 & 1 \end{bmatrix}$ **22.** $\begin{bmatrix} 4 & 7 & -2 \\ 2 & -1 & -1 \\ 5 & 8 & 0 \end{bmatrix}$

23. $\begin{bmatrix} 4 & -2 & 3 \\ 3 & -2 & 1 \\ 2 & 0 & 3 \end{bmatrix}$ **24.** $\begin{bmatrix} 1 & 2 & 3 \\ -2 & 0 & -5 \\ 0 & 3 & -1 \end{bmatrix}$

25. Smith Pharmaceutical and Philips Laboratories both manufacture a new drug for headaches. Each company decides to put its advertising money into one of the following media: radio, television, and newspapers. The amount of extra revenue, in millions of dollars, generated by Smith is given in the following payoff matrix:

		Philips Laboratories		
		Radio	TV	Newspapers
Smith Pharm.	Radio	−0.5	−1.0	0
	TV	−1.5	0	2.0
	Newspapers	1.5	1.0	1.3

Find the optimum strategy for each company.

26. Bank 1 and Bank 2 have offices in Detroit and each plans to expand into one of the following three cities: Ann Arbor, East Lansing, and Royal Oak. If Bank 1 opens branches in Ann Arbor and Bank 2 also opens branches in Ann Arbor, then Bank 1 stands to break even; if Bank 1 opens branches in Ann Arbor and Bank 2 opens branches in East Lansing, then Bank 1 stands to lose $500,000; and if Bank 1 opens branches in Ann Arbor and Bank 2 opens branches in Royal Oak, then Bank 1 stands to lose $200,000. If Bank 1 opens branches in East Lansing and Bank 2 opens branches in Ann Arbor, then Bank 1 stands to make $1,000,000, and so on, as shown in the following payoff matrix. The entries in the payoff matrix represent revenue in millions of dollars to Bank 1.

		Bank 2		
		Ann Arbor	East Lansing	Royal Oak
Bank 1	Ann Arbor	0	−0.5	−0.2
	East Lansing	1.0	1.2	0.2
	Royal Oak	0.3	0	−0.3

Find the optimum strategy for each bank and the payoffs for the strategies.

7.4 NON-STRICTLY DETERMINED GAMES: MIXED STRATEGIES

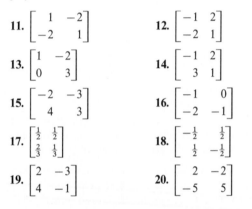

In the last section we considered games in which the strategies of playing the games were pure, but also somewhat uninteresting because there was never a variation. However, not all games are strictly determined. Games that are not strictly determined (that is, games that do not have a saddle point) are said to be **non-strictly determined.** As the next example shows, the matching-pennies game is a non-strictly determined game. In a non-strictly determined game, it can be shown that your best strategy is not a pure strategy.

Example 1 Show that the 2×2 matrix game corresponding to matching pennies is not a strictly determined game.

Solution We have repeated the payoff matrix from the previous section and added the row minima and column maxima.

$$\begin{array}{lcc} & & \text{Row minima} \\ & \begin{bmatrix} 1 & -1 \\ -1 & 1 \end{bmatrix} & \begin{array}{c} -1 \\ -1 \end{array} \\ \text{Column maxima} & \begin{array}{cc} 1 & \;\;1 \end{array} & \end{array}$$

The indicated row minima and column maxima show that there is no saddle point. Therefore, the matching-pennies game is not a strictly determined game. ▪

What is the best strategy for the matching-pennies game? Suppose Player A always plays row 1. Then Player B will eventually detect this and start to play column 2. If A does not change strategy, B will collect one penny every time the game is played. Thus, a pure (row 1) strategy will result in A's paying B a penny each time they play. Player A is better off employing a **mixed strategy,** in which row 1 is played part of the time and row 2 is played part of the time. The question is "What percentage of the time should we choose row 1 and what percentage of the time should we choose row 2?" If Player A decides to randomly choose the row to be played, based upon some percentage, then Player A will be using a mixed strategy.

As an example of a mixed strategy, suppose that in the matching-pennies game the row player randomly plays the first row $\frac{2}{3}$ of the time and the second row $\frac{1}{3}$ of the time. (This strategy might be implemented by tossing a die and choosing the first row if the die lands with 1, 2, 3, or 4 face-up and choosing the second row if a 5 or 6 is showing.) Similarly, a column strategy might be to randomly choose the first column $\frac{1}{4}$ of the time and the second column $\frac{3}{4}$ of the time. We describe the row strategy with the **row probability vector** $[\frac{1}{3}\ \frac{2}{3}]$ and the column strategy with the **column probability vector** $[\frac{1}{4}\ \frac{3}{4}]$. All strategies can be written in terms of row or column probability vectors.

COMMENT: *The column strategy for a strictly determined game can also be written as probability vectors. For example, to express the fact that the second row should always be chosen from a* 3×3 *matrix, we would write* $[0\ 1\ 0]$.

Using the row and column probability vectors we can now assign a probability to each of the possible row-column combinations. For example, if the row and column probability vectors for the matching-pennies matrix game are $[\frac{1}{3}\ \frac{2}{3}]$ and $[\frac{1}{4}\ \frac{3}{4}]$, the probability of the row player's choosing the first row and the column player's choosing the first column is $(\frac{1}{3})(\frac{1}{4}) = \frac{1}{12}$. Similarly, the probability of the row player's choosing the first row and the column player's choosing the second column is $(\frac{1}{3})(\frac{3}{4}) = \frac{1}{4}$; the probability of the row player's choosing the second row and the column player's choosing the first column is $(\frac{2}{3})(\frac{1}{4}) = \frac{1}{6}$; and the probability of the row player's choosing the second row and the column player's choosing the second column is $(\frac{2}{3})(\frac{3}{4}) = \frac{1}{2}$. The expected value of the matching-pennies game for the indicated mixed strategies is then obtained by multiplying each of these probabilities by the corresponding entries from the payoff matrix for the game. Thus,

$$E = 1 \cdot \left(\frac{1}{12}\right) + (-1) \cdot \left(\frac{1}{4}\right) + (-1) \cdot \left(\frac{1}{6}\right) + 1 \cdot \left(\frac{1}{2}\right) = \frac{1}{6}$$

This can be interpreted to mean that if this strategy is used, in the long run Player B will pay Player A $\frac{1}{6}$ of a penny per game.

Expected Value of a Mixed Strategy as a Matrix Product

The procedure employed above to calculate the expected value of the matching-pennies game can get tedious for larger payoff matrices. A more direct approach is to use the formula given below, whose variables are matrices. We will use this formula without demonstrating its validity.

Expected Value of a Matrix Game

The expected value E of a matrix game is

$$E = RPC^t \tag{7.1}$$

where

R is the probability vector representing the row strategy

C^t is the transpose of the probability vector representing the column strategy

P is the payoff matrix

Example 2 Use Formula 7.1 to compute the expected value of the matching-pennies matrix game in which the payoff matrix is

$$P = \begin{bmatrix} 1 & -1 \\ -1 & 1 \end{bmatrix}$$

and the row and column probability vectors are $R = [\frac{1}{3}\ \frac{2}{3}]$ and $C = [\frac{1}{4}\ \frac{3}{4}]$, respectively.

Solution By Formula 7.1, the expected value is

$$E = [\tfrac{1}{3} \;\; \tfrac{2}{3}]\begin{bmatrix} 1 & -1 \\ -1 & 1 \end{bmatrix}[\tfrac{1}{4} \;\; \tfrac{3}{4}]^t [\tfrac{1}{3} \;\; \tfrac{2}{3}]\begin{bmatrix} 1 & -1 \\ -1 & 1 \end{bmatrix}\begin{bmatrix} \tfrac{1}{4} \\ \tfrac{3}{4} \end{bmatrix}$$

$$= \left[\left(\frac{1}{3} - \frac{2}{3}\right) \;\; \left(-\frac{1}{3} + \frac{2}{3}\right)\right]\begin{bmatrix} \tfrac{1}{4} \\ \tfrac{3}{4} \end{bmatrix} = \frac{1}{4}\left(\frac{1}{3} - \frac{2}{3}\right) + \frac{3}{4}\left(-\frac{1}{3} + \frac{2}{3}\right) = \frac{1}{6}$$

The expected value of $\frac{1}{6}$ agrees with the result obtained above. ∎

Example 3 Consider the game whose payoff matrix is given by

$$P = \begin{bmatrix} 2 & 4 & 0 \\ 1 & -2 & 5 \\ -3 & 4 & 1 \end{bmatrix}$$

Determine the value of this game if the probability vector representing the row strategy is given by $R = [\frac{1}{3} \; \frac{2}{3} \; 0]$, and that representing the column strategy is $C = [\frac{1}{2} \; 0 \; \frac{1}{2}]$.

Solution Using Formula 7.1, we have

$$E = [\tfrac{1}{3} \;\; \tfrac{2}{3} \;\; 0]\begin{bmatrix} 2 & 4 & 0 \\ 1 & -2 & 5 \\ -3 & 4 & 1 \end{bmatrix}\begin{bmatrix} \tfrac{1}{2} \\ 0 \\ \tfrac{1}{2} \end{bmatrix} = \tfrac{7}{3}$$

That is, the expected value of this game is $\frac{7}{3}$. ∎

But what is the *best* mixed strategy for a game? That is, given that the game is not strictly determined, we want to determine a strategy R^* that will maximize the expected value for the row player *regardless of the strategy of the column player*. Likewise, we want to determine a strategy C^* that will minimize the expected value for the column player *regardless of the strategy of the row player*. The strategies R^* and C^* are called **optimal strategies**. The fact that such optimal strategies exist for non-strictly determined games is the fundamental theorem of game theory. It can be shown that the expected values of R^* and C^* are equal. We will show how to obtain optimal strategies for any 2×2 non-strictly determined matrix game, but first we will show how this is done for the matching-pennies game.

Example 4

(a) Find the optimal mixed strategy for each player in the matching-pennies game.
(b) Compute the value of the game for each of these strategies.

Solution

(a) Suppose that you are playing the mixed row strategies in the matching-pennies game, and that for each game, you randomly choose the row that you will play. Assume that p is the probability that you will play row 1; then the probability that you will play row 2 is $1 - p$. Thus, the row probability vector is $[p \quad 1 - p]$. From the fundamental theorem of game theory, the expected value of the game will be the same regardless of what column strategies are used. Since the expected value of the game is independent of your opponent's strategy, assume your opponent decides to play column 1 all the time. Then the column probability vector is $[1 \ 0]$ and your expected value for the game is

$$E = [p \quad 1 - p]\begin{bmatrix} 1 & -1 \\ -1 & 1 \end{bmatrix}\begin{bmatrix} 1 \\ 0 \end{bmatrix} = 2p - 1$$

Similarly, if your opponent plays column 2 all of the time, the column probability vector is $[0 \ 1]$ and the expected value is

$$E = [p \quad 1 - p]\begin{bmatrix} 1 & -1 \\ -1 & 1 \end{bmatrix}\begin{bmatrix} 0 \\ 1 \end{bmatrix} = 1 - 2p$$

These two expectations must be equal for the optimal strategy. Using this fact, we have

$$2p - 1 = -2p + 1$$

Solving for p, we get

$$p = \frac{1}{2}$$

Thus, the row probability vector for optimal strategy is $R^* = [\frac{1}{2} \ \frac{1}{2}]$, which means that the best strategy for the row player in the matching-pennies game is to *randomly* pick row 1 one-half of the time and row 2 one-half of the time.

A similar analysis reveals the best mixed strategy for the column player. If q is the probability that the column player will pick column 1, and $1 - q$ is the probability that the column player will pick column 2, then we obtain the expected value of the game for *any* two different row strategies. For the sake of convenience, we use $R = [1 \ 0]$ to obtain

$$E = [1 \quad 0]\begin{bmatrix} 1 & -1 \\ -1 & 1 \end{bmatrix}\begin{bmatrix} q \\ 1 - q \end{bmatrix} = 2q - 1$$

and $R = [0 \ 1]$ to obtain

$$E = [0 \quad 1]\begin{bmatrix} 1 & -1 \\ -1 & 1 \end{bmatrix}\begin{bmatrix} q \\ 1 - q \end{bmatrix} = -2q + 1$$

Equating the two expected values, we have

$$2q - 1 = -2q + 1$$
$$q = \frac{1}{2}$$

Therefore, the best strategy for the column player is described by the probability vector $C^* = [\frac{1}{2}\ \frac{1}{2}]$, which means that column 1 should be randomly chosen one-half of the time and column 2 one-half of the time.

(b) The value of the game for these optimal strategies is given by

$$E = R^*PC^{*t}$$
$$E = [\tfrac{1}{2}\ \tfrac{1}{2}]\begin{bmatrix} 1 & -1 \\ -1 & 1 \end{bmatrix}\begin{bmatrix} \frac{1}{2} \\ \frac{1}{2} \end{bmatrix} = 0$$

Since $E = 0$, the game is a fair game. ∎

The procedure outlined in Example 4 can be used to find the optimal strategies for any 2×2 non-strictly determined matrix game. Let

$$P = \begin{bmatrix} a & b \\ c & d \end{bmatrix}$$

be the payoff matrix and let $R^* = [p \quad 1 - p]$ be a row probability vector that represents the optimal row strategy. We compute the value of $E = R^*PC^t$ for both $C = [1\ \ 0]$ and $C = [0\ \ 1]$. Thus, if $C = [1\ \ 0]$, we have

$$E = R^*PC^t = [p \quad 1 - p]\begin{bmatrix} a & b \\ c & d \end{bmatrix}\begin{bmatrix} 1 \\ 0 \end{bmatrix}$$
$$= ap + c(1 - p)$$

and if $C = [0\ \ 1]$,

$$E = R^*PC^t = [p \quad 1 - p]\begin{bmatrix} a & b \\ c & d \end{bmatrix}\begin{bmatrix} 0 \\ 1 \end{bmatrix}$$
$$= bp + d(1 - p)$$

Equating these two expressions for E and solving for p, we obtain

$$p = \frac{d - c}{a - b - c + d}$$

from which we get

$$1 - p = \frac{a - b}{a - b - c + d}$$

Substituting these values into the assumed expression for R^* yields

$$R^* = \begin{bmatrix} \dfrac{d-c}{a-b-c+d} & \dfrac{a-b}{a-b-c+d} \end{bmatrix}$$

Next, we assume that $C^* = [q \quad 1-q]$, and by a similar series of steps show that C^* is given by

$$C^* = \begin{bmatrix} \dfrac{d-b}{a-b-c+d} & \dfrac{a-c}{a-b-c+d} \end{bmatrix}$$

The value of the game for these optimal strategies is

$$E = R^*PC^{*t} = \begin{bmatrix} \dfrac{d-c}{a-b-c+d} & \dfrac{a-b}{a-b-c+d} \end{bmatrix} \begin{bmatrix} a & b \\ c & d \end{bmatrix} \begin{bmatrix} \dfrac{d-b}{a-b-c+d} \\ \dfrac{a-c}{a-b-c+d} \end{bmatrix}$$

which simplifies, after some effort, to

$$E = \frac{ad-bc}{a-b-c+d}$$

We summarize the results for a 2×2 non-strictly determined game:

Optimal Strategies for a 2 × 2 Non-Strictly Determined Matrix Game

If $P = \begin{bmatrix} a & b \\ c & d \end{bmatrix}$ is the payoff matrix for a non-strictly determined game, then the optimal strategies are given by

$$R^* = \begin{bmatrix} \dfrac{d-c}{a-b-c+d} & \dfrac{a-b}{a-b-c+d} \end{bmatrix}$$

and

$$C^* = \begin{bmatrix} \dfrac{d-b}{a-b-c+d} & \dfrac{a-c}{a-b-c+d} \end{bmatrix}$$

The expected value E, which is called the **value** of the game, is

$$E = \frac{ad-bc}{a-b-c+d}$$

COMMENT: *These formulas cannot be used for strictly determined games. If the game is strictly determined, strange things happen. For instance, the values of p and q may turn out to be greater than 1, which is impossible since p and q are probabilities.*

Example 5 The Celtics and the Bulls are preparing for the NBA championship game. The Bulls expect to average 0.5 point per possession when Michael Jordan is in the game for the Bulls and Larry Bird is in the game for the Celtics, and 0.7 point per possession when Jordan is in the game and Bird is not. If Jordan is on the bench and Bird is in the game, the Bulls expect to average 0.6 point per possession, and if both players are on the bench they expect to average 0.4 point per possession. Treating this as a two-person matrix game,

(a) Write the payoff matrix.
(b) Find the optimal strategy for each coach.
(c) Calculate the value of the game.

Solution

(a) The payoff matrix P is

$$\begin{array}{ccc} & & \text{Larry Bird} \\ & & \begin{array}{cc}\text{In} & \text{Out}\end{array} \\ \text{Michael Jordan} & \begin{array}{c}\text{In} \\ \text{Out}\end{array} & \begin{bmatrix} 0.5 & 0.7 \\ 0.6 & 0.4 \end{bmatrix} \end{array}$$

(b) This is a non-strictly determined game with $a = 0.5$, $b = 0.7$, $c = 0.6$, and $d = 0.4$. Using the formula given above for R^*, the optimal strategy for the Bulls is

$$R^* = \begin{bmatrix} \dfrac{0.4 - 0.6}{0.5 - 0.7 - 0.6 + 0.4} & \dfrac{0.5 - 0.7}{-0.4} \end{bmatrix} = \begin{bmatrix} \dfrac{1}{2} & \dfrac{1}{2} \end{bmatrix}$$

We conclude that the Bulls coach should play Jordan one-half of the time.

Using the formula for C^*, the optimal strategy for the Celtics is

$$C^* = \begin{bmatrix} \dfrac{0.4 - 0.7}{-0.4} & \dfrac{0.5 - 0.6}{-0.4} \end{bmatrix} = \begin{bmatrix} \dfrac{3}{4} & \dfrac{1}{4} \end{bmatrix}$$

Thus, the Celtics coach should play Bird three-quarters of the time.

(c) The value for this game is

$$E = \frac{(0.5)(0.4) - (0.7)(0.6)}{-0.4} = \frac{-0.22}{-0.4} = 0.55$$

That is, by following the optimal strategies, the Bulls can expect to average 0.55 point per possession. ∎

EXERCISES SECTION 7.4

Find the value of the games in Exercises 1–10, with the given strategies.

1. $P = \begin{bmatrix} 2 & -1 \\ 3 & -2 \end{bmatrix}$
   $R = [0.3 \quad 0.7] \qquad C = [0.9 \quad 0.1]$

2. $P = \begin{bmatrix} 2 & -3 \\ 4 & -1 \end{bmatrix}$
   $R = [1 \quad 0] \qquad C = [0 \quad 1]$

3. $P = \begin{bmatrix} 2 & -3 \\ -4 & 5 \end{bmatrix}$
   $R = [0.4 \quad 0.6] \qquad C = [0.2 \quad 0.8]$

4. $P = \begin{bmatrix} -2 & -3 \\ 4 & 3 \end{bmatrix}$
   $R = [0.4 \quad 0.6] \quad C = [0 \quad 1]$

5. $P = \begin{bmatrix} 1 & 3 \\ 2 & -1 \end{bmatrix} \qquad R = [1 \quad 0] \qquad C = [1 \quad 0]$

6. $P = \begin{bmatrix} 1 & 1 & 0 \\ 5 & 6 & 2 \\ -1 & 2 & 1 \end{bmatrix}$
   $R = [0.8 \quad 0.1 \quad 0.1] \qquad C = [0.5 \quad 0.5 \quad 0]$

7. $P = \begin{bmatrix} 4 & 7 & -2 \\ 2 & -1 & -1 \\ 5 & 8 & 0 \end{bmatrix}$
   $R = [1 \quad 0 \quad 0] \qquad C = [0.5 \quad 0.5 \quad 0]$

8. $P = \begin{bmatrix} 4 & -2 & 3 \\ 3 & -2 & 1 \\ 2 & 0 & 3 \end{bmatrix}$
   $R = [0 \quad 0.5 \quad 0.5] \qquad C = [0.2 \quad 0 \quad 0.8]$

9. $P = \begin{bmatrix} 4 & -2 & 3 \\ 3 & -2 & 1 \\ 2 & 0 & 3 \end{bmatrix}$
   $R = [0 \quad 0.5 \quad 0.5] \qquad C = [1 \quad 0 \quad 0]$

10. $P = \begin{bmatrix} 1 & -1 & 0 \\ -1 & 1 & 1 \\ 1 & -1 & 1 \end{bmatrix}$
    $R = [0.5 \quad 0.5 \quad 0] \qquad C = [0 \quad 0.5 \quad 0.5]$

11. Consider the 2×2 matrix game with payoff matrix
    $P = \begin{bmatrix} 2 & 3 \\ 3 & 1 \end{bmatrix}$
    (a) Find the value of the game for $R = [0.5 \; 0.5]$, $C = [0.5 \; 0.5]$.
    (b) Find the value of the game for $R = [\frac{2}{3} \; \frac{1}{3}]$, $C = [\frac{2}{3} \; \frac{1}{3}]$.
    (c) Which is the better strategy?

12. Consider the matrix game with payoff matrix
    $P = \begin{bmatrix} 1 & 2 & 3 \\ -2 & 0 & -5 \\ 0 & 3 & -1 \end{bmatrix}$
    (a) Find the value of the game for the strategy $R = [0.2 \; 0.2 \; 0.6]$, $C = [0 \; 1 \; 0]$.
    (b) Find the value of the game for $R = [0 \; 1 \; 0]$ and $C = [0 \; 1 \; 0]$.
    (c) Compare the value of the game at its saddle point with its value given in (a).

In Exercises 13–20, the 2×2 matrices represent two-person games. In each game, find
(a) The optimal strategy for the row player.
(b) The optimal strategy for the column player.
(c) The value of the game.
(Note: Some of these games are strictly determined. In this case, give the strategy and value of the game.)

13. $\begin{bmatrix} 1 & -2 \\ -2 & 1 \end{bmatrix}$

14. $\begin{bmatrix} -1 & 2 \\ -2 & 1 \end{bmatrix}$

15. $\begin{bmatrix} 1 & -2 \\ 0 & 3 \end{bmatrix}$

16. $\begin{bmatrix} -1 & 2 \\ 3 & 1 \end{bmatrix}$

17. $\begin{bmatrix} -2 & -3 \\ 4 & 3 \end{bmatrix}$

18. $\begin{bmatrix} -1 & 0 \\ -2 & -1 \end{bmatrix}$

19. $\begin{bmatrix} \frac{1}{2} & \frac{1}{2} \\ \frac{2}{3} & \frac{1}{3} \end{bmatrix}$

20. $\begin{bmatrix} -\frac{1}{2} & \frac{1}{2} \\ \frac{1}{2} & -\frac{1}{2} \end{bmatrix}$

21. $\begin{bmatrix} 2 & -3 \\ 4 & -1 \end{bmatrix}$

22. $\begin{bmatrix} 2 & -3 \\ -4 & 5 \end{bmatrix}$

23. Two football coaches are planning their strategies for the big game on Saturday. If Coach A plans a passing attack and Coach B plans to defend against a passing attack, then A can expect to gain three yards per play. If A plans a passing attack and B decides to defend against a running attack, then A can expect to gain ten yards per play. If A plans a running attack and B plans to defend against a running attack, then A can expect to gain two yards per play. Finally, if A plans a running attack and B decides to defend against a passing attack, then A can expect to gain four yards per play. Treating this situation as a two-person matrix game,
(a) Write the payoff matrix.
(b) Find the optimal strategy for each coach.
(c) Find the expected yardage per play.

24. Two basketball coaches are planning their strategies for the final game in the NCAA tournament. If Coach A uses a three-guard offense and Coach B uses a zone defense, then A can expect to score 0.8 point per possession. If A uses a double-post offense and B uses a zone defense, then A can expect to score 0.7 point per possession. If A uses a three-guard offense and B uses a man-to-man defense, then A can expect to score 0.5 point per possession. If A uses a double-post offense and B uses a man-to-man defense, then A can expect to score 1.1 points per possession. Treating this as a two-person matrix game,
(a) Write the payoff matrix.
(b) Find the optimal strategy for each coach.
(c) Find the expected points per possession.

25. Suppose you are playing a game in which you hide a nickel under one cup and a dime under another. After you have mixed them up, your opponent is to pick a cup and guess the coin's value. If your opponent is correct, you pay your opponent the value of the coin under the cup; if your opponent is wrong, he or she pays you the value of the coin under the cup.
(a) Write the payoff matrix.
(b) Find the optimal strategy for each player.
(c) Find the value of the game.

26. Tom and Ann play a game in which, at the same time, each player shows either one or two fingers. If there is a match, Tom wins as many dollars from Ann as there are fingers showing. If there is no match, Ann wins as many dollars from Tom as there are fingers showing.
(a) Write the payoff matrix.
(b) Find the optimal strategy for each player.
(c) Find the value of the game.

IMPORTANT WORDS AND PHRASES CHAPTER 7

The terminology of this chapter is "modern" in the sense that most of these terms were unheard of 100 years ago.

expected value of a game
game theory
  value of a game
Markov chain
  regular Markov chain
matrix games
  non-strictly determined game
  strictly determined game
  zero-sum games
payoff matrix
state vector
  equilibrium-state vector
  initial-state vector
strategy
  mixed
  optimal
  pure
transition matrix
  regular transition matrix

REVIEW EXERCISES CHAPTER 7

In Exercises 1 and 2, find the next state vector for the given state vector.

$$\begin{bmatrix} 0.2 & 0.8 \\ 0.9 & 0.1 \end{bmatrix}$$

1. $[0.6 \quad 0.4]$ **2.** $[0.2 \quad 0.8]$

In Exercises 3 and 4, find the next state vector for the given state vector.

$$\begin{bmatrix} 0.1 & 0.1 & 0.8 \\ 0.7 & 0.2 & 0.1 \\ 0 & 0.7 & 0.3 \end{bmatrix}$$

3. $[0.2 \quad 0.3 \quad 0.5]$ **4.** $[0.1 \quad 0.7 \quad 0.2]$

5. A video game will show two different targets on the screen, but only one at a time. If Target 1 is on the screen when a game ends, there is a probability of 0.3 that it will remain on the screen to start the next game and a probability of 0.7 that the screen will change to Target 2 at the start of the next game. If Target 2 is on the screen when a game ends, there is a probability of 0.5 that it will remain on the screen and a probability of 0.5 that the screen will change to Target 1. Write the transition matrix for this game.

6. A video game shows a frog sitting on one of two lily pads in a pond. If the frog is sitting on Pad 1 when the game starts, there is a probability of 0.3 that it will remain on Pad 1 and a probability of 0.7 that it will jump to Pad 2. If the frog is on Pad 2 when the game starts, there is a probability of 0.6 that it will remain on Pad 2 and a probability of 0.4 that it will jump to Pad 1. Write a transition matrix for the state of the frog.

7. A reporter for *Political News* magazine surveys voters in the last presidential election and finds that 65% of those who voted for the Republican candidate say they will do so again, 25% say they will vote for the Democratic candidate next time, and 10% say they will not vote next time. Of those who voted for the Democratic candidate in the last election, 70% say they will vote the same way next time, 15% say they will vote for the Republican candidate, and 15% say they will not vote. Finally, 25% of those who did not vote in the last election say they will vote for the Republican candidate, 20% say they will vote for the Democratic candidate, and 55% say they will not vote.
(a) Write the transition matrix for voter preference.
(b) Write an initial-state vector to reflect the fact that of those surveyed, 45% voted Republican, 40% voted Democrat, and 15% did not vote.
(c) Find the percentage in each category after the next election.

8. A sociologist finds that there is a constant movement of people between the poor class, the middle class, and the wealthy class. Census data show that 80% of the poor in 1980 remained poor in 1990, 19% moved up to middle class, and 1% moved to the wealthy class. Of those classified as middle class in 1980, 70% remained middle class in 1990, 10% dropped into the poor class, and 20% moved up to the wealthy class. Finally, 92% of the wealthy remained in that class, 7% dropped to middle class, and 1% dropped to the poor class.
(a) Write the transition matrix for this process.
(b) Write the initial-state vector to reflect the fact that of those considered in the study, 10% were from the wealthy class, 50% from the middle class, and 40% from the poor class.
(c) Find the percentage in each class when the 2000 census is taken.

In Exercises 9–12, explain why each transition matrix is regular.

9. $\begin{bmatrix} 0 & 1 \\ 0.3 & 0.7 \end{bmatrix}$ **10.** $\begin{bmatrix} 0.4 & 0.6 \\ 1 & 0 \end{bmatrix}$

11. $\begin{bmatrix} 0.3 & 0.7 & 0 \\ 0.3 & 0.4 & 0.3 \\ 0 & 0.8 & 0.2 \end{bmatrix}$ **12.** $\begin{bmatrix} 0 & 0.4 & 0.6 \\ 0.1 & 0.4 & 0.5 \\ 0.5 & 0.1 & 0.4 \end{bmatrix}$

In Exercises 13–16, use Theorem 7.3 to find the equilibrium-state vector for each regular transition matrix.

13. $\begin{bmatrix} 0.5 & 0.5 \\ 0.2 & 0.8 \end{bmatrix}$ **14.** $\begin{bmatrix} 0.2 & 0.8 \\ 0.6 & 0.4 \end{bmatrix}$

15. $\begin{bmatrix} 0.1 & 0.3 & 0.6 \\ 0.3 & 0.4 & 0.3 \\ 0.4 & 0.4 & 0.2 \end{bmatrix}$ **16.** $\begin{bmatrix} 0.4 & 0.6 & 0 \\ 0.2 & 0.1 & 0.7 \\ 0.7 & 0 & 0.3 \end{bmatrix}$

17. A psychologist studies the traffic pattern of people as they enter a department store by one of three doors and then leave the store by one of the same three doors. As a result of the study, the psychologist sets up the following transition matrix to represent the process:

		Exit Door 1	Exit Door 2	Exit Door 3
Enter	Door 1	0.7	0.2	0.1
	Door 2	0.4	0.3	0.3
	Door 3	0.2	0.6	0.2

Use Theorem 7.3 to find the percentage of customers using each door.

18. A car rental company has offices in Atlanta, Birmingham, and Chattanooga. A person renting a car in one of the three cities can return it to an office in any one of the three. An accountant in the controller's office of the company reports that 85% of the rentals from Atlanta are returned in Atlanta, 10% are returned in Birmingham, and 5% are returned in Chattanooga. From Birmingham, 75% of the rentals are returned in Birmingham, 12% in Atlanta, and 13% in Chattanooga. Of those cars rented in Chattanooga, 90% are returned in Chattanooga, 5% in Birmingham, and 5% in Atlanta. Set up a transition matrix for this process. Use Theorem 7.3 to find the percentage of cars that will end up in each city if the company does not transfer cars between cities.

19. You have the opportunity to choose between two sales jobs. If you take the job in Chicago, there is a 50% chance that you will earn \$45,000 and a 50% chance that you will earn \$30,000. If you take the job in Los Angeles, there is a 70% chance that you will earn \$20,000 and a 30% chance that you will earn \$60,000. If you base your decision on expected value, which job should you take?

20. A contractor bids on the job of constructing a new motel. There is a probability of 0.8 that the profit on the job will be \$25,000, a probability of 0.1 that there will be no profit, and a probability of 0.1 that there will be a loss of \$10,000 (due to bad weather, strikes, and so on). Calculate the expected profit to the contractor.

In Exercises 21–26, identify each matrix game as strictly determined or non-strictly determined. If the game is strictly determined, indicate the saddle point(s). If the game is non-strictly determined, find the optimum strategy for both players and the value of the game.

21. $\begin{bmatrix} 2 & -3 \\ -3 & 2 \end{bmatrix}$ **22.** $\begin{bmatrix} -1 & 3 \\ -3 & 1 \end{bmatrix}$

23. $\begin{bmatrix} 2 & -2 \\ 0 & 3 \end{bmatrix}$ **24.** $\begin{bmatrix} -1 & 3 \\ 2 & 1 \end{bmatrix}$

25. $\begin{bmatrix} 3 & -3 \\ 4 & -2 \end{bmatrix}$ **26.** $\begin{bmatrix} 3 & -3 \\ -5 & 5 \end{bmatrix}$

In Exercises 27 and 28, find the saddle point(s) for each strictly determined matrix game.

27. $\begin{bmatrix} 4 & 7 & -2 \\ 2 & -1 & -1 \\ 5 & 8 & 0 \end{bmatrix}$ **28.** $\begin{bmatrix} 1 & 1 & 0 \\ 5 & 6 & 2 \\ -1 & 2 & 1 \end{bmatrix}$

29. Bank 1 and Bank 2 have offices in Detroit and each plans to expand into one of the following three cities: Ann Arbor, East Lansing, and Royal Oak. If Bank 1 opens branches in Ann Arbor and Bank 2 also opens branches in Ann Arbor, Bank 1 stands to break even; if Bank 1 opens branches in Ann Arbor and Bank 2 opens branches in East Lansing, Bank 1 stands to lose \$500,000; and if Bank 1 opens branches in Ann Arbor and Bank 2 opens branches in Royal Oak, Bank 1 stands to lose \$200,000. If Bank 1 opens branches in East Lansing and Bank 2 opens branches in Ann Arbor, Bank 1 stands to lose \$400,000, and so on, as shown in the payoff matrix below. The entries in the payoff matrix represent revenue in millions of dollars to Bank 1.

		Bank 2 Ann Arbor	Bank 2 East Lansing	Bank 2 Royal Oak
Bank 1	Ann Arbor	0	−0.5	−0.2
	East Lansing	−0.4	−1.2	0.2
	Royal Oak	0.5	0.6	0.8

Find the optimum strategy for each bank.

30. Two football coaches are planning their strategies for the conference title game on Saturday. If Coach A plans a passing attack and Coach B plans to defend against a passing attack, then *A* can expect to gain two yards per play. If *A* plans a passing attack and *B* decides to defend against a running attack, then *A* can expect to gain eight yards per play. If *A* plans a running attack and *B* plans to defend against a running attack, then *A* can expect to lose one yard per play. Finally, if *A* plans a running attack and *B* decides to defend against a passing attack, then *A* can expect to gain three yards per play. Treating this situation as a two- person matrix game,

(a) Write the payoff matrix.

(b) Find the optimum strategy for each coach.

CHAPTER 8

THE MATHEMATICS OF FINANCE

The heart of our economic system is the payment of interest for the temporary use of someone else's money. This can be viewed in two ways: (1) A bank pays us interest when we have money in a savings account, or the federal government pays us interest on savings bonds (technically, the bank or the federal government has borrowed our money), and (2) we pay interest to the bank when we borrow money from the bank to buy a car. In this chapter we discuss various ways of computing interest on savings and on debts.

8.1 Simple Interest

Money paid for the use of another person's or institution's money is called **interest.** A bank may pay you interest if you invest, or deposit, money in the bank. Or the bank may charge you interest if you borrow money from the bank. The amount of money invested, or deposited, is called the **principal,** and the money borrowed is called a **loan.** The terms *principal* and *value of the loan* are used interchangeably, depending upon the viewpoint from which the financial transaction is considered.

The interest paid on an amount of money is determined by the rate of interest. For a given unit of time, the **rate of interest** is the ratio of the interest paid to the principal. Thus, the rate is given by

$$\text{rate of interest} = \frac{\text{interest}}{\text{principal}}$$

The rate of interest is expressed as a percentage, and is usually quoted as an annual, or yearly, rate. For instance, your savings account might pay interest at a rate of 5.6% per year. It is common practice, where it is understood that the interest rate is yearly, to drop the phrase *per year* and simply give the rate. Interest rates for other time periods are easily determined if the annual rate is known. For example, the monthly rate is calculated by dividing the annual rate by 12; the quarterly rate, by dividing the annual rate by 4; and the daily rate, by dividing the annual rate by 360 or 365. Some financial institutions use 360 to simplify their calculations; others use 365. We will use 365 for daily interest calculations. The interest itself is computed in one of two ways: simple and compound.

COMMENT: *All interest rates used in the examples and problems in this text are assumed to be* **annual rates** *unless stated otherwise.*

Simple interest, I_s, on a principal (or loan) is found by multiplying the principal P by the annual rate R and the number of years t.

Simple Interest

The formula for computing the amount of **simple interest** I_s is

$$I_s = PRt \tag{8.1}$$

where P = principal (amount borrowed or loaned)
R = interest rate per year (annual rate)
t = time in years

Example 1 How much simple interest does $1000 earn in one year if the interest rate is 8.5%? How much does it earn in three years?

Solution The principal is $P = \$1000$, and the interest rate is $R = 8.5\%$, or 0.085. [Note that the interest rate must be expressed as a decimal fraction before it is substituted in Formula 8.1.] For one year, $t = 1$, so the interest is

$$\begin{aligned} I_s &= PRt \\ &= 1000(0.085)(1) \\ &= 85 \end{aligned}$$

The amount of interest earned in one year is $85.

The amount of interest earned in three years is found by letting $t = 3$ in Formula 8.1.

$$\begin{aligned} I_s &= PRt \\ &= 1000(0.085)(3) \\ &= 255 \end{aligned}$$

The amount of interest earned in three years is $255, which is three times the interest earned in one year. ■

Example 2 Compute the simple interest on a $10,000 note that is held for five years at 8% paid quarterly. How much interest is paid each quarter?

Solution Even though the interest is paid quarterly, the total interest paid during the five years is still determined by

$$I_s = PRt = \$10{,}000 \times 0.08 \times 5 = \$4000$$

The amount of interest paid each quarter is the total interest divided by the total number of quarters. In this case there are $5 \times 4 = 20$ quarters.

$$\frac{I_s}{\text{no. of quarters}} = \frac{\$4000}{20} = \$200$$

■

Example 3 Jim earns \$216 simple interest on a \$5000 investment during the first half of the year. What is his annual rate of interest?

Solution Using the formula for interest with $I_s = \$216$, $P = \$5000$, and $t = \frac{1}{2}$, we obtain, after solving Formula 8.1 for R:

$$R = \frac{I_s}{P \cdot t} = \frac{216}{5000(\frac{1}{2})} = 0.0864$$

Thus, the annual interest rate is $R = 8.64\%$. ■

Example 4 Jane has an opportunity for a simple-interest investment at 7.9%. Determine how much she should invest for two years if she wishes to earn \$5000 in interest.

Solution Here we know three of the four variables in Formula 8.1, namely $I_s = 5000$, $R = 0.079$, and $t = 2$. Solving Formula 8.1 for P, we have

$$P = \frac{I_s}{R \cdot t} = \frac{5000}{0.079 \cdot 2} = 31{,}645.57$$

Thus, the required investment is \$31,645.57. ■

Underlying the simple interest formula (Formula 8.1) is the assumption that the principal does not change. Thus, money invested for two years at simple interest earns exactly twice as much as it would in one year. Distinguish carefully between the principal and the total value of the investment at the end of t years. The latter amount, called the **future value** of the principal, is equal to the sum of the principal and the interest earned and is given by

$$S_t = P + I_s$$

Since, for simple interest, $I_s = PRt$, we get

$$S_t = P + PRt$$

By factoring the right-hand side of this expression, we get the future-value formula.

Simple Interest—Future Value

The **future value** S_t of principal P, if simple interest is used, is

$$S_t = P(1 + Rt) \qquad \textbf{(8.2)}$$

where P = principal

R = interest rate per year

t = number of years

The term *future value* applies both to the amount of an investment and to the amount of a loan.

Example 5 The Portland Cement Company borrows \$250,000 for three months at 12.4% simple interest. Determine how much money the company must repay.

Solution We use Formula 8.2 with $P = 250{,}000$, $R = 0.124$, and $t = \frac{3}{12} = 0.25$. Thus,

$$\begin{aligned} S_{1/4} = P(1 + Rt) &= 250{,}000[1 + 0.124(0.25)] \\ &= 250{,}000(1.031) \\ &= 257{,}750 \end{aligned}$$

The total amount the company must repay is \$257,750. ▪

Loans Using Simple Interest

Loan companies or individuals typically use two different methods to compute payments on a loan using simple interest: add-on loans and discounted loans. The amount of money paid in each time period to repay a loan is called an **installment.**

Add-On Loans

In an **add-on loan,** the future value of the principal is calculated for the period of the loan to determine the total amount to be repaid. The size of the installment, L, is then determined by dividing the total amount to be repaid by the number of payments, k.

$$L = \frac{P + I_s}{k} = \frac{P + PRt}{k} = \frac{P(1 + Rt)}{k}$$

Installment Payments for Add-On Loans

The amount L of the installment payments for add-on loans is

$$L = \frac{P(1 + Rt)}{k} \quad \textbf{(8.3)}$$

where P = principal
R = interest rate per year
t = number of years
k = number of payments

Example 6 Nina buys a used car for \$4000 and finances it through a 14% simple-interest add-on loan. She wishes to pay back the loan in monthly payments over a period of three years.

(a) How much will her monthly payments be?
(b) What is the total cost of the car?

Solution

(a) Using Formula 8.3 with $P = 4000$, $R = 0.14$, $t = 3$, and $k = 36$, we have

$$L = \frac{4000(1 + 0.14 \cdot 3)}{36} = 157.78$$

Thus, Nina's monthly payments will be \$157.78.

(b) The total cost of the car is $36 \times 157.78 = \$5680.08$. ▪

Discounted Loans

Another method of loaning money, usually used for short-term loans, is called **discounting.** In a discounted loan, the simple interest is computed as usual, but is then subtracted from the principal before the remainder is transferred to the borrower. The amount the borrower receives, P_D, is called the proceeds of the discounted loan. The **lender's discount** is the interest, I_D, that is deducted from the principal. So the amount of the proceeds is $P_D = P - I_D = P - PRt = P(1 - Rt)$.

Proceeds of a Discounted Loan

The **proceeds** P_D for a discounted loan are

$$P_D = P(1 - Rt) \tag{8.4}$$

where P = principal
R = interest rate per year
t = number of years

Example 7 Suppose Chuck wants to borrow \$1000 for three months at a 16% discount rate.

(a) What is the bank discount?
(b) How much money does Chuck actually receive?

Solution

(a) Here, $P = \$1000$, $R = 0.16$, and $t = \frac{3}{12} = \frac{1}{4}$ year. The lender's discount on the money is

$$I_D = PRt = 1000(0.16)(\tfrac{1}{4}) = \$40$$

(b) The amount of money that Chuck receives is

$$P_D = \$1000 - \$40 = \$960$$ ▪

Example 8 The proceeds of Chuck's loan were actually \$960. What was the actual, or "effective," annual interest rate?

Solution Recall that the rate of interest for a given period of time is the ratio of the interest to the principal. However, in a discounted loan the borrower does not receive the principal, but the proceeds. So the actual interest rate for the term of the loan is

$$\text{rate per quarter} = \frac{I_D}{P_D} = \frac{40}{960} = 0.0417$$

Since there are four quarters per year, the annual rate is

$$\text{rate per year} = 4 \times 0.0417 = 0.1668$$

Therefore, the actual interest rate on the loan is 16.68%. ■

COMMENT: *Note that the actual, or effective, annual interest rate is higher than the stated discount rate.*

Example 9 Suppose Chuck really needed \$1000 and not a penny less! How much would he have to borrow at the same 16% discount rate?

Solution In this instance, $P_D = \$1000$, $R = 0.16$, and $t = \frac{1}{4}$. Solving Formula 8.4 for P, we have

$$P = \frac{P_D}{1 - Rt} = \frac{1000}{1 - (0.16)(\frac{1}{4})} = \$1041.67$$

■

EXERCISES SECTION 8.1

1. If \$3000 is invested at an annual simple-interest rate of 6% for two years, how much interest will be earned, and what will be the amount of the investment at the end of that time?
2. If \$500 is invested at 8% annual simple interest for four years, how much interest will be earned and what will be the amount of the investment at the end of that time?
3. Julie invests \$25,000 in a 9.2% simple-interest certificate, payable in five years. How much interest does the investment earn, and how much is it worth in five years?
4. A couple encourages their 15-year-old son to invest his \$1400 in a 3-year 7.9% simple-interest certificate. How much interest does it earn, and how much is the certificate worth at maturity?
5. Pat borrows \$5200 at 8.6% simple interest. How much does he have to repay in 30 months? How much of this repayment is interest?
6. Jim borrows \$9800 and promises to repay it in eight months. If the simple-interest rate is 12.05%, how much must he repay at that time?
7. Suppose a company needs \$15,000 for some new equipment but wants to repay the loan relatively quickly, in 18 months. If the simple interest is 9%, how much must they repay after 18 months? How much is the interest charge?
8. Sam buys a car for \$10,000 with a down payment of \$4200. Instead of monthly payments, he finances the rest with a simple-interest loan of 10.5% for nine months. How much must he pay at the end of that time?
9. A couple decides to buy a \$145,000 home using the proceeds of the sale of their present home. They estimate it will take 60 days to sell their home. The bank offers them a "bridge" loan for this period, which is a simple-interest loan on the \$145,000. If the interest rate is 11.4%, how much will they actually have to repay after the 60 days?
10. A college student needs \$2200 to pay tuition costs and arranges for a loan of that amount at 16.3% simple interest, to be repaid in 90 days. How much will the student have to repay at that time? How much of that repayment is interest?

11. A company paid a \$46.20 annual dividend on stock worth \$500. What was the simple-interest rate of earnings for that stock?

12. Suppose stock in a newspaper company, valued at \$1500, pays an annual dividend of \$154.20, while stock in the local power company, valued at \$2500, pays a dividend of \$261.43. If your only measure of stock quality is interest rate paid, which is the better buy?

13. To buy a certain kind of \$1000 U. S. Bond, you can invest a certain amount now and the bond will mature in 4.25 years at 7.2% simple interest. How much would you invest now to obtain that bond?

14. Refer again to the U. S. Bonds in Exercise 13. Sarah and Jim wish to give their newborn granddaughter a \$10,000 U. S. Bond that will mature after 18 years. If the simple interest rate is 7.1%, how much will Sarah and Jim have to invest now in order to buy that bond?

15. A student has \$5000 in the bank earning 9.2% simple interest. A tuition-and-fees bill is due on January 1 for that amount. A late fee of \$40 is charged for any payments made between January 1 and January 31. Should she pay the bill on January 1 or January 31? Assume the time is in days.

16. The taxes on your home are due on January 15, and add up to \$750. Payments made after that time are assessed a late fee of 0.5% per month for up to three months. If you have the money earning 15% simple interest, is it best for you to pay immediately, or to wait the three months and pay the \$750 plus the penalty?

17. Compute the monthly installment for an add-on loan of \$8000 at 12.3% simple interest, to be repaid in 60 months.

18. Diane buys a car for \$13,500 and finances it through an add-on loan at 15% simple interest. How much are her monthly payments if she wishes to repay within four years? How much interest does she pay on this loan?

19. Diane borrows \$13,500 to buy a car. She finances it for five years with an add-on loan at 15% simple interest. What are Diane's monthly payments?

20. Paul wants to buy a new car for \$12,000. He wishes to finance it with an add-on loan at 15.3% simple interest for 40 months. If he can afford a monthly payment of \$250, how much of a down payment must he make?

21. Beth needs \$4500 for her trip to Africa. She can borrow money at 14.3% simple interest. She calculates that she can afford monthly installments of \$400 for the coming year. How much of her own money does she need to finance her trip?

22. In Exercise 21, what interest rate would Beth have to get in order not to have to make any down payment?

23. Suppose you borrow \$18,000 at 12.2% for nine months on a discounted loan. Compute:
(a) The bank discount on the loan.
(b) The amount that you receive.
(c) The effective interest rate.

24. A store owner borrows \$28,000 at 12% for six months on a discounted loan.
(a) What are the proceeds of the loan?
(b) What is the bank discount?
(c) What is the effective rate of interest?

25. Paula wants to buy an \$8900 car and has \$900 in ready cash. She wants to take out a one-year discounted loan, for which the immediate proceeds are \$8000. What should the principal of the loan be if the interest rate is 15%?

8.2 COMPOUND INTEREST

By far the most popular method of computing interest on home mortgages, on car loans, on savings accounts, or on certificates of deposit is a process called **compounding.** Recall that with simple interest, the amount of interest earned on \$1000 for a five-year period at 9% is $1000(0.09)(5) = $450. This is the total amount of interest earned on the $1000 whether the interest is payable every year, every quarter, or every month. With compound interest, the total amount of interest earned on $1000 over a five-year period at 9% will change depending on how often the interest is compounded, because interest is paid both on the principal and on the previously earned interest. The value of an investment after n interest periods, denoted by A_n, is called the **future value.** The difference between the original principal and the future value is called the **compound interest,** which we will denote by I_c.

To understand the nature of compound interest, let us consider the specific example of an investment of $1000 at an annual interest rate of 9% for five years, compounded annually. *Compounded annually* means that the interest is computed and added to the principal at the end of each year. The compounding time period is the same as the interest-payment period. Thus, *daily compounding* means that the interest is computed and added to the account at the end of every day; *compounded monthly* means that the interest is computed and added to the account at the end of every month. For a $1000 principal, the interest at the end of the first year is

$$I = \$1000(0.09)(1) = \$90$$

Then the total amount in the account at the end of one year is

$$A_1 = \$1000 + \$90 = \$1090$$

which is the same amount that would be in the account if simple interest were used.

In compound interest, to compute the interest for the second year we use A_1 for the principal, *not* the original principal P. Thus, the interest paid at the end of the second year is

$$I = \$1090(0.09)(1) = \$98.10$$

and the total amount in the account at the end of the second year is

$$A_2 = \$1090 + \$98.10 = \$1188.10$$

Notice that if simple interest were used, the interest would be computed on the original principal, P, so that the interest earned would again be $90, the same as the first year. Then the amount S_2 in the account after two years of simple interest would be

$$S_2 = \$1090 + \$90 = \$1180$$

Continuing in the same way, we determine the amount of compound interest earned in the third year; that is,

$$I = \$1188.10(0.09)(1) = \$106.93$$

Adding this to A_2, we have

$$A_3 = \$1188.10 + \$106.93 = \$1295.03$$

In the same way, the interest at the end of the fourth year is

$$I = \$1295.03(0.09)(1) = \$116.55$$

and

$$A_4 = \$1295.03 + \$116.55 = \$1411.58$$

Finally, the interest at the end of the fifth year is

$$I = \$1411.58(0.09)(1) = \$127.04$$

and

$$A_5 = \$1411.58 + \$127.04 = \$1538.62$$

The chart below compares the amounts A_n and S_n for the first five years.

TABLE 8.1 Comparison of Simple and Compound Interest for an Investment of \$1000 for Five Years at 9%

End of Year	Simple Interest I_s	Principal + Interest S_n	Compound Interest I_c	Principal + Interest A_n
1	90	1090	90.00	1090.00
2	90	1180	98.10	1188.10
3	90	1270	106.93	1295.03
4	90	1360	116.55	1411.58
5	90	1450	127.04	1538.62

Notice that the simple interest remains the same for each of the five years, while the compound interest increases each year. The total simple interest over five years is \$450; the total compound interest earned is \$538.63.

To develop a general formula for the compound amount, A_n, suppose a principal P is invested at an interest rate i *for each period*, compounded over n periods. The total amount A_1 at the end of the first interest period is

$$A_1 = P + iP = P(1 + i)$$

At the end of the second period, the total amount is

$$A_2 = A_1 + iA_1 = P(1 + i) + iP(1 + i) = P(1 + i)^2$$

At the end of the third period, the total amount is

$$A_3 = A_2 + iA_2 = P(1 + i)^2 + iP(1 + i)^2 = P(1 + i)^3$$

By continuing this process for n interest periods, we obtain the formula for the compound amount:

Compound Interest—Future Value

The compound amount, or **future value,** A_n for a given principal is

$$A_n = P(1 + i)^n \qquad \textbf{(8.5)}$$

where $P =$ principal

$i =$ interest rate for each interest period

$n =$ number of interest periods

From the compound-interest formula we can easily compute the compound interest after n periods. Using Formula 8.5 we obtain the formula for the amount of compound interest I_c as follows:

$$\begin{aligned} I_c &= A_n - P \\ &= P(1 + i)^n - P \\ &= P[(1 + i)^n - 1] \end{aligned}$$

Amount of Compound Interest

The formula for computing the amount of **compound interest** I_c is

$$I_c = P[(1 + i)^n - 1] \qquad \textbf{(8.6)}$$

where $P =$ principal

$i =$ interest rate for each interest period

$n =$ number of interest periods

COMMENT: *Formulas 8.5 and 8.6 are not usually treated independently. Either A_n or I_c would be computed from 8.5 or 8.6, respectively, and then the remaining quantity determined from the simple relation $P + I_c = A_n$.*

The value of i used in Formulas 8.5 and 8.6, and in any formula that uses the compound-interest principle, **must be the interest rate for one interest period.** Remember, rates are generally quoted on an annual basis, and therefore must be adjusted to reflect the compounding period. In general, if R is the annual interest rate (also called the nominal interest rate) and m is the number of interest periods per year, then interest rate per period, i, is given by

$$i = \frac{\text{annual interest rate}}{\text{number of interest periods per year}} = \frac{R}{m}$$

COMMENT: *There is one possible exception to the above formula. When the compounding is done daily, some banks compute the daily interest rate at the rate of* $i = \frac{R}{360}$ *instead of* $i = \frac{R}{365}$. *The reason for this is historical. In the days before easy access to calculators and computers, it was easier to do the various computations with a divisor of 360 rather than 365. Of course, such difficulties no longer exist, but the custom of using* $i = \frac{R}{360}$ *instead of* $i = \frac{R}{365}$ *continues in some places. The number of days used is either 360 or 365—exceptions are not made for leap year! In this book, when compounding is done daily, we will assume* $i = \frac{R}{365}$.

Example 1 A sum of money is invested at 12% compound interest for three years. Determine the values of i and n to be used in Formulas 8.5 and 8.6 if the interest is to be compounded:

(a) Semiannually.
(b) Quarterly.

Solution We note that the interest rate of 12% is assumed to be the annual rate.

(a) Semiannual compounding means two interest periods per year, one every six months. Thus,

$$i = \frac{0.12}{2} = 0.06 \qquad \text{and} \qquad n = 2(3) = 6$$

(b) Quarterly compounding means four interest periods per year, one every three months. Thus,

$$i = \frac{0.12}{4} = 0.03 \qquad \text{and} \qquad n = 4(3) = 12$$

■

Example 2

(a) Compute the amount of an investment of $10,000 after five years if the interest is 8% compounded quarterly.
(b) What is the total interest earned?
(c) Compare this to the same investment with simple interest.

Solution The quarterly interest rate is $i = \frac{0.08}{4} = 0.02$. The number of interest periods is $n = 5 \cdot 4 = 20$.

(a) To find the compound amount, use Formula 8.5 with these values of i and n and $P = 10{,}000$. Thus,

$$A_{20} = 10{,}000(1 + 0.02)^{20}$$

Using a calculator, we find that the future value of \$10,000 is \$14,859.47.

(b) The interest earned is $I_c = A_{20} - P$, or

$$I_c = \$14{,}859.47 - \$10{,}000 = \$4859.47$$

(c) In Example 2 of Section 8.1, we determined the simple interest on such an investment for five years to be \$4000. There is \$859.47 more interest earned over the five years with quarterly compounding than with simple interest. ▪

Closely related to the concept of future value is the present value of an amount A_n. The **present value** P of an amount A_n is the principal that must be invested today in order to equal A_n after n interest periods. To determine present value, we solve for P in the future-value formula, Formula 8.5, by dividing both sides of the equation by $(1 + i)^n$.

Compound Interest—Present Value

The formula for computing the **present value** P of a given amount is

$$P = A_n(1 + i)^{-n} \qquad \textbf{(8.7)}$$

where A_n = future value of the principal
i = interest rate for each interest period
n = number of interest periods

Example 3 Sylvia wants to make a substantial down payment on a car in five years. She can get 10% interest compounded monthly on her money now. How much should she invest now in order to have \$10,000 in five years for the down payment?

Solution Here, $i = \dfrac{0.10}{12}$, $n = 60$, and $A_{60} = 10{,}000$. Therefore,

$$P = \left(1 + \frac{0.10}{12}\right)^{-60} \times 10{,}000 = \$6077.89$$

Therefore, Sylvia should invest a little over \$6000 now to have the necessary down payment in five years. ▪

WARNING: For the examples and exercises in this text, we will not round off or use approximations when entering the numbers into the calculator. For instance, if you use $i = \dfrac{0.10}{12} \approx 0.0083$ *in the above example instead of* $i = \dfrac{0.10}{12}$, *the amount you get for P is \$6089.95.*

When financial institutions report the annual interest rate they are paying on savings accounts they also often include another rate, called the annual yield. The **annual yield** is the interest rate that would be paid to give the same amount of interest if the compounding period were annual. An annual rate of 6.5% compounded daily, for example, has an annual yield of 6.72%. To determine the annual yield for daily compounding, we consider the interest on $1 from a savings account which nominally pays i% compounded daily:

$$\text{annual yield} = \frac{\text{compound interest}}{\text{principal}} = \frac{I_c}{1}$$

Using Formula 8.6 we obtain the formula for annual yield:

Annual Yield

The formula for the annual yield is

$$\textbf{Annual Yield} = (1 + i)^n - 1 \qquad (8.8)$$

where i = interest rate for each interest period
n = number of periods per year

Table 8.2 gives values for annual yield for some typical interest rates compounded daily. You will notice these and other values on the financial pages of your daily newspaper.

TABLE 8.2 Annual Yield for Interest Rates Compounded Daily

Annual Rate	Annual Yield
5.5	5.65
6.5	6.72
7.0	7.25
7.5	7.79
8.0	8.33
9.5	9.96
10.0	10.52
10.5	11.07

COMMENT: *The values shown in Table 8.2 are computed by comparing the yield of daily compounding with that of annual compounding. A table could also be constructed comparing monthly or quarterly compounding with annual compounding, but this kind of table is not usually done.*

EXERCISES SECTION 8.2

In Exercises 1–6, compute the future value of each amount after one year if the interest is compounded monthly for one year. Give the compound interest earned and compare it to simple interest.

1. \$3000 at 6%.
2. \$500 at 8%.
3. \$4000 at 6.5%.
4. \$4200 at 8.5%.
5. \$10,000 at 9.462%.
6. \$10,000 at 11.012%.

In Exercises 7–10, compute the present value of the given amount at the end of one year if the interest is compounded monthly.

7. \$10,000 at 8.6%.
8. \$5000 at 10%.
9. \$8500 at 9.5%.
10. \$120,000 at 7.8%.

In Exercises 11–15, determine the present value of each amount.

11. \$10,000 at 9% compounded annually for five years.
12. \$2000 at 8% compounded monthly for two years.
13. \$5000 at 7% compounded quarterly for two years.
14. \$10,000 at 9% compounded daily for one year.
15. \$2000 at 8% compounded daily for 90 days.
16. Using Formula 8.8, compute the values in Table 8.2.
17. Tim decides to invest \$500 for five years. He has two choices: 10% simple interest or 9% compounded monthly. Which is his best investment?
18. Kim takes out a \$2000 loan from a bank at 15% compounded monthly. How much does she owe the bank after 30 months?
19. A couple takes out a five-year certificate of deposit for \$10,000. The terms of the certificate are 7% semiannual compounding. What is the value of the CD at maturity?
20. Suppose you have \$2000 to invest for five years. Compare the following terms:
    (a) 10.5% simple interest.
    (b) 10% interest, compounded annually.
    (c) 9.8% interest, compounded semiannually.
21. At the birth of their son, Mr. & Mrs. Profitt invest \$1000 for him at 8% compounded quarterly. How much will this investment be worth when their son is 21 years old?
22. When the Profitts' child is eight years old, they decide that the child will need \$25,000 for his college education 10 years later. How much should they invest in a trust fund that yields 7% compounded semiannually?
23. How much would you have to pay for a \$5000 U. S. Series E Bond that matures in 6.75 years if the interest is 7.2% compounded quarterly?
24. In 1803 the Louisiana Territory was purchased for 3 million dollars. If, instead, the government had put this money in a trust fund that paid 10% compounded annually, what would that trust fund be worth in the year 2003?
25. Sarah and Bill wish to prepare for their retirement by having \$15,000 available for travel ten years from now. How much should they put into a CD that pays 8.5% compounded monthly?
26. What minimum rate of interest should Sarah and Bill get if they have \$7000 to invest? (See Exercise 25.)
27. What minimum rate of interest should you receive on a \$10,000 investment that you wish to double in eight years? Assume semiannual compounding.
28. What annual interest rate will double an investment in eight years if the interest is compounded quarterly?
29. Suppose your home is worth \$100,000 when you buy it and that home market values increase an average of 5% per year. How much is your home worth after ten years?
30. Inflation causes the cost of a college education to increase at roughly 5% per year. If the tuition and board at State U. are \$9000 per year in 1990, how much will tuition and board cost in the year 2000?
31. Suppose the average salary of college professors was \$8000 per year in 1960. If the average inflation rate since then has been about 6% per year, what should the approximate yearly income for college professors have been in 1990?
32. The average salary for college professors was \$42,000 in 1990. Assuming a 6.5% inflation rate, what should their average salary be in 2000? When will the projected average salary reach \$100,000?

8.3 ARITHMETIC SEQUENCES AND ANNUAL PERCENTAGE RATE

The formulas for future value of principal obtained in Section 8.1 for simple interest and in Section 8.2 for compound interest will be used in the next two sections as examples of a more general concept called a sequence. For convenience, we recall Formulas 8.2 and 8.6:

$$S_n = P(1 + Rn) \quad \text{Simple Interest}$$
$$A_n = P(1 + i)^n \quad \text{Compound Interest}$$

where we have slightly altered the simple-interest formula, Formula 8.2, by writing n, instead of t, to indicate that we wish to restrict the time variable to positive integers as we do in the formula for A_n. The future-value formulas, when n is a positive integer, are examples of sequences.

A **sequence** is a set of numbers arranged in a fixed order—for example, 2, 4, 6, 8 or 1, 11, 21, 31, 41, ... Each number in a sequence is called a **term** of the sequence. The first term in the sequence is associated with the number 1, the second with the number 2, the third with the number 3, and so on. In this way, the sequence can be considered a function in which the domain is the set of positive integers and the numbers in the sequence make up the range. The notation we use for a general sequence is

$$a_1, a_2, a_3, a_4, \ldots, a_n, \ldots$$

The subscripts indicate the positions of the terms in the sequence. Thus, a_1 denotes the first term, a_2 denotes the second term, and so on. The term in the nth position is denoted by a_n and is called the **general term.** A sequence that has a finite number of terms is called a **finite sequence**; a sequence with an infinite number of terms is called an **infinite sequence.**

A sequence can be defined either by listing the terms or by giving a formula for the general term a_n. For instance, the sequence of odd numbers is 1, 3, 5, 7, ... This sequence can also be defined by $a_n = 2n - 1$, where n is a positive integer. Thus, $a_1 = 2(1) - 1 = 1$, $a_2 = 2(2) - 1 = 3$, $a_3 = 2(3) - 1 = 5$, and so on.

Example 1

(a) Write the first four terms of the sequence whose general term is $a_n = 3n + 2$.

(b) Write the first four terms of the sequence in which $a_n = \frac{1}{n}$.

Solution

(a) The first term in the sequence is $a_1 = 3(1) + 2 = 5$, the second term is $a_2 = 3(2) + 2 = 8$, the third term is $a_3 = 3(3) + 2 = 11$, and the fourth term is $a_4 = 3(4) + 2 = 14$. So the sequence is 5, 8, 11, 14, ...

(b) Similarly, for $a_n = \frac{1}{n}$, we have $a_1 = \frac{1}{1} = 1$, $a_2 = \frac{1}{2}$, $a_3 = \frac{1}{3}$, $a_4 = \frac{1}{4}$. So the sequence is $1, \frac{1}{2}, \frac{1}{3}, \frac{1}{4}, \ldots$. ■

Example 2 Write the first four terms of the sequence obtained by using the future-value formula for simple interest with $P = 100$ and $R = 0.1$.

Solution Using the formula $S_n = P(1 + Rn)$ with $P = 100$ and $R = 0.1$, we obtain $S_1 = 110$, $S_2 = 120$, $S_3 = 130$, and $S_4 = 140$. We can write these values as the four-term sequence 110, 120, 130, 140. Notice that the terms of this sequence represent the values of an original principal of \$100 invested at a simple-interest rate of 10% after the first four years. ■

Example 3 Find a formula for the general term of the sequence 5, 9, 13, 17, ...

Solution Inspection of this sequence reveals that each term is 4 more than the previous term. Notice also that the first term can be expressed as $5 = 4 + 1$, the second as $9 = 4(2) + 1$, the third as $13 = 4(3) + 1$, and so on. From these observations, we conclude that the nth term for this sequence is $a_n = 4n + 1$. ■

ARITHMETIC SEQUENCES

Examples 2 and 3 are examples of sequences in which the terms are equally spaced; that is, each term differs from the previous term by the same constant amount. In Example 2 each term is ten greater than the previous term and in Example 3 each term is four greater than the previous term. A sequence in which each term differs from the preceding one by the same constant amount is called an **arithmetic sequence.** The constant amount is called the **common difference.** Thus, the sequences in Examples 2 and 3 are arithmetic sequences with common differences 10 and 4, respectively. A formula for a general arithmetic sequence with a common difference d is

$$a_1 = a_1, \qquad a_n = a_{n-1} + d$$

Because the formula above defines a_n in terms of the preceding term a_{n-1}, not directly in terms of n, it is called a **recursive** formula.

The nth term of an arithmetic sequence can be defined explicitly in terms of the first term a_1, the common difference d, and the term number n, by observing that the terms of the general arithmetic sequence are

$$a_1, a_1 + d, a_1 + 2d, a_1 + 3d, \ldots$$

Notice that the coefficient of the common difference d is always 1 less than the term number. So the general term of an arithmetic sequence can be computed as follows:

General Term of an Arithmetic Sequence

The formula for the general term a_n of an arithmetic sequence is

$$a_n = a_1 + (n - 1)d$$

where a_1 = first term of the sequence
d = common difference
n = term number

Example 4 A parachutist falls approximately 16 ft during the first second of free fall, 48 ft during the second second, 80 ft during the third second, and 112 ft during the fourth second.

(a) Show that these distances form an arithmetic sequence.
(b) Give a recursive formula for the sequence.

Solution

(a) The distances form the sequence 16, 48, 80, 112. Notice that the difference between each pair of consecutive numbers is 32, so this is an arithmetic sequence.
(b) The recursive formula for this sequence is $a_n = a_{n-1} + 32$, and $a_1 = 16$. ▪

Whether defined by listing the terms, by giving an explicit formula for the nth term, or by supplying a recursive formula, a sequence may be considered a function whose domain is some subset of the positive integers. In fact, a sequence is sometimes called a **sequence function.** The range values are the terms of the sequence.

Example 5 Give the first four terms of the sequence defined by $a_n = (-1)^n$. What is the range of this sequence function? Is this an arithmetic sequence?

Solution The first four terms of the sequence are $a_1 = (-1)^1 = -1$, $a_2 = (-1)^2 = 1$, $a_3 = (-1)^3 = -1$, $a_4 = (-1)^4 = 1$. The range is $\{1, -1\}$. To determine whether this is an arithmetic sequence, consider $a_1 - a_2 = -1 - 1 = -2$, and $a_2 - a_3 = 1 - (-1) = 2$. Since the differences between two pairs of consecutive terms are not the same, this is not an arithmetic sequence. ▪

SEQUENCES OF PARTIAL SUMS

Associated with each sequence is another sequence, whose nth term is the sum of all terms of the first sequence up to and including a_n. This new sequence is called the **sequence of partial sums,** and is denoted by $\mathbf{S}_n$. Thus, if $a_1, a_2, a_3, \ldots, a_n, \ldots$ are the terms of a sequence, then the terms in the sequence of partial sums are

$$\mathbf{S}_1 = a_1$$
$$\mathbf{S}_2 = a_1 + a_2$$
$$\mathbf{S}_3 = a_1 + a_2 + a_3$$
$$\mathbf{S}_4 = a_1 + a_2 + a_3 + a_4$$
$$\mathbf{S}_n = a_1 + a_2 + a_3 + a_4 + \cdots + a_n = \sum_{k=1}^{n} a_k$$

Thus, this is the explicit formula for the general term of the sequence of partial sums:

$$\mathbf{S}_n = \sum_{k=1}^{n} a_k$$

Example 6 In Example 4, the distance the parachutist fell in each second is given. Determine the sequence of partial sums showing the total distance the parachutist has fallen at the end of each second.

Solution The total distance the parachutist has fallen at the end of each second is the sequence of partial sums of the sequence 16, 48, 80, 112, ... given in Example 4. Thus, $\mathbf{S}_1 = 16$ ft, the distance fallen at the end of the first second; $\mathbf{S}_2 = 16 + 48 = 64$ ft, the total distance fallen at the end of the second second; $\mathbf{S}_3 = 16 + 48 + 80 = 144$ ft; and $\mathbf{S}_4 = 16 + 48 + 80 + 112 = 256$ ft. Therefore, the sequence of partial sums showing the total distance fallen at the end of each second is

$$16, 64, 144, 256, \ldots$$

COMMENT: *Note that the nth term of the sequence of partial sums, $\mathbf{S}_n$, may be found by adding the previous term of the sequence of partial sums, $\mathbf{S}_{n-1}$, to the nth term of the sequence itself. Thus, $\mathbf{S}_n = \mathbf{S}_{n-1} + a_n$ is the recursive formula for the nth term in the sequence of partial sums.*

The sequence of partial sums corresponding to an arithmetic sequence is a_1, $a_1 + a_2 = 2a_1 + d$, $a_1 + a_2 + a_3 = 3a_1 + 3d$, and so on. The formula for the nth term of the sequence of sums is derived by a method attributed to Karl Gauss for computing the sum of the first 100 integers. Gauss recognized the simple but significant fact that the sum is the same regardless of the order in which the terms are written. Thus, for the sum of the first 100 positive integers,

$$\mathbf{S}_{100} = 1 + 2 + 3 + \cdots + 98 + 99 + 100$$

and, in reverse order,

$$S_{100} = 100 + 99 + 98 + \cdots + 3 + 2 + 1$$

Adding these two equalities, we have

$$\begin{aligned} 2S_{100} &= 101 + 101 + 101 + \cdots + 101 + 101 + 101 \\ 2S_{100} &= 100(101) \\ S_{100} &= 5050 \end{aligned}$$

This technique can be used to derive a formula for the nth term of the sequence of sums corresponding to *any* arithmetic sequence. We proceed as follows:

$$\begin{aligned} S_n &= a_1 + (a_1 + d) + (a_1 + 2d) + \cdots + (a_n - 2d) + (a_n - d) + a_n \\ S_n &= a_n + (a_n - d) + (a_n - 2d) + \cdots + (a_1 + 2d) + (a_1 + d) + a_1 \\ \hline 2S_n &= (a_1 + a_n) + (a_1 + a_n) + (a_1 + a_n) + \cdots + (a_1 + a_n) + (a_1 + a_n) + (a_1 + a_n) \end{aligned}$$

$$S_n = \frac{n(a_1 + a_n)}{2}$$

Thus, the sum of n terms of an arithmetic sequence is the average of the first and last terms times the number of terms. Since $a_n = a_1 + (n - 1)d$, we may express this formula in terms of a_1 and d:

$$S_n = \frac{n(a_1 + a_1 + (n - 1)d)}{2} = a_1 \cdot n + \frac{n(n - 1)}{2} d$$

The Sum of n Terms of an Arithmetic Sequence

The formula for the sum of the first n terms of an arithmetic sequence is

$$S_n = a_1 n + \frac{n(n - 1)d}{2} \qquad \textbf{(8.9)}$$

where a_1 = first term of the sequence
n = number of terms
d = common difference

HISTORICAL COMMENT: *Karl Friedrich Gauss (1777–1855), mentioned previously in connection with the sum of the terms of an arithmetic sequence, is often ranked with Archimedes and Isaac Newton as one of the three greatest mathematicians of all time.*

Example 7 Find the sum of the first 50 terms of the arithmetic sequence whose first term is 2 and whose common difference is 3.

Solution Using Formula 8.9, we have

$$S_{50} = 2(50) + \frac{50(49)(3)}{2} = 100 + 3675 = 3775$$

■

ANNUAL PERCENTAGE RATE (APR)

One application of summing arithmetic sequences is to compute the actual annual interest rate paid on add-on installment loans. For example, suppose Rafael and Maria take out a \$6000 small-business loan for one-half year at 12% simple interest, to be repaid in six monthly installments. The add-on interest is $\$6000(0.12)(\frac{1}{2}) = \$360$. Thus, the total amount to be paid back is $\$6000 + \$360 = \$6360$ and the monthly payment is $\frac{\$6360}{6} = \$1060$. However, Rafael and Maria have the use of the entire \$6000 for only one month. After they make their first payment of \$1060 (\$1000 toward the principal and \$60 interest), they have only \$5000 of the original loan left. Yet the \$60 interest was based on \$1000 at 12% for the *entire* half-year. The question is "What is the actual, or true, annual rate of interest they paid for this loan?" The actual rate is called the **annual percentage rate,** or **APR.** We note that the APR is not equal to 12%, because the computation of the \$360 add-on interest is made on the principal for the entire payment period, even though part of the principal has been repaid every month.

To calculate the APR, begin by letting j be the actual APR. We compute the total amount of interest due in terms of j by considering how much interest is due on the unpaid portion of the loan each month. Since j is the APR, it follows that $\frac{j}{12}$ is the monthly interest rate. Then the total interest on the loan for the first month is

$$I_1 = 6000 \cdot \frac{j}{12} \cdot 1$$

The first payment reduces the principal by \$1000. Thus, the interest for the second month is

$$I_2 = 5000 \cdot \frac{j}{12} \cdot 1$$

The interest payments for the remaining months, computed similarly, are

$$I_3 = 4000 \cdot \frac{j}{12} \cdot 1$$

$$I_4 = 3000 \cdot \frac{j}{12} \cdot 1$$

$$I_5 = 2000 \cdot \frac{j}{12} \cdot 1$$

$$I_6 = 1000 \cdot \frac{j}{12} \cdot 1$$

The sum of I_1, I_2, I_3, I_4, I_5, and I_6 is the total amount of interest due and is equal to the actual interest charged on the loan, \$360. Thus,

$$I_1 + I_2 + I_3 + I_4 + I_5 + I_6 = 360$$

But

$$I_1 + I_2 + I_3 + I_4 + I_5 + I_6 = \frac{j}{12}(6000 + 5000 + 4000 + 3000 + 2000 + 1000)$$

The quantity within the parentheses may be summed directly, or may be recognized as the sum of the terms of an arithmetic sequence. In either case, the sum is 21,000; hence,

$$\frac{j}{12} \cdot 21{,}000 = 360$$

Solving for j, we obtain

$$j = \frac{360 \cdot 12}{21{,}000} = 0.2057 = 20.6\%$$

Thus, the APR is over 20 percent, even though the annual rate quoted is 12%. Federal law requires that the APR be disclosed for every loan.

We can derive the general formula for the computation of the APR for add-on loans. Suppose I is the total interest paid on a loan of P dollars, whose payments are to be made k times per year, and suppose a total of n payments are to be made. Then, as in the example above, the total interest I is equal to the sum of the interest charges for each period. The principal is reduced each period by $\frac{P}{n}$ so that interest is charged only on the unpaid balance.

$$I_1 = P \cdot \left(\frac{j}{k}\right)$$
$$I_2 = \left(P - \frac{P}{n}\right) \cdot \left(\frac{j}{k}\right)$$
$$I_3 = \left(P - \frac{2P}{n}\right) \cdot \left(\frac{j}{k}\right)$$
$$\vdots$$
$$I_n = \frac{P}{n} \cdot \left(\frac{j}{k}\right)$$

The sum of these interest charges is equal to I, so we have

$$\left(\frac{j}{k}\right)\left[P + \left(P - \frac{P}{n}\right) + \left(P - \frac{2P}{n}\right) + \left(P - \frac{3P}{n}\right) + \cdots + \frac{2P}{n} + \frac{P}{n}\right] = I$$

After factoring $\left(\frac{P}{n}\right)$ out of the expression in brackets, we obtain

$$\left(\frac{j}{k}\right) \cdot \left(\frac{P}{n}\right) \cdot [n + (n - 1) + (n - 2) + (n - 3) + \cdots + 2 + 1] = I$$

or

$$\frac{jP}{kn} \cdot [n + (n - 1) + (n - 2) + (n - 3) + \cdots + 2 + 1] = I$$

The expression in brackets is the sum of the first n integers. Substituting $a_1 = 1$ and $d = 1$ into Formula 8.9, the expression in brackets may be replaced by $\frac{n(n+1)}{2}$.

Hence,

$$\frac{jP}{kn} \cdot \frac{n(n + 1)}{2} = I$$

Solving this expression for j, we obtain the expression for the APR.

APR for Add-On Loans

The formula for computing the annual percentage rate, APR, for add-on loans is

$$\text{APR} = \frac{2Ik}{P(n + 1)} \qquad \textbf{(8.10)}$$

where I = total interest paid
k = number of payments per year
P = amount of the loan
n = total number of payments

Example 8 To finance a car, Carl and Mai take out a 36-month add-on loan of \$8000 at 15%. Payments are to be made monthly. Determine their monthly payments and the APR.

Solution The interest on the loan is \$8000 · 3 · 0.15 = \$3600. Hence, the monthly payments are

$$\frac{\$8000 + \$3600}{36} = \$322.22$$

The APR may be computed from Formula 8.10 with $I = \$3600$, $k = 12$, $n = 36$, and $P = 8000$:

$$\text{APR} = \frac{2 \cdot 3600 \cdot 12}{8000 \cdot 37} = 0.292 = 29.2\%$$

■

EXERCISES SECTION 8.3

In Exercises 1–10, determine which of the given sequences are arithmetic. Give the formula for the general term for each sequence.

1. 1, 2, 3, 4, 5, ...

2. 2, 5, 8, 11, ...

3. 3, 6, 9, 12, 15, ...

4. 1, −1, 1, −1, 1, −1, ...

5. $5, 1, \frac{1}{5}, \frac{1}{25}, \ldots.$

6. $\frac{1}{2}, \frac{1}{4}, \frac{1}{8}, \frac{1}{16}, \frac{1}{32}, \ldots.$

7. $\frac{1}{2}, \frac{1}{4}, 0, -\frac{1}{4}, -\frac{1}{2}, \ldots.$

8. 3, 9, 27, 81, ...

9. 2, 4, 16, 256, ...

10. 5, 1, −3, −7, ...

11. Find a formula for the general term of an arithmetic sequence whose first term is 2 and whose common difference is 3.

12. Find the common difference of the terms of an arithmetic sequence whose first term is 3 and whose seventh term is 23.

13. Find the common difference of the terms of an arithmetic sequence whose first term is 2 and whose seventh term is 35.

14. Find the common difference of the terms of an arithmetic sequence whose first term is −3 and whose fifth term is 5.

15. Find the sum of the first eight terms of an arithmetic sequence whose first term is −1 and whose common difference is 3.

16. Find the sum of the first ten terms of an arithmetic sequence whose second term is 5 and whose common difference is −4.

17. Jim wants to invest a total of $570 over a period of six months. If he knows that each month he will be able to invest $30 more than he did the previous month, what should his initial investment be? (Ignore any interest accumulations.)

18. A company runs a contest in which it sends the winner some money every day for a month. The amounts are $50 the first day, $100 the second day, $150 the third day, and so on. In a 30-day month, how much will the winner collect?

19. Find the monthly payments for an add-on loan of $10,000 at 14% if the loan is for 48 months. What is the APR?

20. Find the monthly payments for an add-on loan of $1000 at 12% if the loan is for six months. What is the APR?

21. Find the monthly payments for an add-on loan of $6000 at 10% if the loan is for 30 months. What is the APR?

22. Find the monthly payments for an add-on loan of $11,000 at 6.9% if the loan is for 60 months. What is the APR?

8.4 GEOMETRIC SEQUENCES

José has the good fortune to obtain a 14.9% interest rate on some special bonds. At this rate his investment doubles about every five years. If he invests $100, he will have $200 at the end of five years, $400 at the end of ten years, $800 at the end of fifteen years, and so on. This is a sequence whose terms may be written as 100, 200, 400, 800, and so on, and whose general term is $a_n = 2^{n-1} \cdot 100$. Notice that in this sequence, the ratio of succeeding terms is always 2 to indicate the doubling every five years. Such a sequence is an example of a geometric sequence.

Geometric Sequences

A sequence such as the one above, which is constructed by successively multiplying by the same number, is called a **geometric sequence.** The constant multiplier is called the **common ratio** and is usually denoted by r. The recursive definition of a geometric sequence is

$$a_1 = a_1$$
$$a_n = ra_{n-1}$$

Example 1 The sequence giving the number of your direct ancestors in preceding generations is 2, 4, 8, 16, ... Each term is obtained by multiplying the previous term by 2; therefore this is a geometric sequence. ■

A geometric sequence is completely defined by its first term, a_1, and the common ratio r. To determine the general term, the nth term, we note that

$$\begin{aligned} a_1 &= a_1 \\ a_2 &= a_1 r \\ a_3 &= a_2 r = a_1 r^2 \\ a_4 &= a_3 r = a_1 r^3 \\ &\vdots \\ a_n &= a_{n-1} r = a_1 r^{n-1} \end{aligned}$$

The General Term of a Geometric Sequence

The formula for the general term a_n of a geometric sequence is

$$a_n = a_1 r^{n-1} \tag{8.11}$$

where a_1 = first term of the sequence
r = common ratio
n = term number

COMMENT: *Notice that the ratio of any term to the preceding term in the geometric sequence is a constant; in particular, it is equal to r, the common ratio.*

Example 2 What is the tenth term of a geometric sequence whose first term is 500 and whose common ratio is 1.2?

Solution Using Formula 8.11 for the nth term, we have

$$a_{10} = 500 \cdot (1.2)^9 = 2579.89$$ ■

Example 3 Show that the sequence 150, 165, 181.5, 199.65, 219.615 is a geometric sequence and determine the common ratio.

Solution In order to show that the sequence is a geometric one, we consider the ratio of two successive terms, for instance a_2 and a_3. Thus,

$$\frac{a_3}{a_2} = \frac{181.5}{165} = 1.1$$

Also, $\frac{a_5}{a_4} = \frac{219.615}{199.65} = 1.1$. So, we see that the sequence is geometric and the common ratio is 1.1. ▪

Example 4 Write the first four terms of the geometric sequence formed by the future value of \$1000 compounded monthly at 6%. What is the common ratio?

Solution In Formula 8.5, $A_n = P(1 + i)^n$, we let $P = 1000$ and $i = \frac{0.06}{12} = 0.005$. Then $A_1 = 1000(1 + 0.005)$, $A_2 = 1000(1 + 0.005)^2$, $A_3 = 1000(1 + 0.005)^3$, and $A_4 = 1000(1 + 0.005)^4$. This forms the sequence \$1005.00, \$1010.03, \$1015.08, \$1020.15, where $r = 1.005$. ▪

In Section 8.2 we computed the future value of one single compound-interest investment. However, this is not the way many people save money. Instead of making one lump-sum deposit or payment, they invest money or pay back loans with a sequence of (usually equal) periodic payments. Saving money by making a sequence of periodic payments or deposits is referred to as an **annuity.** When loans are repaid in this manner, the equal periodic payments are known as **mortgage payments.** Annuities and mortgage payments are closely related concepts; both require an understanding of geometric sequences. However, to discuss these topics we must first consider summing the terms of a geometric sequence. That is, for the geometric sequence, $a, ar, ar^2, ar^3, \ldots, ar^{n-1}$, we want to compute the sum of the first n terms,

$$S_n = a + ar + ar^2 + ar^3 + \cdots + ar^{n-1}$$

To find a formula for this sum, multiply both sides of the equality by r:

$$rS_n = ar + ar^2 + ar^3 + ar^4 + \cdots + ar^n$$

Now subtract this expression from the one for S_n, to get

$$\begin{aligned} S_n - rS_n = &(a + ar + ar^2 + ar^3 + \cdots + ar^{n-1}) \\ &- (ar + ar^2 + ar^3 + ar^4 + \cdots + ar^n) \end{aligned}$$

Simplifying,

$$S_n - rS_n = a - ar^n$$
$$S_n(1 - r) = a(1 - r^n)$$

From this, if $r \neq 1$, we get the formula for the sum of the first n terms of a geometric sequence whose first term is a.

Sum of the First *n* Terms of a Geometric Sequence

The formula for the sum, S_n, of the first n terms of a geometric sequence is

$$S_n = \frac{1 - r^n}{1 - r} \cdot a_1 \qquad \textbf{(8.12)}$$

where a_1 = first term

r = common ratio

Example 5 Jean deposits \$1 in her account and then doubles the deposit each month for a year. What is her last deposit? What is the total amount of her deposits?

Solution The deposits form the geometric sequence 1, 2, 4, 8, 16, 32, 64, 128, 256, 512, 1024, 2048. Her last deposit is \$2048. The sum of the numbers of this sequence is given by Formula 8.12, with $a_1 = 1$, $r = 2$, and $n = 12$. Thus,

$$S_n = \frac{1 - 2^{12}}{1 - 2} \cdot 1 = \frac{1 - 4096}{-1} = 4095$$

The total amount of Jean's deposits is \$4095. ■

Geometric sequences can be used to describe a wide variety of physical problems. For instance, a geometric sequence can be used to describe the size of a population Y_0, which increases in size at the rate of r percent each year. Thus, the population at the end of the first year is $Y_1 = Y_0(1 + r)$, at the end of the second year it is $Y_2 = Y_1(1 + r) = Y_0(1 + r)^2$, and so on. Therefore, the size of the population at the end of n years is

$$Y_n = Y_0(1 + r)^n$$

COMMENT: *Note the similarity between the future value of a principal with compounding, $A_n = P(1 + i)^n$, and the population equation, $Y_n = Y_0(1 + r)^n$. Two different phenomena described by essentially the same equation! In fact, there are many other practical applications related to the concept of compounding.*

Example 6 The world population in 1978 was approximately 4 billion. The population is increasing at the rate of 2% each year.

(a) Estimate the world population in the year 2000.
(b) Estimate the doubling period of the population. (The **doubling period** is the time it takes the population to double in size.)

Solution

(a) Letting $Y_0 = 4 \times 10^9$, $r = 0.02$, and $n = 22$, we have

$$Y_{22} = 4 \times 10^9(1.02)^{22} = 6.18 \times 10^9$$

Thus, based on the compounding model for population growth, the world population should be a little over 6 billion by the turn of the century.

(b) The doubling period depends only on the rate of growth, r, and not on the value of Y_0. We let $Y_n = 2Y_0$ in the population equation to obtain

$$2Y_0 = Y_0(1.02)^n$$

from which we get

$$2 = (1.02)^n$$

We use a calculator to estimate n by trying values. Try, for example, 30, 32, and so on, until $(1.02)^n$ is approximately equal to 2. We find, in this way, that $n = 35$ years; that is, the world population will double every 35 years if the population growth rate is 2%. ▪

COMMENT: *The doubling period in Example 6 can also be found by taking the logarithm of both sides of* $2 = (1.02)^n$. *Thus,*

$$\log 2 = \log(1.02)^n$$
$$= n \log 1.02$$

Solving for n, we get

$$n = \frac{\log 2}{\log(1.02)} \approx 35$$

EXERCISES SECTION 8.4

In Exercises 1–8, give the formula for the general term for each sequence. Identify which are geometric, and for each of those give the common ratio.

1. 1, 2, 4, 8, 16, ...

2. 1, 3, 5, 7, ...

3. 2, 4, 6, 8, ...

4. $\frac{1}{2}, \frac{1}{4}, \frac{1}{8}, \frac{1}{16}, \ldots .$

5. $\frac{1}{3}, -\frac{2}{3}, \frac{4}{3}, -\frac{8}{3}, \ldots .$

6. $-2, 2, -2, 2, \ldots .$

7. $2, 1, \frac{1}{2}, \frac{1}{4}, \ldots .$

8. $2, \frac{4}{3}, \frac{8}{9}, \frac{16}{27}, \ldots .$

9. Find the sum of the first four terms of the geometric sequence whose first term is 3 and whose common ratio is 5.

10. Find the sum of the first ten terms of the geometric sequence whose second term is 2 and whose common ratio is 2.

11. What is the fifth term of the geometric sequence whose first term is 1000 and whose common ratio is 1.15?

12. What is the tenth term of the geometric sequence whose first term is 500 and whose common ratio is 0.73?

13. Between 1985 and 1990, the population of a certain city grew approximately 6% per year. If the population in 1990 was 200,000, what was it in 1985?

14. An old story has it that a king wanted to reward one of his subjects by placing one grain of wheat on the first square of a checkerboard, two on the second, four on the third, eight on the fourth, and continuing to double the amount until all 64 squares on the checkerboard were filled. How many grains of wheat would the king have placed on the 64th square?

15. A problem similar to that posed in Exercise 14 asks how you would react to having your monthly salary paid as follows: 1 cent the first day, 2 cents the second day, 4 cents the third day, 8 cents the fourth day, and so forth, with your salary doubling each day for 30 days.
(a) How much would you be paid on the 30th day?
(b) What would be your total pay for the month?

16. A rubber ball, when dropped from a height, rebounds to three-fourths of that height. On what bounce will the ball fail to reach one-fourth of the original height?

17. With each stroke, a vacuum pump removes one-fifth of the air in a container. After how many strokes will it have removed three-fourths of the original air?

18. A particular radioactive substance decays at a rate such that it loses one-eighth of its weight each day. How long will it be until it loses half its initial weight? (Note: This length of time is called the *half-life* of the substance.) In how many days will it have lost 90% of its initial weight?

19. The speed of a chemical reaction approximately doubles each time the temperature increases by 10°C. Write a geometric sequence to show how many times faster paper will burn if the temperature rises by 10°C, 20°C, 30°C, and 40°C.

20. Inflation in Argentina is such that it has the effect of reducing one's income by $\frac{1}{4}$ every six months. If the income of a farm laborer in Argentina is $400 per month today, what will his monthly income be worth 18 months from now? Assume he does not get a raise.

21. Jim invests $2000 at 8% compounded monthly. Each month he receives a statement on which is printed the value of his investment. Give the values shown on the statements for the first five months.

22. Your stockbroker says that a growth stock increases by 10% per year. The stock sells today at $48 per share. Based upon your stockbroker's prediction, what will be the selling price of the stock in five years?

23. Luis is told that his new automobile will depreciate at the rate of 20% each year. If he bought the car in 1990 for $12,000, what will be its value in the year 2000?

24. Sociologists predict that the inner-city population in a large city will decrease by about 10% per year. If the inner-city population was 500,000 in 1990, what will it be in 1995?

25. Rob must pay off an interest-free debt by making monthly payments of 12% of the unpaid balance. What percentage of the debt does he pay off in one year?

26. A personal computer is said to depreciate at 25% of its value annually. If Jane bought a computer for $3500 in 1992, what will be its value by 1996?

27. A chain letter is a (now illegal) "game" in which you are sent a letter with, say, five names on it. You mail a dollar to the name at the top of the list and cross out that name. Then you put your name at the bottom and mail the letter to five of your friends. If your friends continue this, and their friends continue, and so forth, how much do you stand to gain?

28. In a certain court proceeding involving an unpaid debt, an individual is asked to make monthly payments of 5% of the outstanding debt for the next 24 months. If the debt now stands at $150,000, what percentage of the debt would be paid off in that time? (Assume no interest is added to the debt during this period.)

8.5 ANNUITIES AND MORTGAGES

One of the most popular methods of saving money is by making a sequence of regular payments. As we noted in the previous section, such savings methods are called annuities. In this section, we develop the mathematical formulas you need to understand annuities and make appropriate calculations.

Ordinary Annuity

An **annuity** is any series of equal payments or deposits made at regular intervals. The present value of an annuity is computed by a systematic application of the future-value and present-value formulas for compound interest, as well as by the use of the formula for the sum of the first n terms of a geometric sequence. We will restrict discussion to the most common kind of annuity, called an **ordinary annuity.** In an ordinary annuity:

- The payments are equal.
- The payments are made at the end of each payment interval.
- The payment intervals coincide with the compounding periods.

The **future value** of an annuity, sometimes called the *face value,* is the sum of the principal and the compound interest that will accumulate if all stipulated payments are made for the complete term of the annuity. For example, consider an annuity for which monthly payments of \$100 are made for two years at a nominal rate of 7%. This annuity would have a future value equal to the sum of the \$100 principals plus the accumulated compound interest. In this case, there are 24 principals to consider. The \$100 payment made at the end of the first month will earn interest for 23 months at a monthly interest rate of $\frac{0.07}{12}$. Thus, the future value of this \$100 payment, using Formula 8.5, is given by

$$V_1 = 100\left(1 + \frac{0.07}{12}\right)^{23}$$

The second \$100 monthly payment will earn interest for 22 months, so its future value is

$$V_2 = 100\left(1 + \frac{0.07}{12}\right)^{22}$$

Similarly, the other \$100 payments will earn interest for 21 months, 20 months, and so on. The twenty-third payment will earn interest for one month, and the very last payment will earn no interest at all. Thus,

$$V_3 = 100\left(1 + \frac{0.07}{12}\right)^{21}$$

$$V_4 = 100\left(1 + \frac{0.07}{12}\right)^{20}$$

$$\vdots$$

$$V_{23} = 100\left(1 + \frac{0.07}{12}\right)^{1}$$

$$V_{24} = 100\left(1 + \frac{0.07}{12}\right)^{0}$$

The total value V of the annuity is thus the sum of these 24 amounts:

$$V = V_1 + V_2 + V_3 + V_4 + \cdots + V_{23} + V_{24}$$

By examining the terms of this sum, we see that this is a geometric sequence with $a_1 = V_{24} = 100$ and common ratio $r = \left(1 + \frac{0.07}{12}\right) \approx 1.0058333$. Using Formula 8.12 to calculate the sum of the first 24 terms, we find

$$V = 100\left(\frac{1 - 1.005833^{24}}{1 - 1.005833}\right) = 2568.09$$

Hence, the future value of the annuity is \$2568.09.

This same approach can be used to obtain a formula for the future value of an ordinary annuity with n payments. If i is the interest rate per payment period, and there are n payments of P dollars each, then the amount of the annuity is given by the sum,

$$V = \sum_{k=0}^{n-1} P(1 + i)^k$$

The n terms of this sum form a geometric sequence with common ratio $r = (1 + i)$ and $a_1 = P$. Hence, using Formula 8.12, we have

$$V = P\frac{1 - (1 + i)^n}{1 - (1 + i)}$$

Algebraically simplifying this expression, we get the formula for the future value of an annuity:

Future Value of an Annuity

The formula for the future value of an annuity V, at maturity, is

$$V = P \cdot \frac{(1 + i)^n - 1}{i} \qquad \textbf{(8.13)}$$

where P = amount of each payment

i = interest rate per payment period

n = total number of payment periods

Example 1 To save toward the down payment on their next car, Madeline and Alex invest in an annuity with quarterly payments of \$150 for five years at a nominal annual interest rate of 8%. Find the value of the annuity at maturity.

Solution We use Formula 8.13 with $P = \$150$, $i = \dfrac{0.08}{4} = 0.02$, and $n = 5 \cdot 4 = 20$.

$$V = 150 \cdot \frac{(1 + 0.02)^{20} - 1}{0.02} = 3644.61$$

Hence, the value of the annuity at the end of five years is \$3644.61. ▪

Sinking Funds

Businesses and individuals are often faced with the prospect of a major purchase in the future. A school district wants to invest money regularly over the next three years in order to upgrade its computer. A shopping center's directors plan to put aside money each quarter to defray the cost of replacing the roof in five years. A sculptor, who knows she wants to take an extended tour of Europe for study in four years, is trying to decide how much to save weekly to meet that expense. These all are examples of regular saving for a future expense. Money saved systematically for use at a future date is called a **sinking fund.** If we consider the amount in the sinking fund to be the future value V of an annuity of n payments invested at $i\%$ interest per payment period, then to determine the amount of the payment we solve Formula 8.13 for P.

DEFINITION Sinking Fund Payments

The formula for the amount of regular sinking fund payments P is

$$P = V \cdot \frac{i}{(1 + i)^n - 1} \qquad \textbf{(8.14)}$$

where V = amount of the sinking fund
i = interest rate per payment period
n = number of payment periods

Example 2 The DeSmet School District wants to upgrade the computer labs in its schools in three years. The estimated cost for this upgrading is \$35,000. How much should the board invest each quarter at 7.5% annual interest if it wants to have \$35,000 in three years?

Solution Using Formula 8.14, the amount of the sinking fund is $V = 35{,}000$, with $i = \dfrac{0.075}{4}$ and $n = 4 \cdot 3 = 12$. Hence,

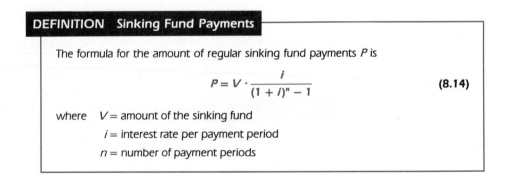

$$P = \frac{35{,}000\left(\frac{0.075}{4}\right)}{\left(1 + \frac{0.075}{4}\right)^{12} - 1} = 2627.98$$

So the school board needs to invest approximately \$2630 each quarter. ▪

Example 3 Claudia would like to invest in a 9% annuity for her European trip in four years. How much should she deposit at the end of each month for four years if she wishes to have \$10,000 for the trip?

Solution We use Formula 8.14 with $V = 10{,}000$, $i = \frac{0.09}{12} = 0.0075$, and $n = 4 \cdot 12 = 48$. Thus,

$$P = \frac{10{,}000 \cdot 0.0075}{(1.0075)^{48} - 1} = 173.85$$

Thus, Claudia should invest about \$175 per month for four years to accumulate \$10,000 in her European trip fund. ▪

PRESENT VALUE OF AN ANNUITY

Let's shift our perspective. Up to this point we have discussed annuities from the view of *saving* a fixed amount regularly to build up a lump sum for the future. But suppose you already have a sum of money (for example, an inheritance or lottery winnings) and you want it paid out to you in periodic payments over a fixed period of time, while the balance continues to earn interest. The amount of money to be invested is called the **present value of the annuity.** The periodic payments are also known as **withdrawals.** For instance, if you were to invest \$220,000 in a 9% annuity today, you could receive (or withdraw) almost \$2000 per month for 20 years. That is almost \$480,000, or more than twice what you originally invested. Let's consider how we can determine the amount of money needed to make regular withdrawals over a fixed period of time.

Let A denote the amount of money to be initially deposited. From A, we expect a payment of Q dollars at the end of each of n payment periods. Assume the interest rate for each payment period is i. Consider the total value of the money paid out, T, in two different ways:

- $T =$ the total value of an annuity with Q dollars per payment period for n payment periods. From this viewpoint, we use Formula 8.12 to obtain

$$T = Q \cdot \frac{(1 + i)^n - 1}{i}$$

- $T =$ the future value of A compounded at $i\%$ per pay period. From this viewpoint, we use Formula 8.5 to obtain

$$T = A(1 + i)^n$$

Equating these two expressions for T, the total value of the annuity at the end of n periods, we get

$$Q \cdot \frac{(1 + i)^n - 1}{i} = A(1 + i)^n$$

Solving for A by multiplying both sides by $(1 + i)^{-n}$, we obtain the formula for the present value of an annuity:

Present Value of an Annuity

The formula for the present value A of an annuity is

$$A = Q\left(\frac{1 - (1 + i)^{-n}}{i}\right) \qquad \textbf{(8.15)}$$

where $Q =$ payment per period
$i =$ interest rate per payment period
$n =$ number of payment periods

Example 4 Use Formula 8.15 to determine the present value of the annuity discussed earlier. In that case, we were considering a 9% annuity that would pay \$2000 per month for 20 years.

Solution Here, we have $i = \frac{0.09}{12} = 0.0075$, $Q = 2000$, and $n = 20 \cdot 12 = 240$, so that

$$A = 2000\left(\frac{1 - 1.0075^{-240}}{0.0075}\right) = 222{,}289.91$$

Thus, the present value for this annuity is \$222,289.91. ▪

Mortgages

When a debt is paid off by a sequence of equal payments, the debt is said to be **amortized** by an annuity; the debt is known as a **mortgage,** and includes both the amount of money borrowed (the principal) and the interest on the loan. These payments are *mortgage payments*. The payments required to amortize a debt are determined from the equation for the present value of an annuity, Formula 8.15. We simply identify the debt with the quantity A and the mortgage payment with the periodic payment Q and solve for Q. So we have

$$A = Q\left(\frac{1-(1+i)^{-n}}{i}\right)$$

Solving for Q yields the mortgage payment formula:

Mortgage Payments

The formula for computing the mortgage payment is

$$Q = A\left(\frac{i}{1-(1+i)^{-n}}\right) \tag{8.16}$$

where A = debt (principal)
i = interest rate per payment period
n = total number of payments

Example 5 Janan and Brent King purchase a \$70,000 home with a down payment of \$15,000 and a \$55,000 loan to be amortized over 30 years at 9.5%. Determine the size of the monthly payment. What is the total amount of money paid on the \$55,000 mortgage?

Solution Using Formula 8.16 with $A = 55{,}000$, $i = \dfrac{0.095}{12}$, and $n = 12 \cdot 30 = 360$, we have

$$Q = \frac{55{,}000\left(\dfrac{0.095}{12}\right)}{1-\left(1+\dfrac{0.095}{12}\right)^{-360}} = 462.47$$

Hence, the monthly mortgage payment is \$462.47. The total amount paid is the amount of a mortgage payment times the total number of payments; in this case, that is

$$\$462.27 \cdot 360 = \$166{,}417.20$$

Each of the equal mortgage payments combines payment of the interest due for that payment period and a portion of the principal to reduce the debt. That is, if P is the principal at the beginning of the payment period and if i is the interest per payment period, then iP is the interest for that period. The debt reduction is then $Q - iP$. A listing of the debt reductions for all the mortgage payments is called an *amortization schedule*.

Example 6 A debt of \$10,000 with interest at 12% is amortized by quarterly payments over two years. Determine the quarterly payments and construct an amortization schedule.

Solution The quarterly payments are calculated using Formula 8.16 with $A = 10{,}000$, $i = \frac{0.12}{4}$, and $n = 2 \cdot 4 = 8$. Thus,

$$Q = \frac{10{,}000(0.03)}{1 - (1.03)^{-8}} = 1424.56$$

The quarterly payment, then, is \$1424.56. The amortization schedule is given in Table 8.3. The principal is reduced each payment period by the amount of that payment that exceeds the interest due. That is, if P_n is the principal at the beginning of a payment period and I_n is the interest due at the end of that period, and Q is the mortgage payment, then the principal at the beginning of the next payment period is

$$P_{n+1} = P_n - (Q - I_n)$$

The interest due at the end of a given period is

$$I_n = P_n \cdot i$$

TABLE 8.3 **Amortization Schedule**

Payment Number	Principal at Beginning of Each Payment Period (P)	Interest Due at End of Period ($I_n = 0.03 \times P$)	Payment Amount (Q)	Amount Paid on Principal ($Q - I$)
1	\$10,000.00	\$300.00	\$1424.56	\$1124.56
2	8875.44	266.26	1424.56	1158.30
3	7717.14	231.51	1424.56	1193.05
4	6524.09	195.72	1424.56	1228.84
5	5295.25	158.86	1424.56	1265.70
6	4029.55	120.89	1424.56	1303.67
7	2725.88	81.78	1424.56	1342.78
8	1383.10	41.49	\$1424.59*	1383.10

* The last payment is \$0.03 higher than the others so that the debt can be paid off in eight payments.

EXERCISES SECTION 8.5

In Exercises 1–6, find the future value of each annuity. Assume that the compounding period equals the payment period.

1. $5 per day for 5 years at 10%.
2. $360 semiannually for 20 years at 8%.
3. $60 monthly for 20 years at 8%.
4. $2 daily for 20 years at 8%.
5. $100 monthly for 10 years at 10%.
6. $600 semiannually for 10 years at 10%.
7. Christine decided to invest weekly to save for her wedding. How much should she invest each week at 6.2% interest in order to have $3500 in two years?
8. Luis wants to save for a down payment for a Corvette. How much should he invest each week at 5.7% interest in order to have $10,500 in four years?
9. Sam and Mary want to save for the down payment on a home. How much should they invest each month at 8.15% interest if they want to have $20,000 in five years?
10. The Dorans want to invest in a quarterly annuity to supplement their retirement income. How much should be invested each quarter at 6.5% interest for them to have $135,000 in seven years?
11. Marlene wants to make a down payment on a car in three years. How much should she invest each month at 7.1% interest in order to have $5000 at that time? What is the total amount invested and how much interest is earned?
12. Basil wants to invest money weekly at 6.8% interest to have $3500 in two years. What is the total amount that he will invest and how much interest will be earned?
13. A company needs to have $400,000 in ten years to help pay the cost of medical insurance for its retirees. How much money must the company invest annually at 6% in order to have $400,000 at the end of ten years?
14. How much should you invest quarterly at 6.5% annual interest to be sure that you can buy a new car (costing approximately $16,000) in five years without buying on credit?

In Exercises 15–18, determine the present value of an annuity required to meet each payment schedule. Assume the compounding period is equal to the payment period.

15. An 8.5% 10-year annuity with monthly payments of $500.
16. A 7.2% 15-year annuity with monthly payments of $1000.
17. An 8.2% 15-year annuity with annual payments of $10,000.
18. A 5.9% 25-year annuity with quarterly payments of $1500.
19. Sam is offered, on retirement from his company, a 5.6% annuity with monthly payments of $200 for 10 years or a cash payment equal to the value of the annuity. What is the value of the lump-sum payment? If he were to choose the annuity, what would be the total of the payments he received?
20. Upon retirement, Pat Flaherty is offered a choice: she can accept an 8.5% annuity with quarterly payments of $500 for 15 years, or she can take the equivalent lump-sum value of the annuity now. What lump sum would she get? If she were to choose the annuity, what would be the total amount she received?
21. Marvin Wasson won the state lottery, which will pay him $30,000 per year for the next 20 years. How much must the state invest now at 7.5% in order to fulfill its obligation to this lottery winner?
22. Grandma and Grandpa Heatherton wish to provide money for their three grandchildren's college educations. How much money must they have in an annuity that pays 8.2%, if they plan on giving each child $5000 per year for five years?
23. Mookie Johnson buys a house and finances $20,000 at a fixed interest rate of 8% for 20 years. Calculate his monthly payments.
24. The loan officer at First Federal offers the Pryors a 20-year fixed rate 9% second mortgage of $25,000

on their home. Determine the amount of the Pryors' monthly payment.

25. Suppose you are going to borrow \$25,000 for 20 years at 10% interest.
    (a) Calculate your monthly payment.
    (b) Calculate the total amount you would pay back.
    (c) Calculate the total amount of interest you would pay.

26. Repeat Exercise 25 with \$35,000, 10%, and 30 years.

27. A debt of \$2000 with a nominal annual interest rate of 10% is amortized by monthly payments over one year. Determine the equal monthly payments and make up an amortization schedule.

28. Create an amortization schedule for retiring a debt of \$3000 with 12% nominal annual interest if payments are to be quarterly for two years.

IMPORTANT WORDS AND PHRASES CHAPTER 8

From the single word *interest* flow many important financial terms. The financial terms *simple interest* and *compound interest* give rise to arithmetic and geometric sequences. This chapter has provided a rich array of applications to the everyday problems of banking, loans, mortgage payments, and annuities. You should have become familiar with the following terms:

add-on loan
amortization
annual yield
annual percentage rate (APR)
annuity, ordinary
arithmetic sequence
common difference, ratio
compound interest
discounting
doubling period
future value
  annuity
  compound interest
  simple interest
general term
  arithmetic sequence
  geometric sequence
geometric sequence
installment
interest
  compound
  simple
loan
mortgage
present value
  of an annuity
  of a compound amount
rate of interest
sequence
  of partial sums
sinking fund
sum
  of an arithmetic sequence
  of a geometric sequence
withdrawals

REVIEW EXERCISES CHAPTER 8

In Exercises 1–6, compute the simple interest for each set of conditions.

1. \$470 at 5.5% for three years.
2. \$750 at 9.2% for four years.
3. \$84,000 at 8.7% for 18 months.
4. \$1850 at 10.4% for 16 months.
5. \$685 at 14.5% for 90 days.
6. \$105,500 at 6.8% for 60 days.
7. What principal, invested at 9% simple interest for two years, has a future value of \$9800?

8. What principal, invested at 13.1% simple interest for 18 months, has a future value of $25,785?

9. What is the interest rate if a principal of $1000 earns $25 simple interest in six months?

10. What is the interest rate if a principal of $3500 earns $100 simple interest in four months?

11. Sarah takes out a short-term add-on loan of $800 for eight months. At an interest rate of 9.5%, what are the monthly payments?

12. Akim needs a short-term loan for tuition. What monthly payments would he have to make for an add-on loan of $1250 at 8.8% for 12 months?

13. A couple borrows $7000 at 12.2% simple interest for nine months on a discounted loan. Compute
(a) The interest on the loan.
(b) The amount of money they receive.
(c) The effective interest rate.

14. Repeat the computations for the discounted loan in Exercise 13 with a loan period of 15 months.

15. If $650 is invested at 8.9% compounded quarterly, how much interest will be earned in three years?

16. If $3400 is invested at 10.5% compounded monthly, how much interest will be earned in two years?

17. Compute the present value of a bond that will be worth $10,000 in 15 years, assuming it pays 8% interest compounded annually.

18. A bond will be worth $5000 in ten years. If the bond pays 7.8% interest compounded annually, what is its present value?

19. What interest rate will double your money in seven years if the interest is compounded semiannually?

20. What interest rate will double your money in six years if the interest is compounded monthly?

In Exercises 21–26, give the annual yield for each compound interest rate.

21. 8.7% compounded daily.

22. 7.9% compounded daily.

23. 10.5% compounded monthly.

24. 11.2% compounded monthly.

25. 6.4% compounded semiannually.

26. 8.1% compounded semiannually.

For Exercises 27–34, determine whether each sequence is geometric or arithmetic, and find the common ratio or common difference.

27. 1, 3, 5, 7, 9, ...

28. 10, 8, 6, 4, 2, ...

29. 2, 4, 8, 16, 32, ...

30. $\frac{2}{3}, \frac{2}{9}, \frac{2}{27}, \frac{2}{81}, \ldots$

31. $-5, 25, -125, 625, \ldots$

32. $-1, 4, 9, 14, 19, \ldots$

33. 3, 9, 15, 21, ...

34. $8, 5, 2, -1, \ldots$

For Exercises 35–42, find the sum of the first 20 terms of each of the sequences in Exercises 27–34.

43. What is the APR for an add-on loan of $5000 at 8.0% with monthly payments for one year?

44. What is the APR for an add-on loan of $750 at 6.8% with monthly payments for eight months?

In Exercises 45–48, find the future value of each annuity.

45. $100 quarterly for 20 years at 9% compounded quarterly.

46. $500 semiannually for 20 years at 10% compounded semiannually.

47. $25 monthly for 10 years at 9.8% compounded monthly.

48. $2000 annually for 25 years at 8.6% compounded annually.

49. The Powells wish to save for a substantial down payment on a home six years from now. How much should they invest each month at 7.2% interest in order to have $35,000 at that time?

50. The Rios want to anticipate the education expenses of their new daughter. How much should they invest each month at 5.8% interest in order to have $150,000 available to them in 20 years?

51. José is offered a retirement annuity with monthly payments of $250 for seven years. If the current interest rate is 7%, what lump sum could he take instead of the annuity? Compare this amount with the total amount he would receive if he were to accept the monthly payments.

52. Nicole is offered a settlement on an estate of a lump-sum payment or a 4.7% annuity with monthly payments of \$100 for 15 years. What lump-sum payment should she expect? Compare this amount with the total amount she would receive with the annuity.

53. Compute the monthly payment for a loan of \$25,000 at 12.4% for 30 years.

54. Compute the monthly payment for a loan of \$40,000 at 11.8% for 20 years.

55. The Steiners decide to buy a \$100,000 home and make a \$25,000 down payment. Compute the monthly payments for the \$75,000 mortgage if the loan is obtained at 9.7% for 25 years. How much interest is paid on this loan if it is paid off on schedule?

56. Mr. and Mrs. Larkin buy a \$200,000 home and make a \$50,000 down payment. Compute the monthly payments for a mortgage on their home if the loan is obtained at 11.1% for 30 years. If the loan will be entirely paid off over 30 years, how much interest will be paid?

57. Create an amortization schedule for retiring a debt of \$3000 at 8% interest if payments are made quarterly for two years.

CHAPTER 9

BOOLEAN ALGEBRA

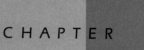

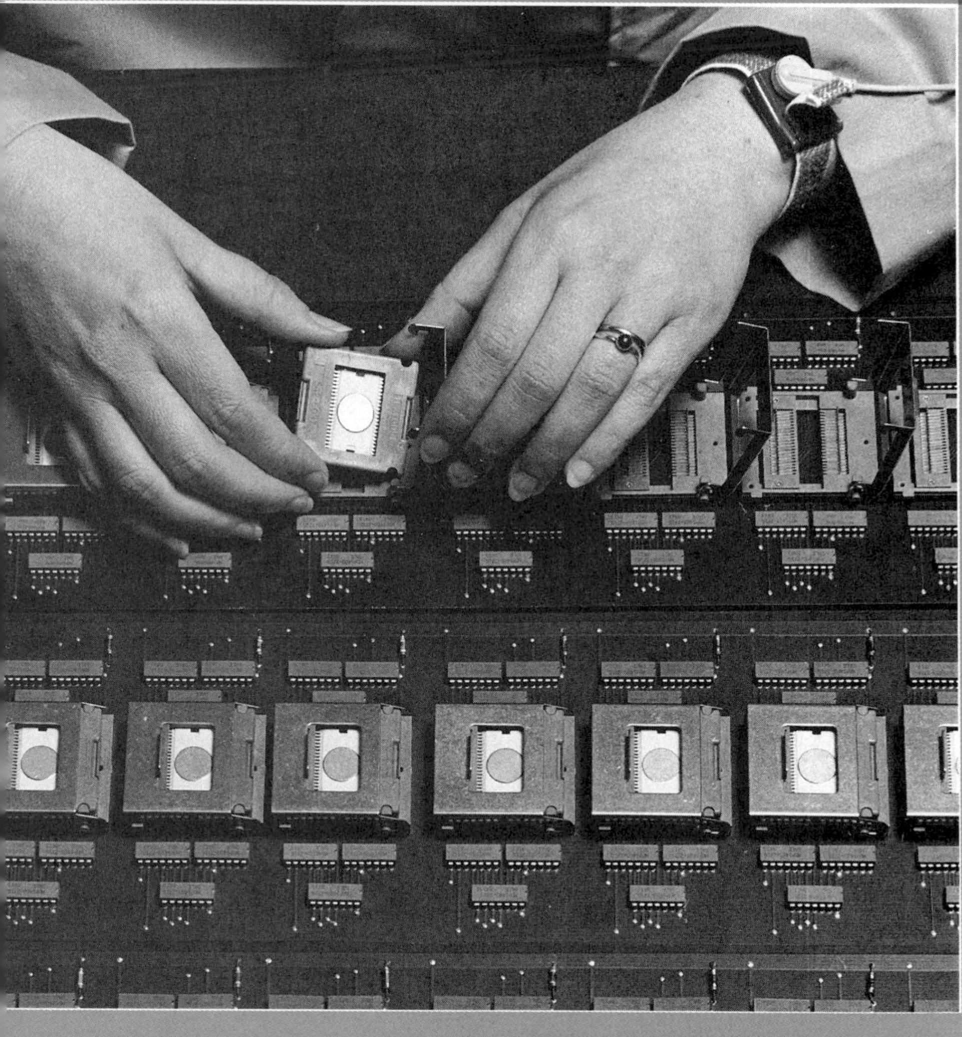

Algebra had its origin in the study of numbers. The idea that *symbols* could be used to represent numbers led to the discovery of basic relationships between numbers. As the centuries passed, mathematicians discovered that an *unknown number* could often be found if its relation to *known numbers* could be discovered. The study of the relationships between known and unknown numbers, and of the algorithms used in the process of finding the unknown numbers, was developed in the ninth century as a part of Arab culture. Originally this study was called *al-jebr*, and eventually this Arab word became our word *algebra.*

Today algebra has a wider meaning; it refers to the study of the structures of mathematical systems. *Mathematical system* is a broad term, referring to a set of objects connected by some operations. The structure of such a system is loosely defined as the properties of the operations that can be performed on the members of the system. Thus, the *algebra of real numbers* is the study of the properties of the operations of addition, subtraction, multiplication, and division on the real numbers. However, there are algebras other than the algebra of real numbers. In this chapter we look at an algebra called Boolean algebra.

HISTORICAL COMMENT: *Boolean algebra was introduced by George Boole, an English logician, in 1840 as a tool for analyzing logical statements. This system has since found applications in several other areas. In 1938, Claude E. Shannon published a milestone paper that used Boolean algebra in the analysis of switching circuits. In this chapter, we will first describe Shannon's switching algebra and then move to the abstract Boolean algebra.*

9.1 ALGEBRA OF SWITCHES

We begin our study of Boolean algebra with a mathematical analysis of electric switching circuits. For our purposes an electric switch, or relay, is a device that either permits, or does not permit, electricity to flow in a circuit. No need to worry about how much or how little flows, just whether it flows or not. In fact, any device that has two possible states may be thought of as a switch. For instance, an electric light bulb that is on or off, a door that is open or closed, a valve that permits water to flow or not, may all be represented by switches. In general, this kind of switch is said to be **bistable.** Figure 9.1 shows a schematic diagram of a switch in two different states, or *positions:* open and closed. An open switch prevents electricity from flowing from one terminal to another; a closed switch permits the electricity to flow.

Figure 9.1

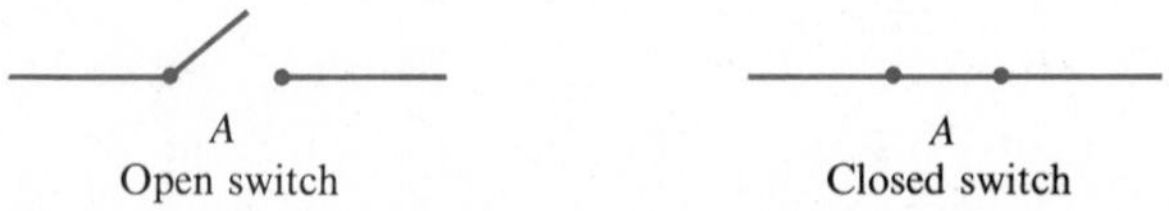

We usually represent switches in circuits without indicating their states of closure; that is, rather than show the switch as either open or closed, we simply denote it as Switch A, as shown in Figure 9.2. We assume that Switch A may be open or closed.

Figure 9.2

The **complement** of Switch A is one that is open when A is closed and closed when A is open. We denote the complement of A by A'. See Figure 9.3 for the schematic diagram of Switches A and A'.

Figure 9.3

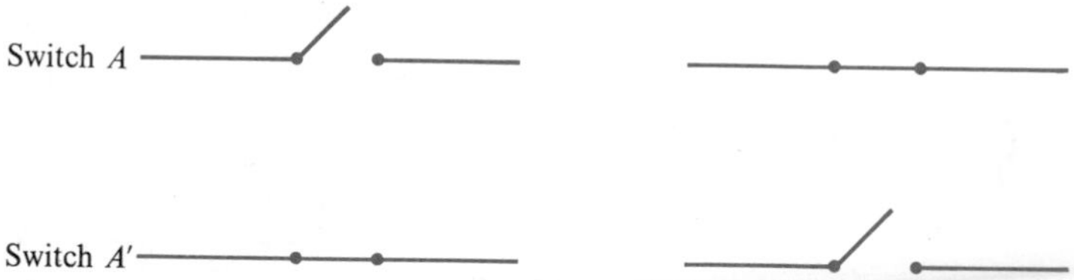

If Switches A and B are connected end-to-end, like those in Figure 9.4(a), then we say they are **connected in series.** For electric current to go from one side of a series connection to the other, it must pass through both Switch A and Switch B. A series circuit is closed only if all of the switches in the series connection are closed, and open if any or all of the switches in the circuit are open. If Switches A and B are connected like those in Figure 9.4(b), then we say that they are **connected in parallel.** A parallel circuit is closed if any or all of the switches in the circuit are closed, and is open only if all of the switches are open.

Figure 9.4

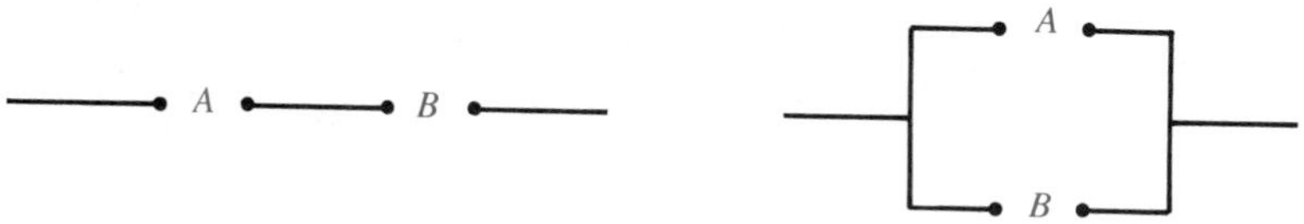

(a) Series circuit (b) Parallel circuit

Suppose that we represent Switch A and Switch B in series by AsB; then AsB is closed only if both A and B are closed; otherwise, AsB is open. The following table shows the closure of AsB for the four possible conditions. Similarly, if we represent Switch A and Switch B in parallel by ApB, then ApB is open only if both A and B are open; otherwise, ApB is closed. The table for the closure of ApB is shown beside the table for AsB.

TABLE 9.1 Tables of Closure for Series and Parallel Circuits

	Series			Parallel	
A	*B*	*AsB*	*A*	*B*	*ApB*
closed	closed	closed	closed	closed	closed
closed	open	open	closed	open	closed
open	closed	open	open	closed	closed
open	open	open	open	open	open

Sometimes these tables of closure are written in operational form, in which the placing of two switches in series or parallel is represented as an *operation* involving the two switches. Then the result of the operation is shown in a table such as those in Table 9.2, where *open* is represented by O and *closed* by C. The state of closure of Switch A is given in the first column, and that of Switch B in the first row. The state of closure of the combination is then the intersection of the appropriate column and row. So, if A is open (O) and B is closed (C), then Table 9.2 shows that a series combination of A and B is open (O).

TABLE 9.2 **Tables of Closure for Series and Parallel Operations**

Series					Parallel			
		Switch B					Switch B	
	s	*C*	*O*			*p*	*C*	*O*
Switch A	*C*	*C*	*O*		Switch A	*C*	*C*	*C*
	O	*O*	*O*			*O*	*C*	*O*

An Algebra of Switches

Now we will describe an algebra of switches; that is, we will show how to use symbols to represent switching circuits. We make the following definitions:

- *An open switch or circuit will be identified by the symbol* 0 (*zero*), *and a closed switch or circuit by the symbol* 1 (*one*). These are the same symbols that represent the real numbers zero and one but, in this case, the symbols represent states of switches and not numerical quantities.
- *A switch will be considered to be a variable, represented by a letter such as A or B, whose "value" is the state of the switch.* Thus, if A is a variable representing a switch, its value is either 0 (open) or 1 (closed).
- *The series arrangement of two switches A and B will be denoted by* $A \cdot B$, *instead of AsB*. The $(\cdot)$ is the same symbol used to mean the multiplication of real numbers. The dot is often omitted, so that $A \cdot B$ is written AB.
- *The parallel arrangement of two switches A and B will be denoted by* $A + B$, *instead of ApB*. The $(+)$ is the same symbol normally associated with the addition of real numbers.
- *The complement of a switch A will be denoted by* A'. Thus, A' is a switch whose value is 0 when $A = 1$ and 1 when $A = 0$.

Using 0 for *open* and 1 for *closed*, and $(\cdot)$ for *series* and $(+)$ for *parallel*, Table 9.1, the table of closure for the parallel and series connection of two switches, can be rewritten as Table 9.3. We will use Table 9.3 in all future references to switching circuits.

TABLE 9.3 Algebraic Description of Parallel and Series Circuit

A	B	Series $A \cdot B$	Parallel $A + B$
1	1	1	1
1	0	0	1
0	1	0	1
0	0	0	0

Table 9.2, rewritten with these new symbols, becomes Table 9.4.

TABLE 9.4 Tables for the Operations (·) and (+)

Series

	$(\cdot)$	Switch B: 1	Switch B: 0
Switch A	1	1	0
	0	0	0

Parallel

	$(+)$	Switch B: 1	Switch B: 0
Switch A	1	1	1
	0	1	0

COMMENT: *Taken out of the context of switching circuits, one of the results of Table 9.4 would seem to be incorrect. Note that the (+) operation (for parallel circuits) for closed relays reads* $1 + 1 = 1$*. Interpreted as a real-number statement, this is obviously false. However, as a switching-algebra statement, it simply means that if two closed relays are connected in parallel, the circuit is closed.*

Using switching algebra, we can write expressions involving variables and the operations of (+) and (·) to represent complex arrangements of series and parallel switches.

Example 1 Draw a circuit represented by the algebraic expression $AB + A'$.

Solution The product AB represents a series arrangement of Switches A and B. The sum $AB + A'$ represents a parallel arrangement of AB and A'. The circuit is shown in Figure 9.5.

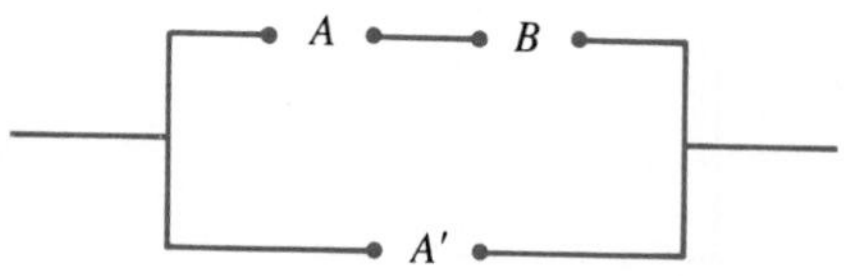

Figure 9.5

■

Example 2 Write a switching-algebra expression for the following circuit:

Figure 9.6

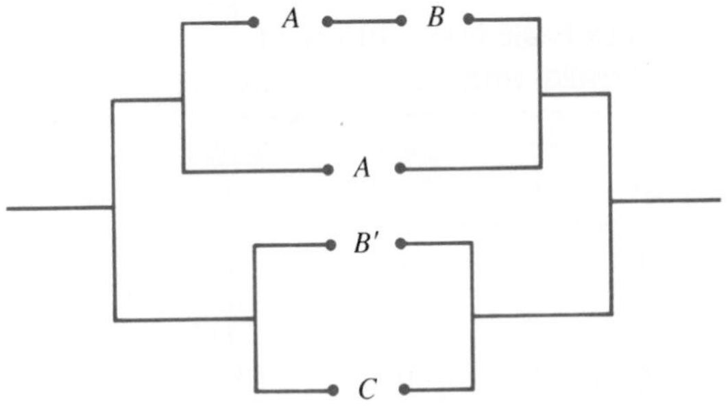

Solution We note that Switch A is in parallel with the series connection represented by AB, which, in turn, is in parallel with the parallel connection $B' + C$. Hence, the expression that represents the circuit is

$$A + AB + (B' + C)$$

EQUIVALENT CIRCUITS

Combinations of series and parallel arrangements of switches are used to build other circuits. We say that two circuit arrangements are *equivalent* if their closure properties are the same for any given position of the relays making up the circuits. We write "Circuit 1 $\sim$ Circuit 2" to indicate equivalence. If two circuits are equivalent, their switching-algebra expressions are equal. Although these expressions look like real-number-algebra expressions, they are switching-algebra expressions for some series/parallel arrangement of switches, and must be evaluated using Table 9.3.

To show that two circuits are equivalent, we need only show that both circuits produce the same sequence of 0s and 1s when all possible cases are examined. We can demonstrate the equivalence of two circuits by representing them with switching-circuit algebraic expressions and then constructing a closure table to show the possible outcomes. If the outcomes for each circuit are identical, the circuits are equivalent.

Example 3 Show that the following two circuits are equivalent:

Figure 9.7

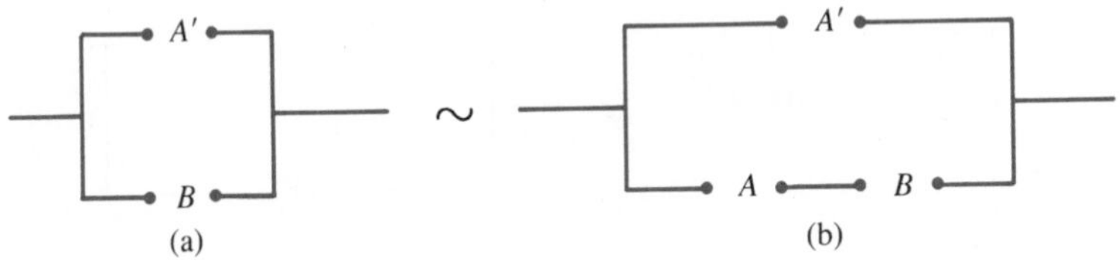

Solution The circuit in (a) is represented by the expression $A' + B$, and the circuit in (b) is represented by $A' + AB$. In switching algebra, there are only two possible values for a variable A: either $A = 1$ or $A = 0$. Therefore, there are four possible arrangements of Switches A and B in combination, as shown in the following table:

A	B	A'	$A' + B$	$A' + AB$
1	1	0	$0 + 1 = 1$	$0 + 1 \cdot 1 = 0 + 1 = 1$
1	0	0	$0 + 0 = 0$	$0 + 1 \cdot 0 = 0 + 0 = 0$
0	1	1	$1 + 1 = 1$	$1 + 0 \cdot 1 = 1 + 0 = 1$
0	0	1	$1 + 0 = 1$	$1 + 1 \cdot 0 = 1 + 0 = 1$

Inspection of this table reveals that the sequence of 0s and 1s in the fourth column $(A' + B)$ is identical to that in the fifth column $(A' + AB)$. Since $A' + B$ and $A' + AB$ have identical values for all possible values of A and B, we conclude that the two circuits represented by these expressions are equivalent. We indicate the equivalence in switching algebra by writing $A' + B = A' + AB$. ■

DUALITY

Switching-circuit algebra exhibits a property that does not exist for real-number algebra, called the **principle of duality.** To understand the principle of duality, we first learn how to find the dual of a switching circuit.

Dual of a Circuit

To find the dual of a given circuit:

- Change all parallel connections to series connections.
- Change all series connections to parallel connections.
- Change any switches that are open to switches that are closed.
- Change any switches that are closed to switches that are open.

Algebraically, the dual of a circuit is found by changing $(+)$ to $(\cdot)$ and $(\cdot)$ to $(+)$, also interchanging 0 and 1 if they occur in the expression. For example, the dual of the parallel circuit represented by $A + B$ is the series circuit whose representation is AB.

Example 4

(a) The dual of $A + AB$, obtained by the interchange $(+) \leftrightarrow (\cdot)$, is $A \cdot (A + B)$.

(b) The dual of $A + 1$, obtained by $(+) \leftrightarrow (\cdot)$ and $1 \leftrightarrow 0$, is $A \cdot 0$. ■

The concept of the dual of a circuit is used extensively in circuit algebra. The importance of duality comes from the following principle:

Principle of Duality

If two switching circuits are equivalent, then their dual circuits are also equivalent.

Example 5

(a) Since $A + 1 = 1$, by the principle of duality, $A \cdot 0 = 0$.
(b) It can be shown that $AB + A'B = B$. Therefore, by the principle of duality, $(A + B)(A' + B) = B$. ■

Example 6 Show that the circuit represented by $A + AB$ is equivalent to its dual.

Solution From Example 4(a), the dual of $A + AB$ is $A(A + B)$. The following table of closure shows that these two expressions are equal for all possible values of A and B:

A	B	$A + AB$	$A(A + B)$
1	1	1	$1(1 + 1) = 1(1) = 1$
1	0	1	$1(1 + 0) = 1(1) = 1$
0	1	0	$0(0 + 1) = 0(1) = 0$
0	0	0	$0(0 + 0) = 0(0) = 0$

Since the sequences of 0s and 1s in the third and fourth columns are identical, we conclude that $A + AB = A(A + B)$; that is, $A + AB$ and its dual $A(A + B)$ are equivalent. ■

The equivalence of certain circuits is so basic as to be considered to be part of the laws of our switching algebra. Figure 9.8 shows an important pair of equivalent circuits, whose equivalence we verify by appealing to known physical conditions. For example, the two circuits in Figure 9.8(a) are equivalent, since when two switches are wired in series it makes no difference which is "first" and which is "second." We conclude from this argument that $A \cdot B = B \cdot A$, and we say that the switching operation $(\cdot)$ is **commutative.** Similarly, we can argue that the two circuits in Figure 9.8(b) are equivalent and, therefore, the switching operation of $(+)$ is also commutative; that is, $A + B = B + A$. Equalities such as $A \cdot B = B \cdot A$ and $A + B = B + A$, which are valid for all values of A and B, are called *identities.*

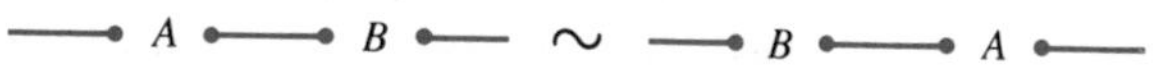

(a) $A \cdot B = B \cdot A$

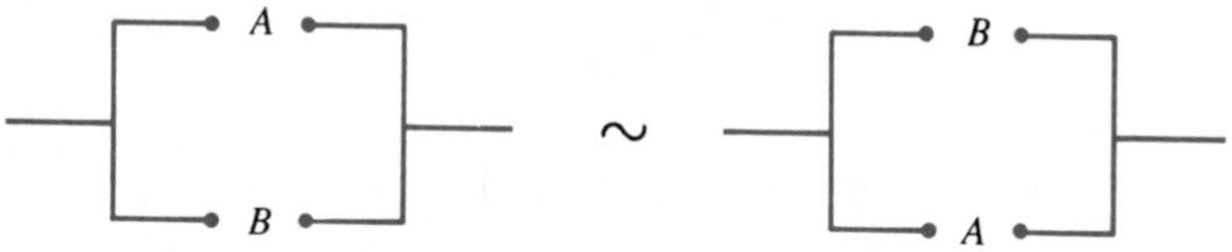

(b) $A + B = B + A$

Figure 9.8 Commutative Laws

The identity $AB = BA$ is also called the *commutative law for multiplication* and is listed in the summary table, below, with the other laws of switching algebra. We have left the verification of most of these laws for the Exercises. Notice that each law is stated in conjunction with its dual.

TABLE 9.5 **Properties of 0 and 1**

If A is a bistable switch, then

$A + 0 = A$	and	$A \cdot 1 = A$
$A \cdot 0 = 0$	and	$A + 1 = 1$
$AA' = 0$	and	$A + A' = 1$
$1' = 0$	and	$0' = 1$

Algebraic Laws

If A and B are bistable switches, then

Commutative Laws	$AB = BA$	$A + B = B + A$
Associative Laws	$A(BC) = (AB)C$	$A + (B + C) = (A + B) + C$
Distributive Laws	$A(B + C) = AB + AC$	$A + BC = (A + B)(A + C)$
Idempotent Laws	$AA = A$	$A + A = A$
Absorption Laws	$A(A + B) = A$	$A + AB = A$
Complementation Laws	$AA' = 0$	$A + A' = 1$
DeMorgan's Laws	$(AB)' = A' + B'$	$(A + B)' = A'B'$

COMMENTS:

- *Several of these laws have "look-alike" counterparts in real-number algebra and some are peculiar to switching algebra. However, even those with counterparts have totally different meanings in the two algebras.*
- *Try to understand the meaning of these laws. For example, the first idempotent law says that a circuit in which a switch is wired in series with itself is equivalent to a circuit consisting of the switch itself.*

- *Notice that a direct consequence of idempotency is that there are no multiples or powers in switching algebra; that is, there is no need to think of 3A or A^2, since $3A = A + A + A = A$ and $A^2 = A \cdot A = A$.*
- *The second distributive law is the* ***distributive law of addition over multiplication.*** *This law may seem a bit odd, because there is no corresponding law in real-number algebra.*

The laws, or identities, in Table 9.5 may henceforth be used to prove other identities. Consequently, there are two ways to prove the validity of an identity: it may be proved by appeal to the listed algebraic properties or by constructing a table of closure. Both techniques are demonstrated in the next example.

Example 7

(a) Sketch the circuit diagrams for $(A + B) + AC$ and $A + B$.

(b) Use a table of closure to show that $(A + B) + AC = A + B$ is an identity.

(c) Use the laws of switching algebra to show that $(A + B) + AC = A + B$ is an identity.

Solution

(a) The circuit for $A + B + AC$ is shown in Figure 9.9(a) and that for $A + B$ is shown in Figure 9.9(b).

Figure 9.9

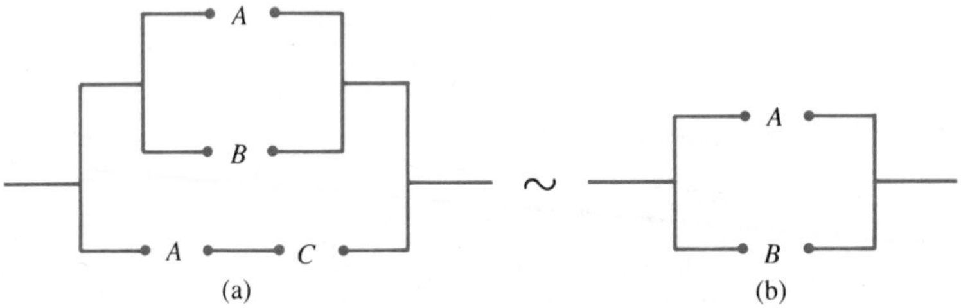

$(A + B) + AC = A + B$

(b) The table of closure for these circuits is shown below. Notice that for three switches there are eight possible arrangements of 1s and 0s.

A	B	C	$A+B$	AC	$A+B+AC$
1	1	1	1	1	1
1	1	0	1	0	1
1	0	1	1	1	1
1	0	0	1	0	1
0	1	1	1	0	1
0	1	0	1	0	1
0	0	1	0	0	0
0	0	0	0	0	0

Since the table of closure for $A + B$ is identical to that for $(A + B) + AC$, the two circuits are equivalent.

(c) We will now show that $A + B = A + B + AC$ is an identity by using the laws of switching algebra.

$(A + B) + AC = (B + A) + AC$	Commutative property of (+)
$= B + (A + AC)$	Associative property of (+)
$= B + A(1 + C)$	Distributive property of (·) over (+)
$= B + A$	$1 + C = 1$ and $A \cdot 1 = A$ (properties of 1)
$= A + B$	Commutative property of (+) ■

COMMENT: *Example 7 is an example of an application of switching algebra. When scientists build complicated circuits involving millions of switches, algebraic simplifications such as those in (c) are used to minimize the number of switches and simplify the design. The savings in circuit design are evidenced not only in smaller numbers of switches, and therefore lower costs, but also in faster operating speeds.*

Example 8 Write the switching-algebra expression that represents the circuit arrangement shown in Figure 9.10(a). Use the laws in Table 9.5 to simplify the expression. Sketch the circuit represented by the simplified expression.

Figure 9.10

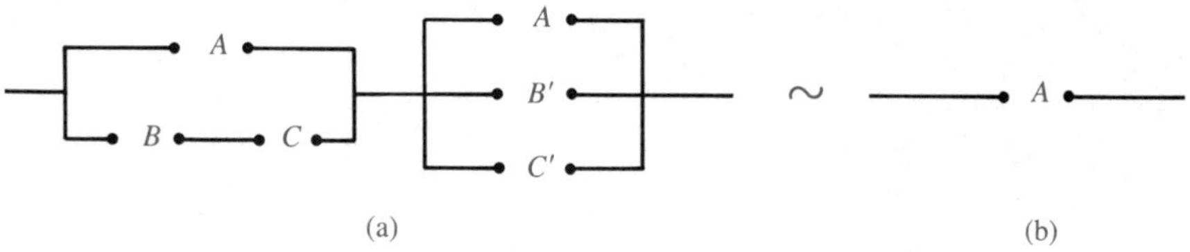

Solution The circuit in Figure 9.10(a) may be represented by $(A + BC)(A + B' + C')$. The following steps show the simplification:

$(A + BC)(A + B' + C')$	
$= (A + BC)(A + (BC)')$	DeMorgan's law for $B' + C'$
$= AA + A(BC)' + BCA + BC(BC)'$	Distributive property of (·) over (+)
$= A + A((BC)' + BC) + 0$	$AA = A$ (Idempotent Law) Distributive property of (·) over (+) $BC(BC)' = 0$ (Complementation Law)
$= A + A \cdot 1$	$BC + (BC)' = 1$ (Complementation Law)
$= A + A$	$A \cdot 1 = A$ (property of 1)
$= A$	Idempotent Law

Thus, the given circuit is actually nothing more than Switch A. This is shown in Figure 9.10(b). ■

EXERCISES SECTION 9.1

The expressions in Exercises 1–10 correspond to switching circuits. Sketch each circuit and use switching algebra to show that it may be reduced to one relay, an open circuit, or a closed circuit.

1. $A(A + B)$
2. $AB + A$
3. $(A + B)(A' + B)$
4. $A + B + A'$
5. $AB + A(A + B)$
6. $ABC + AB + A$
7. $AB + BC' + A'B$
8. $AA' + A + AB$
9. $A(A + B) + A(A + C)$
10. $C + BC + C' + AC'$

For Exercises 11–20, write the dual of each of the expressions in Exercises 1–10.

In Exercises 21–30, prove the given identity by using the laws in Table 9.5 and also by constructing a table of closure. Use the principle of duality to write another identity.

21. $(A + B)(A' + B) = B$
22. $(A + B)' = A'B'$
23. $AB(A + C) = AB$
24. $AB + B + BC' + C = B + C$
25. $(A + B + C)(ABC + D) = ABC + (A + B + C)D$
26. $(A + B')(A' + B) = 1$
27. $(AB + AB' + A'B')' = A'B$
28. $AB' + AC + AB = A$
29. $ABC + (A + B)(A + C) = A + BC$
30. $(A' + C)(A + B)(A + B') = AC$
31. Use switching algebra or a table of closure to show that

$$(A + B)(A' + C)(B + C) = (A + B)(A' + C)$$

In Exercises 32–37, write and simplify algebraic expressions that describe each of the switching circuits. Sketch equivalent simplified circuits.

32.

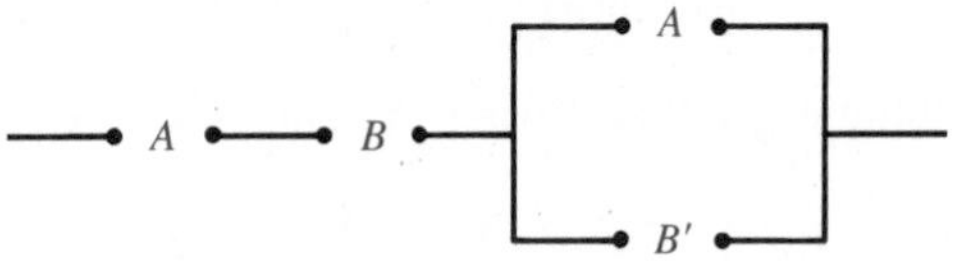

33.

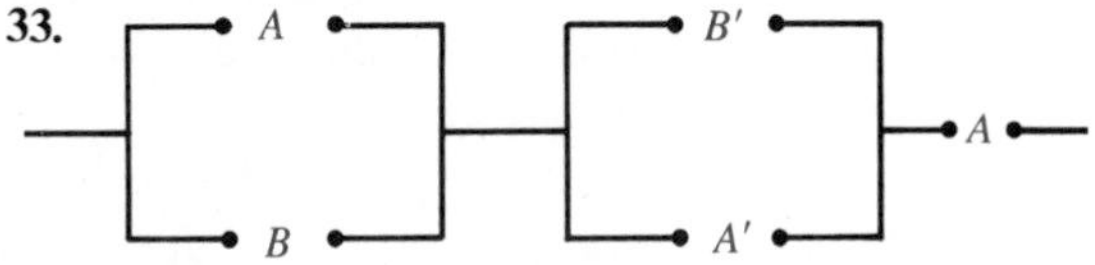

34.

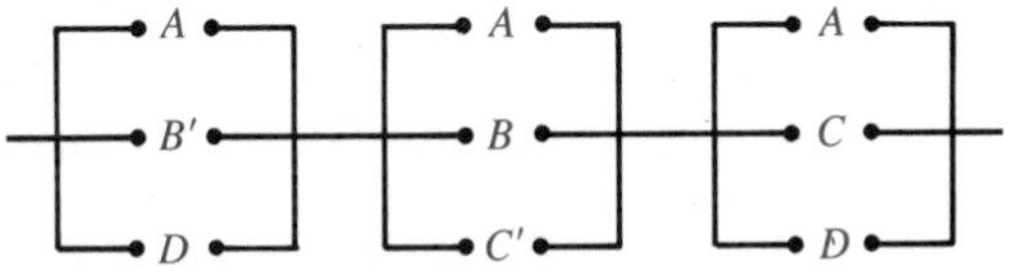

35.

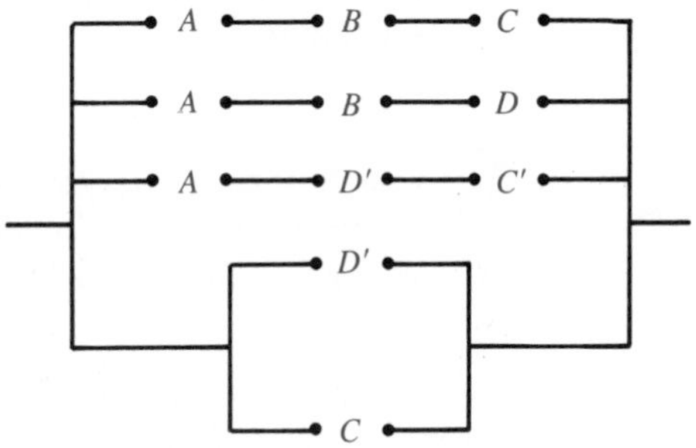

36.

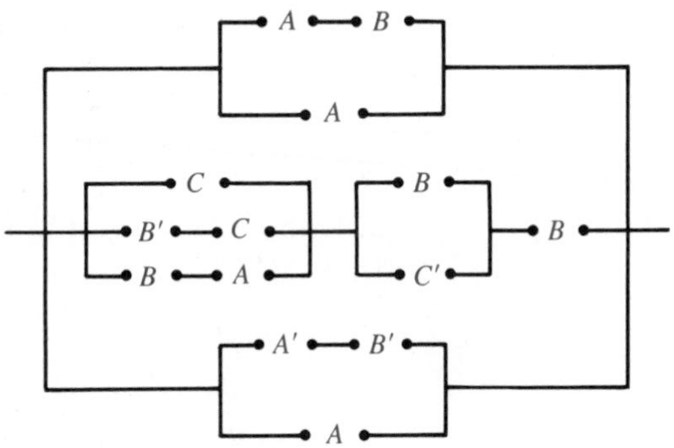

37.

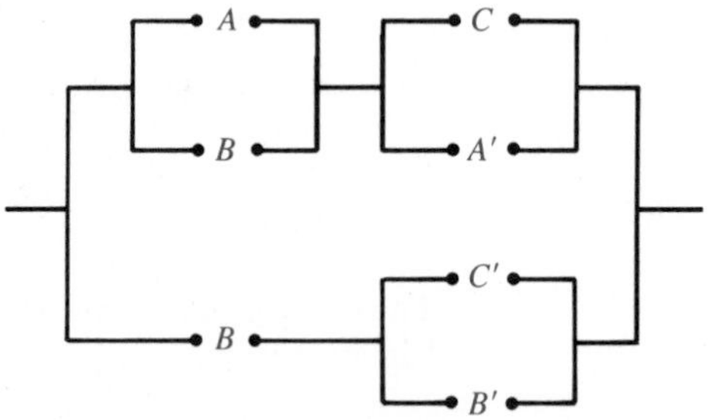

9.2 BOOLEAN ALGEBRA: AN ABSTRACT SYSTEM

We may use the algebra of switches developed in Section 9.1 to motivate the definition of the more general algebra called *Boolean algebra.* This process of considering the laws and properties of a mathematical system independent of any application is called *abstraction,* and is at the heart of many mathematical innovations. Whereas the laws derived and discussed in Section 9.1 had to do with electric switches, the process of abstraction allows us to study these laws independent of switching circuits and to identify applications in areas unrelated to switches.

BOOLEAN ARITHMETIC

Recall that a symbol in the algebra of real numbers may take on any real value defined by the domain of the variable. In Boolean algebra, the symbols represent only two values—namely, 0 and 1. These numbers are called **Boolean numbers** and should not be confused with real numbers. The arithmetic of Boolean numbers involves only the operations of addition $(+)$ and multiplication $(\cdot)$. The operations of subtraction and division are not used in Boolean arithmetic. The following laws of Boolean arithmetic are valid for the operations of addition and multiplication:

Laws of Boolean Arithmetic

1. $0 + 0 = 0$ $\quad 1 \cdot 1 = 1$
2. $1 + 1 = 1$ $\quad 0 \cdot 0 = 0$
3. $0 + 1 = 1 + 0 = 1$ $\quad 0 \cdot 1 = 1 \cdot 0 = 0$

COMMENT: *Only one of the six laws that govern Boolean arithmetic seems to be unusual. When you first see the identity* $1 + 1 = 1$*, you may think there has been a typographical error. However, this is not a misprint;* $1 + 1 = 1$ *is an identity in binary Boolean arithmetic. It may help to recall from Section 9.1 that 1 represents a closed circuit and* $(+)$ *represents a parallel arrangement. In the context of circuits,* $1 + 1 = 1$ *means simply that if both switches in a parallel circuit are closed, the parallel circuit is closed.*

These laws of Boolean arithmetic can also be expressed in terms of addition and multiplication tables, similar to those used in ordinary arithmetic. One major difference, however, is that Boolean arithmetic tables are complete with only the numbers 0 and 1, whereas real-number addition and multiplication tables are incomplete because they only include a few of the positive integers.

Addition

$(+)$	1	0
1	1	1
0	1	0

Multiplication

$(\cdot)$	1	0
1	1	0
0	0	0

Equations and Identities

In Boolean algebra, as in real-number algebra, we can construct conditional statements, called **equations,** involving unknown quantities. Thus, $X + 0 = 1$ is a Boolean equation. Since there are only two Boolean numbers, 0 and 1, it is easy to determine whether there are any values for which the statement is true. In the case of the equation $X + 0 = 1$, the equation is true for $X = 1$ but not true for $X = 0$. An equation such as $X + 1 = 0$ has no solution, because neither $X = 0$ nor $X = 1$ will make this statement true. A Boolean equation that is true for all values of the variable is called an **identity.**

Example 1 Show that $X + 1 = 1$ is a Boolean identity; that is, show that $X + 1 = 1$ is true for all possible values of the variable.

Solution In Boolean algebra, there are only two possible values for X, either $X = 1$ or $X = 0$. If we substitute $X = 1$, we have $1 + 1 = 1$, which is true from the second rule of Boolean arithmetic. If we substitute $X = 0$, we have $0 + 1 = 1$, which is true from the third rule of Boolean arithmetic. Since these are the only two possibilities, we can say $X + 1 = 1$ is true for all possible values of X and is, therefore, an identity. ■

A Boolean **dual** is obtained from a given Boolean statement by interchanging $(+)$ with $(\cdot)$ and 0 with 1. Notice that the six laws of Boolean arithmetic are written as three pairs of laws that are duals of one another. Thus, $0 + 0 = 0$ and $1 \cdot 1 = 1$ are dual statements obtained by applying $(+) \leftrightarrow (\cdot)$ and $(0) \leftrightarrow (1)$ at the same time. Identities can be obtained from one another using the principle of duality.

Principle of Duality

The dual of a Boolean identity is also an identity.

COMMENT: *The principle of duality is not an assumed principle, but one that we conclude must exist because of the symmetry of Boolean algebra. Notice that the principle of duality does* not *apply to the algebra of real numbers. (Try it!)*

Example 2 Since $X + 1 = 1$ is a Boolean identity, by the principle of duality, its dual $X \cdot 0 = 0$, obtained by $(+) \leftrightarrow (\cdot)$ and $(0) \leftrightarrow (1)$, is also an identity. ■

Proving Boolean Identities

If X and Y represent Boolean variables, then $X = X + XY$ is an identity, in which X and Y can take on only the values 0 and 1. Proving the validity of this identity is a relatively straightforward process. Since X and Y may take on only the values 0 and 1, we can prove a Boolean identity by enumerating all possible cases and examining the results to see whether both sides of the equality produce the same sequence of 0s and 1s. We can demonstrate the validity of identities by constructing a table to show all possible outcomes.

Example 3 Prove the Boolean identity $X = X + XY$.

Solution We examine the four possible cases in the following table:

X	Y	$X + XY$
1	1	$1 + 1 \cdot 1 = 1$
1	0	$1 + 1 \cdot 0 = 1$
0	1	$0 + 0 \cdot 1 = 0$
0	0	$0 + 0 \cdot 0 = 0$

Inspection of this table reveals that the sequence of 0s and 1s in the first column (X) is identical to the third column ($X + XY$). Since X and $X + XY$ have identical values for all possible values of X and Y, we conclude that $X = X + XY$ is a Boolean identity. ■

COMMENT: *Notice that because of the principle of duality, $X = X(X + Y)$, the dual of the identity in Example 3, is also an identity. This, of course, can be proved by constructing a table of values showing that X and $X(X + Y)$ have identical values for all possible values of X and Y.*

COMPLEMENTS

We encountered the idea of the complement of a switch in Section 9.1: if Switch A is open, its complement, Switch A′, is closed and vice versa. We generalize the complement operator (′) in Boolean algebra and define it as follows:

$$1' = 0 \qquad \text{and} \qquad 0' = 1$$

Thus, the complement of 1 is 0 and the complement of 0 is 1. The next example shows that the complement of a complement is the original variable.

Example 4 Show that $(X')' = X$.

Solution We demonstrate the validity of this statement in the following table:

X	X'	$(X')'$
1	0	1
0	1	0

Since the first and third columns are identical, we conclude that $(X')' = X$. ■

We are now ready to list the basic laws of Boolean algebra. These laws are generalizations of the laws introduced in Section 9.1 for switching algebra. The validity of each can be checked by the method used in Example 3 to verify Boolean identities. Each law consists of a statement and its corresponding dual.

TABLE 9.6 **Basic Laws of Boolean Algebra**

Commutative Laws	$XY = YX$	$X + Y = Y + X$
Associative Laws	$X(YZ) = (XY)Z$	$X + (Y + Z) = (X + Y) + Z$
Distributive Laws	$X(Y + Z) = XY + XZ$	$X + YZ = (X + Y)(X + Z)$
Idempotent Laws	$XX = X$	$X + X = X$
Absorption Laws	$X(X + Y) = X$	$X + XY = X$
Complementation Laws	$XX' = 0$	$X + X' = 1$
DeMorgan's Laws	$(XY)' = X' + Y'$	$(X + Y)' = X'Y'$
Identity Laws	$1 \cdot X = X$	$0 + X = X$
Dominance Laws	$0 \cdot X = 0$	$1 + X = 1$

COMMENTS:

1. *Because of the idempotent law, there are no powers or multiples of a Boolean variable.*
2. *The distributive law of* $(+)$ *over* $(\cdot)$ *seems strange because there is no such law in real-number algebra.*
3. *The Boolean number* 1 *has the interesting property that* $X + 1 = 1$. *This result follows from the fact that* $1 + 1 = 1$ *and* $0 + 1 = 1$.

SIMPLIFYING ALGEBRAIC EXPRESSIONS

The basic laws in Table 9.6 may be used to simplify a Boolean expression—that is, an expression, consisting of a finite number of unknowns, that involves the operations of $(+)$, $(\cdot)$, and $(')$. The ability to simplify Boolean expressions is sometimes more than just an academic exercise, as we saw in the discussion of switching algebra.

DEFINITION Boolean Algebra Expressions

- A **monomial** is either a single variable or a product of single variables. For example, X, X', XY, and $X'YZ$ all are monomials.
- A **polynomial** is an indicated sum of monomials. Each of the monomials is called a term of the polynomial. $X + X' + XY + X'YZ$ is an example of a polynomial.
- A **factor** is a monomial or polynomial that multiplies another monomial or polynomial. For example, in the expression $X(X + Y)$ the factors are X and $X + Y$.
- An expression is **linear** if it consists of a single variable (with or without a prime) or a sum of single variables. For example, X and $X' + Y$ are linear expressions, but YZ and $XY + Z$ are not linear.

The expansion and simplification of Boolean algebraic expressions is based upon the laws in Table 9.6. The following example shows the process:

Example 5 Expand and simplify $(X + Y)(X' + Y)$.

Solution

$(X + Y)(X' + Y)$	
$= X(X' + Y) + Y(X' + Y)$	Distributive property of $(\cdot)$ over $(+)$
$= XX' + XY + YX' + YY$	Distributive property of $(\cdot)$ over $(+)$
$= XY + X'Y + Y$	$XX' = 0$ (Complementation Law) $YY = Y$ (Idempotent Law)
$= Y(X + X') + Y$	Factoring (Distributive property of $(\cdot)$ over $(+)$)
$= Y \cdot 1 + Y$	$X + X' = 1$ (Complementation Law)
$= Y + Y$	$Y \cdot 1 = Y$ (Multiplicative Identity)
$= Y$	$Y + Y = Y$ (Idempotent Law) ▪

COMMENT: *As in real-number algebra, the word* simplify *can be ambiguous. As a general rule, the* simplest form *is the one that requires the least number of symbols, where each variable, parenthesis, or operation constitutes a symbol. Thus, the expression $(X + Y)(X' + Y)$ has eleven symbols and Y has one symbol.*

Example 6 Show that $X + X'Y = X + Y$.

Solution

$X + X'Y = (X + X')(X + Y)$	Distributive property of $(+)$ over $(\cdot)$
$= 1(X + Y)$	$X + X' = 1$
$= X + Y$	Multiplicative identity ▪

COMMENT: *The result of Example 6 is sometimes listed as a basic law of Boolean algebra because it can be very useful in simplifying other Boolean expressions. A similar rule, $X' + XY = X' + Y$, is also very useful. Its proof is left for the Exercises at the end of this section.*

Example 7 Simplify $X'(X + Y) + Y'(Y' + Z) + Y$.

Solution

$X'(X + Y) + Y'(Y' + Z) + Y$	
$= X'X + X'Y + Y'Y' + Y'Z + Y$	Distributive property of $(\cdot)$ over $(+)$
$= X'Y + Y' + Y'Z + Y$	$X'X = 0$ and $Y'Y' = Y'$
$= Y' + X' + Y + Z$	Result of Ex. 6 in form $Y' + YX' = Y' + X'$ and $Y + Y'Z = Y + Z$
$= 1 + X' + Z$	$Y' + Y = 1$
$= 1$	Dominance Law ▪

Example 8 Factor the polynomial $XW + XZ + YW + YZ$ into linear factors.

Solution Proceed exactly as in real-number algebra:

$XW + XZ + YW + YZ$	
$= X(W + Z) + Y(W + Z)$	Distributive property of $(\cdot)$ over $(+)$
$= (X + Y)(W + Z)$	Distributive property of $(\cdot)$ over $(+)$ ▪

COMMENT: *It can be shown that every Boolean polynomial can be factored into a product of linear factors. To carry out this factorization, you should learn to use* both *distributive laws, as illustrated in the next example.*

Example 9 Factor the binomial $XY + WZ$ into linear factors.

Solution

$XY + WZ = (XY + W)(XY + Z)$	Distributive property of $(+)$ over $(\cdot)$
$= (W + XY)(Z + XY)$	Commutative Law
$= (W + X)(W + Y)(Z + X)(Z + Y)$	Distributive property of $(+)$ over $(\cdot)$

■

Finally, we wish to show an example in which DeMorgan's laws are used. In passing, we note without proof that DeMorgan's laws can be expanded to a sum or product of more than two terms; that is,

$$(X + Y + Z)' = X'Y'Z'$$

and

$$(XYZ)' = X' + Y' + Z'$$

Example 10 Simplify $(X + YZ)(X + Y' + Z')$.

Solution

$(X + YZ)(X + Y' + Z')$	
$= (X + YZ)(X + (YZ)')$	DeMorgan's law on $Y' + Z'$
$= XX + X(YZ)' + YZX + YZ(YZ)'$	Distributive property of $(\cdot)$ over $(+)$
$= X + X((YZ)' + YZ) + 0$	$XX = X$ (Idempotent Law) Distributive property of $(\cdot)$ over $(+)$ $YZ(YZ)' = 0$ (Complementation Law)
$= X + X \cdot 1$	$YZ + (YZ)' = 1$ (Complementation Law) Additive Identity
$= X + X$	Multiplicative Identity
$= X$	Idempotent Law

■

EXERCISES SECTION 9.2

In Exercises 1–6, prove each law by examining all possible cases.

1. Absorption Law: $X(X + Y) = X$
2. Absorption Law: $X + XY = X$
3. DeMorgan's law: $(XY)' = X' + Y'$
4. DeMorgan's law: $(X + Y)' = X'Y'$
5. Distributive Law: $X + YZ = (X + Y)(X + Z)$
6. Distributive Law: $X(Y + Z) = XY + XZ$

Expand and simplify the expressions in Exercises 7–16.

7. $(X + Y)(Y' + X)$
8. $(X + Y')(X' + Y')$
9. $X(X' + Y)$
10. $(XY' + Y')(X + Y')$
11. $(X + Y)(X' + Y)(X + Y')$
12. $(X' + Y')(X + Y)(X' + Y)(X + Y')$
13. $(X + Y' + Z)(XZ'(Y + Z)')$
14. $(X + Y + X'Y')(XY + Y'X)'$
15. $(X + Y + Z)'XY'$
16. $(XYZ)'(X + Y')'$

In Exercises 17–26, factor each expression into linear factors, or reduce it to a single term.

17. $X + Y'W$
18. $X + YZ$
19. $X + YZ + XZ$
20. $X + XY + Y$
21. $XZ + XY + YZ$
22. $XYZ + X'Z$
23. $XYZ + X'W$
24. $XY' + XW(Y + Z)$
25. $XY + X' + Y'$
26. $(XYZ)' + (X' + Y' + Z')'$

For Exercises 27–36, write the dual of each expression in Exercises 17–26. Expand each dual and compare it to the corresponding answer you found in Exercises 17–26.

Simplify the following expressions:

37. $XY'X'Y'$
38. $XY + XY' + X'Y + X'Y'$
39. $XYZ + X' + Y' + Z'$
40. $(XY + X'Y + X'Y')'$
41. $XYZ + XY'Z + Z'XYZ$
42. $XY(XY' + XZ + XYZ)$
43. $XY + X'Y'W'Z + XYWZ'$
44. $(X + Y' + X'Y)(X + Y + X'Y')$
45. $X'Z + Y'Z + XYZW'$
46. $(XY + X'Y' + X'Y)'[(X' + Y')(X' + Y)]'$
47. Use Boolean algebra to show that $(X + Y)(X' + Z)(Y + Z) = (X + Y)(X' + Z)$.
48. Use Boolean algebra to show that $XY(X + Y)(X' + Y)(Y + Z)(Y' + Z') = XYZ'$.

9.3 THE ALGEBRA OF STATEMENTS

The mathematics department at Conservative University is going to hire a new faculty member. The search committee is meeting to decide on the desirable qualifications for the position. Each of the five members of the committee brings a list of recommended qualifications to the meeting. The five recommendations are:

- A Ph.D. in mathematics, a published textbook, and at least five years of teaching experience.
- A Ph.D. in mathematics, a published textbook, and experience in grant-writing.
- No Ph.D., a published textbook, and less than five years of teaching experience.
- Experience in grant-writing and at least five years of teaching experience.
- A published textbook but no experience in grant-writing.

Can these qualifications be reconciled in any way to give a coherent description of the desired faculty member? There appears to be some overlap in the desirable qualifications as submitted by the committee members. However, without some systematic method of reconciling their differences, the faculty seems headed for a long and argumentative afternoon. In this section, we will show how simple statements and combinations of statements may be logically analyzed. We will also show how to solve the dilemma of the qualifications for a new faculty member!

The mathematical analysis of such statements can be traced to the origins of Boolean algebra. In fact, George Boole invented his algebra primarily as a tool for studying logic. In this form, Boolean algebra is called **symbolic logic** or the *algebra of statements.* The role of symbolic logic in some areas of computer science has provided motivation for its study by persons other than logicians. We have taken a slightly different path in this chapter; that is, we have shown how the laws of Boolean algebra can be viewed as simple restatements of the algebra of switches. Consequently, we must show how "statements" can be mathematized.

In symbolic logic, a **statement** is a declarative sentence that can be judged to be either true or false, but not both. Statements are the *variables* of symbolic logic.

Example 1

(a) The statement "The person is 30 years of age or older" is true or false depending on the age of the person. Hence, it is an acceptable variable.

(b) The statement "Today it is raining" can be determined to be true or false merely by looking outside; therefore, it is a valid variable.

(c) The statement "The statement you are reading is false" is self-contradictory and, hence, is not a valid variable.

(d) The question "Do you have a dog?" is not a declarative sentence and, hence, is not a valid variable.

(e) The statement "She is the best basketball player" does not qualify as a statement because it involves a value judgment; that is, it is true or false depending upon who is making the evaluation. ▪

Statements in symbolic logic are represented by lowercase letters such as p, q, r, and s. Simple statements are combined into compound statements by using the connectors **and** and **or**. Thus, if

p: The car is red

and

q: The car is a convertible

are simple statements, then we can use the *and* connector to make the compound statement "The car is red and the car is a convertible." In symbolic logic the *and* connector is denoted by $\wedge$, so "p and q" is written "$p \wedge q$". The compound statement "The car is red or the car is a convertible," obtained by using the *or* connector, means "*the car is red or it is a convertible or it is a red convertible.*" Here, as in set theory, we use the inclusive *or*. The *or* connector is denoted by $\vee$, so "p or q" is written "$p \vee q$".

Example 2

(a) The compound statement "Birds fly and fish swim" consists of the two simple statements:

p: Birds fly

q: Fish swim

joined by the *and* connector. In symbolic form, we write $p \wedge q$ to represent such a compound statement.

(b) The compound statement "His name is Sam or he plays the trumpet" consists of the two simple statements

r: His name is Sam

s: He plays the trumpet

joined by the *or* connector. Symbolically, we write $r \vee s$ to represent such statements. ▪

The *negation* of a statement p is denoted by $\sim p$. The negation of

p: The sky is blue

is the statement

$\sim p$: The sky is not blue

Example 3 Consider the statements

p: It is raining outside

and

q: I have my umbrella with me

(a) The statement "It is raining outside *and* I have my umbrella with me" is considered to be true if and only if both statements p and q are true.

(b) The statement "It is raining outside *or* I have my umbrella with me" is true if either p is true or q is true or both are true.

(c) The statement "It is *not* raining outside" is considered to be the negation of statement p; that is, it is true if p is false, and false if p is true. ▪

As we noted earlier, statements are declarative sentences that can be judged to be true or false. The truth or falsity of a statement is called its **truth value.** The truth value of a compound statement may be determined by writing all of the possible truth-value combinations for a given connector or series of connectors. For instance, the truth values of $p \wedge q$ and $p \vee q$ are shown in the truth tables below. Notice that $p \wedge q$ is true if and only if both p and q are true, and $p \vee q$ is true if either p is true or q is true.

p	q	$p \wedge q$	p	q	$p \vee q$
T	T	T	T	T	T
T	F	F	T	F	T
F	T	F	F	T	T
F	F	F	F	F	F

Next, we show the tables for $(\cdot)$ and $(+)$ from Boolean algebra. The similarity between these sets of tables is obvious. If we substitute 1 for T and 0 for F, the truth table for *and* becomes the Boolean multiplication table and the truth table for *or* becomes the Boolean addition table.

X	Y	$X \cdot Y$	X	Y	$X + Y$
1	1	1	1	1	1
1	0	0	1	0	1
0	1	0	0	1	1
0	0	0	0	0	0

Thus, symbolic logic and Boolean algebra are abstractly identical under the following correspondence:

Symbolic Logic		Boolean Algebra
T	$\leftrightarrow$	1
F	$\leftrightarrow$	0
$\wedge$	$\leftrightarrow$	$(\cdot)$
$\vee$	$\leftrightarrow$	$(+)$
$\sim$	$\leftrightarrow$	$(')$

Consequently, we may think of symbolic logic as an **algebra of statements,** in which a *variable* is any declarative sentence that can be determined to be true or false.

COMMENT: *Two mathematical systems that are abstractly identical are said to be* ***isomorphic.*** *Thus the algebra of statements and Boolean algebra are isomorphic, as are the algebra of switches and Boolean algebra.*

In most applications of Boolean algebra, values are assigned to the variables. In the algebra of statements, the possible values that may be assigned to the variables are $T \rightarrow 1$ and $F \rightarrow 0$. Thus, the value of a variable p in the algebra of statements is either 1 or 0. Keep in mind that the values 1 and 0 have no numerical significance; they are simply symbols that represent truth and falsity.

Since the algebra of statements and Boolean algebra are isomorphic, the laws given in Table 9.6 of Section 9.2 for Boolean algebra apply directly to the algebra of statements. Of course, the laws are interpreted in the context of the statements that they represent. Thus, the commutative property of $(+)$ says that the truth value of "p or q" is identical to that of "q or p." The other laws should be interpreted in the same way. The next example shows how compound statements may be simplified using the algebra of statements in Boolean form.

Example 4 Use the laws of the algebra of statements, from Table 9.6, to show that $p \wedge (p \vee q) = p$. This is one of the **absorption laws.**

Solution First make the interchange $\wedge \to (\cdot)$ and $\vee \to (+)$ in the given statement to obtain the statement we wish to prove, $p \cdot (p + q) = p$.

$p \cdot (p + q) = p \cdot p + p \cdot q$	Distributive property of $(\cdot)$ over $(+)$
$= p + p \cdot q$	$p \cdot p = p$ (Idempotent Law)
$= p \cdot (1 + q)$	Factoring (Distributive property of $(\cdot)$ over $(+)$)
$= p \cdot 1 = p$	$1 + q = 1$ (Dominance)

We have shown that $p \wedge (p \vee q)$ has the same truth value as p. ■

Example 5 Use the laws of the algebra of statements to simplify the compound statement $(p \vee q) \wedge (p' \vee q)$.

Solution After the interchange, we have

$(p + q)(p' + q) = p(p' + q) + q(p' + q)$	Distributive property of $(\cdot)$ over $(+)$
$= pp' + pq + p'q + qq$	Distributive property of $(\cdot)$ over $(+)$
$= pq + p'q + q$	$pp' = 0$ and $qq = q$
$= q(p + p') + q$	Factoring
$= q + q$	$p + p' = 1$
$= q$	$q + q = q$

We have shown that $(p + q)(p' + q) = q$; that is, the truth values of $(p \vee q) \wedge (p' \vee q)$ and q are identical. ■

A table of values, or truth table, may also be used to show that Boolean statements are identical without relying upon algebraic simplification. The next example involves an identity that is rather difficult to prove algebraically. See how easy it is using a table of values.

Example 6 Show that $(q \wedge r') \vee (r \wedge s) \vee (q \wedge s') = q \vee (r \wedge s) \vee (q \wedge s')$ by using a table of values.

Solution First write the given expression as a Boolean statement. Then the truth table tells the story. The two last columns are identical, and we have considered all eight possible cases.

q	r	s	qr'	rs	qs'	$qr' + rs + qs'$	$q + rs + qs'$
1	1	1	0	1	0	1	1
1	1	0	0	0	1	1	1
1	0	1	1	0	0	1	1
1	0	0	1	0	1	1	1
0	1	1	0	1	0	1	1
0	1	0	0	0	0	0	0
0	0	1	0	0	0	0	0
0	0	0	0	0	0	0	0

Since the entries for $qr' + rs + qs'$ and $q + rs + qs'$ are identical for all eight cases, we conclude that $qr' + rs + qs' = q + rs + qs'$. ▪

COMMENT: *Notice that in Example 6 $qr' + rs + qs' = q + rs + qs'$ is an identity and, thus, "cancellation" may not be used to prove identities; otherwise, we could conclude that $qr' = q$,* and this is not true.

We now show some examples of translating English statements into Boolean algebra symbols.

Example 7 Translate "Jim will go for a walk and mow the grass or go to the movies" into Boolean algebra notation.

Solution There are three activities here: going for a walk, mowing the grass, and going to the movies. We make the following assignments:

p: Jim takes a walk q: Jim mows the grass r: Jim goes to the movies

Notice that the sentence describing Jim's activities is ambiguous. Does it mean Jim will choose between (walking and mowing grass) and (going to a movie)? If this is the interpretation we take, then the description of what Jim is going to do is given by (p and q) or r. Substituting ($\cdot$) for *and* and ($+$) for *or*, we have $p \cdot q + r$.

However, if the sentence means that Jim will choose between (walking) and (mowing grass or going to a movie), then the description of what Jim is going to do is $p(q + r)$. In either case, the Boolean notation leaves no room for ambiguity. ▪

Example 8 Judy and Mary are trying to decide what to do. Judy says she wants to eat lunch and go to the movies. Mary says she wants to eat lunch or go to the movies. Describe both their desires symbolically and help them decide what to do.

Solution We make the assignment

p: We eat lunch q: We go to the movies

Then Judy's choice is (pq) and Mary's choice is ($p + q$). Now if we want to satisfy Judy *and* Mary, we want the expression $pq(p + q)$. We can show that this reduces to pq. This means that the most logical choice in this case would be for the two to have lunch *and* go to the movies.

Notice, however, that if we wish to satisfy Judy *or* Mary, we have the expression $pq + (p + q)$, which reduces to $p + q$. In this case, the most logical choice would be for the two to have lunch *or* go to the movies. ▪

Example 9 Let us return to the problem posed at the beginning of this section with regard to the new faculty position at Conservative U. We make the following assignments:

p: A person has a Ph.D. in mathematics
q: A person has published a textbook
r: A person has five years of teaching experience
s: A person has experience in grant-writing

Thus, the first committee member's recommended qualifications may be described as pqr, those of the second as pqs, those of the third as $p'qr'$, those of the fourth as rs, and those of the fifth as qs'. Thus, the opinions of the committee as to the necessary qualifications for the position can be written in Boolean algebra notation as

$$pqr + pqs + p'qr' + rs + qs'$$

In one of the exercises, you are asked to show that this Boolean expression can be reduced to $q + rs$. Since this can be done, the job description may be written simply as "The candidate must be a textbook author, or else have five years of teaching experience and also experience in grant writing." ▪

EXERCISES SECTION 9.3

In Exercises 1–10, rewrite each logical statement using algebra-of-statements notation. Use truth tables to show the equivalence of the two statements.

1. $p \vee q$; $q \vee p$

2. $(p \vee q) \wedge p$; $p \vee (p \wedge q)$

3. $(p \vee q) \wedge p$; p

4. $p \vee (q \wedge r)$; $(p \vee q) \wedge (p \vee r)$

5. $\sim p \vee p$; 1

6. $p \wedge \sim p$; 0

7. $q \vee (q \wedge \sim r)$; q

8. $(p \wedge q) \vee (\sim p \wedge r) \vee (q \wedge r)$; $(p \wedge q) \vee (\sim p \wedge r)$

9. The dual of Exercise 8

10. $(p \vee q) \wedge (p \vee r)$; $(p \vee r) \wedge (p \vee q \vee \sim r)$

For Exercises 11–20, let p: I will study, q: I will watch TV, r: I will sleep, s: I will eat. *Describe in words each of the following:*

11. pq

12. $p + q$

13. $pq + r$

14. $pq + rs$

15. $p + p'$

16. pp'

17. $s + r + s' + pq$

18. $p' + q' + r'$

19. $(pqr)'$

20. $sr'q$

For Exercises 21–30, translate each statement into symbols:

21. I will study or watch TV.

22. I will eat and watch TV.

23. I will not sleep and study.

24. I will study or watch TV and eat.

25. I will study and not sleep, or watch TV, or eat.

26. I will not eat, and I will study, and I will not watch TV.

27. I will watch TV and do none of the following: sleep or eat or study.

28. I will study while eating and watching TV.

29. I will not study while watching TV.

30. I will not sleep, watch TV, eat, or study.

31. Three friends are trying to decide what to do: Jane wants to eat hamburgers or eat hot dogs. Jim wants to eat hamburgers and eat hot dogs and go home. Sally wants to eat hot dogs and go home. Resolve their dilemma by using Boolean logic, assuming you wish to satisfy Jane and Jim and Sally.

32. Do Example 6 algebraically by justifying the following steps:

$$\begin{aligned} qr' + rs + qs' &= qr' + rs + qrs + qs' + qs'r \\ &= qr' + rs + qs' + qr(s + s') \\ &= q(r + r') + rs + qs' \\ &= q + rs + qs' \end{aligned}$$

Remember, no "cancellations" are permitted.

33. A technician at a pharmaceutical plant is testing tablets of a drug for impurities. She has two trays. One contains tablets that have been found to be up to specifications and the other contains tablets that are considered defective. In a moment of confusion, she puts one good tablet in the tray of rejects and puts one defective one in the good tray. Then another technician moves the unlabeled trays. Now the first technician decides to test two samples from one of the trays. Consider the following propositions:

p: The technician takes a tablet from the tray of good tablets

q: The first tablet meets specifications

r: The second tablet meets specifications

(a) Is q equivalent to p?
(b) Is $q \wedge r$ equivalent to p?

34. Show the validity of the identity in Example 9.

9.4 COMPUTER LOGIC CIRCUITS

In this section, we will show how switching-circuit elements are used to represent basic logic functions. In this context, a logic element is an electrical device with two input leads and one output lead. We use two arrows leading to a logic element to denote wires, or leads, where voltages are applied; we use one arrow leading from the logic element to represent a wire on which the output voltage is sensed. The applied input voltages and the output voltage are commonly called **signals.** These signals may be of long duration or may be brief changes in voltage levels, or pulses, as they are called.

Digital computer circuits are designed to respond to two different voltage levels; a high voltage (or voltage pulse) on a lead constitutes one signal on that lead, and a low voltage constitutes the opposite signal. The precise voltage levels are of no interest to us but, for the purposes of our analysis, we assign to the high-voltage signal on a lead the symbol "1" and, to the low-voltage signal the symbol "0." The basic elements, or switches, that direct the flow of logic signals in a digital logic circuit are called **logic gates.** We now want to develop an algebra of logic gates and show it to be identical to Boolean algebra.

Consider any circuit with two input leads such that there is a "1" on the output lead if and only if there is a 1 on both input leads. Such an element is called an **AND-gate,** and is represented symbolically in Figure 9.11(a). Consider a circuit in which there is a 1 on the output lead if and only if there is a 1 on either or both of the two input leads. Such an element is called an **OR-gate,** and its schematic symbol is shown in Figure 9.11(b). Finally, we have a circuit with only one input lead, as shown in Figure 9.11(c), with the property that if the voltage on the input lead is high—that is, 1—then the voltage on the output lead is low—that is, 0—and vice versa. This circuit is called a **NOT-gate,** or sometimes an *inverter.*

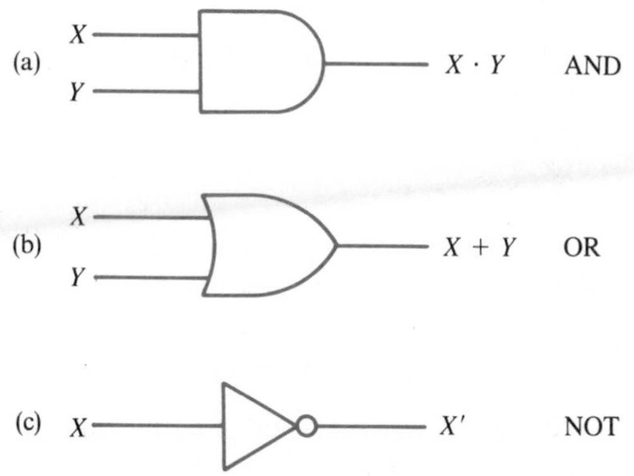

Figure 9.11

An OR-gate corresponds to the *or* connector for two statements in logic. To write the logic statement for an OR-gate, let X represent the voltage on one lead, Y represent the voltage on the other lead, and Z represent the output voltage. Then the output of an OR-gate is represented by

$$Z = X + Y$$

Similarly, an AND-gate corresponds to the *and* connector in logic, so the output of an AND-gate is represented by

$$Z = XY$$

The combination of an OR-gate followed by a NOT-gate is called a **NOR-gate,** and the combination of an AND-gate followed by a NOT-gate is called a **NAND-gate.** Thus, if X, Y, and Z have the same meanings as before, then a NAND-gate is represented by $Z = (XY)'$. Similarly, the output of a NOR-gate is given by $Z = (X + Y)'$. Figure 9.12 shows commonly used symbols for the NAND-gate and the NOR-gate. ☒

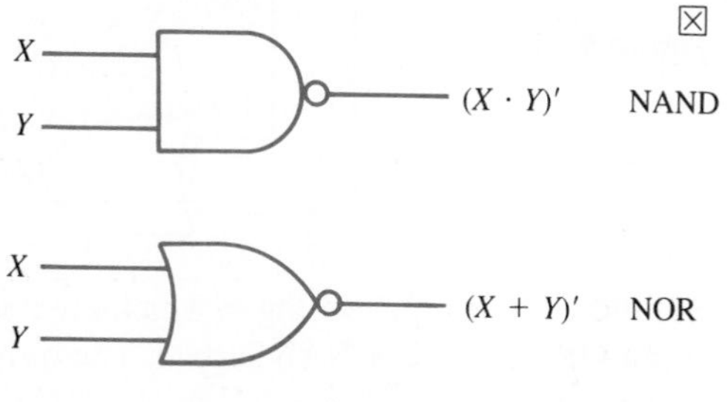

Figure 9.12

We note that an AND-gate corresponds to Boolean multiplication in the sense that the output of an AND-gate is a 1 only if both of the input voltages have the value "1"; otherwise, the output will be a 0. Similarly, the output of an OR-gate corresponds to Boolean addition, because the output of an OR-gate will be a 0 only if both voltage levels are 0; otherwise, the output will be a 1. The NOT-gate acts as the complement. With these correspondences, the logic-gate circuitry is another specific example of Boolean algebra. The table below shows the correspondence between Boolean algebra and the most important logic gates.

Gate Name	Boolean Notation
AND	$X \cdot Y$
OR	$X + Y$
NOT	X'
NAND	$(X \cdot Y)'$
NOR	$(X + Y)'$

The next example shows the working tables for AND-, NAND-, OR-, and NOR-gates using Boolean notation.

Example 1 Develop the tables for AND-, NAND-, OR-, and NOR-gates.

Solution A NAND-gate is equivalent to $(X \cdot Y)'$ and a NOR-gate to $(X + Y)'$.

X	Y	AND $X \cdot Y$	NAND $(X \cdot Y)'$	OR $(X + Y)$	NOR $(X + Y)'$
1	1	1	0	1	0
1	0	0	1	1	0
0	1	0	1	1	0
0	0	0	1	0	1

■

The next two examples show some simple logic diagrams. The output voltages for all possible input voltages are shown in the accompanying tables.

Example 2 Determine the Boolean expression for the logic circuit in Figure 9.13 and make a table showing all possible input and output voltages.

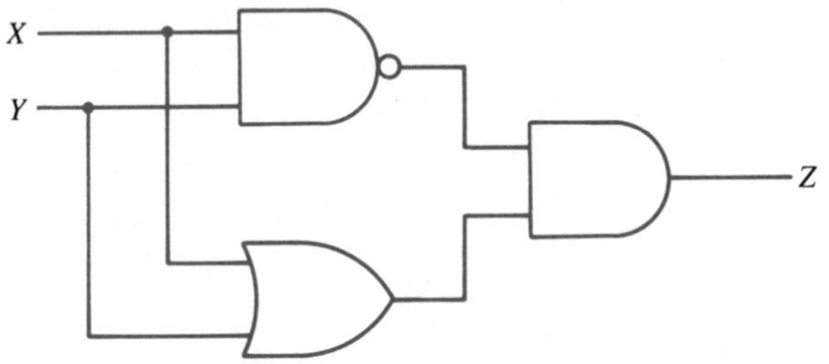

Figure 9.13

Solution The input voltages, which are denoted by X and Y, are applied directly to an OR-gate and a NAND-gate. The output of both gates then forms the input of an AND-gate. Thus, the Boolean representation of this logic circuit is

$$(X + Y) \cdot (XY)'$$

The table for this circuit is then

X	Y	$X+Y$	XY	$(XY)'$	$(X+Y)(XY)'$
1	1	1	1	0	0
1	0	1	0	1	1
0	1	1	0	1	1
0	0	0	0	1	0

■

Example 3 A logic diagram is given in Figure 9.14.

(a) Determine the Boolean expression for the logic circuit.
(b) Show a table of all possible input and output voltages.
(c) Use Boolean algebra to simplify the circuit.

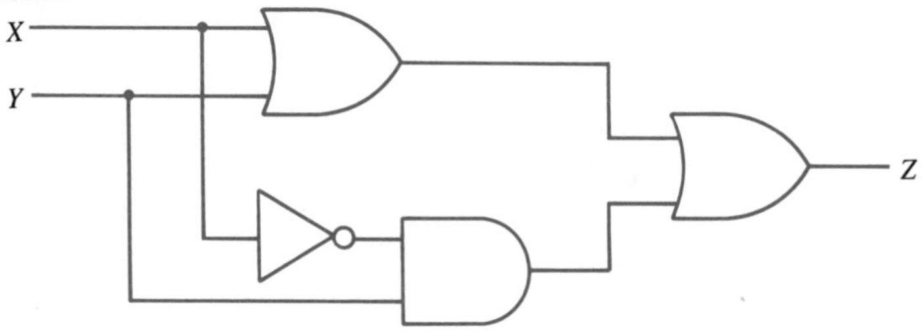

Figure 9.14

Solution

(a) We note that X and Y are input voltages for the left-hand OR-gate, and X' and Y are input voltages for the AND-gate. Therefore, the output of the OR-gate is denoted by $X + Y$ and the output of the AND-gate by $X'Y$. Since these two output values are used as input to the OR-gate, the desired expression for the output Z is

$$(X + Y) + X'Y$$

(b) The output values for this circuit are shown in the table below for all possible input values.

X	Y	$X+Y$	$X'Y$	$(X+Y)+X'Y$
1	1	1	0	1
1	0	1	0	1
0	1	1	1	1
0	0	0	0	0

Inspection of the columns of the table reveals that the $X + Y$ column and the $(X + Y) + X'Y$ column are identical. We can therefore conclude that the given circuit can be replaced with a single AND-gate.

(c) To simplify the given expression using the Boolean laws in Table 9.6, we note that

$$\begin{aligned}(X+Y)+X'Y &= X+Y+X'Y \\ &= X+Y(1+X') && \text{Distributive property of } (\cdot) \text{ over } (+) \\ &= X+Y && 1+X'=1 \text{ (Dominance)}\end{aligned}$$

This agrees with our observation in (b). ▪

Example 4 Sketch a diagram of a logic circuit corresponding to the Boolean expression $(X+Y')(XY')$.

Solution Since Y' occurs in both expressions, we first have Y as an input to a NOT-gate. The lead from the output of the NOT-gate representing Y' is used along with the lead for X as inputs to both an AND-gate and an OR-gate. Then these outputs are used as inputs to an AND-gate. (See Figure 9.15.)

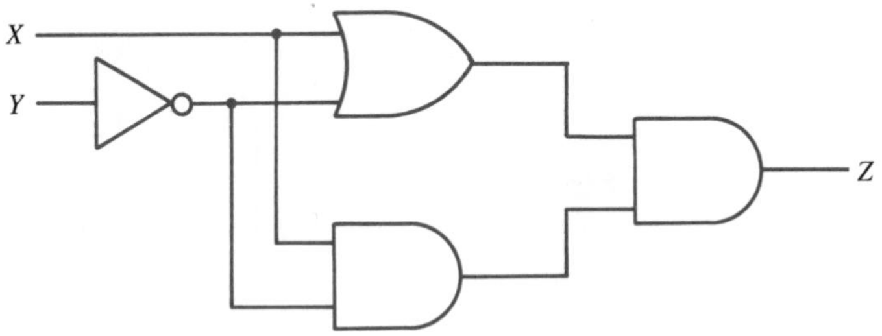

Figure 9.15

▪

EXERCISES SECTION 9.4

In Exercises 1–10, each expression represents a logic circuit. Draw the logic circuit and use Boolean algebra to determine a simplified expression for the output.

1. $X+X$ **2.** $X \cdot 0$

3. $X+1$ **4.** $X+0$

5. $X \cdot 1$ **6.** $X+XY$

7. $X(1+Y)$ **8.** $X(X+Y)$

9. $(X+Y')(XY)$ **10.** $(X'+Y')'$

In Exercises 11–20, write the Boolean expression for each logic circuit and make a table of input and output voltages.

11.

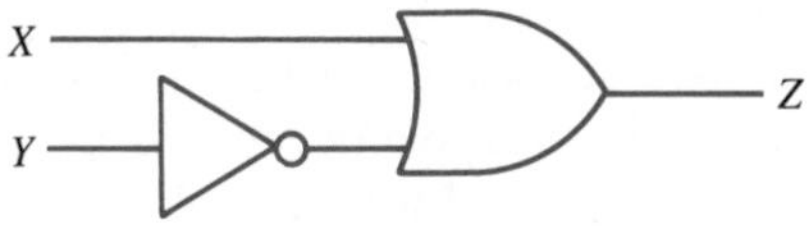

12.

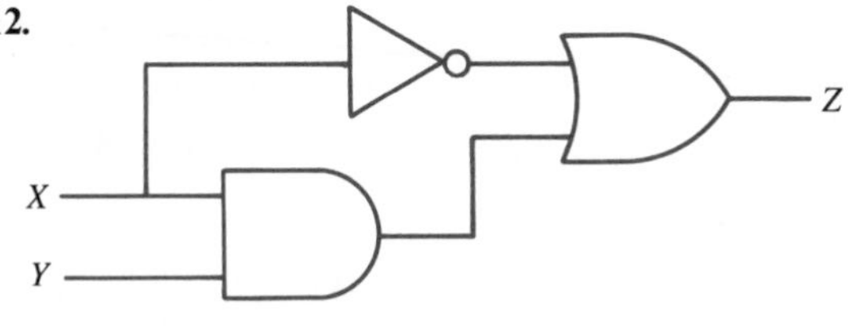

13.

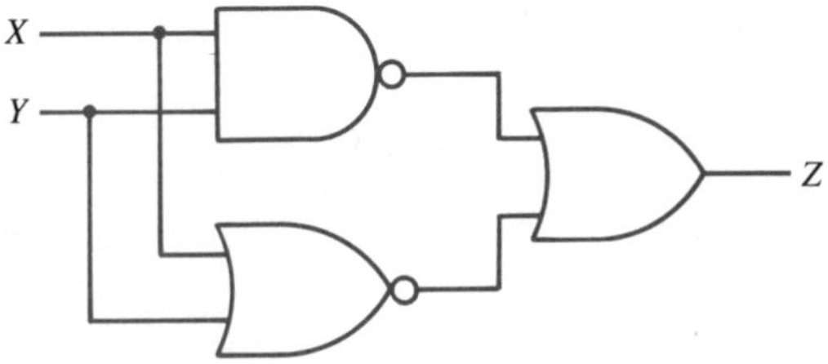

14.

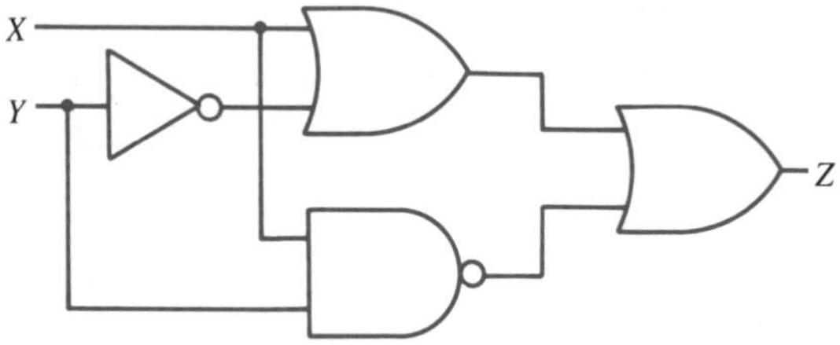

15.

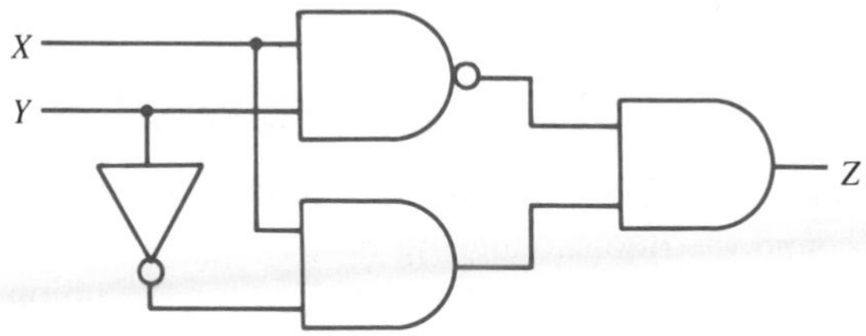

16.

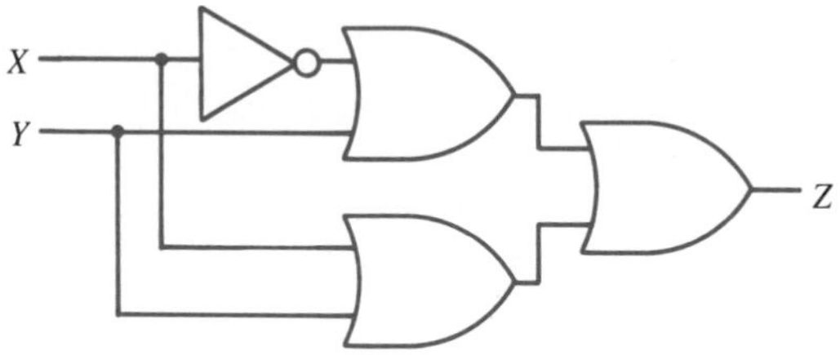

17.

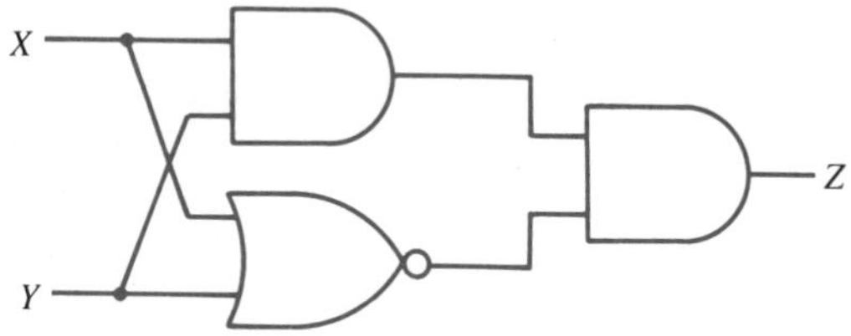

18.

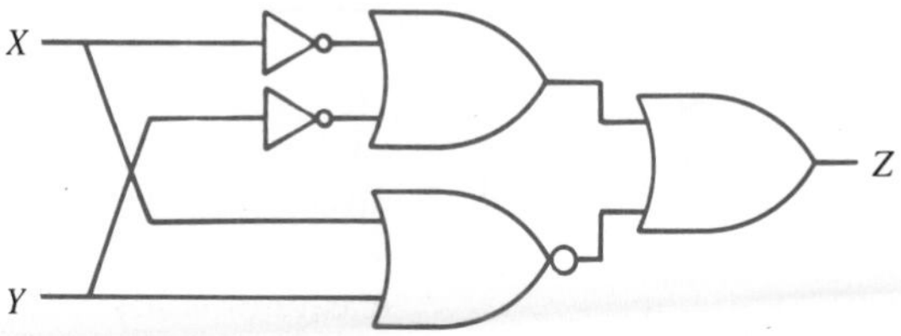

19.

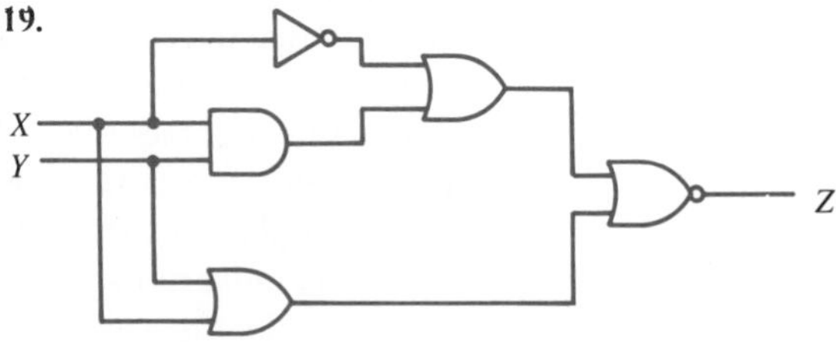

20.

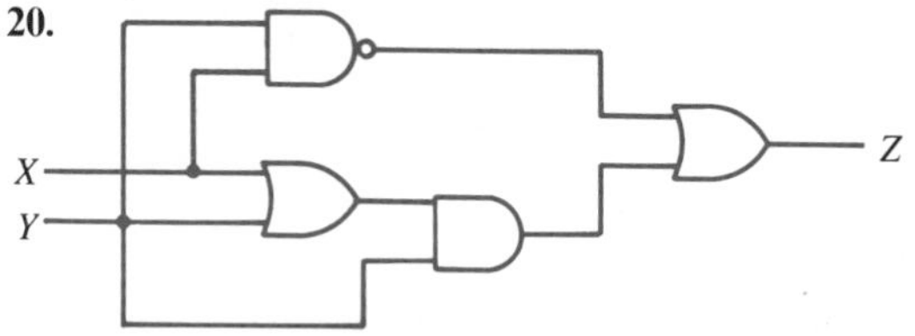

9.5 THE ALGEBRA OF SETS

We discussed set notation in Chapter 4. Now we will take another look at what we know of sets and see what, if anything, sets have in common with Boolean algebra. Suppose we examine the set operations of union and intersection. Recall that the union of set A and set B is denoted $A \cup B$ and is defined as the set of elements in A or in B. The intersection of A and B is denoted $A \cap B$ and is defined as the set of elements in A and in B. The fact that an element a is in set A is denoted by $a \in A$, and the fact that a is not in set A is denoted by $a \notin A$. The following tables are intended to show the possible outcomes for the operations of union and intersection. The entry $\in$ in a column means that an element is in the set indicated by the column heading. Similarly, the entry $\notin$ means that the element is not in the set.

A	B	$A \cup B$	A	B	$A \cap B$
$\in$	$\in$	$\in$	$\in$	$\in$	$\in$
$\in$	$\notin$	$\in$	$\in$	$\notin$	$\notin$
$\notin$	$\in$	$\in$	$\notin$	$\in$	$\notin$
$\notin$	$\notin$	$\notin$	$\notin$	$\notin$	$\notin$

Next, we show the tables for the Boolean operations of $(+)$ and $(\cdot)$.

X	Y	$X+Y$	X	Y	$X \cdot Y$
1	1	1	1	1	1
1	0	1	1	0	0
0	1	1	0	1	0
0	0	0	0	0	0

A careful study of these four tables reveals a similarity between the tables for the set operations of $\cup$ and $\cap$ and the Boolean algebra tables for $(+)$ and $(\cdot)$. As a matter of fact, if we substitute 1 for $\in$ and 0 for $\notin$, the table for the union of two sets becomes the Boolean addition table and the table for the intersection of two sets becomes the Boolean multiplication table. This suggests that set theory is abstractly the same as Boolean algebra, which is in fact the case. We say that an **isomorphism** exists between the two systems.

The universal set U is identified with the Boolean number 1 and the empty set $\varnothing$ is identified with 0. Using this notation, an isomorphism exists between Boolean algebra and set theory, under the following correspondence:

Set Theory		Boolean Algebra (Algebra of Sets)
U	$\leftrightarrow$	1
$\varnothing$	$\leftrightarrow$	0
$\cup$	$\leftrightarrow$	$(+)$
$\cap$	$\leftrightarrow$	$(\cdot)$

OPERATIONS FOR AN ALGEBRA OF SETS ■

- The *universal set,* which was previously represented by U, is represented by the symbol 1 (the same symbol that we use in arithmetic for the first natural number). Thus, 1 consists of all elements under discussion and it follows that for any set A, $A \subseteq 1$.
- The *empty set* or *null set,* which was previously represented by $\varnothing$, is represented by the numeral 0 (the same symbol that we use for zero in arithmetic).
- The *union* of set A and set B is the set of elements in A or in B. The union operator, which was previously represented by the symbol $\cup$, is represented by the $(+)$ symbol. Thus, in our algebra of sets, the union of A and B is denoted by $A + B$.
- The *intersection* of set A and set B is the set of elements in both A and B. The intersection operator, which was previously represented by $\cap$, is represented by the $(\cdot)$ symbol. Thus, the intersection of A and B is denoted $A \cdot B$.

- The *complement* of a set A, which can be represented by $\bar{A}$, is represented in set algebra by A'. The complement of a set A consists of all elements in the universal set 1 that are not in A. Thus, A' is the set of all elements that are in 1, but not in A.

Warning: *We remind you that as they are used here, the operations* $(+)$ *and* $(\cdot)$ *do not represent addition and multiplication, but are symbols for union and intersection. Similarly,* 0 *and* 1 *are* not *numbers, but symbols for the empty set and the universal set, respectively. Recall that we did essentially the same thing in Section 9.1 when describing the algebra of switches.*

As an immediate consequence of the above definitions of $(+)$, $(\cdot)$, $(')$, 0, and 1, the elementary rules from set theory can be transformed into rules of set algebra. The first column gives the rule in ordinary set notation and the second column gives the notation for the corresponding rule in the algebra of sets.

Set Notation		Set Algebra Notation
$A \cup \varnothing = A$	$\leftrightarrow$	$A + 0 = A$
$A \cap \varnothing = \varnothing$	$\leftrightarrow$	$A \cdot 0 = 0$
$A \cup U = U$	$\leftrightarrow$	$A + 1 = 1$
$A \cap U = A$	$\leftrightarrow$	$A \cdot 1 = A$
$A \cup \bar{A} = U$	$\leftrightarrow$	$A + A' = 1$
$A \cap \bar{A} = \varnothing$	$\leftrightarrow$	$AA' = 0$

COMMENT: *Notice that in set algebra it is not so strange that* $A + 1 = 1$*, since this property is just a restatement of the fact that the union of any set with the universal set is the universal set.*

Example 1 Write the following set expressions using set-algebra notation:

(a) $\overline{(A \cup B)} \cup B$
(b) $(A \cap B) \cup \varnothing$

Solution Since $\cup \to (+)$, $\cap \to (\cdot)$, $\overline{\quad\quad} \to (')$, and $\varnothing \to 0$, we have:

(a) $\overline{(A \cup B)} \cup B \to (A + B)' + B$
(b) $(A \cap B) \cup \varnothing \to AB + 0$ ■

Example 2 Write the following set-algebra expressions using set notation:

(a) $A(A + B)$
(b) $A + B' + A'B$

Solution

(a) $A(A + B) \to A \cap (A \cup B)$
(b) $A + B' + A'B \to A \cup \bar{B} \cup (\bar{A} \cap B)$ ■

BASIC PROPERTIES

When two systems are isomorphic, the laws that govern the operations defined for one system are identical to those for the other. Thus, the laws of Boolean algebra and those of set theory are identical. The algebraic properties of union and intersection are restated here using the notation of the algebra of sets.

Laws of Set Algebra

	Union	Intersection
Commutative Laws:	$A + B = B + A$	$AB = BA$
Associative Laws:	$(A + B) + C = A + (B + C)$	$(AB)C = A(BC)$
Distributive Laws:	$A(B + C) = AB + AC$ (intersection over union)	$A + (BC) = (A + B)(A + C)$ (union over intersection)
Absorption Laws:	$A(A + B) = A$	$A + AB = A$
DeMorgan's Laws:	$(A + B)' = A'B'$	$(AB)' = A'B'$

Each of these properties may be justified by identifying the statement with the set-theoretic notation developed earlier.

Example 3 Show that $A \cup (\bar{A} \cap B) = A \cup B$.

Solution First we change from set theory notation to set algebra notation. Thus, the given expression is $A + A'B = A + B$.

$$A + A'B = (A + A')(A + B) \qquad \text{Distributive property of } (+) \text{ over } (\cdot)$$
$$= 1 \cdot (A + B) \qquad A + A' = 1$$
$$= A + B \qquad 1 \cdot A = A$$

So we have shown that $A \cup (\bar{A} \cap B) = A \cup B$. ■

As we did for switching algebra, we observe that *some* of the laws for the algebra of sets are also true for the algebra of real numbers. However, some laws are distinctly different. For example, in the algebra of sets there is a second distributive property, $A + (BC) = AB + AC$, that has no parallel in the algebra of real numbers.

EXERCISES SECTION 9.5

For Exercises 1–14, interpret each expression in set-theoretic notation.

1. AB

2. $A + B$

3. $A(B + A)$

4. $A + BA$

5. AA

6. $A + A$

7. $1 \cdot 0$

8. $1 + 0$

9. $1 \cdot A$

10. $A + 0$

11. A'

12. $A(A + B)'$

13. $A + A'$

14. AA'

For Exercises 15–24, write each expression in set algebra notation.

15. $A \cap B$

16. $(A \cap B) \cap B$

17. $\overline{A \cup B}$

18. $\bar{A} \cup \bar{B}$

19. $A \cap \varnothing$

20. $A \cap A$

21. $A \cup A$

22. $A \cup U$

23. $B \cup \bar{B}$

24. $\bar{A} \cap \bar{B}$

For Exercises 25–32, write each expression in set-algebra notation and use the laws of set algebra to simplify the expression.

25. $A \cap A$

26. $A \cup A$

27. $A \cap (A \cup B)$

28. $A \cap \overline{(A \cup B)}$

29. $(A \cup B) \cap (\bar{B} \cup A)$

30. $(A \cup \bar{B}) \cap (\bar{A} \cup \bar{B})$

31. $A \cap (\bar{A} \cup B)$

32. $((A \cap \bar{B}) \cup \bar{B}) \cap (A \cup \bar{B})$

33. How would you state the principle of duality in set algebra? Is the principle valid in set algebra?

34. Does set algebra have an idempotent law? State both forms of the law in set notation.

For Exercises 35–44, write the dual of each of the statements in Exercises 15–24:

(a) In set-theoretic notation.
(b) In algebraic notation.

IMPORTANT WORDS AND PHRASES CHAPTER 9

Boolean algebra has many applications, so the terms associated with this chapter vary widely. Since, in the abstract, all the systems considered in this chapter are identical, it is easy to forget that many of these terms have special meanings. Initially, the arithmetic of Boolean algebra may seem strange, but when you have learned some of its applications, you should be comfortable with a mathematical system in which $1 + 1 = 1$.

absorption laws
algebra of statements
algebra of switches
and-connector
AND-gate
associative property
bistable devices
Boolean algebra
Boolean arithmetic
Boolean expression
commutative property
complementation
DeMorgan's laws
distributive property
  of (+) over (·)
  of (·) over (+)
duality
equivalent circuits
idempotency
isomorphism
NAND-gate
NOR-gate
or-connector
OR-gate
parallel connection
principle of duality
series connection
set algebra
table of closure

REVIEW EXERCISES CHAPTER 9

In Exercises 1–4, prove each given Boolean law by examining all possible cases.

1. Idempotent Law: $XX = X$
2. Idempotent Law: $X + X = X$
3. Complementation Law: $XX' = 0$
4. Complementation Law: $X + X' = 1$

Expand and simplify the Boolean expressions in Exercises 5–12.

5. $X(X + Y)$

6. $X(X + X')$

7. $Y(X + X')$

8. $(X + Y')(X + Y')$

9. $X(X + Y)(X + Y)$

10. $(XY)(XY + X)$

11. $(X + Y + Z')XY'$

12. $X'Z(X + Y + Z')$

Factor each Boolean expression in Exercises 13–20 into linear factors, or reduce it to a single term.

13. $Y + YX$

14. $Y + Y' + YZ$

15. $Y + YZ + Z$

16. $XYZ + Y'Z$

17. $ZY + Z' + Y'$

18. $(XY)' + (X' + Y')$

19. $XW' + X'W + X'W'$

20. $XZ' + XY' + XYZW'$

In Exercises 21–28, write each expression in set-algebra notation and use the laws of set algebra to simplify the expression.

21. $A \cup (A \cup B)$

22. $A \cup \bar{A}$

23. $A \cup \overline{(A \cap B)}$

24. $(A \cup B) \cap (\bar{A} \cup B)$

25. $(\bar{A} \cup B) \cap (A \cup \bar{B})$

26. $\bar{A} \cap (A \cap \bar{B})$

27. $(A \cap B) \cup \bar{A} \cup \bar{B}$

28. $A \cup \bar{A} \cup (A \cap B)$

Write each logic statement in Exercises 29–32 in terms of the algebra of statements. Use a truth table to show that the two statements are equivalent.

29. $\sim(p \wedge q)$; $\sim p \vee \sim q$

30. $\sim(p \vee q)$; $\sim p \wedge \sim q$

31. $p \vee (p \wedge q)$; $p \wedge (p \vee q)$

32. $(r \wedge \sim t) \vee (r \wedge s') \vee (r \wedge s \wedge t \wedge u)$; $r \wedge (\sim s \vee \sim t \vee u)$

In Exercises 33–36, write each statement in symbolic-logic form.

33. It is raining and the temperature is above 80°.

34. I will watch the football game or I will not watch TV.

35. The flag is red and white or it is not my flag.

36. The dog is brown and wears a collar.

In Exercises 37–42, sketch the switching circuit corresponding to each Boolean expression.

37. $A + AB$

38. $(A + B')(A + B)$

39. $A(AB + B)$

40. $AB(A + B)$

41. $AB'C$

42. $ABC + A' + C'$

In Exercises 43–46, write and simplify the Boolean expressions that describe each switching circuit. Sketch the equivalent simplified circuit.

43.

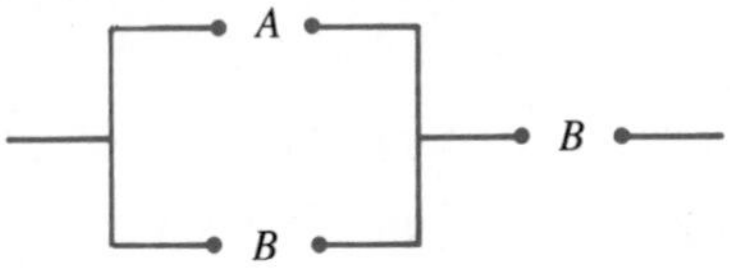

44.

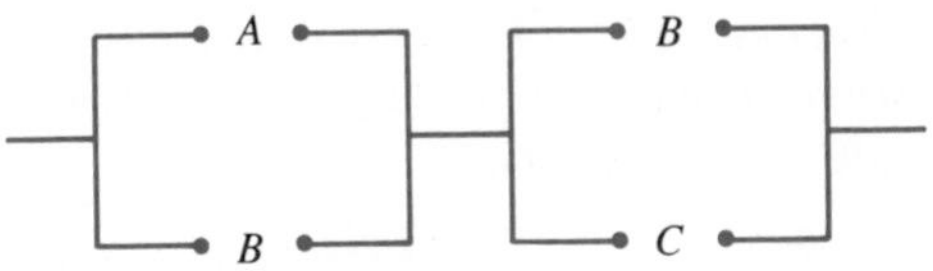

45.

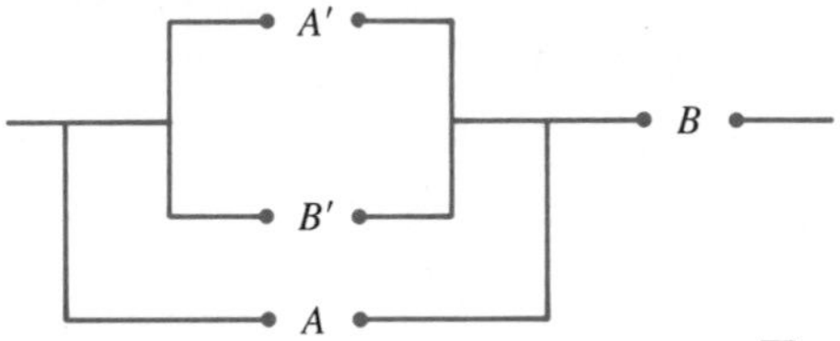

46.

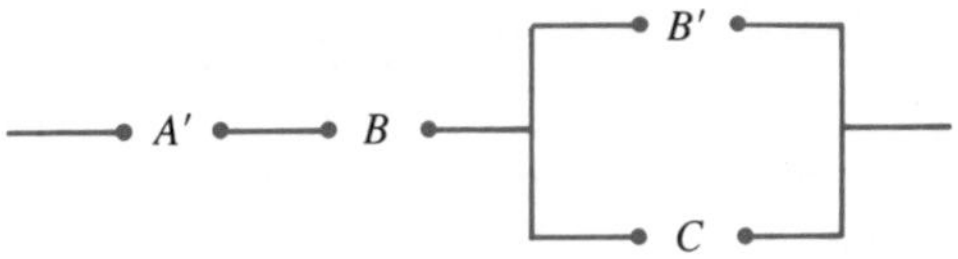

In Exercises 47–50, write and simplify the Boolean expression for the output of each logic circuit and make a table of input and output voltages.

47.

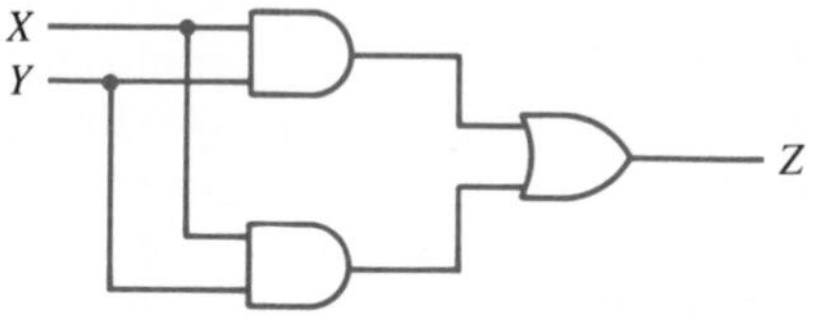

48.

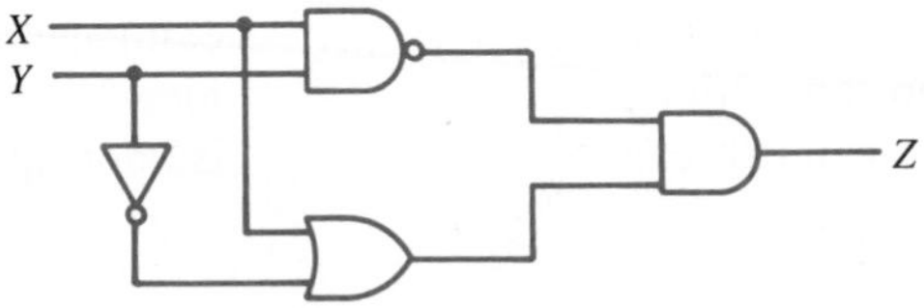

49.

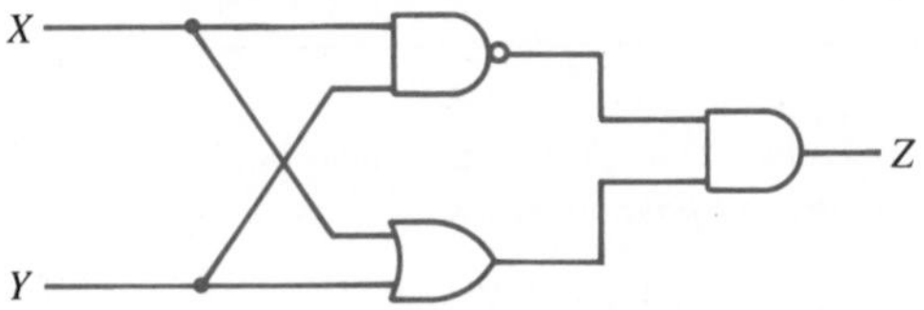

50.

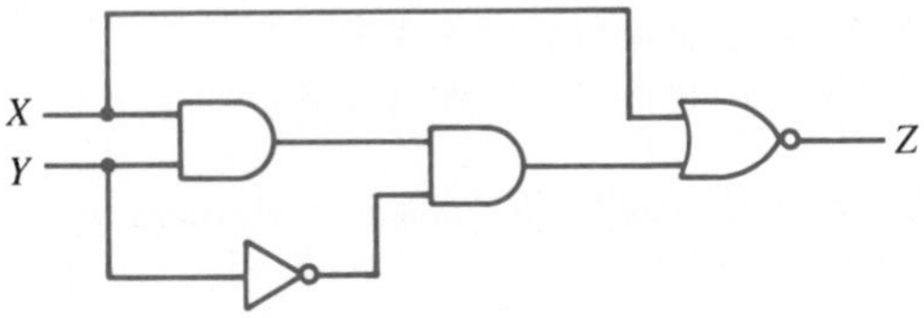

Answers to Odd-Numbered Problems

Chapter 1

Section 1.1

1. $-\frac{3}{2}$ **3.** $\frac{2}{3}$ **5.** 13 **7.** $\frac{18}{25}$ **9.** 7 **11.** $\frac{1}{2}$ **13.** -20 **15.** $\frac{1}{3}$ **17.** 6 **19.** 2 **21.** $-\frac{22}{3}$

23. $\frac{4}{5}$ **25.** $\frac{5}{3}$ **27.** $\frac{p}{q}$ **29.** $R = \frac{R_1R_2}{R_1 + R_2}$ **31.** $P = A(1 + r)^{-n}$ **33.** $a = \frac{v^2 - v_0^2}{2s}$ **35.** $n = \frac{2D}{d - D}$

Section 1.2

1. $A = \frac{1}{2}bh$ **3.** $P = 4s$ **5.** $a = \frac{v}{t}$ **7.** $x = y - 8$ **9.** $a = 5(b + 3)$ **11.** $0.2(x - 12{,}000)$

13. $\frac{1}{3}(x + y + z)$ **15.** $i = pr$ **17.** $2(x + y)$ **19.** $A = 6s^2$ **21.** \$8411 **23.** 102, 104, 106

25. 0.7 mi **27.** 220 mi **29.** 133 mph **31.** 0.5 hr

33. four touchdowns, four extra points, four field goals, and two safeties

35. 8 ft from the boy and 12 ft from the girl **37.** 1.785 hr

Section 1.3

1. Function **3.** Function **5.** Function **7.** Not a function **9.** Function **11.** Not a function

13. $D: -\infty < x < \infty$, $R: -\infty < y < \infty$ **15.** $D: -\infty < x < \infty$, $R: -\infty < y < \infty$ **17.** $D: x \le 0$, $R: y \ge 0$

19. $D: -\infty < x < \infty$, $R: -\infty < y < \infty$ **21.** D: all x except $x = -3$, R: all y except $y = 0$

23. $D: \{0, 2, 5\}$, $R: \{1, -1, 7\}$ **25.** $1 \le y \le 11$ **27.** $\frac{15}{2} \le x \le \frac{45}{2}$

29. (a) 10 (b) $3\pi + 1$ (c) $3z + 1$ (d) $3(x - h) + 1$ (e) $3x - 3h$ (f) $x = 3$ (g) $-\infty < y < \infty$

31. (a) 3 (b) -3 (c) Undefined (d) $\{3, 9, -3\}$ (e) $\{2, 5, 7\}$
(f) $H(2 + 5) = H(7) = -3$, $H(2) + H(5) = 3 + 9 = 12$; $H(2 + 5) \ne H(2) + H(5)$

33. $4x^2 - 48x + 135$ **35.** $9x^4 - 36x^3 + 36x^2$

37. $f(x) + g(x) = 2x$, $f(x) - g(x) = -6$, $f(x) \cdot g(x) = (x - 3)(x + 3)$, $\frac{f(x)}{g(x)} = \frac{x - 3}{x + 3}$

39. $f(x) + g(x) = x^2 + x$, $f(x) - g(x) = x^2 - x - 2$, $f(x) \cdot g(x) = (x^2 - 1)(x + 1)$, $\frac{f(x)}{g(x)} = \frac{x^2 - 1}{x + 1}$

41. $f(g(x)) = -2x - 2$, $g(f(x)) = -2x + 13$ **43.** $f(g(x)) = 6x + 4$, $g(f(x)) = 6x + 8$

45. $f(g(x)) = x$, $g(f(x)) = x$ **47.** $f(g(x)) = x$, $g(f(x)) = x$

Section 1.4

1.

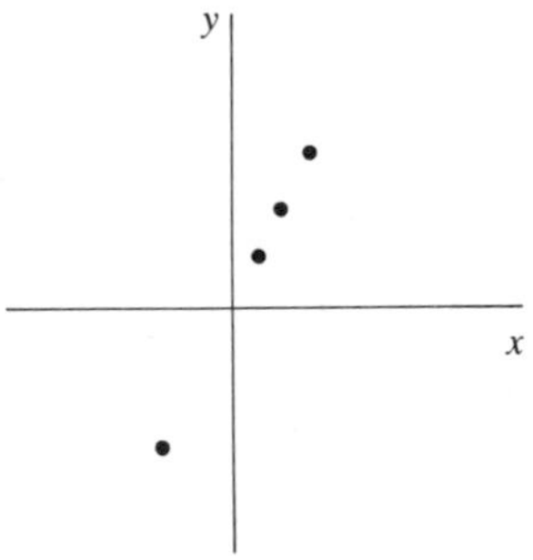

3.

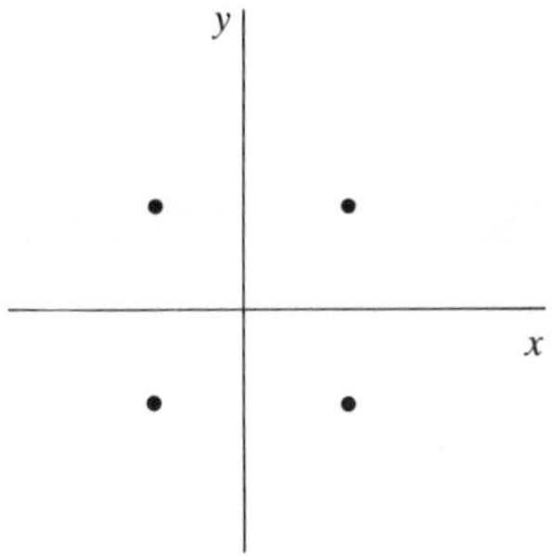

5.

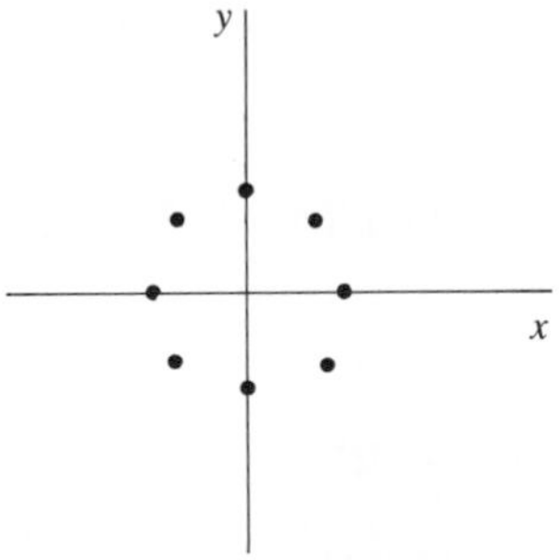

7.

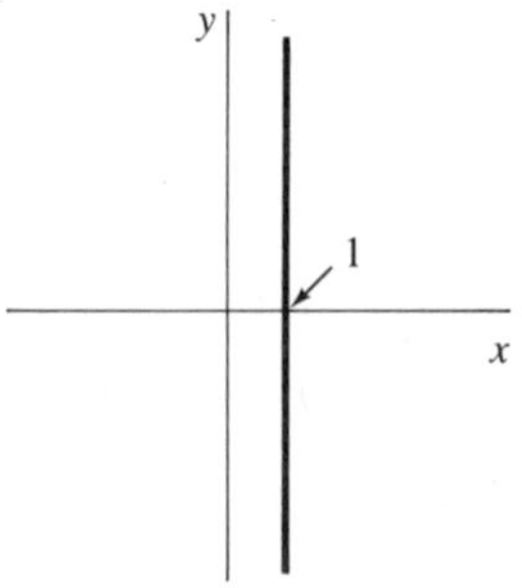

9.

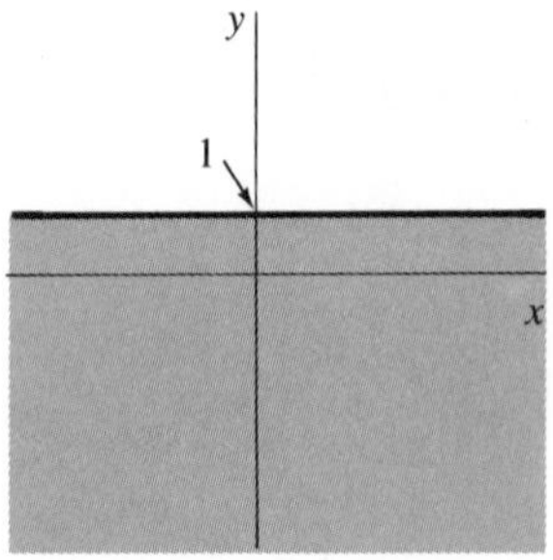

11.

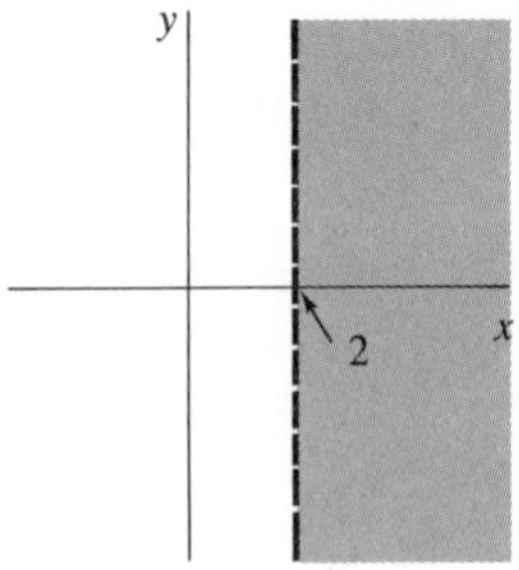

13.

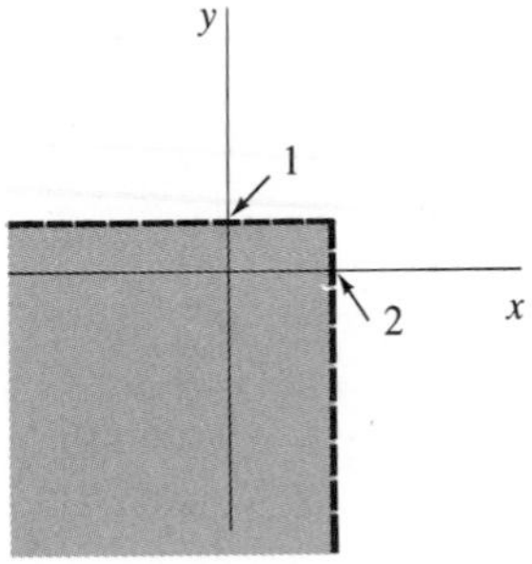

15. No points in common

17. (a) Function (b) Not a function

19. (a) Not a function (b) Function

21. $D: -\infty < x < \infty$, $R: -\infty < y < \infty$

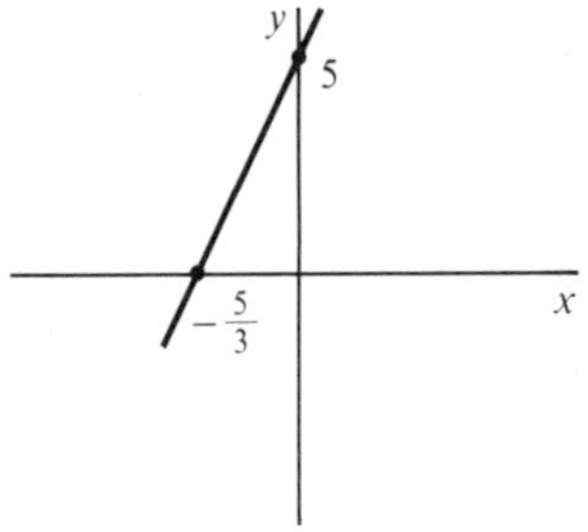

23. $D: -\infty < x < \infty$, $R: -\infty < y < \infty$

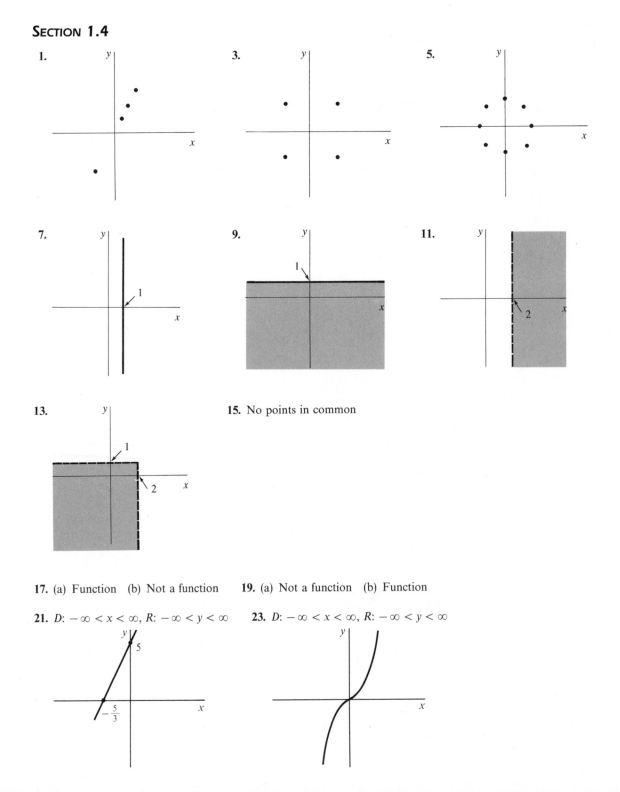

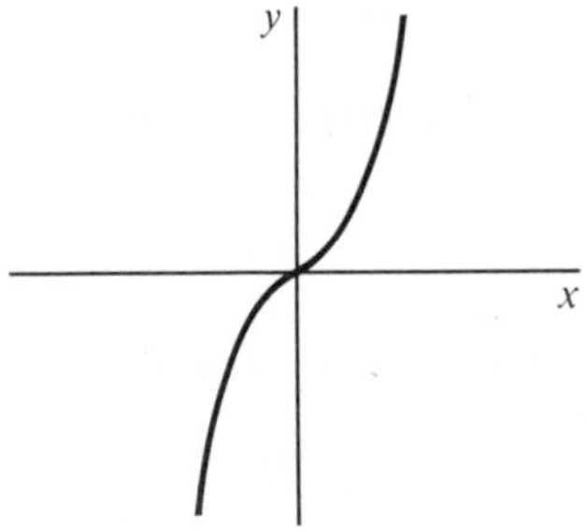

25. D: $-\infty < t < \infty$, R: $z \geq 4$

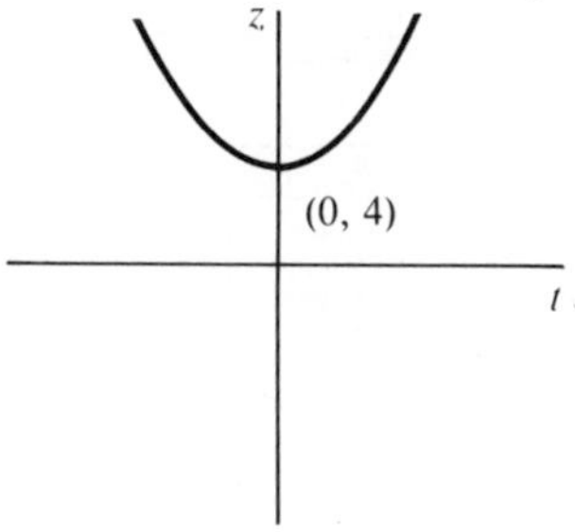

27. D: $x \geq 0$, R: $y \geq 0$

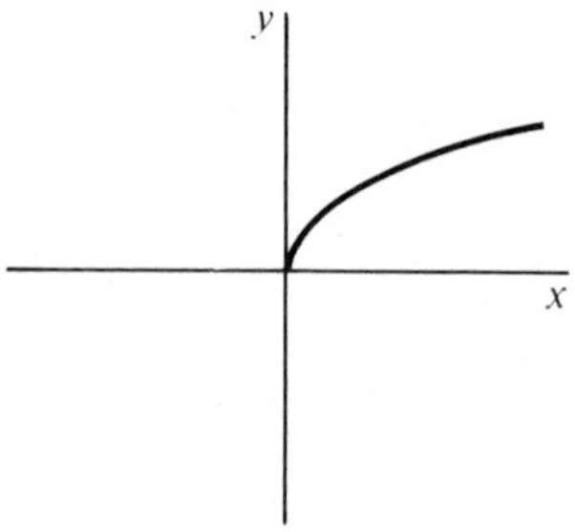

29.

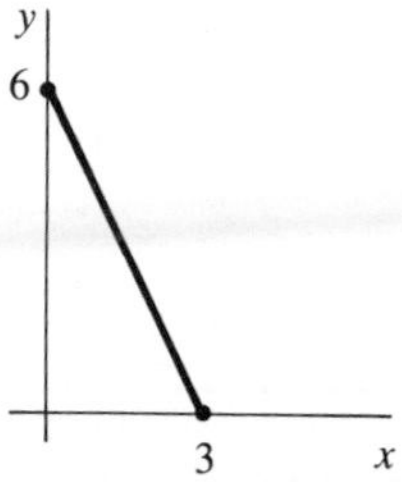

31.

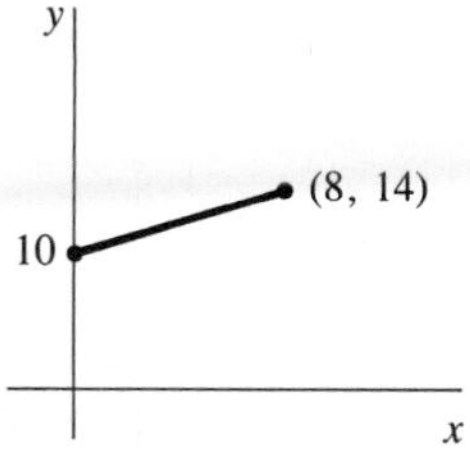

33.

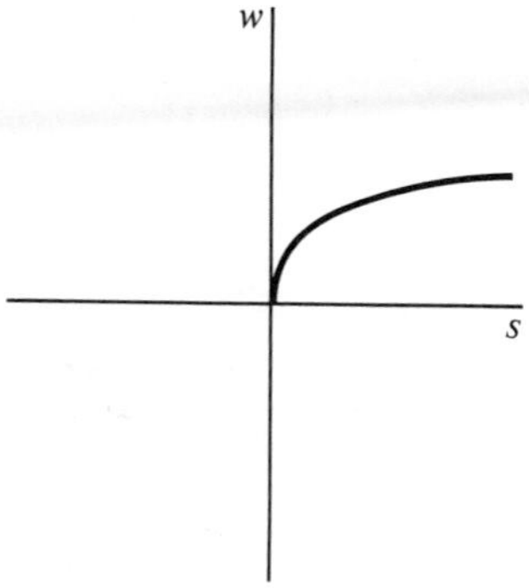

35.

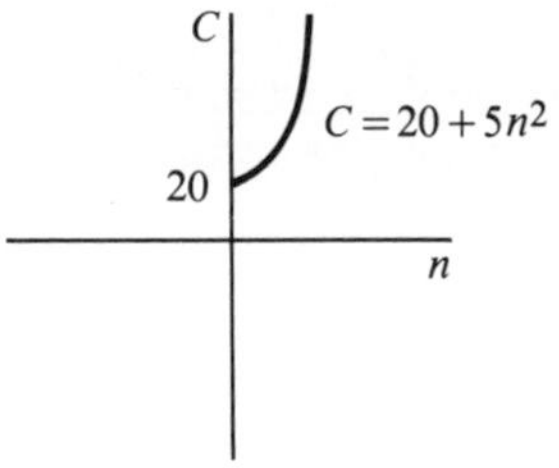

37.

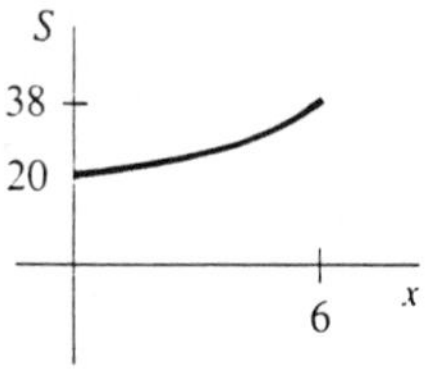

39.

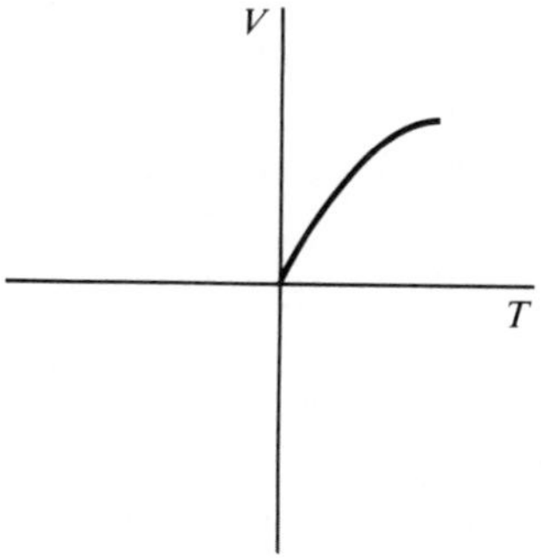

41. (a) \$3418 (b)

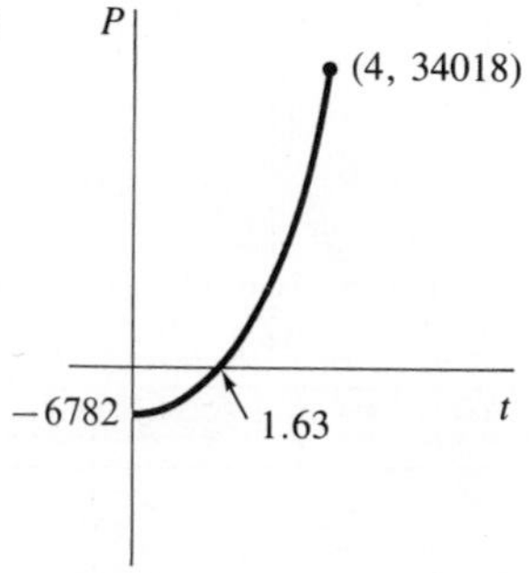

(c) 1.63 years

CALCULATOR EXERCISES

1.

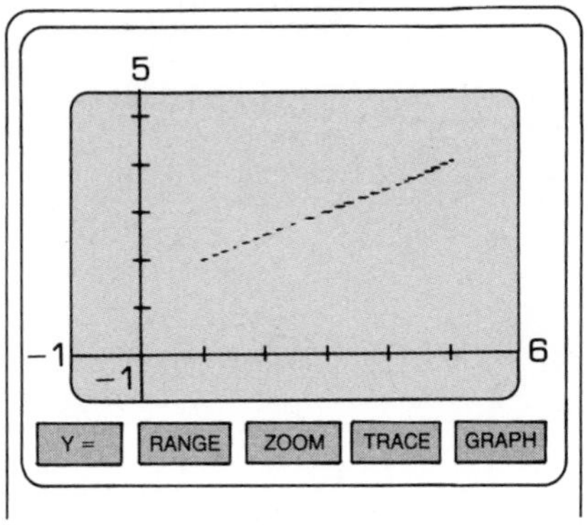

3.

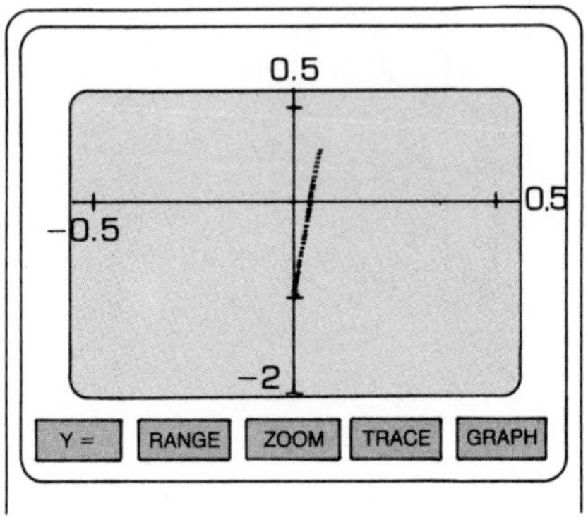

5.

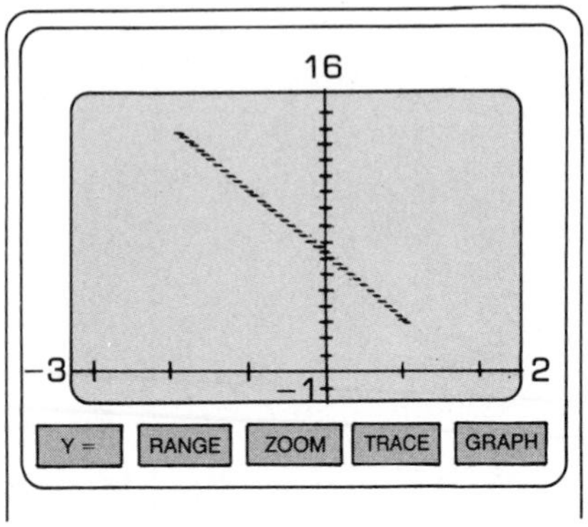

7.

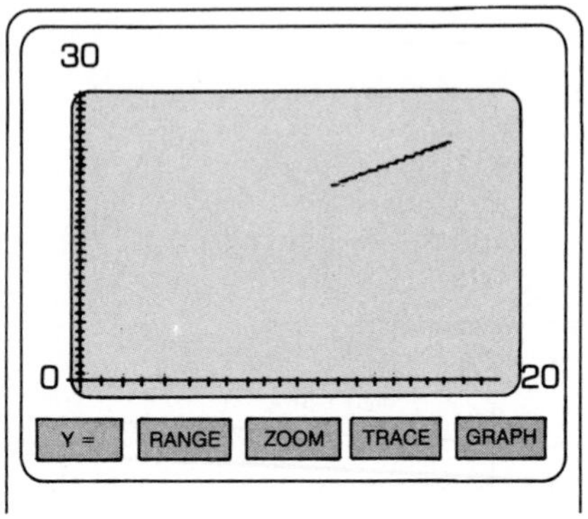

9.

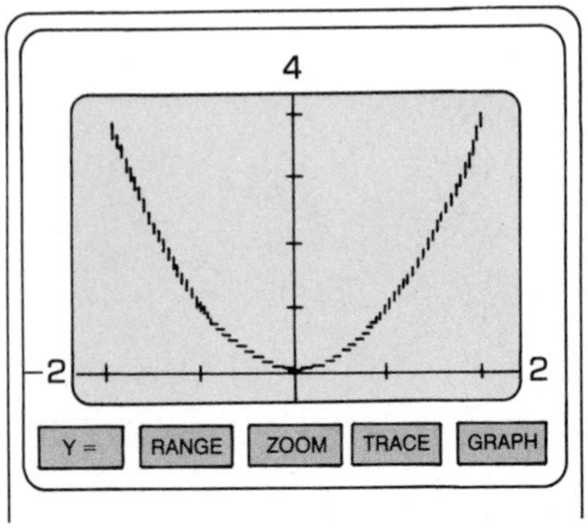

11.

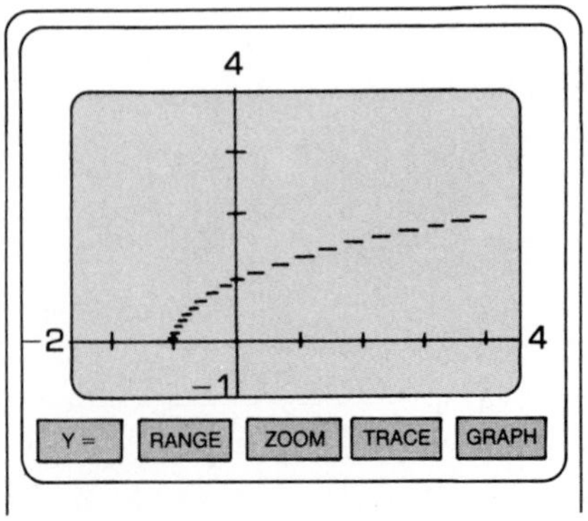

13.

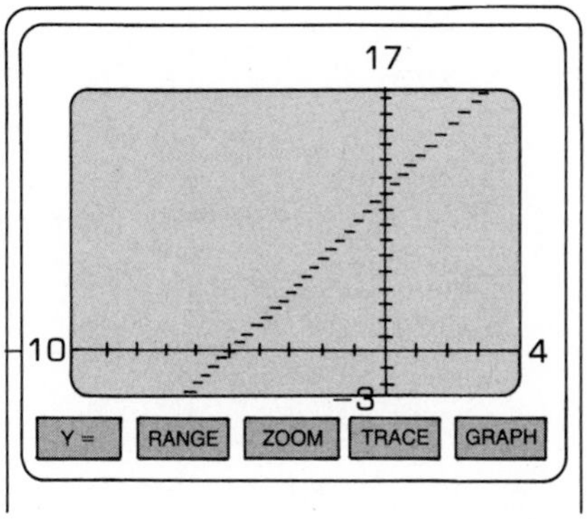

15.

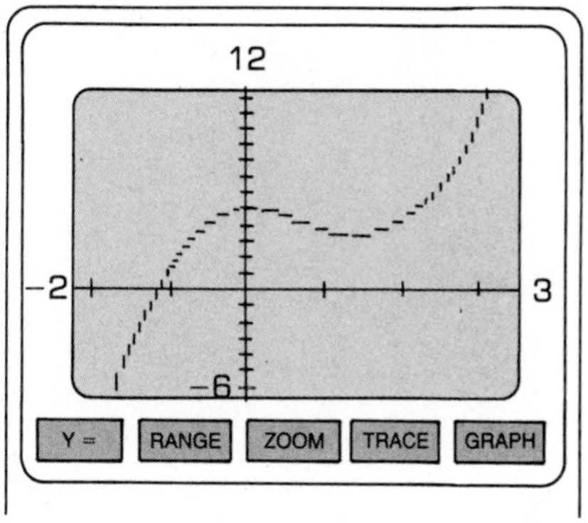

17.

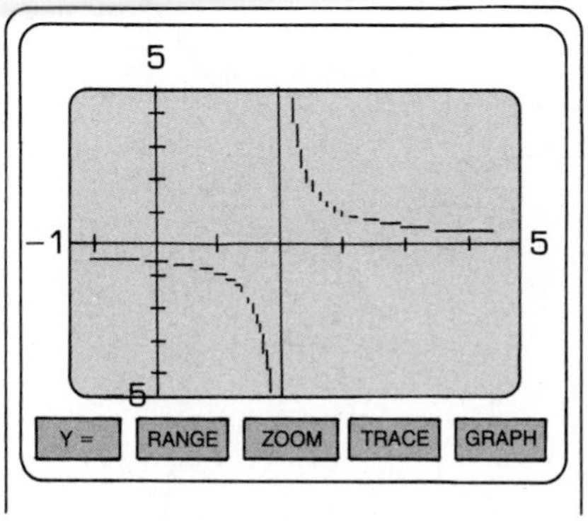

SECTION 1.5

1.

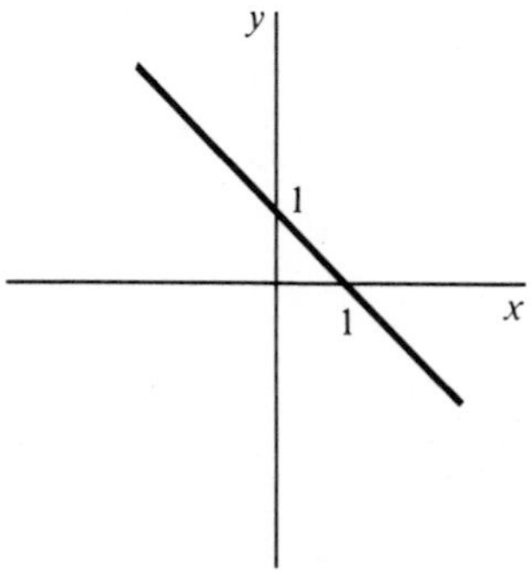

3.

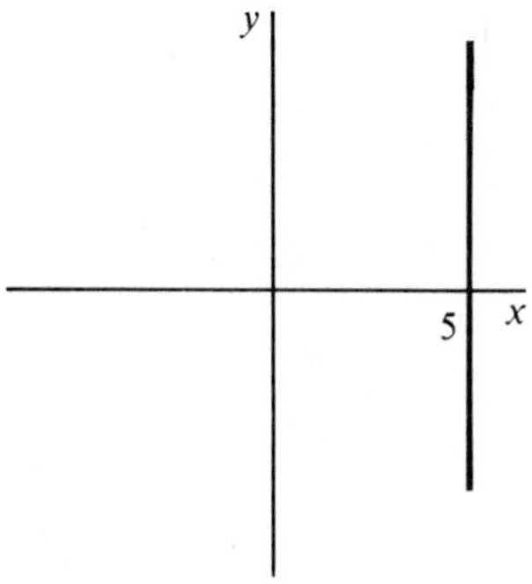

5.

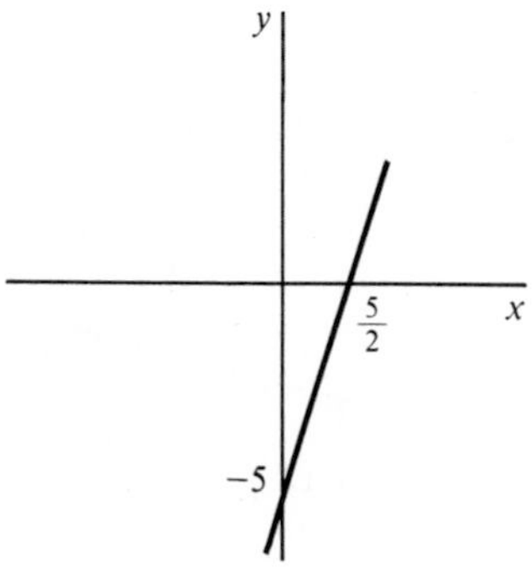

7.

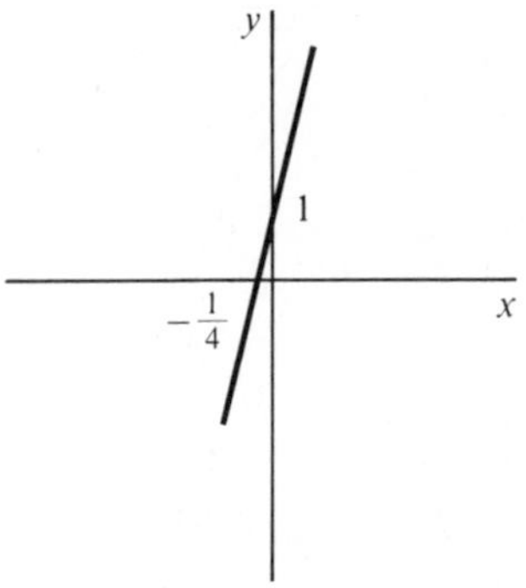

9.

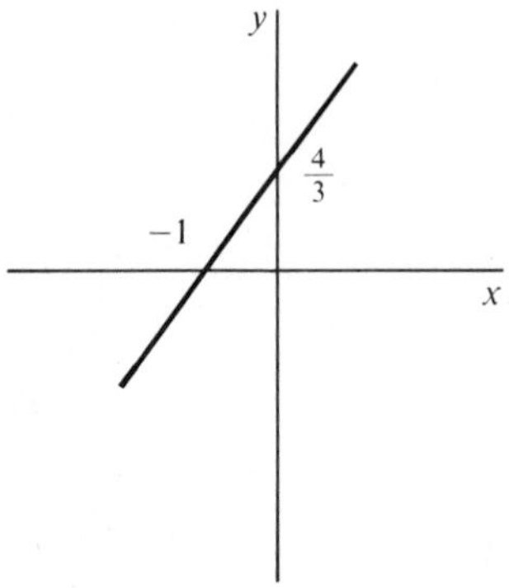

11.

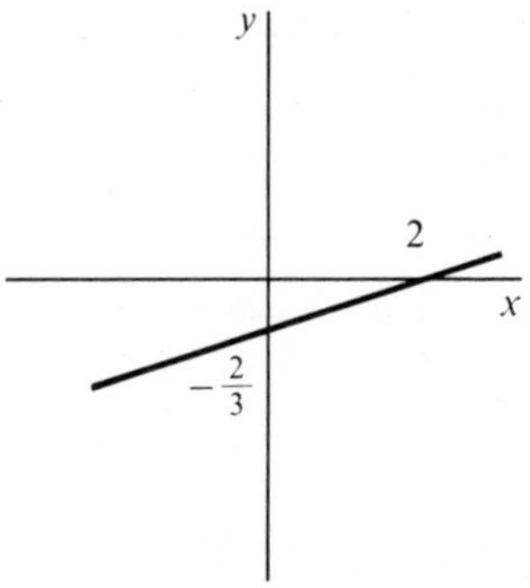

13. $m = \frac{1}{2}$

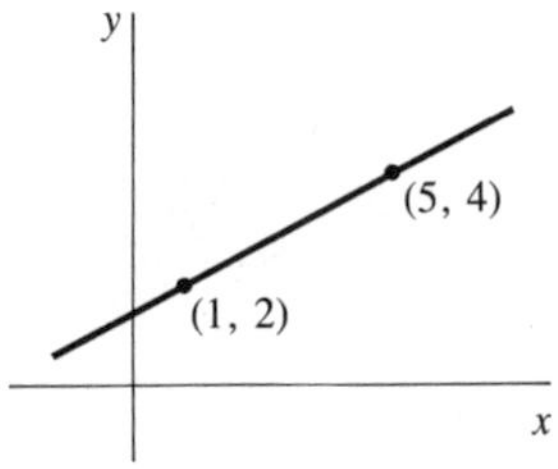

15. $m = -\frac{5}{4}$

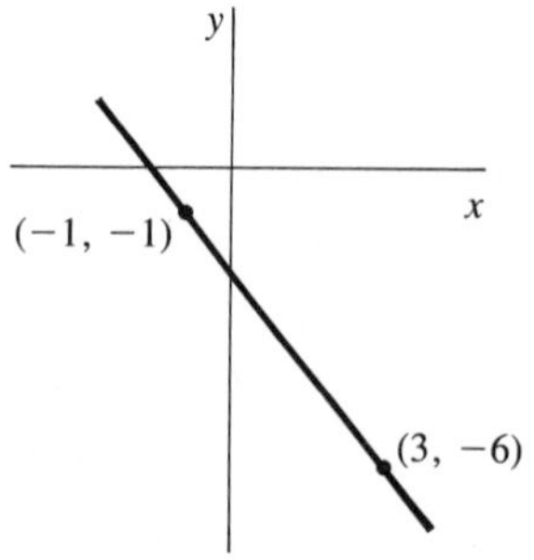

17. $m = \frac{4}{3}$

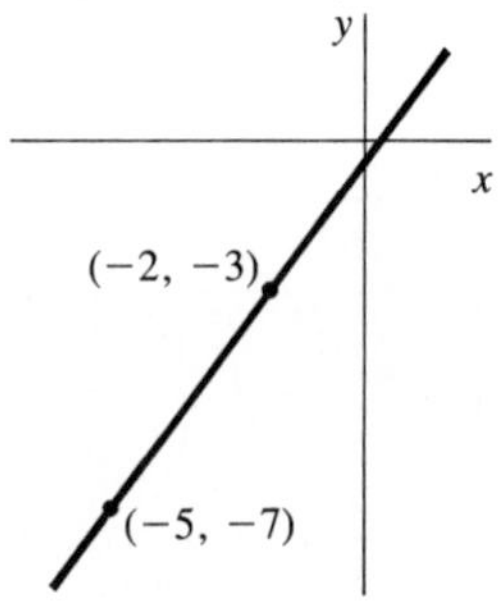

19. $m = 0$

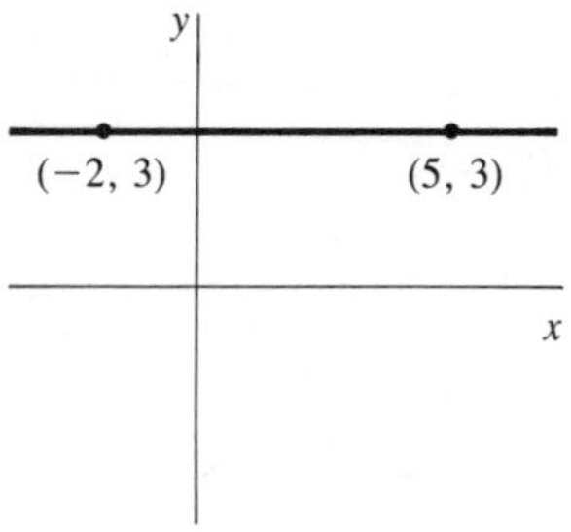

21.

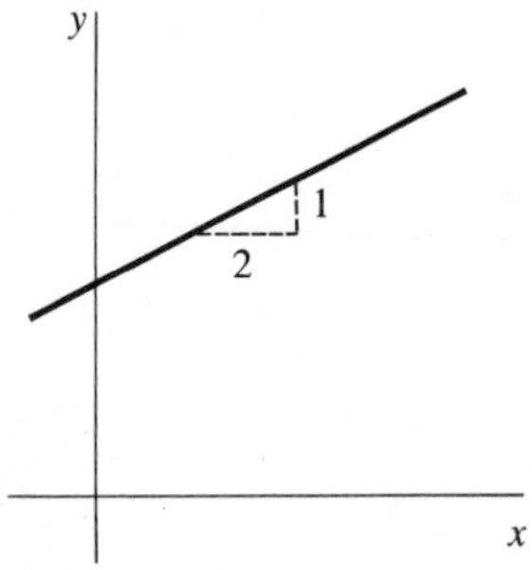

23.

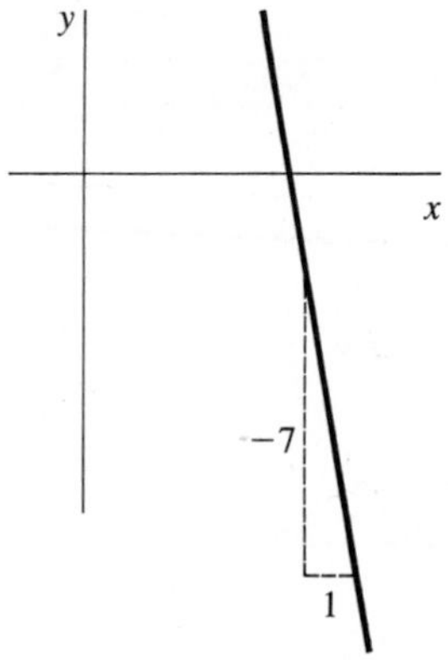

25. 2 **27.** 1.83; No

29.

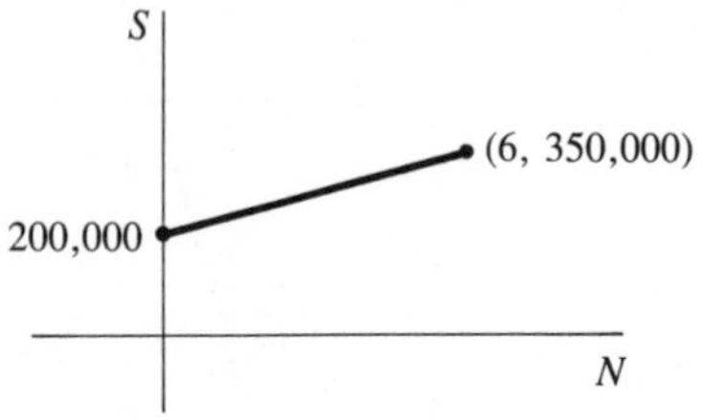

31. −2400 residents per year. The negative sign indicates that the population is decreasing.

33. (a) $150x + 250y = 750{,}000$ (b)

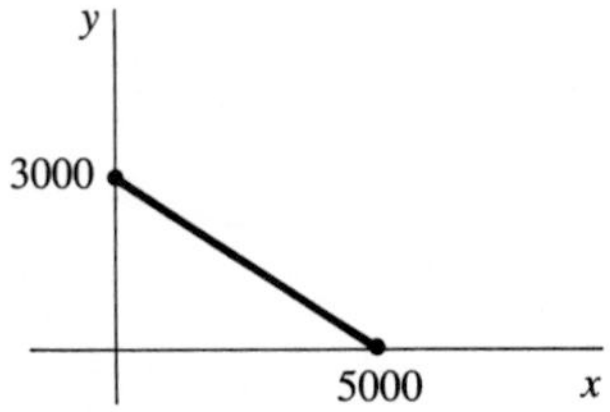

(c) 2500

SECTION 1.6

1. $x - 2y = -8$

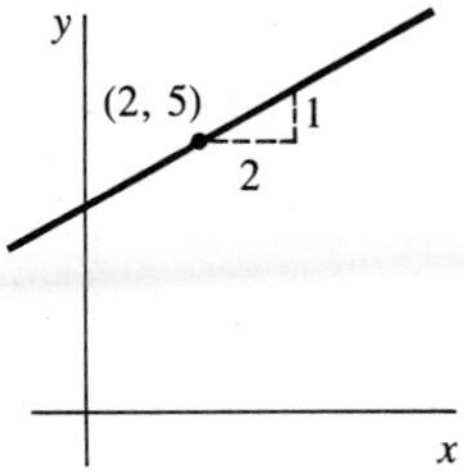

3. $7x + y = 33$

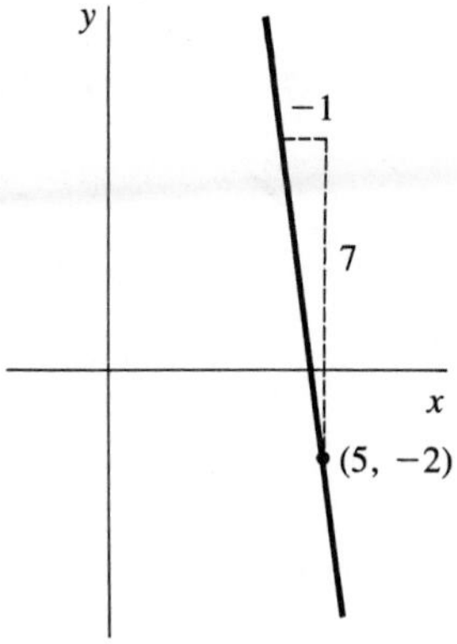

5. $x + 5y = 16$

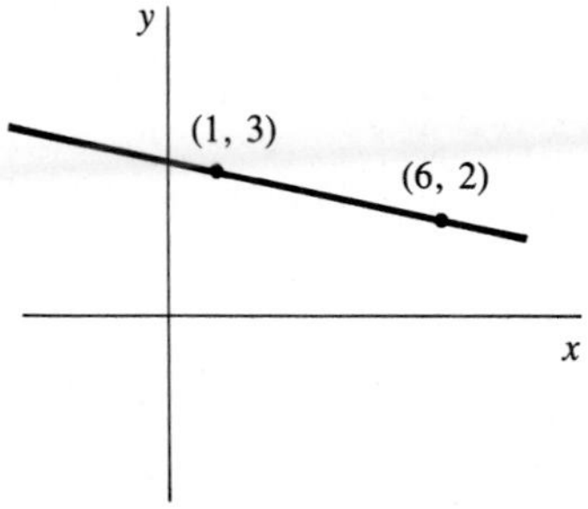

7. $2y - 3x = 1$

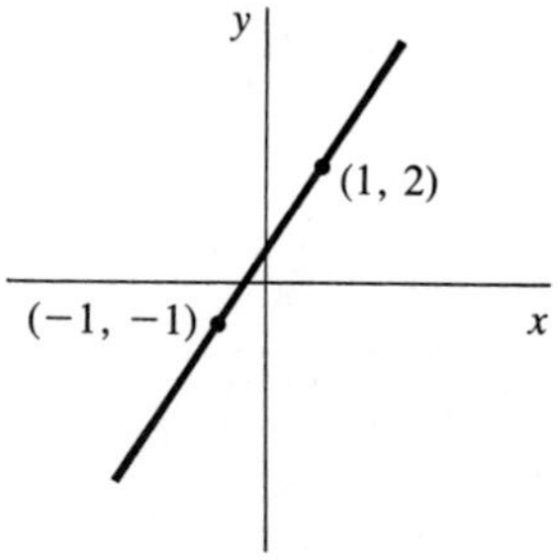

9. $5y - 2x = 10$

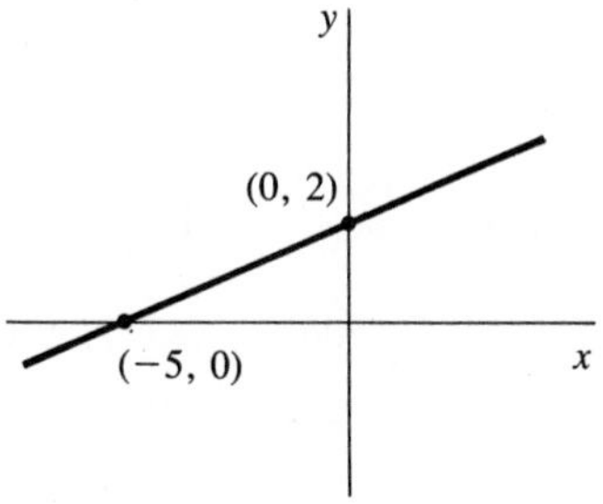

11. $m = \frac{2}{3}, b = -\frac{5}{3}$

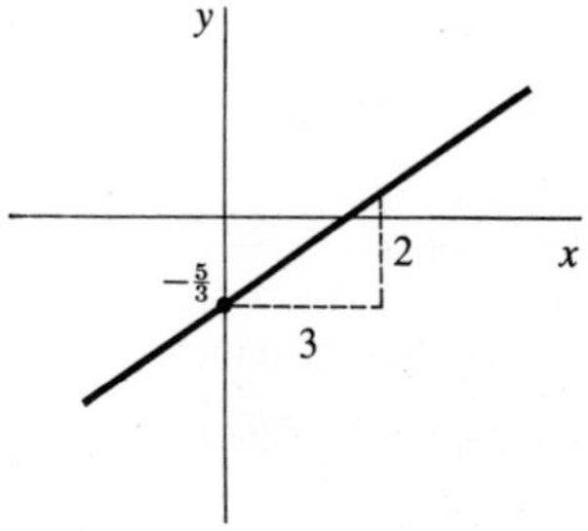

13. $m = -1, b = 2$

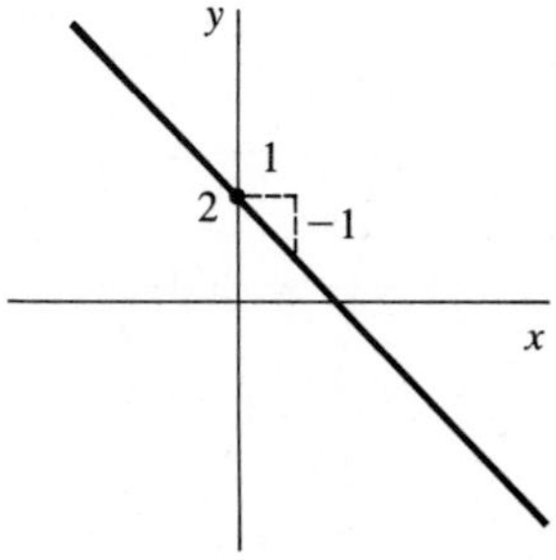

15. $m = \frac{1}{2}, b = -2$

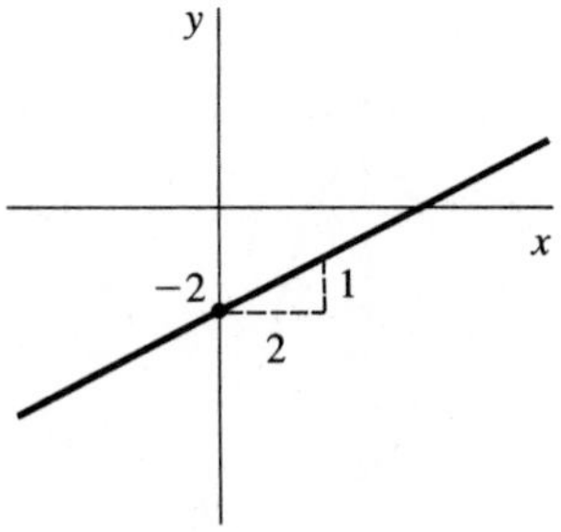

17. $m = \frac{1}{2}, b = \frac{5}{2}$

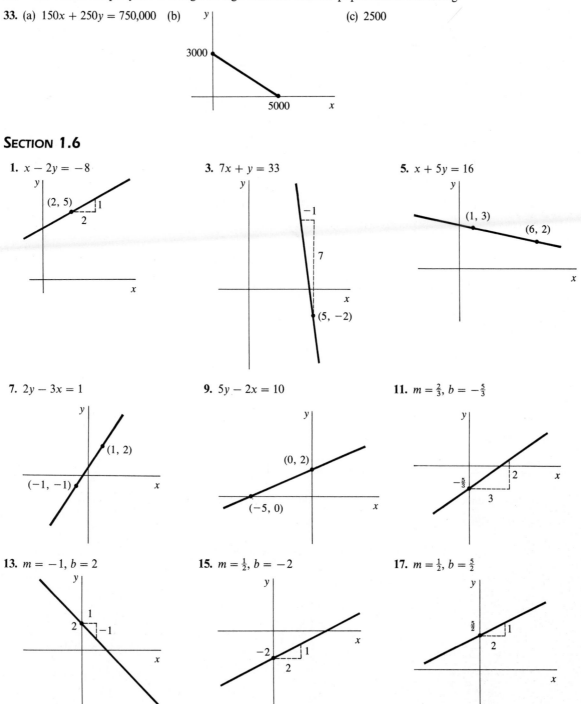

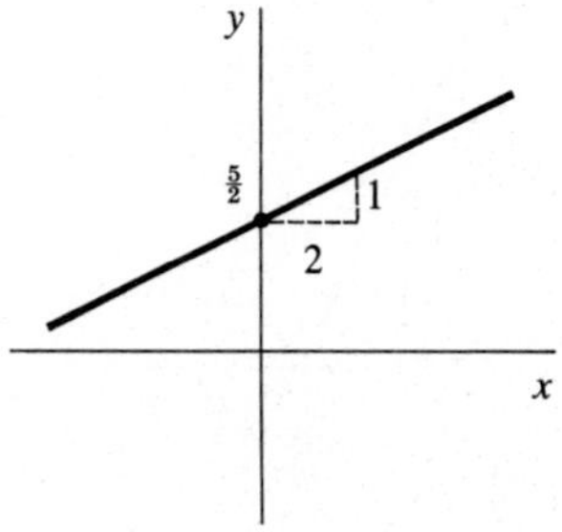

19. $m = 0$, $b = 5$

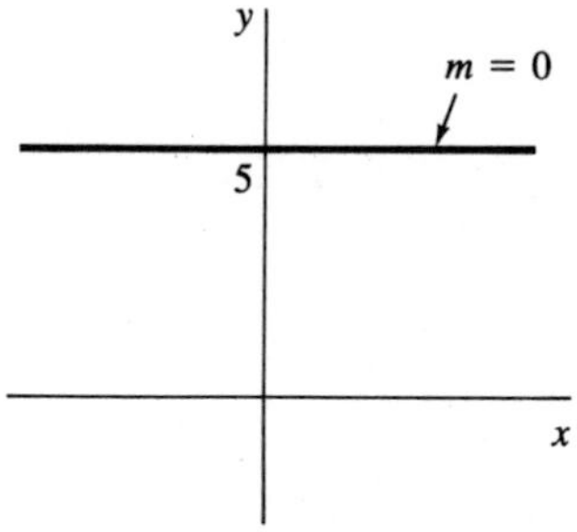

21. $m = \frac{1}{5}$, $b = \frac{7}{5}$

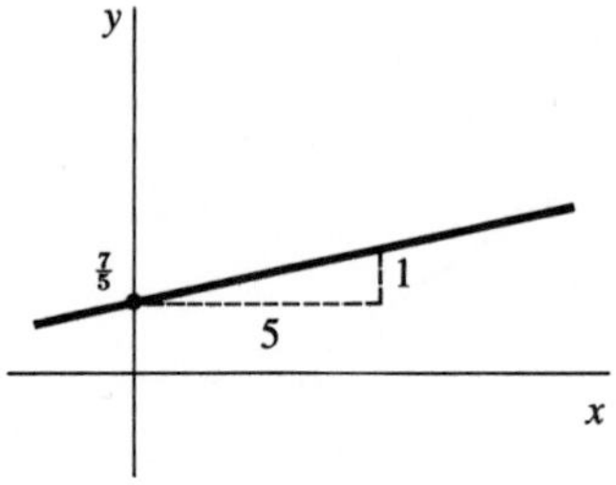

23. $2x - 3y = -4$ **25.** $x + y = 8$ **27.** $2x + 3y = 8$ **29.** $7x + 5y = 0$ **31.** $3x + y = -5$

33. (a) $\dfrac{x}{6} + \dfrac{y}{2} = 1$ (b) $\dfrac{x}{\frac{5}{2}} - \dfrac{y}{5} = 1$ **35.** $I = \dfrac{V}{12}$ **37.** \$225 **39.** \$151.11 **41.** $C = \frac{5}{9}(F - 32)$

43. (a) $C = 45 + 8t$ (b) \$95 (c)

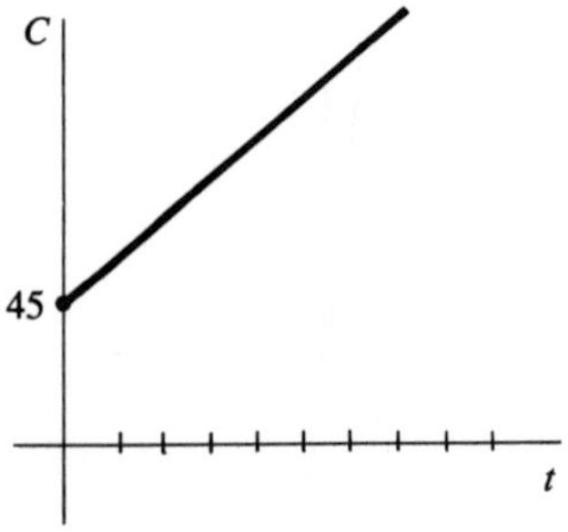

45. (a) $C = 9 + 0.5n$ (b) \$9

Review Exercises Chapter 1

1. $\frac{7}{3}$ **3.** $\frac{11}{14}$ **5.** 3 **7.** $x = \frac{60}{43}$ **9.** $x = \dfrac{18a}{1 + 2a}$ **11.** $x = \frac{7}{10}$ **13.** $y = \frac{23}{8}$ **15.** $m_0 = \dfrac{c^2m - E}{c^2}$

17. 57 nickels, 27 dimes **19.** 10 min **21.** Function **23.** Not a function **25.** Function

27. Function **29.** D: $-\infty < x < \infty$, R: $-\infty < y < \infty$ **31.** D: $x \geq 3$, R: $y \geq 0$

33. $f(g(x)) = -3x + 11$, $g(f(x)) = 5 - 3x$ **35.** $f(g(x)) = \dfrac{1}{x} + 3$, $g(f(x)) = \dfrac{1}{x + 3}$

37.

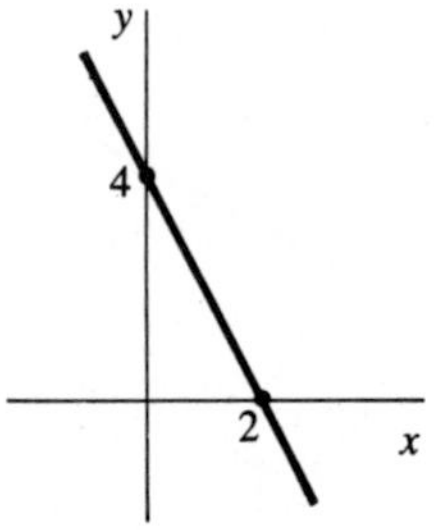

39.

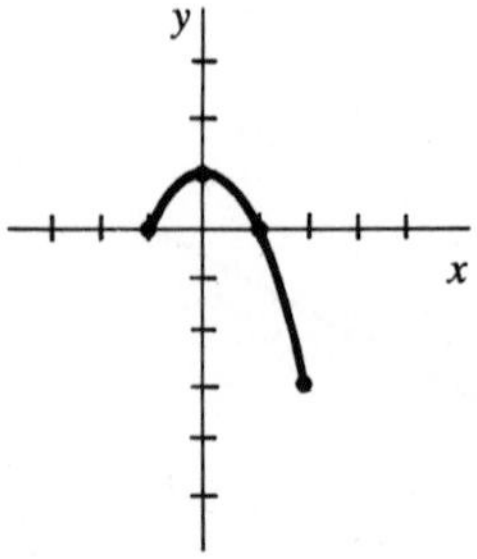

41.

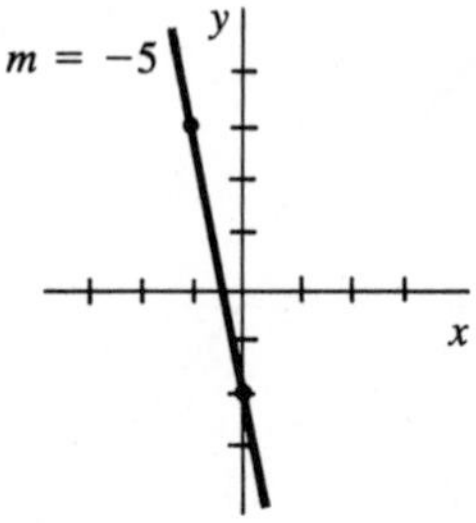

43.

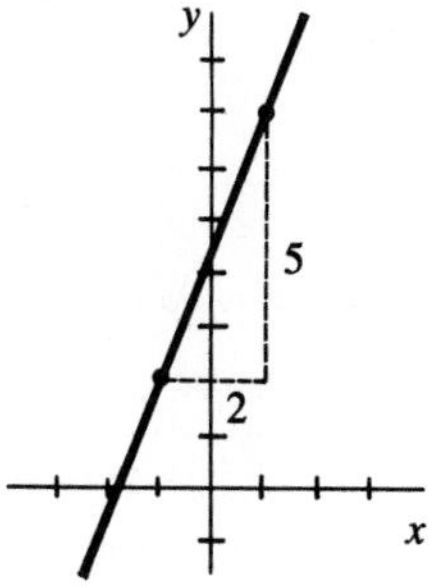

45. $-\frac{6}{5}$ **47.** $x - 3y = -11$ **49.** $4x + 2y = 5$ **51.** $v = \frac{1}{6}(13t + 98)$ **53.** $m = \frac{9}{50}$

55. (a)

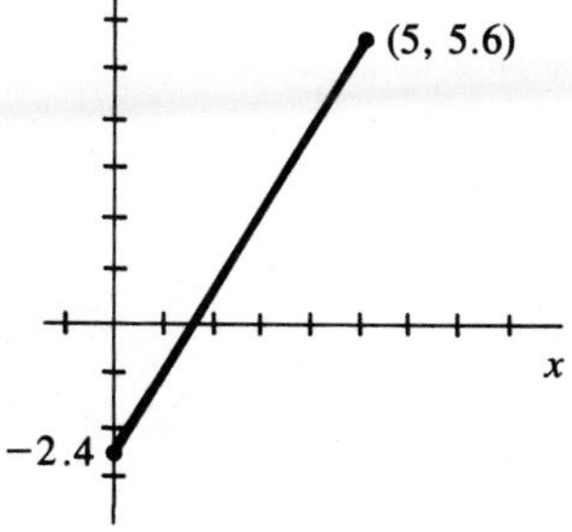

(b) 1.5years

57. (a) $P = 18n - 7500$ (b)

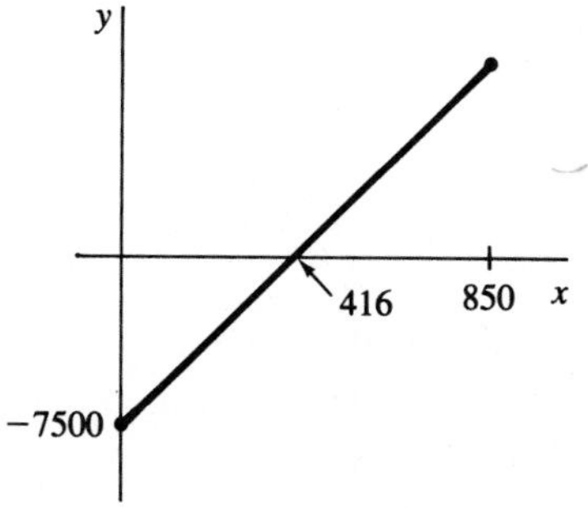

(c) 416 tickets

59. (a)

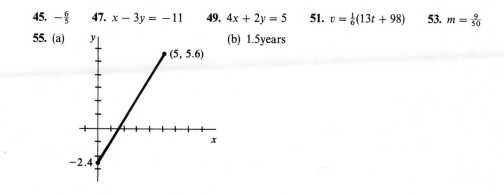

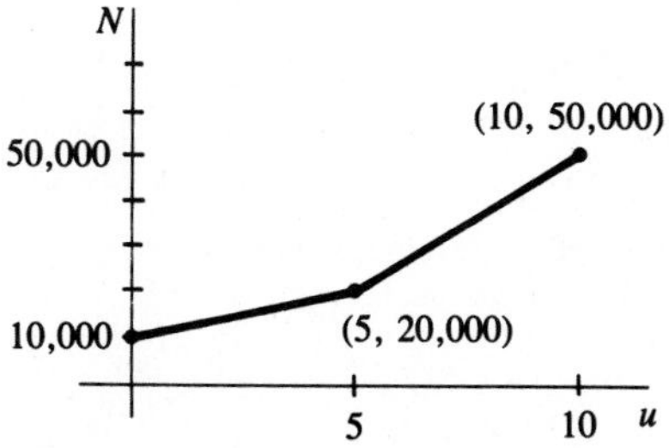

(b) 42,500

CHAPTER 2

SECTION 2.1

1. $(-3, -2)$ **3.** $(5, 2)$ **5.** Dependent **7.** Inconsistent **9.** $(\frac{8}{5}, \frac{14}{5})$ **11.** $(\frac{1}{4}, -\frac{3}{4})$ **13.** $(\frac{10}{9}, \frac{53}{63})$
15. $(-5, 4)$ **17.** $w = 3, z = 4$ **19.** $(-2, -2)$ **21.** $A = 14, B = 11$ **23.** $(8, 4)$ **25.** $a = 0, b = \frac{1}{2}$
27. 6.75 mph and 0.75 mph **29.** $\frac{10}{3}$ lb and $\frac{20}{3}$ lb **31.** 20 n and 15 d **33.** 18 min and 22.5 min
35. 80,000 D and 20,000 R **37.** Demand: $p = -3x + 700$; supply: $p = \frac{2}{3}x + 250$; equilibrium: $x = 123$, $p = \$332$

SECTION 2.2

1. $(2, 4)$ **3.** $(2, 1)$ **5.** $(-\frac{3}{4}, 2, 3)$ **7.** $(8, 5, 6)$ **9.** $(6, \frac{5}{3}, 2)$ **11.** Inconsistent **13.** $(3, 2, -1)$
15. Inconsistent **17.** $x = \frac{1}{7}(t - 9)$, $y = \frac{1}{7}(5t - 3)$, $z = t$ **19.** $(4, -3, 2)$ **21.** $(5.52, 2.28, 2.21)$
23. $x = 1, y = z = t$ **25.** $x = \frac{1}{2}(5 - 2t)$, $y = 2 - 2t$, $z = t$ **27.** $x = 2 - \frac{3}{2}t$, $y = \frac{1}{2}t$, $z = t$
29. $x = 2 - \frac{3}{2}t$, $y = -\frac{1}{4}(t + 10)$, $z = t$ **31.** $(1, -1, 2, 3)$ **33.** $(0, 1, -1, 2)$ **35.** $n = 246, d = 226, q = 428$
37. $I_1 = 15, I_2 = 17, I_3 = 13$ **39.** 25 g of Brand A, 25 g of Brand B, 50 g of Brand C

SECTION 2.3

1. $a_{22} \neq b_{22}$ **3.** A and B are different sizes **5.** $h = 3, k = 3$ **7.** $x = 4, y = \pm 4$

9. $h = -1$ or 2, $k = 0$ or 3 **11.** $\begin{bmatrix} 4 & 2 \\ 2 & 1 \end{bmatrix}; \begin{bmatrix} 6 & -2 \\ 0 & 1 \end{bmatrix}; \begin{bmatrix} 10 & 0 \\ 2 & 2 \end{bmatrix}$ **13.** $\begin{bmatrix} 2 & 2 & 5 \\ 4 & 5 & 7 \end{bmatrix}; \begin{bmatrix} 0 & 2 & 1 \\ 4 & -5 & -5 \end{bmatrix}; \begin{bmatrix} 2 & 4 & 6 \\ 8 & 0 & 2 \end{bmatrix}$

15. $\begin{bmatrix} 8 \\ -1 \\ 7 \end{bmatrix}; \begin{bmatrix} -6 \\ 1 \\ -3 \end{bmatrix}; \begin{bmatrix} 2 \\ 0 \\ 4 \end{bmatrix}$ **17.** $[-1 \quad 5 \quad 1]; [1 \quad 1 \quad 1]; [0 \quad 6 \quad 2]$

19. $A + B$ and $A - B$ cannot be done because A and B are different sizes; $2A = \begin{bmatrix} 2 & 0 \\ 10 & 4 \end{bmatrix}$ **21.** $\begin{bmatrix} 1 & -1 \\ 2 & 5 \end{bmatrix}$

23. $\begin{bmatrix} 2 & 2 \\ -1 & -3 \\ 0 & 3 \end{bmatrix}$ **25.** $\begin{bmatrix} -2 \\ 1 \\ 8 \\ 2 \end{bmatrix}$ **27.** $[17 \quad 3]$ **29.** $\begin{bmatrix} 1 & 0 & 0 \\ 0 & 1 & 0 \\ 0 & 0 & 1 \end{bmatrix}$

33. (a) 243 (b) 80 (c) 371 (d) 493 (e) $\begin{bmatrix} 271 & 493 \\ 175 & 279 \end{bmatrix}$ (f) $\begin{bmatrix} 1 & 23 \\ 5 & 7 \end{bmatrix}$

SECTION 2.4

1. $AB = \begin{bmatrix} -5 & 10 \\ 0 & 2 \end{bmatrix}$; $BA = \begin{bmatrix} -3 & 2 \\ 5 & 0 \end{bmatrix}$ **3.** Cannot be done because size requirements are not met.

5. $AB = \begin{bmatrix} -2 & 0 \\ 10 & 0 \end{bmatrix}$; $BA = [-2]$ **7.** $AB = \begin{bmatrix} 4 \\ -1 \end{bmatrix}$; BA cannot be done

9. $AB = \begin{bmatrix} 12 & 15 \\ 13 & 13 \\ -1 & 4 \end{bmatrix}$; BA cannot be done **11.** $\begin{bmatrix} 2 & 3 \\ 3 & 1 \end{bmatrix}\begin{bmatrix} x \\ y \end{bmatrix} = \begin{bmatrix} 7 \\ 10 \end{bmatrix}$ **13.** $\begin{bmatrix} 5 & 1 \\ 1 & -3 \end{bmatrix}\begin{bmatrix} x \\ y \end{bmatrix} = \begin{bmatrix} 0 \\ 7 \end{bmatrix}$

15. $\begin{bmatrix} 1 & 1 & 1 \\ 2 & -1 & -1 \\ 1 & -1 & 1 \end{bmatrix}\begin{bmatrix} x \\ y \\ z \end{bmatrix} = \begin{bmatrix} 1 \\ 2 \\ 12 \end{bmatrix}$ **17.** $\begin{bmatrix} 1 & 3 & -1 \\ 0 & 4 & -2 \\ 2 & 0 & 1 \end{bmatrix}\begin{bmatrix} x \\ y \\ z \end{bmatrix} = \begin{bmatrix} 0 \\ 1 \\ 13 \end{bmatrix}$ **19.** $\begin{bmatrix} 1 & 1 & -4 \\ 2 & 3 & -1 \\ 7 & 0 & 2 \end{bmatrix}\begin{bmatrix} r \\ s \\ t \end{bmatrix} = \begin{bmatrix} 7 \\ 0 \\ 6 \end{bmatrix}$

31. $(A + B)^2 = A^2 + AB + BA + B^2$

35. (a) 311 business majors will live on campus. (b)

	On	Off	Comm.
Engr	286	200	65
Bus	311	174	84
Educ	189	93	54

37. (a) [500 2000 1500] (b) $\begin{bmatrix} 30 & 20 \\ 20 & 10 \\ 15 & 20 \end{bmatrix}$ (c) [77,500 60,000] (d) $\begin{bmatrix} 20 \\ 10 \end{bmatrix}$ (e) [2,150,000]

Section 2.5

1. $\begin{bmatrix} 1 & 0 \\ 0 & 1 \end{bmatrix}$ **3.** $\begin{bmatrix} 0 & 1 \\ 1 & -2 \end{bmatrix}$ **5.** $\begin{bmatrix} \frac{1}{2} & 0 & 0 \\ 0 & \frac{1}{3} & 0 \\ 0 & 0 & \frac{1}{5} \end{bmatrix}$ **7.** $\begin{bmatrix} 0 & 0 & -1 \\ -2 & 1 & 0 \\ 1 & 0 & 1 \end{bmatrix}$ **9.** No inverse

11. $\begin{bmatrix} \frac{1}{a} & -\frac{b}{ad} \\ 0 & \frac{1}{d} \end{bmatrix}$ **13.** $\begin{bmatrix} -1 & 1 & 5 \\ 0 & \frac{1}{4} & \frac{1}{2} \\ 1 & -1 & -4 \end{bmatrix}$ **15.** $\begin{bmatrix} -\frac{5}{3} & \frac{8}{3} & -1 \\ 0 & 1 & -\frac{1}{2} \\ -\frac{1}{3} & \frac{1}{3} & 0 \end{bmatrix}$ **17.** $x = y = \frac{4}{3}$ **19.** $x = \frac{2}{5}$, $y = \frac{3}{5}$, $z = \frac{1}{5}$

21. $x = -1$, $y = 2$, $z = 1$ **23.** $x = 4$, $y = 2$, $z = 0$ **25.** Inconsistent system

29. (a) 25 −8 63 9 −9 32 12 −19 46 21 −19 81 (b) math counts

31. (a) 48 83 43 50 73 41 11 8 2 56 85 47 22 16 2 (b) help a friend

Section 2.6

1. (a) 20 (b) Fuels (c) $\begin{bmatrix} 70 \\ 30 \end{bmatrix}$ (d) $\begin{bmatrix} 334 \\ 192 \end{bmatrix}$ **3.** (a) $\begin{bmatrix} 65 \\ 80 \\ 0 \end{bmatrix}$ (b) $\begin{bmatrix} 461 \\ 222 \\ 185 \end{bmatrix}$

5. (a) $\begin{bmatrix} 0.1 & 0.3 & 0.1 \\ 0.2 & 0.3 & 0.1 \\ 0.3 & 0.2 & 0.2 \end{bmatrix}$ (b) $\begin{bmatrix} 130 \\ 290 \\ 160 \end{bmatrix}$ (c) $\begin{bmatrix} 243 \\ 317 \\ 233 \end{bmatrix}$

7. (a) $\begin{bmatrix} 0.7 & -0.4 & -0.2 \\ -0.3 & 0.5 & -0.6 \\ -0.4 & -0.1 & 0.8 \end{bmatrix}\begin{bmatrix} x_1 \\ x_2 \\ x_3 \end{bmatrix} = \begin{bmatrix} 0 \\ 0 \\ 0 \end{bmatrix}$ (b) $x_1 = 1.48x_3$; $x_2 = 2.09x_3$

Review Exercises Chapter 2

1. (2, 1) **3.** $s = \frac{41}{7}$, $t = -\frac{22}{7}$ **5.** $x = 3$, $z = -3$ **7.** $(\frac{5}{2}, -1, -\frac{1}{2})$ **9.** $(-13, -2, -6)$ **11.** (1, 2)

13. $(2, \frac{1}{2})$ **15.** (4, 3, 2) **17.** $\begin{bmatrix} 2 & 3 \\ 1 & -1 \end{bmatrix}\begin{bmatrix} x \\ y \end{bmatrix} = \begin{bmatrix} 7 \\ 8 \end{bmatrix}$ **19.** $\begin{bmatrix} 1 & 2 & 5 \\ 0 & 1 & -1 \\ 2 & -1 & 1 \end{bmatrix}\begin{bmatrix} x \\ y \\ z \end{bmatrix} = \begin{bmatrix} 10 \\ 2 \\ -1 \end{bmatrix}$

21. Cannot be done because A and B are not the same size. **23.** $\begin{bmatrix} -2 & 8 & 8 \\ 2 & 1 & 5 \\ -2 & 11 & 0 \end{bmatrix}$

25. $AB = [0]$, $BA = \begin{bmatrix} -2 & -3 & 1 \\ 4 & 6 & -2 \\ 8 & 12 & -4 \end{bmatrix}$ **27.** $AB = \begin{bmatrix} 9 & -13 \\ 15 & -9 \end{bmatrix}$, $BA = \begin{bmatrix} 1 & -5 \\ 23 & -1 \end{bmatrix}$

29. $AB = \begin{bmatrix} 1 & -12 \\ 12 & -4 \end{bmatrix}$, $BA = \begin{bmatrix} -8 & 2 & -5 \\ 10 & 1 & 1 \\ 40 & 4 & 4 \end{bmatrix}$ **31.** $\begin{bmatrix} -\frac{1}{2} & \frac{1}{2} \\ \frac{3}{4} & -\frac{1}{4} \end{bmatrix}$ **33.** $\begin{bmatrix} -\frac{5}{17} & \frac{4}{17} \\ \frac{3}{17} & \frac{1}{17} \end{bmatrix}$

35. The inverse does not exist. **37.** $x = 1$, $y = 2$ **39.** $x = 2$, $y = \frac{1}{2}$ **41.** $a = 4$, $b = 3$, $c = 2$

43. 46.5°, 43.5° **45.** \$125,000 at 9%, \$625,000 at 11%

47. (a) 584 cars (b) 546 (c) $\begin{bmatrix} 405 & 508 \\ 268 & 279 \end{bmatrix}$ (d) $\begin{bmatrix} 5 & 8 \\ 18 & 9 \end{bmatrix}$

49. (a) 35 (b) 80 (c) Manufactured products (d) $\begin{bmatrix} 525 \\ 385 \\ 245 \end{bmatrix}$ (e) $\begin{bmatrix} 25 \\ 115 \\ 55 \end{bmatrix}$ (f) $\begin{bmatrix} 996 \\ 894 \\ 541 \end{bmatrix}$

Chapter 3

Section 3.1

1.
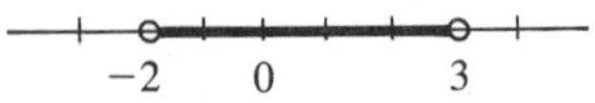

3.

5.

7.
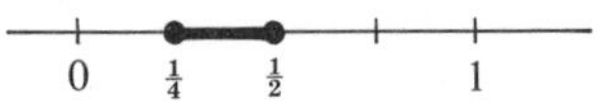

9.

11. $x > -\frac{5}{2}$
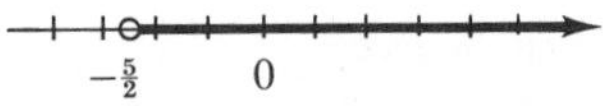

13. $x \geq \frac{1}{4}$

15. $x > \frac{70}{11}$

17. $x \leq \frac{13}{2}$

19. No solution

21. $-8 < x < 2$

23. $\frac{7}{3} \leq x \leq 3$
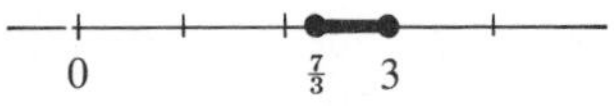

25. $x > 8$

27. $x \geq -\frac{1}{2}$

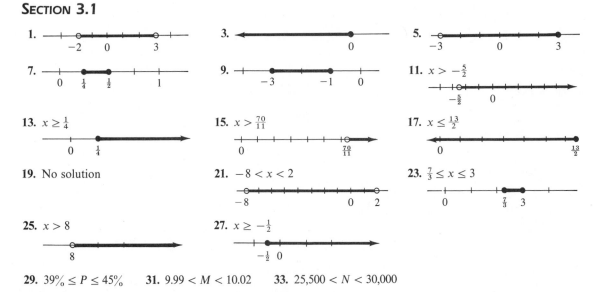

29. $39\% \leq P \leq 45\%$ **31.** $9.99 < M < 10.02$ **33.** $25{,}500 < N < 30{,}000$

Section 3.2

1.

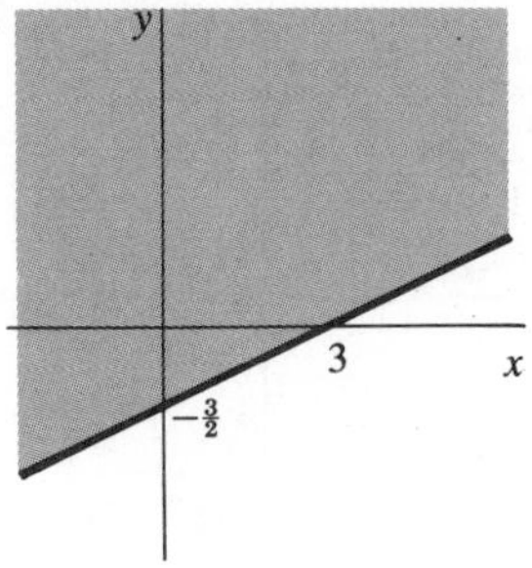

3.

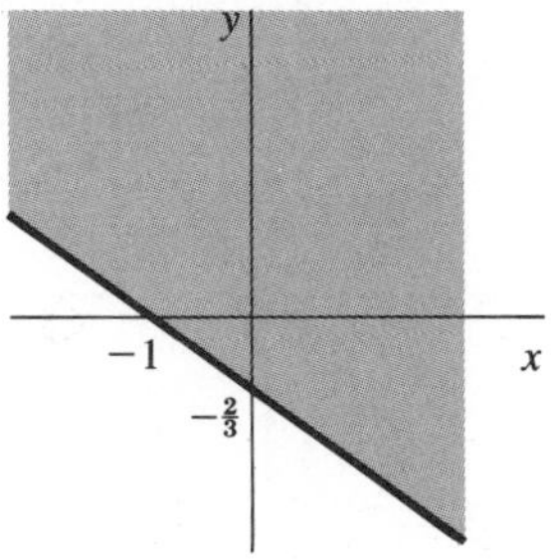

5.

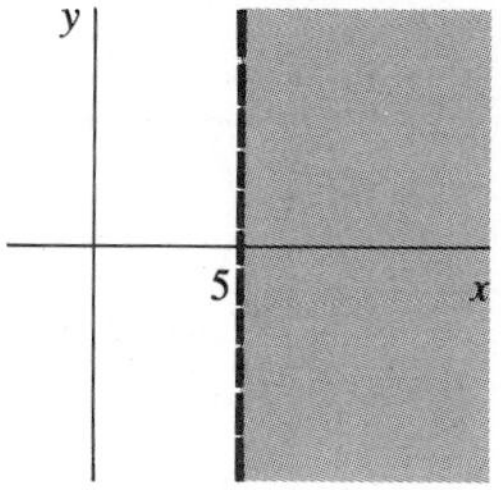

7.

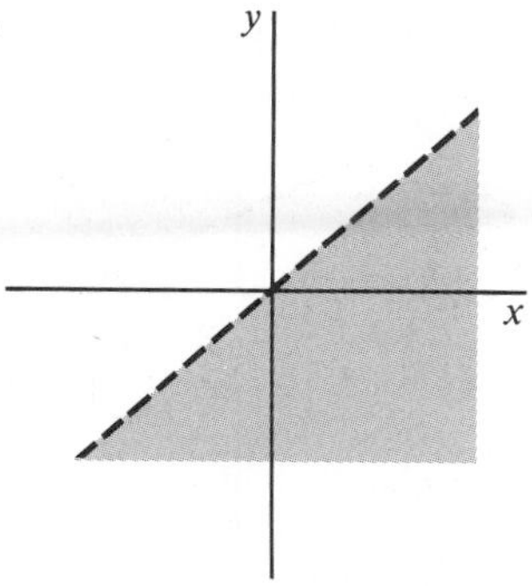

9.

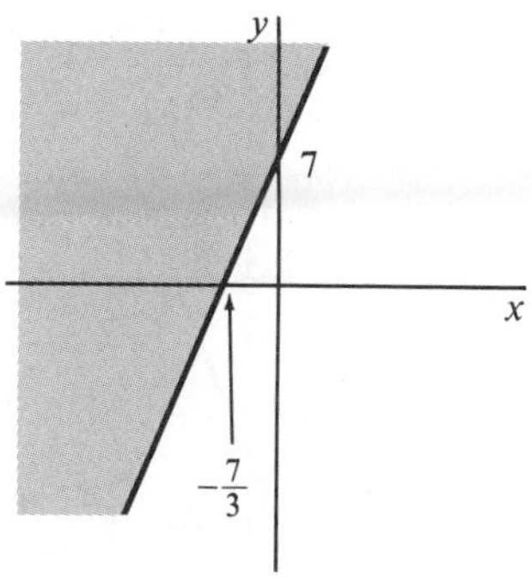

11.

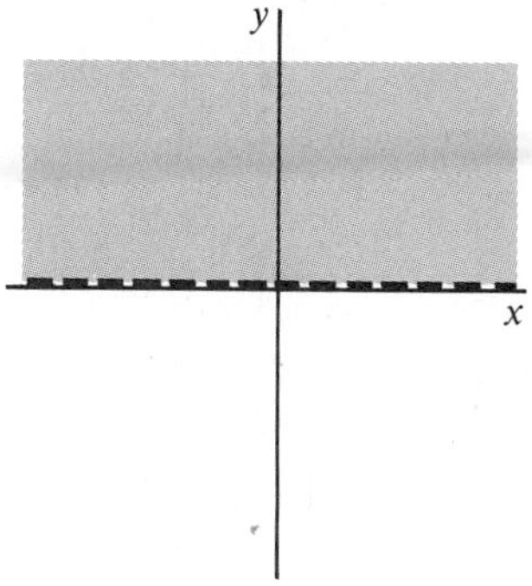

13.

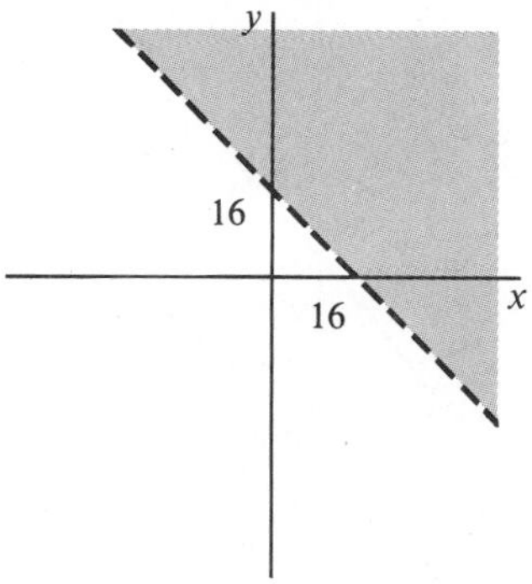

15.

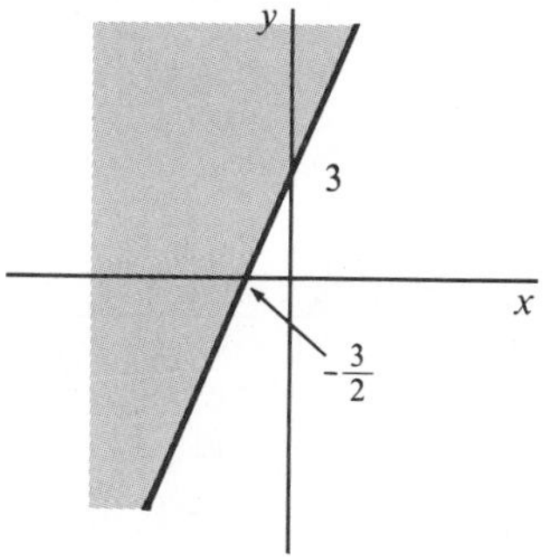

17. $C < -\frac{160}{9}$

19.

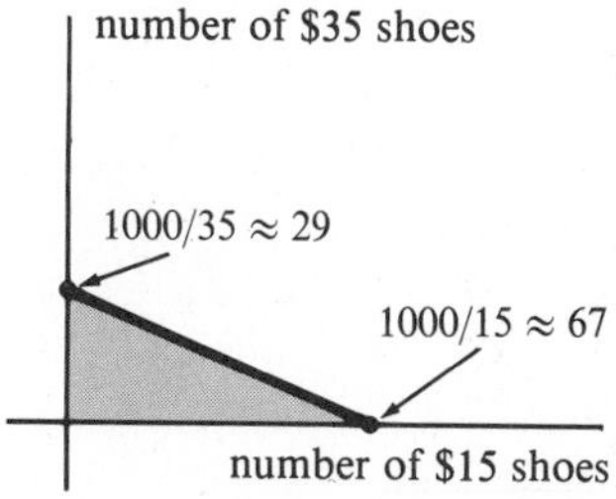

21. $2x + y \leq 1000$

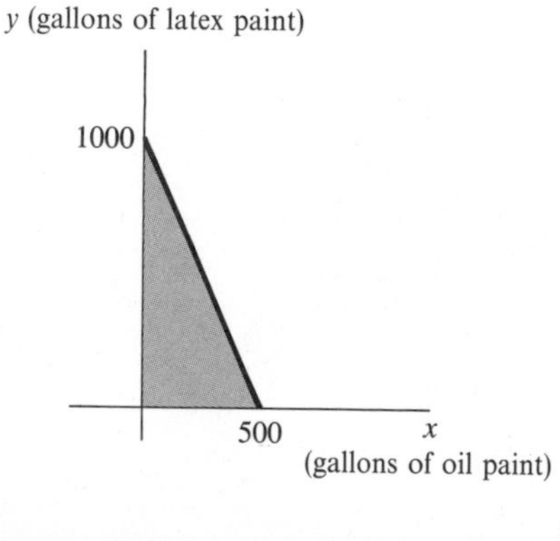

SECTION 3.3

1.

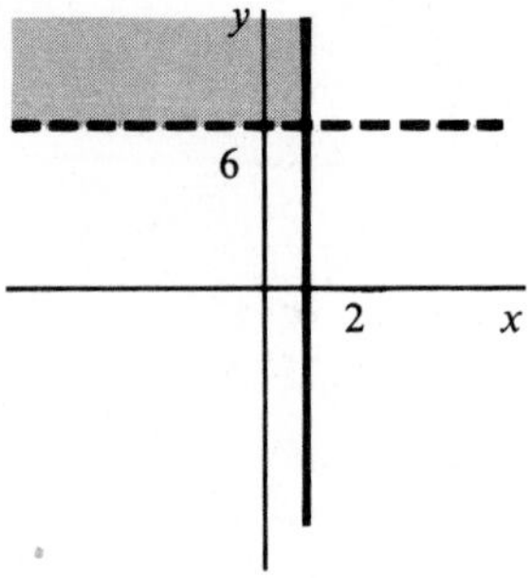

3.

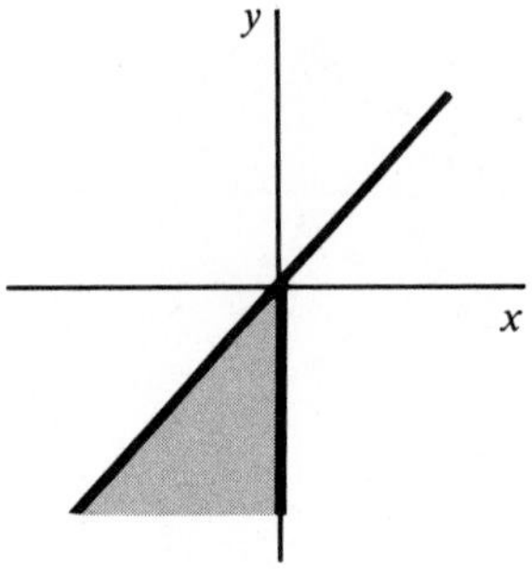

5.

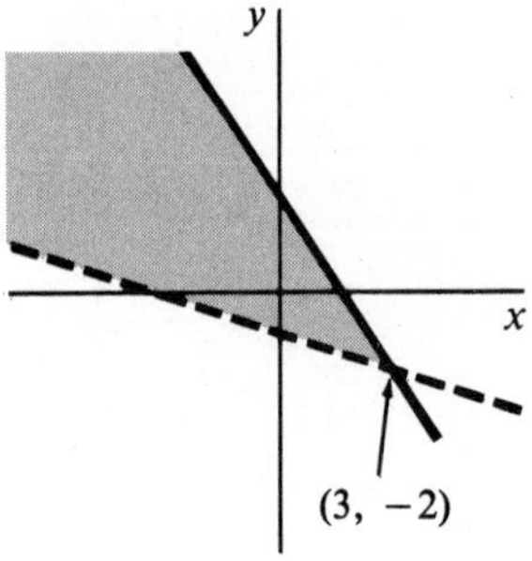

7.

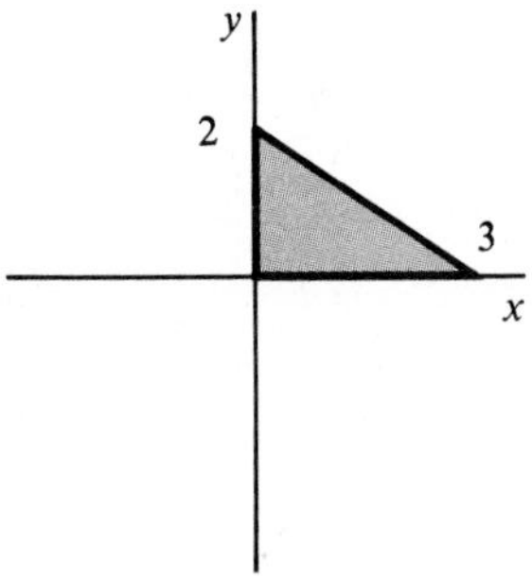

9.

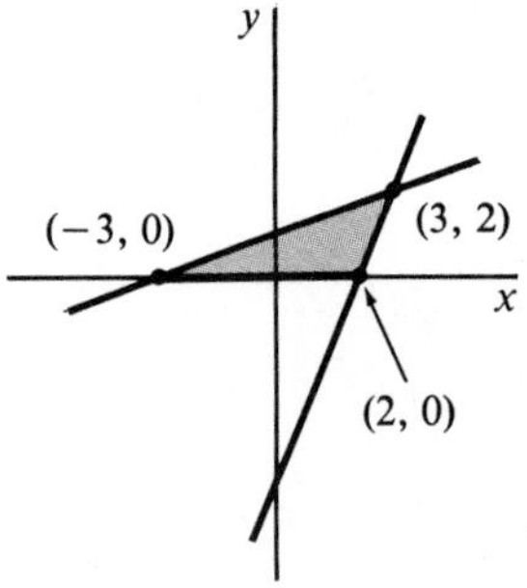

11.

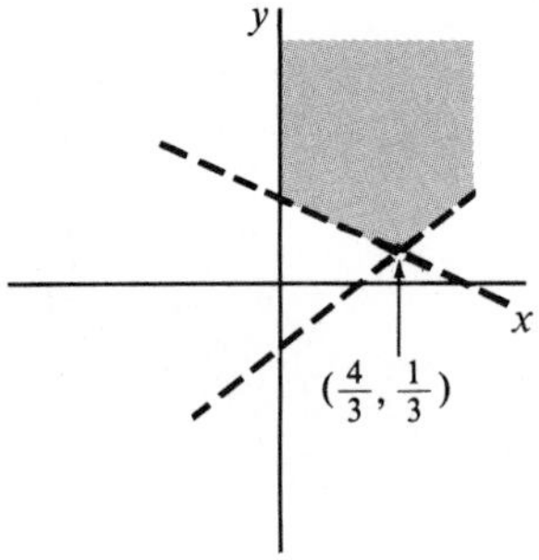

13.

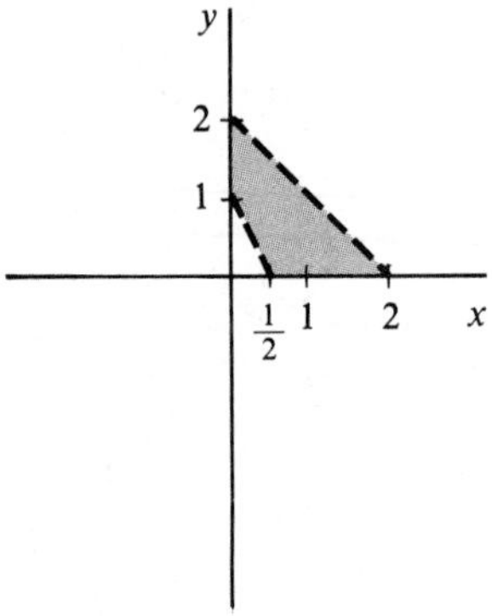

15.

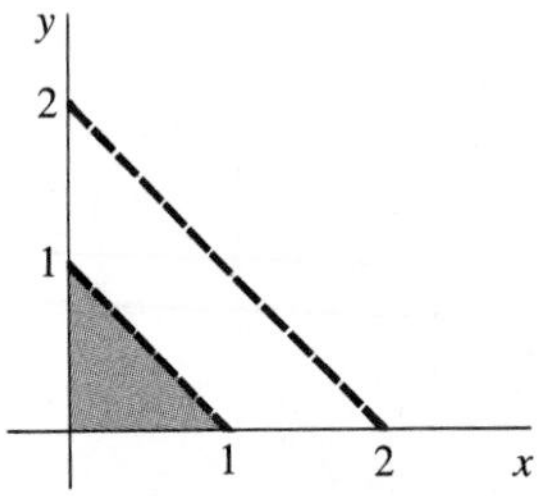

17.

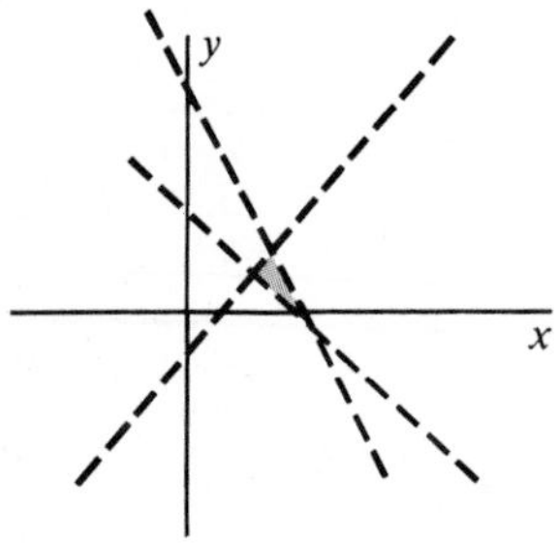

19.

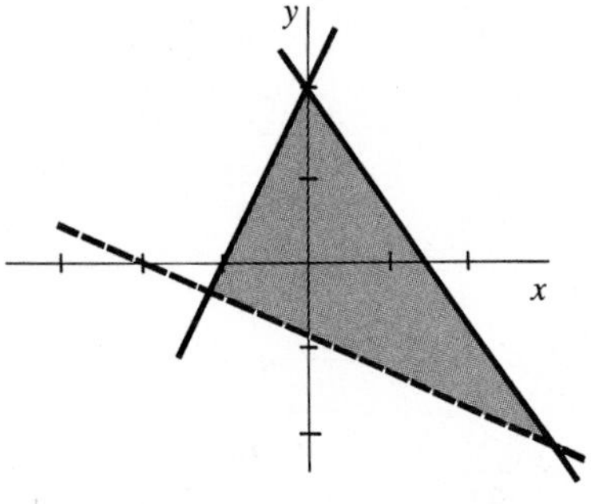

21. $x + y \leq 1000$, $x \leq 800$, $y \leq 700$, $x \geq 0$, $y \geq 0$

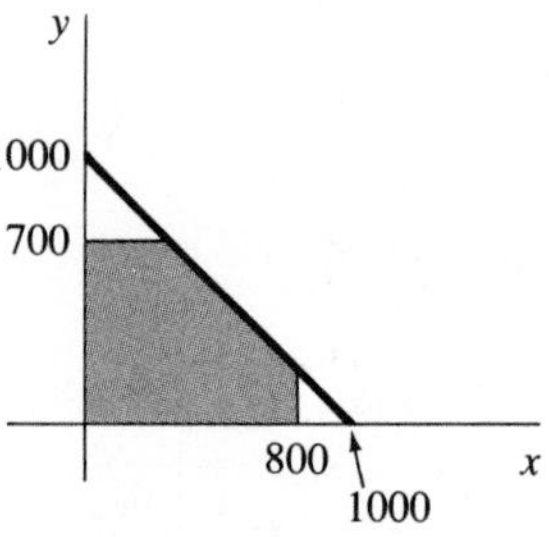

23. $x + y \le 10$, $x \le 5$, $y \le 7$, $x \ge 0$, $y \ge 0$

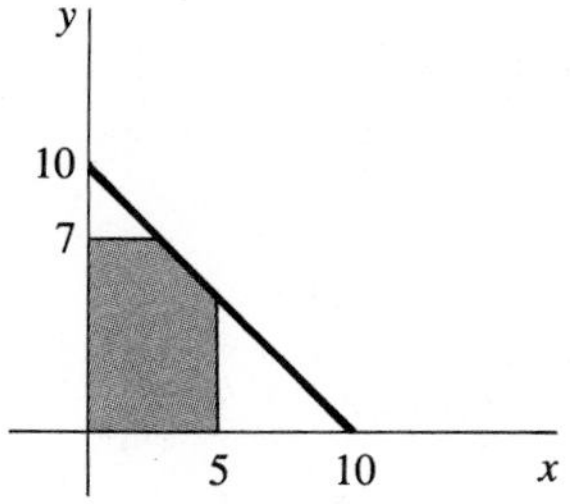

25. $150x + 250y \le 30{,}000$, $x + y \le 160$, $x \ge 0$, $y \ge 0$

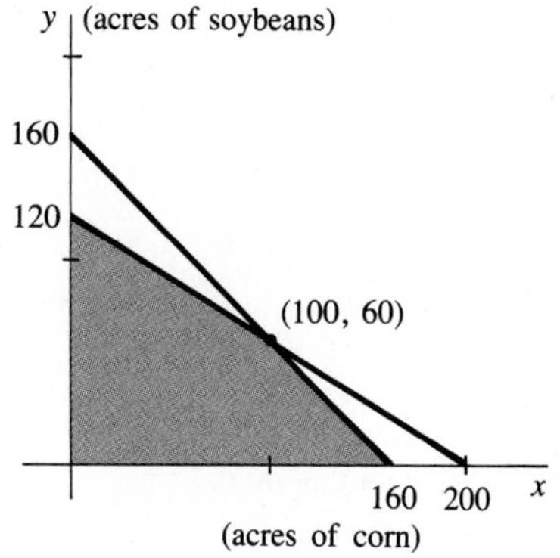

SECTION 3.4

1. max 10 at (4, 2) **3.** max 25 at (5, 0) **5.** (a) max 10 at (2, 2) (b) max 9 at (3, 0) **7.** min 12 at (0, 6)

9. (a) min 10 at (2, 2) (b) min 6 at (0, 6) **11.** No solution **13.** (a) max 16 at (0, 4) (b) min 4 at (0, 1)

15. (a) min 11 at (1, 2) (b) min 6 at (0, 3) **17.** min 14 at (2, 2); $x + y \ge 1$ is a superfluous constraint.

19. $x = 15$, $y = 30$

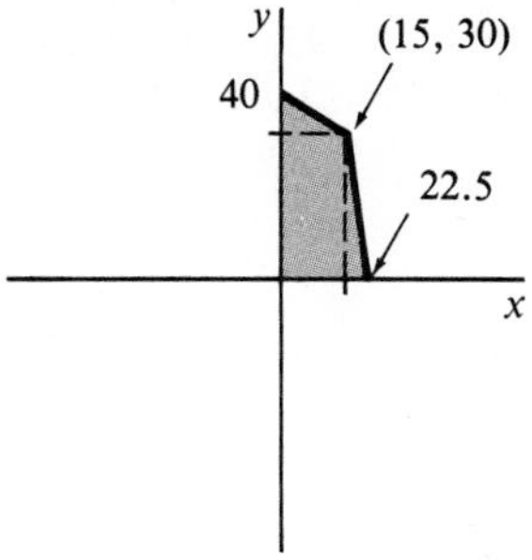

21. \$13,300 at 8.5%
\$6700 at 7.0%

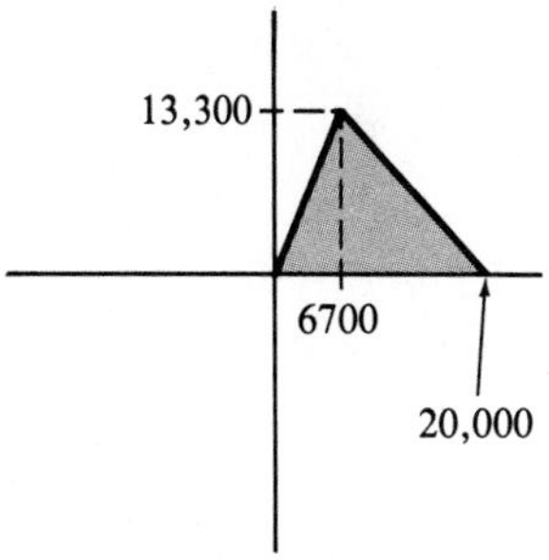

23. min \$15,500; three large planes and eight small planes

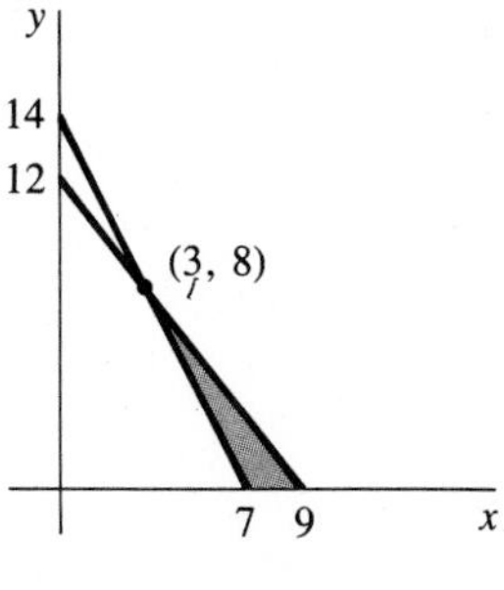

SECTION 3.5

1. $\begin{bmatrix} 3 & 2 & 1 & 0 & 0 & 10 \\ 1 & 4 & 0 & 1 & 0 & 15 \\ -1 & -2 & 0 & 0 & 1 & 0 \end{bmatrix}$ **3.** $\begin{bmatrix} 0 & 1 & 1 & 0 & 0 & 2 \\ 1 & 2 & 0 & 1 & 0 & 8 \\ -1 & -3 & 0 & 0 & 1 & 0 \end{bmatrix}$

5. $\begin{bmatrix} 1 & 1 & 1 & 0 & 0 & 5 \\ 1 & 2 & 0 & 1 & 0 & 8 \\ -5 & -1 & 0 & 0 & 1 & 0 \end{bmatrix}$ **7.** $\begin{bmatrix} 1 & 0 & 0 & 1 & 0 & 0 & 0 & 5 \\ 0 & 1 & 0 & 0 & 1 & 0 & 0 & 8 \\ 0 & 0 & 1 & 0 & 0 & 1 & 0 & 10 \\ -2 & 0 & -5 & 0 & 0 & 0 & 1 & 0 \end{bmatrix}$

9. Pivot is 2 **11.** Either 1 or 2 **13.** $P = 20$ for $x = 18$, $y = 12$

15. $\left[\begin{array}{ccccc|c} 1 & \frac{5}{2} & \frac{1}{2} & 0 & 0 & 5 \\ 0 & -\frac{3}{2} & -\frac{1}{2} & 1 & 0 & 5 \\ 0 & \frac{3}{2} & \frac{11}{2} & 3 & 1 & 25 \end{array}\right]$ using 2 as the pivot. **17.** $P = 250$ for $x = 10$, $y = 18$, $z = 10$

19. $P = \frac{164}{3}$ at $x = 8$, $y = 0$, $z = \frac{4}{3}$ **21.** $P = 40$ at $x = 5$, $y = 0$, $z = 5$

25. 15 steel castings, 30 aluminum castings **27.** \$6700 at 7%, \$13,300 at 8.5%

29. Tables $= \frac{10}{7}$, desks $= \frac{23}{14}$, cabinets $= \frac{9}{7}$. Since we cannot make fractional pieces of furniture, we round off to the nearest unit; that is, one table, two desks, and one cabinet.

Section 3.6

1. max $P = 180$ at $x_1 = 60$, $x_2 = 0$ **3.** max $P = 15$ at $x_1 = 15$, $x_2 = 0$ **5.** No solution

7. max $P = 75$ at $x_1 = 15$, $x_2 = 0$ **9.** No solution

11. Maximize $P = 8y_1 + 15y_2$
Constraints: $2y_1 + 3y_2 \le 3$
$y_1 + 5y_2 \le 2$
$y_1 \ge 0,\ y_2 \ge 0$
min $C = \frac{87}{7}$ at $x_1 = \frac{25}{7}$, $x_2 = \frac{6}{7}$

13. Maximize $P = 8y_1 + 15y_2$
Constraints: $2y_1 + 3y_2 \le 3$
$3y_1 - 5y_2 \le 2$
$y_1 \ge 0,\ y_2 \ge 0$
min $C = 15$ at $x_1 = 5$, $x_2 = 0$

15. Maximize $P = 8y_1 + 15y_2$
Constraints: $2y_1 + 3y_2 \le 3$
$-y_1 + 5y_2 \le -2$
$y_1 \ge 0,\ y_2 \ge 0$
min $C = 15$ at $x_1 = 5$, $x_2 = 0$

17. Maximize $P = 8y_1 + 15y_2 + 12y_3$
Constraints: $2y_1 + 3y_2 + y_3 \le 1$
$5y_2 + 2y_3 \le 2$
$y_1 - y_2 + y_3 \le 1$
$y_1 \ge 0,\ y_2 \ge 0,\ y_3 \ge 0$
min $C = 12$ at $x_1 = 12$, $x_2 = 0$, $x_3 = 0$

19. Maximize $P = 8y_1 + 15y_2 + 12y_3$
Constraints: $2y_1 + 3y_2 + y_3 \le 1$
$y_1 + 5y_2 + 2y_3 \le 2$
$y_1 - y_2 + 3y_3 \le -1$
$y_1 \ge 0,\ y_2 \ge 0,\ y_3 \ge 0$
min $C = 12$ at $x_1 = 12$, $x_2 = 0$, $x_3 = 0$

21. min $\frac{1450}{3}$ calories; $\frac{8}{3}$ oz of Cereal 1, $\frac{14}{3}$ oz of Cereal 2

23. Plant 1 should run $\frac{180}{11} \approx 16.4$ days
Plant 2 should run $\frac{100}{11} \approx 9.1$ days
minimum cost = \$14,363

Review Exercises Chapter 3

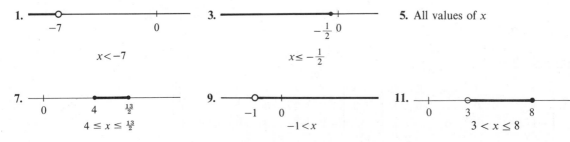

1. 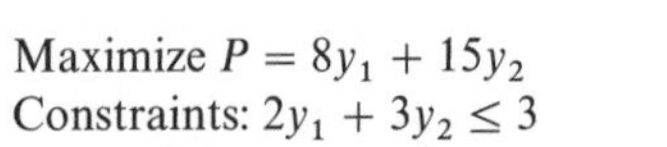

$x < -7$

3. 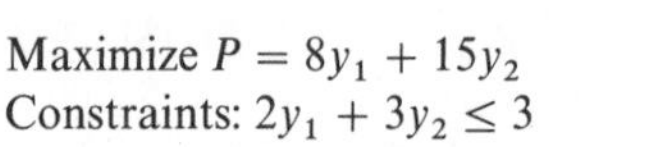

$x \le -\frac{1}{2}$

5. All values of x

7. 

$4 \le x \le \frac{13}{2}$

9. 

$-1 < x$

11. 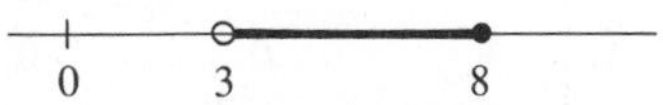

$3 < x \le 8$

13.

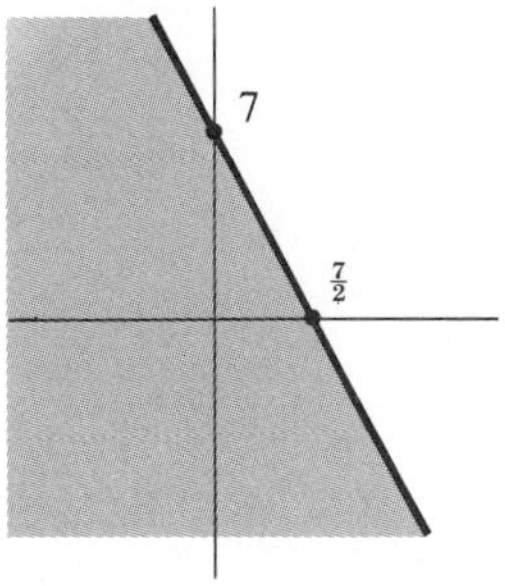

15.

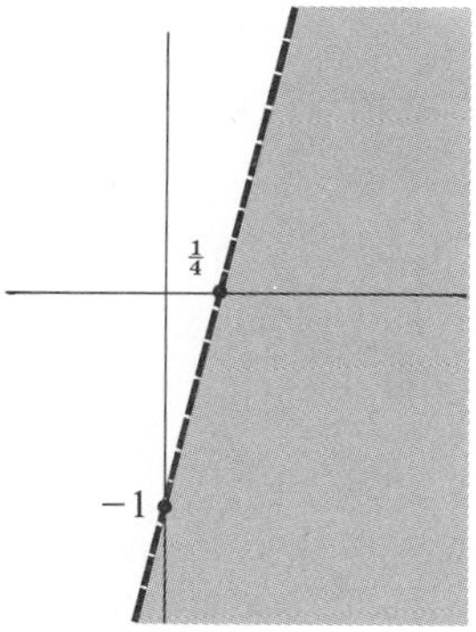

17.

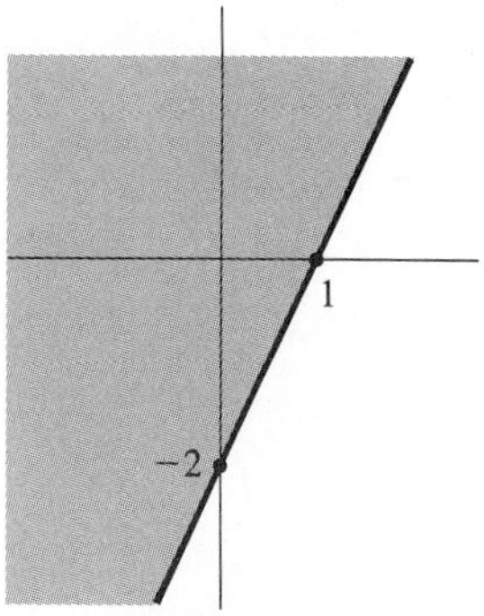

19. $C > 0$

21.

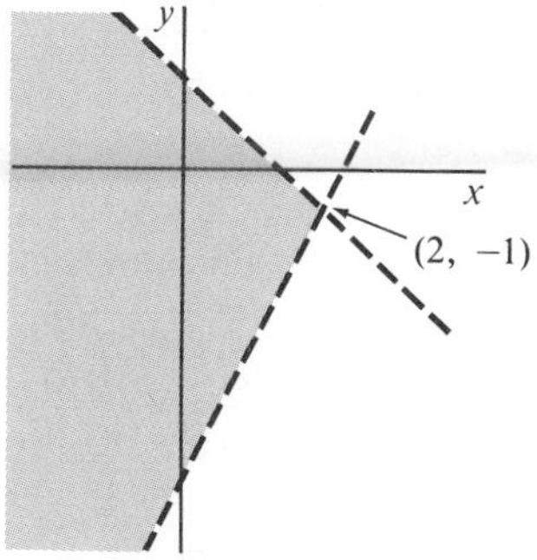

23.

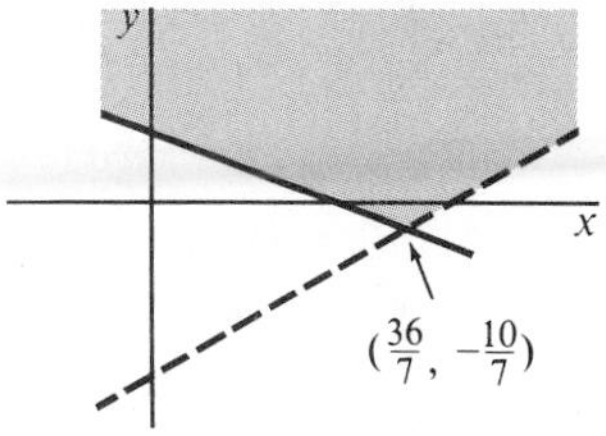

25.

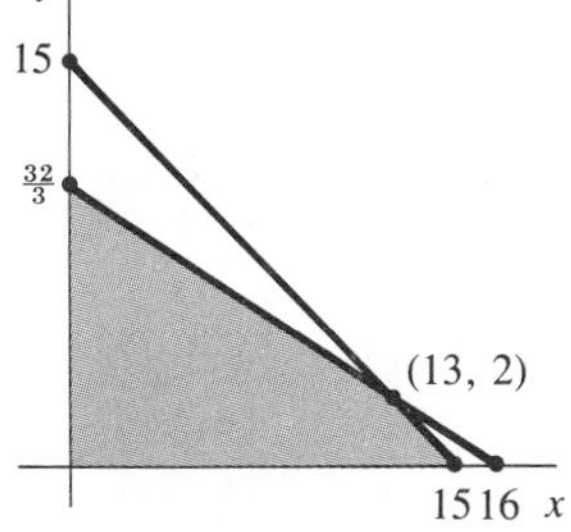

27. max $P = 24$ at $(0, 12)$

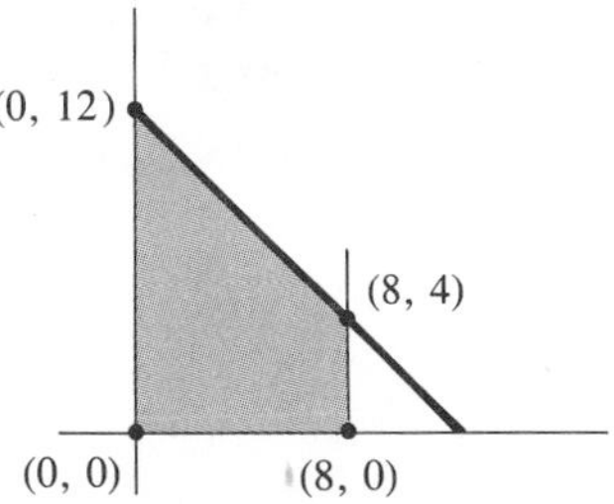

29. min $P = 16$ at $(2, 2)$

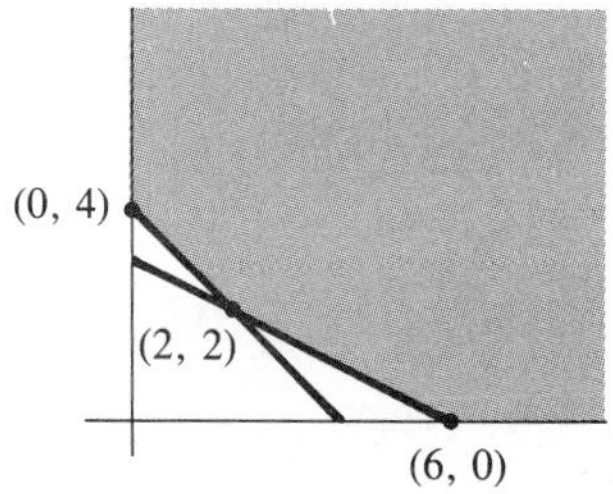

31. max $P = 1000$ at $(500, 0)$

33. $P = 5y_1 + 8y_2$, $y_1 + y_2 \le 7$, $y_1 + y_2 \le 2$
min $C = 16$ at $(0, 8)$

35. $P = 24y_1 + 15y_2$, $2y_1 + 3y_2 \le 3$, $3y_1 + y_2 \le 2$
min $C = 21$ at $(3, 6)$

37. min $C = 6$ at $x_1 = 6$, $x_2 = 0$, $x_3 = 0$

Chapter 4

Section 4.1

1. (a) $c \in A$ (b) $m \in A$ (c) $d \notin A$ **3.** (a) True (b) False (c) False (d) True

5. $A = \{-2, -1, 0, 1, 2, 3, 4, 5\}$ **7.** $B = \{2\}$ **9.** $S = \{x \mid x \text{ is an odd integer between 2 and 10}\}$

11. $C = \{x \mid x \text{ is a person in the U.S. under the age of 25}\}$ **13.** False. $\{x \mid x + 2 = 0\} = \{-2\} \ne \varnothing$.

15. True. Each set is equal to $\{-2, 2\}$.

17. False. The solutions to $x^2 - 3x + 2 = 0$ are $x = 1, 2$, and $\{1, 3\}$ is not a subset of $\{1, 2\}$.

19. $A = \{\text{e, s, v, n, l, m, t}\}$ **21.** (a) $\{2\}$ (b) $\varnothing$ (c) $\varnothing$ (d) $\{-\sqrt{3}, \sqrt{3}\}$

23. $\{5, 6, 8, 9\}, \{5, 6, 8\}, \{5, 6, 9\}, \{6, 8, 9\}, \{5, 9, 8\}, \{5, 6\}, \{5, 8\}, \{5, 9\}, \{6, 8\}, \{6, 9\}, \{8, 9\}, \{5\}, \{6\}, \{8\}, \{9\}, \varnothing$

25. (a) $X = \{2, 4, 6, 8, 10\}$ (b) $Y = \{6, 7, 8, 9, 10\}$ **27.** $A \cup B = \{1, 2, 3, 5, 6, 7, 8\}$ **29.** $A \cap B = \{1, 3, 8\}$

31. $C \cap B = \{3\}$ **33.** $A' = \{1, 3, 5, 7, 9\}$ **35.** $C' = \{1, 2, 3, 5, 6, 8, 10\}$ **37.** $B \cap C' = \{1, 2, 3\}$

39. $(A \cap B)' = \{1, 3, 5, 6, 7, 8, 9, 10\}$

41. (a) A represents the set of Super Bowl winners from 1980 to 1990.
(b) B represents the set of Super Bowl runner-ups from 1980 to 1990.
(c) $A \cup B$ represents the set of teams who either won the Super Bowl or were runner-ups or both from 1980 to 1990.
(d) $A \cap B$ represents the set of teams each of which was both a Super Bowl winner and a Super Bowl runner-up from 1980 to 1990.

43. (a) $A \cap B$ represents U.S. citizens whose first and second marriages both ended in divorce.
(b) $A \cup B$ represents U.S. citizens whose first or second marriages, or both, ended in divorce.

SECTION 4.2

1. $\{1, 2, 3, 4, 5, 6, 7, 9\}$ **3.** $\{1, 2, 3, 4, 5, 6\}$ **5.** $\{4, 5, 6\}$

7.

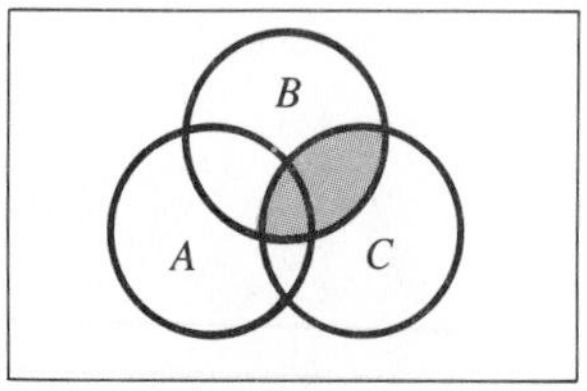

$B \cap C$

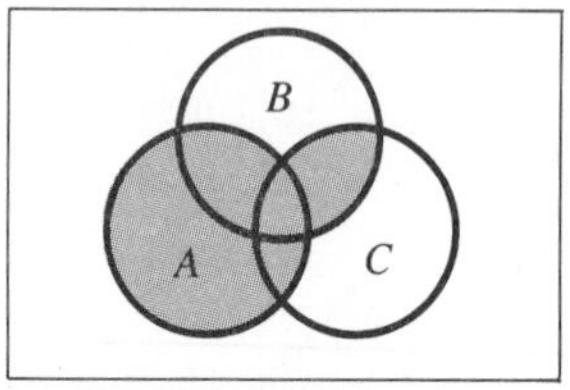

$A \cup (B \cap C)$

9. $U = \{1, 2, 3, 4, 5, 6, 7, 8, 9, 10\}$ **11.** $\{1, 2, 3, 4, 5, 6, 8, 10\}$ **13.** $\{1, 2, 3, 4, 7, 9\}$ **15.** $\{6, 8, 10\}$

17. $\{2\}$ **19.** $\{1, 3, 4, 5, 7, 9\}$

25.

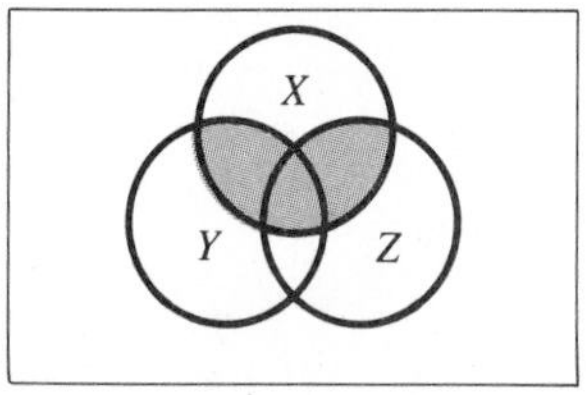

$X \cap (Y \cup Z)$

27.

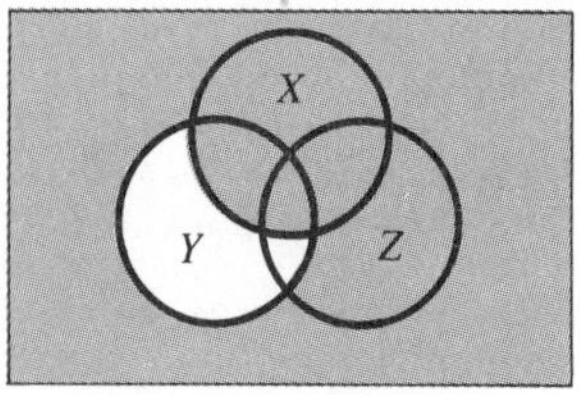

$X \cup Y'$

29.

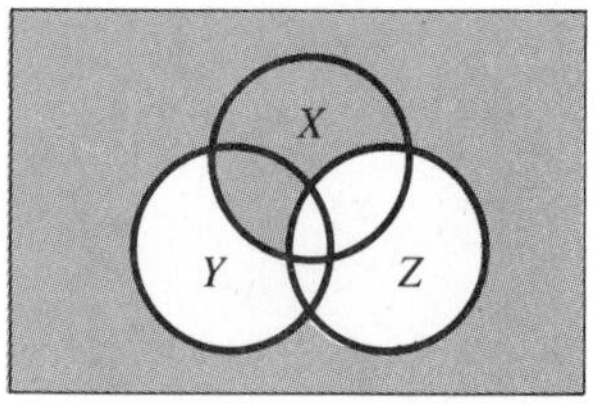

$(X \cup Y') \cap Z'$

31. (a) $M \cap S$ (b) $E' \cap S$ (c) $(M' \cap E) \cup (M' \cap S)$ or $M' \cap (E \cup S)$ (d) $M \cap E \cap S'$ **33.** 6 **35.** 7

37. 661 **39.** 33 **41.**

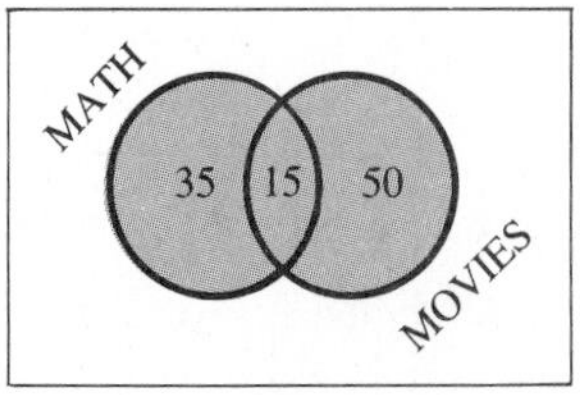

Section 4.3

1. 6, 5 **3.** *wx, xw, wy, yw, wz, zw, xy, yx, xz, zx, yz, zy* **5.** 840 **7.** 30 **9.** $10! = 3{,}628{,}800$

11. $\frac{15!}{6!} = 1{,}816{,}214{,}400$ **13.** 24 **15.** 180 **17.** 40,320 **19.** 20,160 **21.** 11

23. Both are equal to 380.

Section 4.4

1. 6 **3.** 21 **5.** 1365 **7.** 133,784,560 **9.** 1, 6, 15 **11.** 21 **13.** 20 **15.** 300 **17.** 8400

19. 31,500 **21.** 8 **23.** 252

Section 4.5

1. $x^8 + 8x^7y + 28x^6y^2 + 56x^5y^3 + 70x^4y^4 + 56x^3y^5 + 28x^2y^6 + 8xy^7 + y^8$

3. $H^{12} + 12H^{11}T + 66H^{10}T^2 + 220H^9T^3 + 495H^8T^4 + 792H^7T^5 + 924H^6T^6 + 792H^5T^7 + 495H^4T^8 + 220H^3T^9 + 66H^2T^{10} + 12HT^{11} + H^{12}$

5. $35a^4b^3$ **7.** $12{,}376H^6T^{11}$ **9.** 56 **11.** 36 **13.** $n = 13$ **15.** (a) 1 (b) 10 (c) 5 **17.** 9

19. 1941 **21.** 9438 **23.** 125,970

Review Exercises Chapter 4

1. $\{2, 3, 4, 5, 6\}$ **3.** $\{2, 3, 4, 5, 6, 7, 10\}$ **5.** $\{1, 2, 3, 5, 6, 7, 8, 9, 10\}$ **7.** $\{2, 4\}$ **9.** U **11.** 3

13. $\{1, 2, 3\}, \{1, 2\}, \{1, 3\}, \{2, 3\}, \{1\}, \{2\}, \{3\}, \varnothing$ **15.** 5040 **17.** 67,060,224 **19.** 15 **21.** (a) 56 (b) 126

Chapter 5

Section 5.1

1. $\frac{1}{4}$ **3.** $\frac{1}{13}$ **5.** $\frac{1}{52}$ **7.** $\frac{3}{13}$ **9.** 0 **11.** $\frac{82}{500}$ **13.** $\frac{422}{1000}$ **15.** {e, o}

17. (a) $S = \{HHH, HHT, HTH, THH, TTH, THT, HTT, TTT\}$
(b) $E = \{HHH, HHT, HTH, THH, TTH, THT, HTT\}$ (c) $F = \{HHT, HTH, THH\}$

19. (a) $S = \{2, 3, 4, 5, 6, 7, 8, 9, 10, 11, 12\}$ (b) $A = \{2, 4, 6, 8, 10, 12\}$ (c) $B = \{8, 9, 10, 11, 12\}$ (d) No
(e) $C = \{8, 10, 12\}$ **21.** $\{1, 2, 3, 4, 5, 6, 8, 10\}$ **23.** $\{2, 4, 5, 6, 7, 8, 9, 10\}$ **25.** $\varnothing$ **27.** $\{1, 3, 5, 7, 9\}$

29. $\{6, 8, 10\}$ **31.** {John, Jerry, Fred, Pete, Ben, Julie} **33.** $\varnothing$ **35.** {Julie, Kristin, Hylda}

37. {Jerry, Pete, Ben}

Section 5.2

1. $\frac{1}{4}$ **3.** $\frac{1}{52}$ **5.** $\frac{3}{8}, \frac{5}{8}$ **7.** $\frac{1}{3}$ **9.** $\frac{1}{3}$ **11.** $\frac{3}{8}$ **13.** $\frac{6}{13}$ **15.** $\frac{10}{13}$ **17.** $\frac{5}{16} = 0.3125$

19. $\frac{C_{5,3} \cdot C_{8,4}}{C_{13,7}} = 0.408$ **21.** $\frac{C_{15,2}}{C_{26,2}} = 0.323$ **23.** $\frac{C_{4,3}}{C_{52,3}} = 1.8 \times 10^{-4}$ **25.** $\frac{C_{13,3}}{C_{52,3}} = 0.0129$

27. $\frac{C_{13,2} \cdot C_{13,1}}{C_{52,3}} = 0.04588$ **29.** $\frac{C_{4,2} \cdot C_{48,3}}{C_{52,5}} = 0.0399$ **31.** $\frac{C_{4,2} \cdot C_{4,2} \cdot C_{4,1}}{C_{52,5}} = 5.5 \times 10^{-5}$ **33.** $\frac{1}{5}$

35. $\frac{3}{7}$ **37.** $\frac{8}{11}$ **39.** $\frac{7}{13}$ **41.** $\frac{1}{24}$

SECTION 5.3

1. $\frac{1}{3}$ **3.** $\frac{7}{9}$ **5.** $\frac{4}{13}$ **7.** (a) 0.7 (b) 0.3 (c) 0.8 **9.** (a) 0.55 (b) 0.44 (c) 0.06 **11.** $\frac{2}{9}$ **13.** $\frac{1}{9}$

15. $\frac{13}{18}$ **17.** $\frac{4}{13}$ **19.** $\frac{7}{13}$ **21.** (a) 0.61 (b) 0.39 **23.** $\frac{21}{50}$ **25.** $\frac{7}{36}$ **27.** $\frac{13}{36}$ **29.** $\frac{8}{19}$

SECTION 5.4

1. $\frac{1}{16}$ **3.** $\frac{3}{51}$ **5.** $\frac{169}{10{,}200}$ **7.** (a) $\frac{28}{143}$ (b) $\frac{63}{143}$ **9.** $\frac{1}{45}$ **11.** $\frac{5}{21}$ **13.** $\frac{1}{5}$ **15.** 1.2×10^{-5}

17. 0.081 **19.** $\frac{1}{3}$ **21.** $\frac{4}{15}$ **23.** $\frac{16}{45}$ **25.** (a) 0.1864 (b) nine draws **27.** 0.756 **29.** 0.03125

31. (a) 0.25 (b) 0.499 (c) 0.749 (d) 0.998

SECTION 5.5

1.–6.

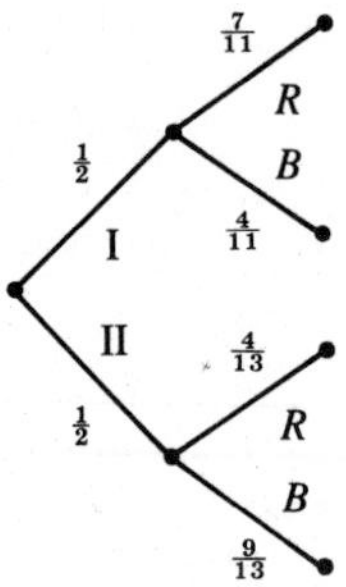

1. 0.318 **3.** 0.472 **5.** 0.674 **7.** $\frac{1}{169}$ **9.** $\frac{4}{663}$ **11.** $\frac{13}{102}$

13. $\frac{1}{3}$ **15.** $\frac{16}{45}$

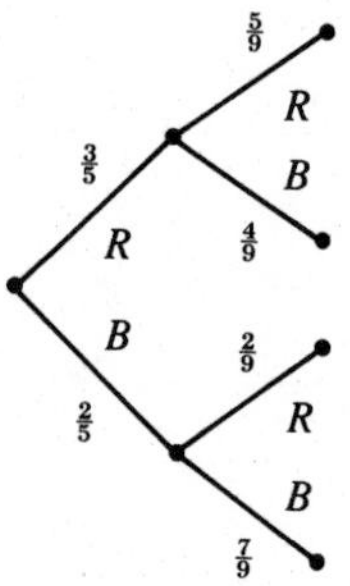

17. $\frac{17}{42}$

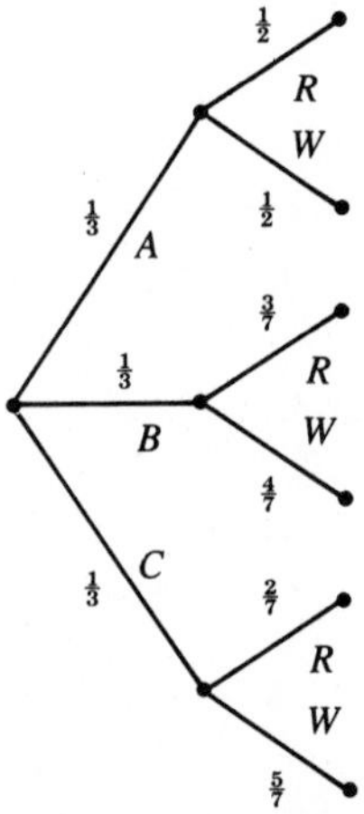

19. 0.6

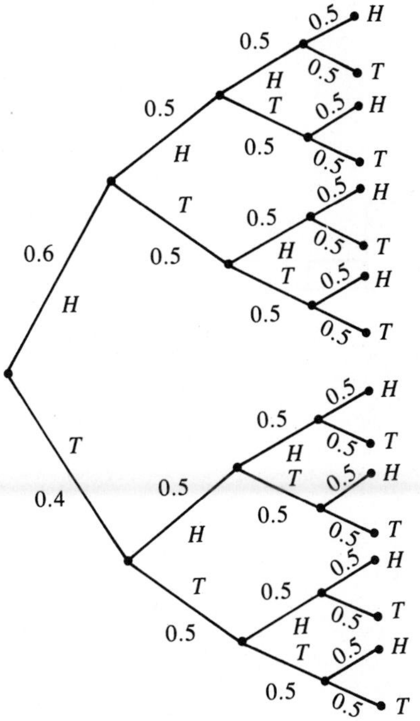

21. $\frac{345}{985}$ **23.** $\frac{250}{997}$

Section 5.6

1. 0.298 **3.** 0.625 **5.** 0.273 **7.** 0.542 **9.** 0.980 **11.** 0.424 **13.** 0.229 **15.** 0.151 **17.** 0.804

Review Exercises Chapter 5

1. $\frac{165}{1000}$

3. (a) {(1, 2), (2, 1), (2, 3), (3, 2), (3, 4), (4, 3), (1, 4), (4, 1), (2, 5), (5, 2), (1, 6), (6, 1), (5, 4), (4, 5), (3, 6), (6, 3), (5, 6), (6, 5)} (b) No (c) {(1, 2), (2, 1)} **5.** (a) $\frac{1}{4}$ (b) $\frac{1}{2}$ (c) $\frac{1}{52}$ **7.** $\frac{1}{4}$ **9.** $\frac{5}{8}$ **11.** $\frac{6}{11}$ **13.** 0.937

15. (a) $\frac{1}{169}$ (b) $\frac{1}{221}$ **17.** 0.871 **19.** $\frac{13}{30}$

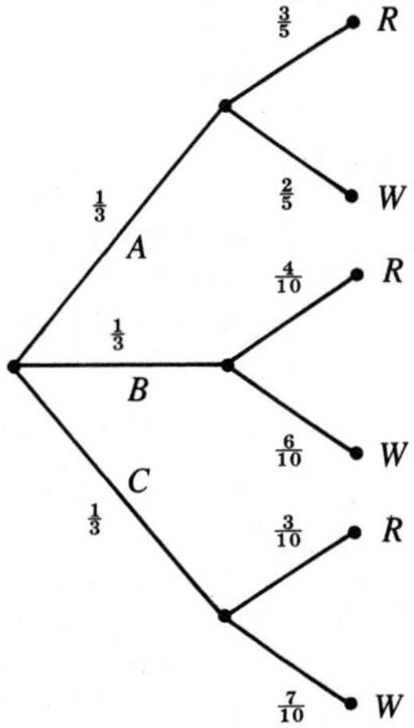

Chapter 6

Section 6.1

1. Yes **3.** Yes **5.** No **7.** Yes

9. (a) $HHH \to 0$,
$HHT, HTH, THH \to 1$
$HTT, THT, TTH \to 2$
$TTT \to 3$

(b) $P(0) = \frac{1}{8}$
$P(1) = \frac{3}{8}$
$P(2) = \frac{3}{8}$
$P(3) = \frac{1}{8}$

(c)

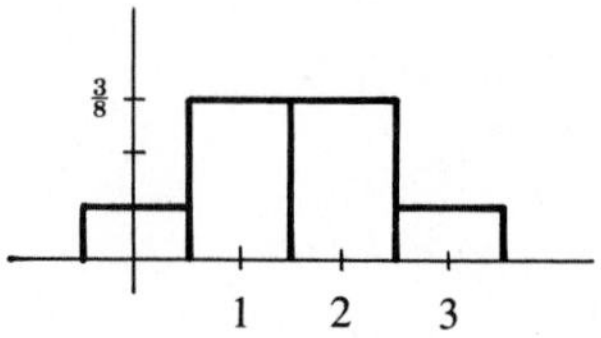

11. (a) $(1, 1) \to 2$
$(1, 2), (2, 1) \to 3$
$(1, 3), (3, 1), (2, 2) \to 4$
$(1, 4), (4, 1), (2, 3), (3, 2) \to 5$
$(1, 5), (5, 1), (2, 4), (4, 2), (3, 3) \to 6$
$(1, 6), (6, 1), (2, 5), (5, 2), (3, 4), (4, 3) \to 7$
$(2, 6), (6, 2), (3, 5), (5, 3), (4, 4) \to 8$
$(3, 6), (6, 3), (4, 5), (5, 4) \to 9$
$(4, 6), (6, 4), (5, 5) \to 10$
$(5, 6), (6, 5) \to 11$
$(6, 6) \to 12$

(b) $P(2) = \frac{1}{36}$
$P(3) = \frac{2}{36}$
$P(4) = \frac{3}{36}$
$P(5) = \frac{4}{36}$
$P(6) = \frac{5}{36}$
$P(7) = \frac{6}{36}$
$P(8) = \frac{5}{36}$
$P(9) = \frac{4}{36}$
$P(10) = \frac{3}{36}$
$P(11) = \frac{2}{36}$
$P(12) = \frac{1}{36}$

(c)

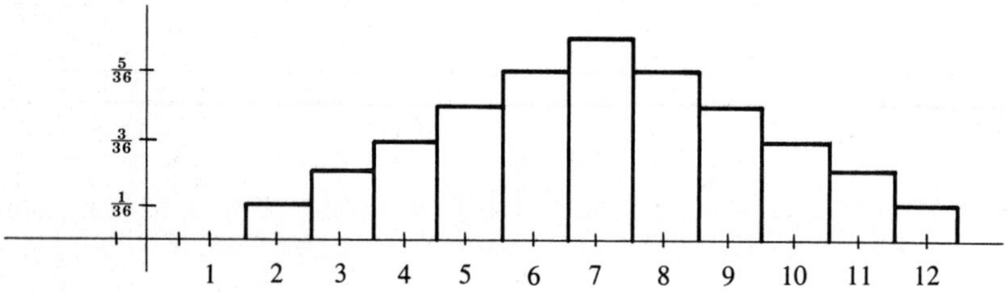

13. (a) Red $\to 2$
Black $\to -1$
Not colored $\to -1$

(b) $P(2) = \frac{18}{38}$
$P(-1) = \frac{20}{38}$

(c)

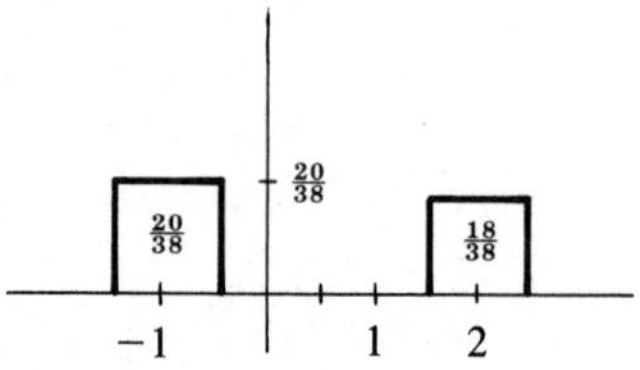

15. (a) Carroll wins $\to 0.50$
Carroll loses $\to -0.50$

(b) $P(0.50) = \frac{1}{2}$
$P(-0.50) = \frac{1}{2}$

(c)

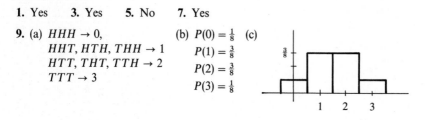

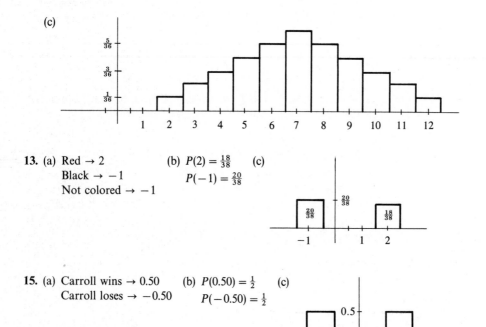

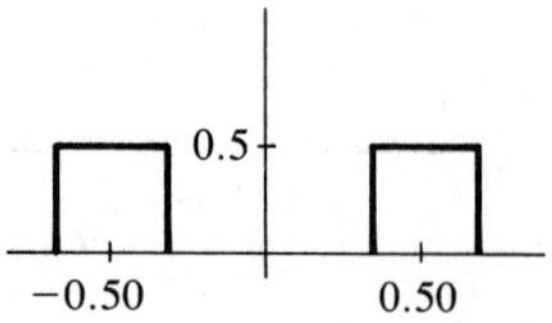

17. (a) $X = \{0, 1, 2\}$ (b) $P(0) = 0.9409$ (c)
$P(1) = 0.0582$
$P(2) = 0.0009$

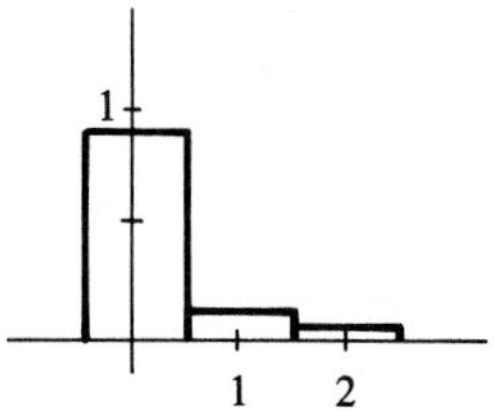

19. (a) 0.2 (b) 0.3 (c) 0.5

SECTION 6.2

1. $2\frac{5}{8}$

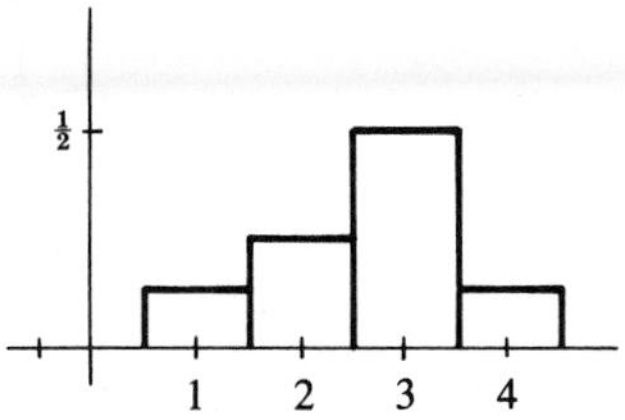

3. 13

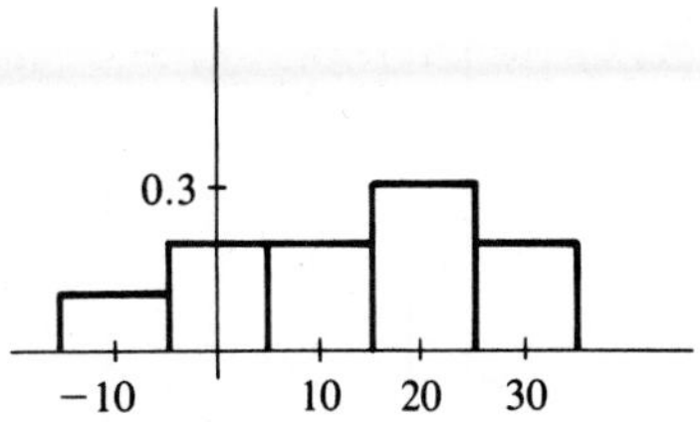

5. 1.2 **7.** 7 **9.** 11.5 **11.** 1.875 **13.** 0.462 **15.** $-\frac{1}{19}$ **17.** (a) −0.65 (b) −3.25
19. $19,000 **21.** Chicago

SECTION 6.3

1. $\mu = 2.625$, $\sigma = 0.857$ **3.** $\mu = 13$, $\sigma = 12.69$ **5.** $\mu = 1.177$, $\sigma = 0.683$ **7.** $\mu = 7$, $\sigma = 2.415$
9. $\mu = 11.481$, $\sigma = 6.779$ **11.** $\mu = 1.875$, $\sigma = 0.817$ **13.** $\mu = 1.6$, $\sigma = 1.2$
15. (a) 0.75 (b) 0.89 (c) 0.69 **17.** $97.5 \le x \le 122.5$, 0.84 **19.** 0.91 **21.** 0.84

SECTION 6.4

1. 0.234 **3.** (a) 0.016 (b) 0.094 (c) 0.984 **5.** 0.078
7. (a) 1.25×10^{-4} (b) 3.57×10^{-6} (c) 1.04×10^{-4} **9.** 0.282 **11.** 0.056 **13.** 0.851 **15.** 0.999
17. 0.410 **19.** 0.012 **21.** 0.287

23. (a)

X	p
0	0.132
1	0.329
2	0.329
3	0.165
4	0.041
5	0.004

(b)

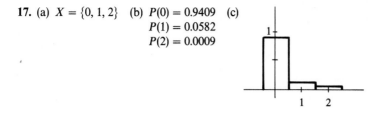

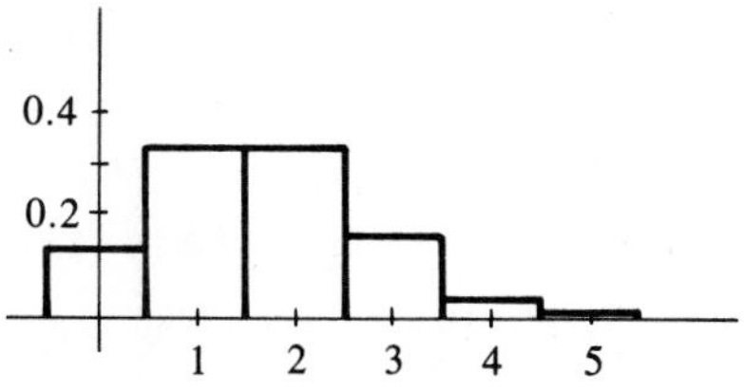

(c) $\mu = 1.667$
$\sigma = 1.054$

25. (a)

X	p
0	0.284
1	0.420
2	0.233
3	0.057
4	0.005

(b)

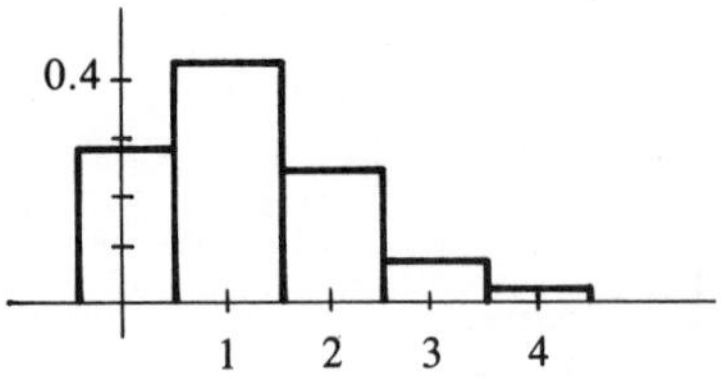

(c) $\mu = 1.08$
$\sigma = 0.888$

27. (a)

X	p
0	0.216
1	0.432
2	0.288
3	0.064

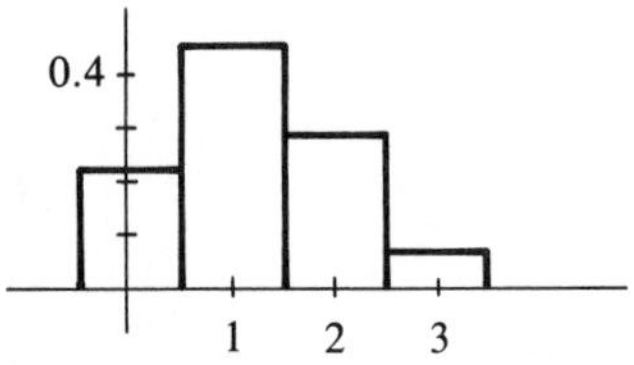

(b)

X	p
0	0.047
1	0.187
2	0.311
3	0.276
4	0.138
5	0.037
6	0.004

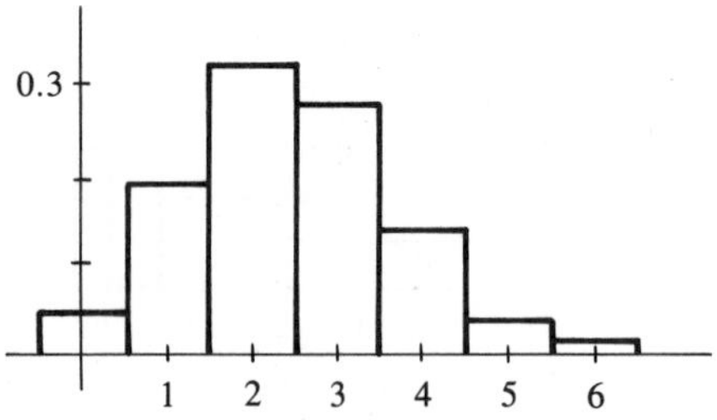

(c)

X	p
0	0.002
1	0.017
2	0.064
3	0.142
4	0.213
5	0.227
6	0.177
7	0.101
8	0.042
9	0.012
10	0.002
11	0.000
12	0.000

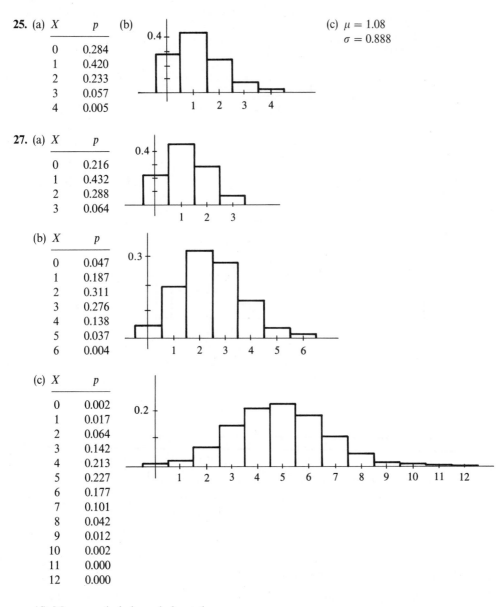

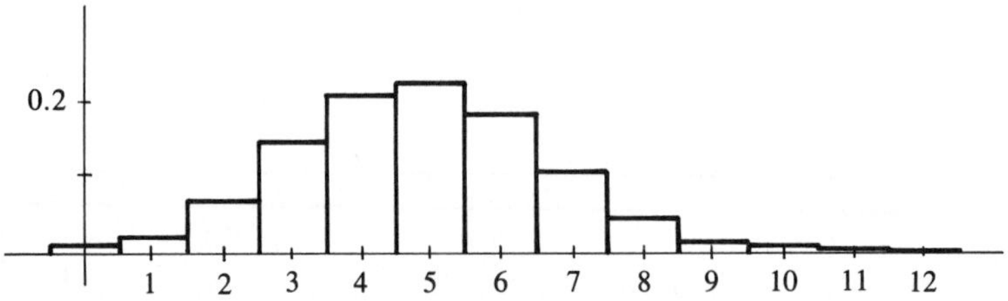

(d) More evenly balanced about the mean

SECTION 6.5

1. 0.8577 **3.** 0.3085 **5.** 0.4332 **7.** 0.2486 **9.** 0.3686 **11.** 0.4908 **13.** 0.1170 **15.** 1.28
17. 1.645 **19.** 0.255 **21.** 0.6915 **23.** 0.7794 **25.** 0.1193 **27.** 0.2486 **29.** 0.5987 **31.** 0.3520
33. 0.2514 **35.** 0.0838 **37.** 0.0359

Section 6.6

1. No **3.** No **5.** Yes **7.** No **9.** 0.3885 **11.** 0.3156 **13.** 0.3085 **15.** (a) 0.118 (b) 0.1214 **17.** 0.1251 **19.** 0.4168

Section 6.7

1. $\bar{x} = 3.11$, median $= 3$, mode $= 5$ **3.** $\bar{x} = 20.66$, median $= 21$, no mode

5. $\bar{x} = 2.78$, median $= 3$, mode $= 2, 3$ **7.** $\bar{x} = 3.11$, $s_x = 1.69$ **9.** $\bar{x} = 20.67$, $s_x = 2.80$

11. $\bar{x} = 2.78$, $s_x = 1.20$ **13.** $\bar{x} = 3.11$, $s_x = 1.69$ **15.** $\bar{x} = 20.67$, $s_x = 2.80$ **17.** $\bar{x} = 2.78$, $s_x = 1.20$

19. $\bar{x} = 27.43$, $s_x = 12.34$ **21.** $\bar{x} = 7.39$, $s_x = 3.77$ **23.** $\bar{x} = 27.43$, $s_x = 12.33$ **25.** $\bar{x} = 7.39$, $s_x = 3.77$

27. (a) $\bar{x} = 40.81$, $s_x = 2.50$ (b)

1000 miles	# of cars
34–36	1
36–38	2
38–40	4
40–42	7
42–44	5
44–46	1

(c)

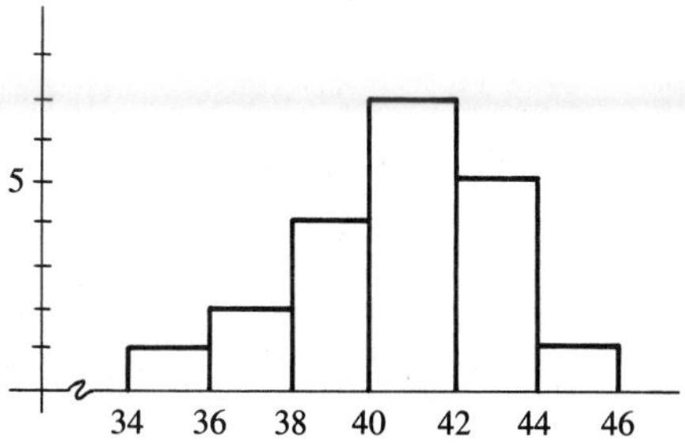

(d) $\bar{x} = 40.60$, $s_x = 2.48$ (e) $\bar{x} = 40.60$, $s_x = 2.48$, grouped
$\bar{x} = 40.81$, $s_x = 2.50$, ungrouped

29. (a) $\bar{x} = 2.32$, $s_x = 0.534$ (b)

ppm	f
1.2–1.5	2
1.5–1.8	1
1.8–2.1	5
2.1–2.4	7
2.4–2.7	5
2.7–3.0	2
3.0–3.3	5

(c)

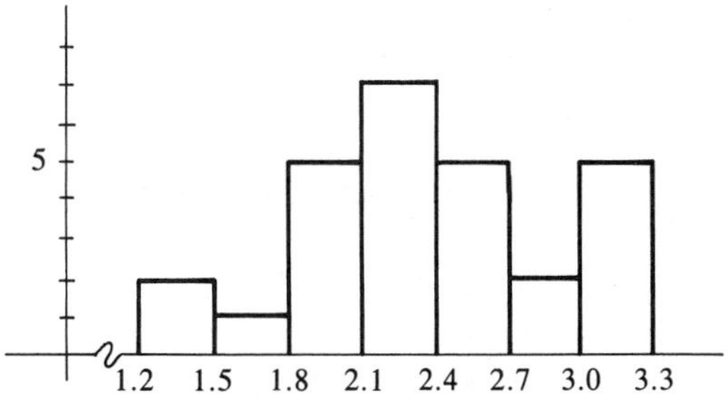

(d) $\bar{x} = 2.37$, $s_x = 0.528$ (e) $\bar{x} = 2.32$, $s_x = 0.534$, ungrouped
$\bar{x} = 2.37$, $s_x = 0.528$, grouped

Review Exercises Chapter 6

1. (a) $\{0, 1, 2\}$ (b) $P(0) = \frac{21}{55}$
$P(1) = \frac{28}{55}$
$P(2) = \frac{6}{55}$
(c)

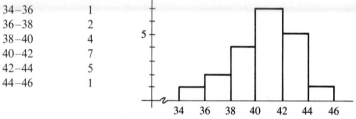

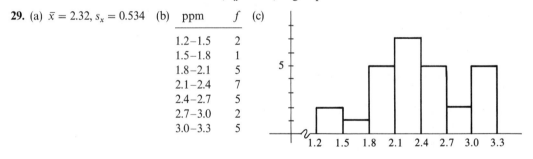

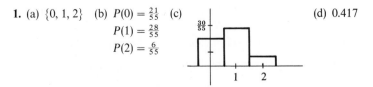

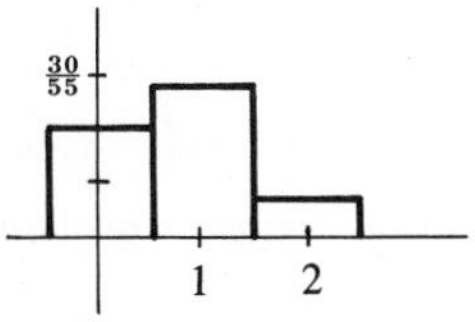

(d) 0.417

3. (a) {0, 1, 2} (b) $P(0) = \frac{2}{15}$, $P(1) = \frac{8}{15}$, $P(2) = \frac{5}{15}$ (c)

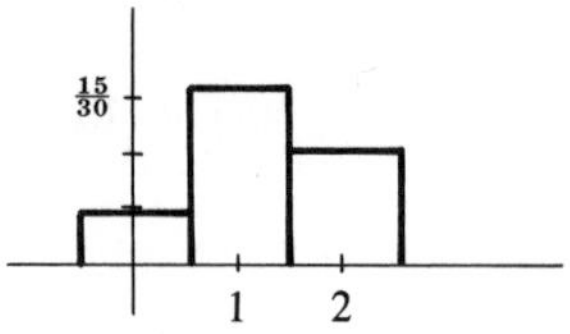

(d) 0.426

5. Buy domestic. **7.** $19 \leq x \leq 31$, 0.56 **9.** 0.93 **11.** (a) 0.276 (b) 0.820 (c) 0.197

13.

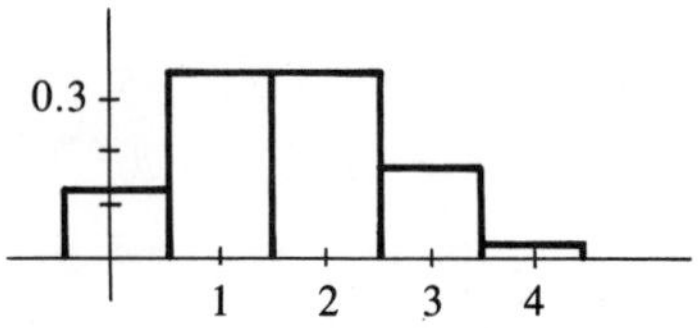

15. 0.9906 **17.** 0.3577 **19.** 0.2825 **21.** 0.8485 **23.** 0.841

25. 0.9332 **27.** 0.1339 **29.** 0.8849 **31.** 0.8708 **33.** 0.6591 **35.** 1

37. (a) $\bar{x} = 25.8$, median = 26, mode = 25, 28, $s_x = 3.12$ (b)

# of cars	f
18–21	2
21–24	2
24–27	7
27–30	7
30–33	2

(c)

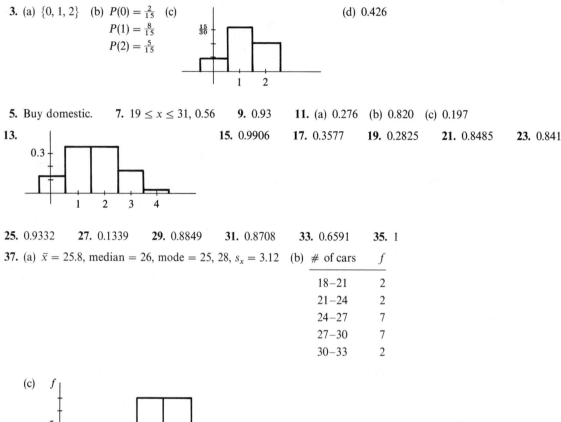

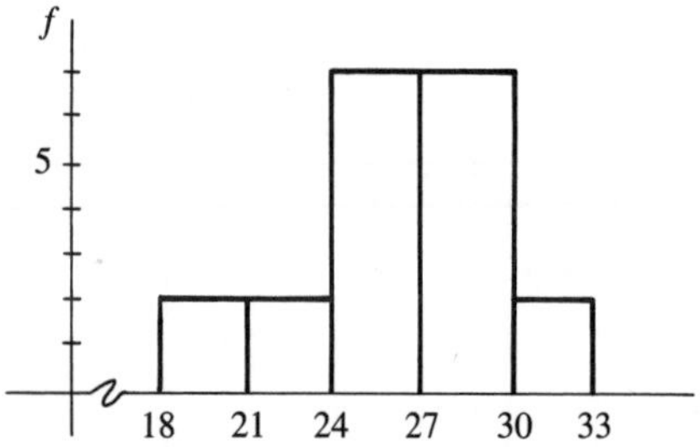

(d) $\bar{x} = 26.25$, $s_x = 3.35$ (e) $\bar{x} = 26.25$, $s_x = 3.35$, grouped
$\bar{x} = 25.80$, $s_x = 3.12$, ungrouped

CHAPTER 7

SECTION 7.1

1. Yes, it is a transition matrix.

3. No, this is not a transition matrix, because the number of rows is not equal to the number of columns.

5. No, this is not a transition matrix, because the sum of the elements of the third row is not equal to 1.

7. Yes, it is a transition matrix.

9. $\begin{bmatrix} 0.4 & 0.6 \\ 0.7 & 0.3 \end{bmatrix}$

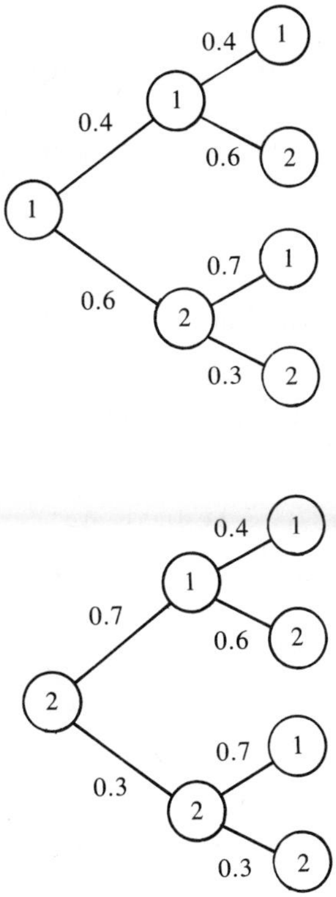

11. $\begin{bmatrix} 0.2 & 0.8 \\ 0.6 & 0.4 \end{bmatrix}$

13. $\begin{bmatrix} 0.7 & 0.2 & 0.1 \\ 0.15 & 0.73 & 0.12 \\ 0.15 & 0.20 & 0.65 \end{bmatrix}$

15. $\begin{bmatrix} 0.80 & 0.19 & 0.01 \\ 0.1 & 0.7 & 0.2 \\ 0.01 & 0.07 & 0.92 \end{bmatrix}$

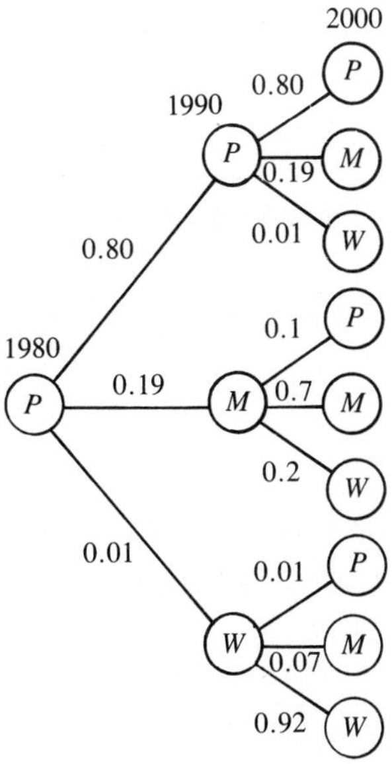

17. [0.5 0.5] **19.** [0.75 0.25] **21.** [0.48 0.21 0.31] **23.** [0.54 0.22 0.24]

25. (a) 0.417 (b) 0.444 **27.** (a) [0.35 0.20 0.45] (b) [0.1835 0.334 0.4825]

29. Transition Matrix $\begin{bmatrix} 0.82 & 0.18 \\ 0.45 & 0.55 \end{bmatrix}$

Among granddaughters of a grandmother whose IQ is greater than 120, 75% will have IQs greater than 120 and 25% will have IQs below 120.

Among granddaughters of a grandmother whose IQ is less than 120, 62% will have IQs greater than 120 and 38% will have IQs below 120.

SECTION 7.2

1. Because T has all positive elements. **3.** Because T^2 has all positive elements.

5. Because T has all positive elements. **7.** Because T^2 has all positive elements.

9. (a) $[0.53 \quad 0.47]$ (b), (c) Same equilibrium-state vector. **11.** (a), (b), (c) $[0.36 \quad 0.64]$

13. $[\frac{5}{11} \quad \frac{6}{11}] \approx [0.45 \quad 0.55]$ **15.** $[\frac{1}{6} \quad \frac{5}{6}] \approx [0.17 \quad 0.83]$ **17.** $[\frac{1}{3} \quad \frac{1}{3} \quad \frac{1}{3}]$

19. $[\frac{36}{91} \quad \frac{45}{91} \quad \frac{10}{91}] \approx [0.396 \quad 0.495 \quad 0.110]$ **21.** (a), (b), (c) $[0.43 \quad 0.57]$ **23.** $[0.53 \quad 0.47]$

25. The equilibrium-state vector is $[\frac{3}{7} \quad \frac{4}{7}]$. In the long run 42.9% of the games will start with Godzilla and 57.1% will start with King Kong.

27. Cincinnati, 32.7%; Dayton, 22.1%; Columbus, 45.1%.

SECTION 7.3

1. $2\frac{5}{8}$ **3.** 13 **5.** \$0.11 **7.** $-$\$0.58

9. Trade it. The expected value of the jackpot is \$9333, which is greater than \$5000. **11.** Not strictly determined.

13. Not strictly determined.

15. Strictly determined. Saddle point is 3. Game is unfair. Row strategy: choose row 2. Column strategy: choose column 2.

17. Strictly determined. Saddle point is $\frac{1}{2}$. Game is unfair. Row strategy: choose row 1. Column strategy: choose column 2.

19. Strictly determined. Saddle point is -1. Game is unfair. Row strategy: choose row 2. Column strategy: choose column 2.

21. 2 **23.** 0 **25.** Smith should choose newspaper and Philips should choose TV. Payoff \$1,000,000.

SECTION 7.4

1. 2.26 **3.** 1.12 **5.** 1 **7.** 5.5 **9.** 2.5 **11.** (a) 2.25 (b) 2.33 (c) Strategy in (b) gives greater payoff.

13. $R^* = [0.5 \quad 0.5]$, $C^* = [0.5 \quad 0.5]$, value $= -0.5$ **15.** $R^* = [0.5 \quad 0.5]$, $C^* = [0.83 \quad 0.17]$, value $= 0.5$

17. $R^* = [0 \quad 1]$, $C^* = [0 \quad 1]$, value $= 3$ **19.** $R^* = [1 \quad 0]$, $C^* = [0.5 \quad 0.5]$, value $= 0.5$

21. $R^* = [0 \quad 1]$, $C^* = [0 \quad 1]$, value $= -1$

23. (a)

		Coach B Pass	Coach B Run
Coach A	Pass	3	10
	Run	4	2

(b) Coach A: pass $\frac{2}{9}$ and run $\frac{7}{9}$; Coach B: defend pass $\frac{8}{9}$ and defend run $\frac{1}{9}$

(c) 3.9 yards per play

25. (a)
$$\begin{array}{c} \\ \text{Cup 1} \\ \text{Cup 2} \end{array} \begin{array}{c} \begin{array}{cc} \text{Cup 1} & \text{Cup 2} \end{array} \\ \begin{bmatrix} -5 & 5 \\ 10 & -10 \end{bmatrix} \end{array}$$

(b) Your opponent should choose Cup 1 about 50% of the time and Cup 2 about 50% of the time. You should place nickel under Cup 1 about 67% of the time and dime under Cup 2 33% of the time. (c) 0

REVIEW EXERCISES CHAPTER 7

1. [0.48 0.52] **3.** [0.23 0.43 0.34] **5.** $\begin{bmatrix} 0.3 & 0.7 \\ 0.5 & 0.5 \end{bmatrix}$

7. (a) $\begin{bmatrix} 0.65 & 0.25 & 0.1 \\ 0.15 & 0.7 & 0.15 \\ 0.25 & 0.2 & 0.55 \end{bmatrix}$ (b) [0.45 0.40 0.15] (c) 39% Republican, 42.25% Democrat, 18.75% did not vote.

9. Matrix is regular because the row elements add to 1, and T^2 has all positive elements.

11. Matrix is regular because the row elements add to 1, and T^2 has all positive elements.

13. $[\frac{2}{7} \;\; \frac{5}{7}] \approx [0.29 \;\; 0.71]$ **15.** $[\frac{12}{43} \;\; \frac{16}{43} \;\; \frac{15}{43}] \approx [0.28 \;\; 0.37 \;\; 0.35]$

17. 52% use door 1, 30% use door 2, and 18% use door 3.

19. E. V. of the Chicago job is $37,500. E. V. of the L. A. job is $32,000. Choose the Chicago job.

21. Not strictly determined. $R^* = [0.5 \;\; 0.5]$, $C^* = [0.5 \;\; 0.5]$, value $= -0.5$.

23. Not strictly determined. $R^* = [\frac{3}{7} \;\; \frac{4}{7}]$, $C^* = [\frac{5}{7} \;\; \frac{2}{7}]$, value $= \frac{6}{7}$.

25. Strictly determined: -2 is a saddle point. Value $= -2$. **27.** 0

29. Bank 1 should open in Royal Oak; Bank 2 should open in Ann Arbor.

CHAPTER 8

SECTION 8.1

1. $360, $3360 **3.** $11,500, $36,500 **5.** $6318, $1118 **7.** $17,025, $2025 **9.** $147,717.26

11. 9.24% **13.** $765.70 **15.** Pay on January 1 because the interest for January would be only $37.81.

17. $215.33 **19.** $393.75 **21.** $300.52 **23.** (a) $1647 (b) $16,353 (c) 13.4% **25.** $9411.76

SECTION 8.2

1. $3185.04, $185.04, $180 **3.** $4267.88, $267.88, $260 **5.** $10,988.33, $988.33, $946.20 **7.** $9178.76

9. $7732.56 **11.** $6499.31 **13.** $4352.06 **15.** $1960.94 **17.** 9% compounded monthly.

19. $14,106 **21.** $5277.33 **23.** $3088.73 **25.** $6430.47 **27.** 8.85% **29.** $163,000 **31.** $45,950

SECTION 8.3

1. Arithmetic: $a_n = n$ **3.** Arithmetic: $a_n = 3n$ **5.** Not arithmetic: $a_n = 5^{2-n}$ **7.** Arithmetic: $a_n = \frac{3}{4} - \frac{1}{4}n$

9. Not arithmetic: $a_n = 2^{(2^n - 1)}$ **11.** $a_n = 3n - 1$ **13.** $d = \frac{11}{2}$ **15.** $S = 76$ **17.** $a_1 = \$20$

19. $325 per month, APR = 27.4% **21.** $250 per month, APR = 19.4%

SECTION 8.4

1. Geometric: $a_n = 2^{n-1}$, $r = 2$ **3.** Not geometric: $a_n = 2n$ **5.** Geometric: $a_n = (-2)^{n-1}/3$; $r = -2$ **7.** Geometric: $a_n = 2^{2-n}$, $r = \frac{1}{2}$ **9.** 468 **11.** 1749 **13.** 149,452 **15.** (a) \$5,368,710 (b) \$10,737,400 **17.** 7 strokes **19.** speed $= 2^{(T-T_0)/10}$ **21.** \$2013.33, \$2026.75, \$2040.26, \$2053.87, \$2067.56 **23.** \$1288.50 **25.** 78.4% **27.** \$3125

SECTION 8.5

1. \$11,837.09 **3.** \$35,341.23 **5.** \$20,484.50 **7.** \$31.63 **9.** \$271.13 **11.** (a) \$125.03 (b) Total invested: \$4501.08 (c) Total interest: \$498.92 **13.** \$30,347.18 **15.** \$40,327 **17.** \$84,559.39 **19.** Present value: \$18,344.83; total payments received: \$24,000 **21.** \$305,834.80 **23.** \$167.29 **25.** (a) \$241.26 (b) \$57,902.40 (c) \$32,902.40 **27.** The monthly payments are \$175.83. The interest payments for each month are \$16.67, \$15.34, \$14.00, \$12.65, \$11.29, \$9.92, \$8.54, \$7.15, \$5.74, \$4.32, \$2.89, and \$1.45.

REVIEW EXERCISES CHAPTER 8

1. \$77.55 **3.** \$10,962 **5.** \$24.49 **7.** \$8305.08 **9.** 5% **11.** \$106.33 **13.** (a) \$640.50 (b) \$6359.50 (c) 13.4% **15.** \$196.44 **17.** \$3152.42 **19.** 10.15% **21.** 9.09% **23.** 11.02% **25.** 6.5% **27.** Arithmetic: $d = 2$ **29.** Geometric: $r = 2$ **31.** Geometric: $r = -5$ **33.** Arithmetic: $d = 6$ **35.** 400 **37.** 2,097,150 **39.** -7.9472×10^{13} **41.** 1200 **43.** 14.8% **45.** \$21,911.76 **47.** \$5062.88 **49.** \$390.08 **51.** \$16,564.33; \$21,000 **53.** \$264.88 **55.** Monthly payment: \$665.73; interest paid: \$124,719 **57.** Quarterly payments are \$1092.08. The interest payments for each month are: \$160.00, \$141.36, \$122.34, \$102.95, \$83.17, \$62.99, \$42.41, and \$21.41.

CHAPTER 9

SECTION 9.1

1.

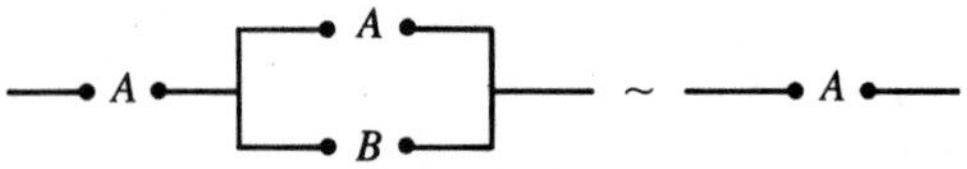

3.

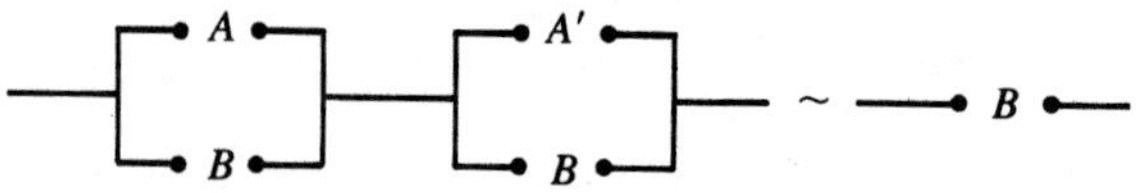

5.

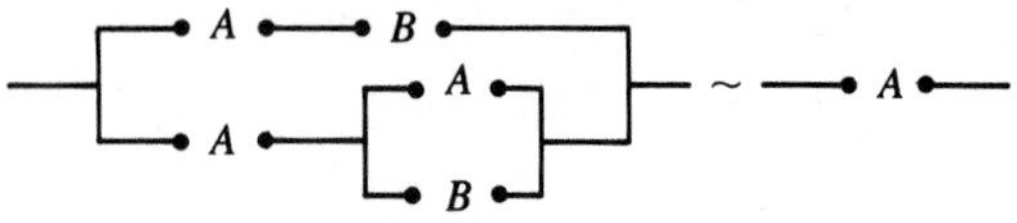

7.

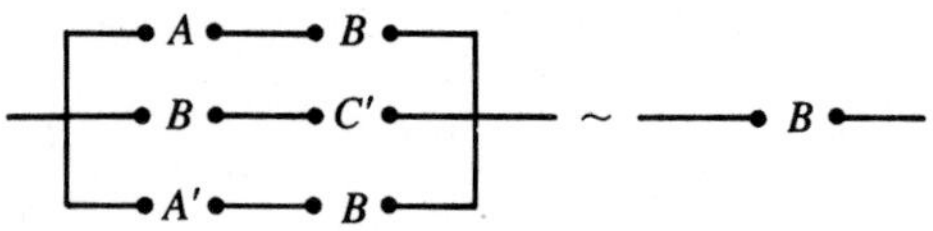

9.

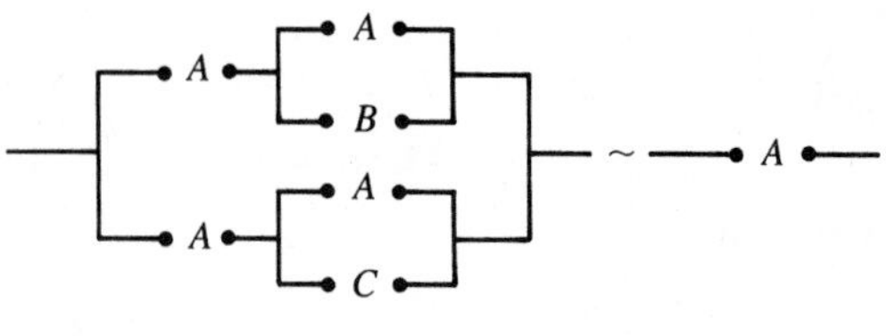

11. $A + AB$ **13.** $AB + AB$

15. $(A + B)(A + AB)$ **17.** $(A + B)(B + C')(A' + B)$ **19.** $(A + AB)(A + AC)$

33. AB'

35. $AB + C + D'$

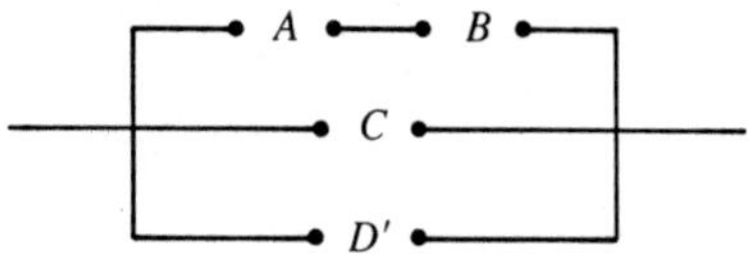

37. $AC + B$

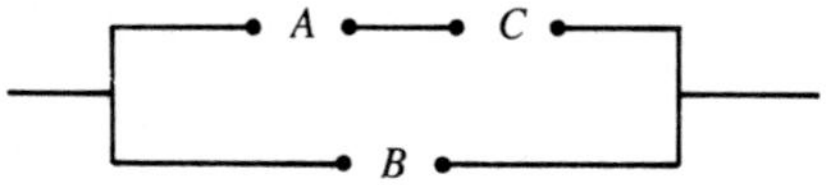

Section 9.2

7. X **9.** XY **11.** XY **13.** $XY'Z'$ **15.** 0 **17.** $(X + Y')(X + W)$ **19.** $(X + Y)(X + Z)$

21. $(X + Y)(X + Z)(Y + Z)$ **23.** $(X' + Y)(X' + Z)(X + W)(Y + W)(Z + W)$ **25.** 1

In the answers to Exercises 27–35, the dual is written first and then expanded.

27. $X(Y' + W) = XY' + XW$ **29.** $X(Y + Z)(X + Z) = XY + XZ$

31. $(X + Y)(X + Z)(Y + Z) = XY + XZ + YZ$ **33.** $(X + Y + Z)(X' + W) = X'Y + X'Z + XW + YW + ZW$

35. $(X + Y)X'Y' = 0$ **37.** 0 **39.** 1 **41.** XZ **43.** $XY + W'Z$ **45.** $Z(XYW)'$

47. $(X + Y)(X' + Z)(Y + Z) = (X + Y)(X'Y + X'Z + YZ + Z) = (X + Y)(X'Y + YZ + Z) = (X + Y)(X'Y + Z)$
$= XZ + X'Y + YZ + XX' = Z(X + Y) + X'(X + Y) = (X' + Z)(X + Y)$

Section 9.3

1. $p + q = q + p$ **3.** $(p + q)p = p$ **5.** $p' + p = 1$ **7.** $q + qr' = q$

9. $(p + q)(p' + r)(q + r) = (p + q)(p' + r)$ **11.** I will study and watch TV.

13. I will study and watch TV or I will sleep. **15.** I will study or not study.

17. I will eat or sleep, or not eat, or study and watch TV. **19.** I will not study, watch TV, and sleep.

21. $p + q$ **23.** $r'p$ **25.** $p(r' + q' + s')$ **27.** $q(r + s + p)'$ **29.** qp'

31. Should eat hamburgers, eat hot dogs, and go home. **33.** (a) No (b) Yes

Section 9.4

1.

3.
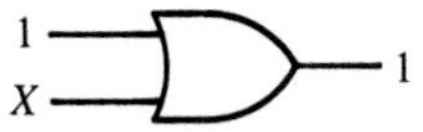

5.
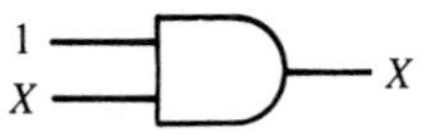

7.
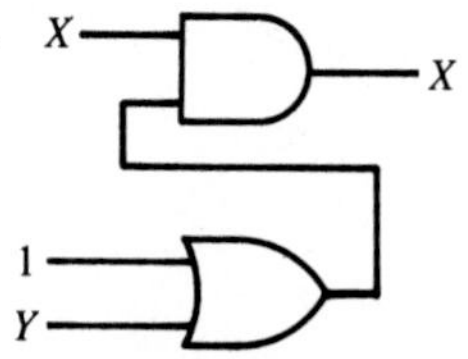

9.
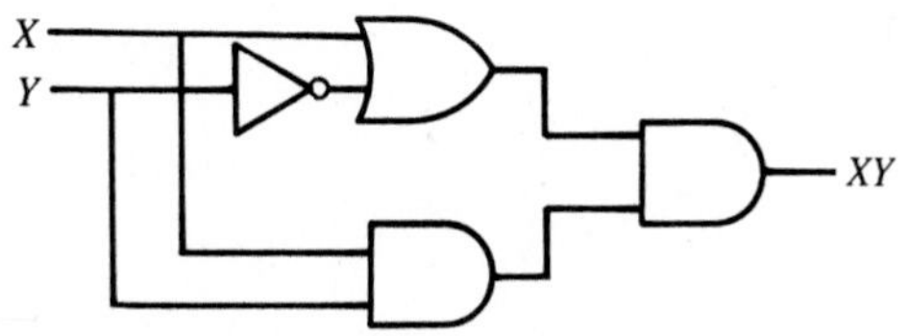

11.

X	Y	Output $= X + Y'$
1	1	1
1	0	1
0	1	0
0	0	1

13.

X	Y	Output $= (X + Y)' + (XY)'$
1	1	0
1	0	1
0	1	1
0	0	1

15.

X	Y	Output $= XY'$
1	1	0
1	0	1
0	1	0
0	0	0

17.

X	Y	Output $= (X + Y)'XY$
1	1	0
1	0	0
0	1	0
0	0	0

19.

X	Y	Output $= 0$
1	1	0
1	0	0
0	1	0
0	0	0

Section 9.5

1. $A \cap B$ **3.** $A \cap (B \cup A)$ **5.** $A \cap A$ **7.** $U \cap \varnothing$ **9.** $U \cap A$ **11.** $\bar{A}$ **13.** $A \cup \bar{A}$ **15.** AB

17. $(A + B)'$ **19.** $A \cdot 0$ **21.** $A + A$ **23.** $B + B'$ **25.** $A \cdot A = A$ **27.** $A \cdot (A + B) = A$

29. $(A + B)(A + B') = A$ **31.** $A(A' + B) = AB$

33. In any set identity, change $\cup$ to $\cap$, $\cap$ to $\cup$, $\varnothing$ to U, and U to $\varnothing$. The new expression will also be an identity. This duality principle is valid in set algebra.

35. (a) $A \cup B$ (b) $A + B$ **37.** (a) $\overline{A \cap B}$ (b) $(AB)'$ **39.** (a) $A \cup U$ (b) $A + 1$

41. (a) $A \cap A$ (b) $A \cdot A$ **43.** (a) $B \cap \bar{B}$ (b) $B \cdot B'$

Review Exercises Chapter 9

1. No answer required **3.** No answer required **5.** X **7.** Y **9.** X **11.** XY' **13.** Y **15.** $Y + Z$

17. 1 **19.** $X' + W'$ **21.** $A + B$ **23.** 1 **25.** $AB + A'B'$ **27.** 1 **29.** $(pq)' = p' + q'$

31. $p + pq = p(p + q)$ **33.** p: It is raining; q: The temperature is above 80°; $p \wedge q$

35. p: The flag is red and white; q: It is my flag; $p \vee q'$

37.
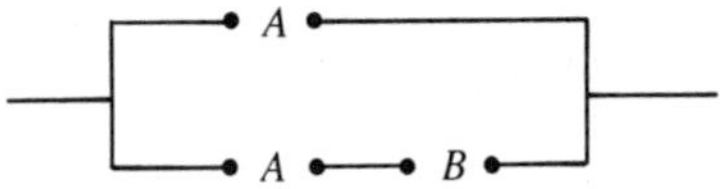

39.
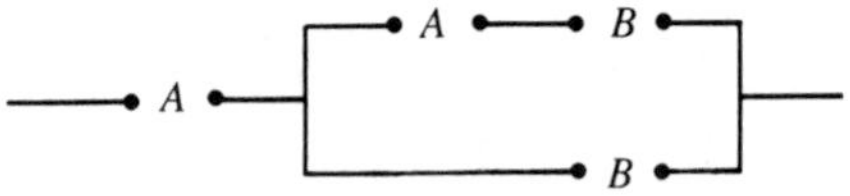

41.
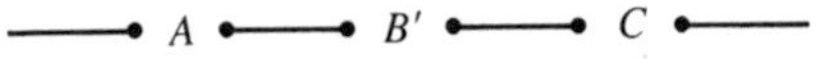

43.
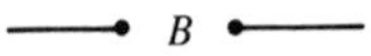

45.
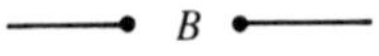

47.

X	Y	Output $= XY$
1	1	1
1	0	0
0	1	0
0	0	0

49.

X	Y	Output $= XY' + YX'$
1	1	0
1	0	1
0	1	1
0	0	0

APPENDIX TABLES

TABLE I Table of Binomial Probabilities

		p													
n	x	.01	.05	.10	.20	.30	.40	.50	.60	.70	.80	.90	.95	.99	x
2	0	.980	.902	.810	.640	.490	.360	.250	.160	.090	.040	.010	.002	.000	0
	1	.020	.095	.180	.320	.420	.480	.500	.480	.420	.320	.180	.095	.020	1
	2	.000	.002	.010	.040	.090	.160	.250	.360	.490	.640	.810	.902	.980	2
3	0	.970	.857	.729	.512	.343	.216	.125	.064	.027	.008	.001	.000	.000	0
	1	.029	.135	.243	.384	.441	.432	.375	.288	.189	.096	.027	.007	.000	1
	2	.000	.007	.027	.096	.189	.288	.375	.432	.441	.384	.243	.135	.029	2
	3	.000	.000	.001	.008	.027	.064	.125	.216	.343	.512	.729	.857	.970	3
4	0	.961	.815	.656	.410	.240	.130	.062	.026	.008	.002	.000	.000	.000	0
	1	.039	.171	.292	.410	.412	.346	.250	.154	.076	.026	.004	.000	.000	1
	2	.001	.014	.049	.154	.265	.346	.375	.346	.265	.154	.049	.014	.001	2
	3	.000	.000	.004	.026	.076	.154	.250	.346	.412	.410	.292	.171	.039	3
	4	.000	.000	.000	.002	.008	.026	.062	.130	.240	.410	.656	.815	.961	4
5	0	.951	.774	.590	.328	.168	.078	.031	.010	.002	.000	.000	.000	.000	0
	1	.048	.204	.328	.410	.360	.259	.156	.077	.028	.006	.000	.000	.000	1
	2	.001	.021	.073	.205	.309	.346	.312	.230	.132	.051	.008	.001	.000	2
	3	.000	.001	.008	.051	.132	.230	.312	.346	.309	.205	.073	.021	.001	3
	4	.000	.000	.000	.006	.028	.077	.156	.259	.360	.410	.328	.204	.048	4
	5	.000	.000	.000	.000	.002	.010	.031	.078	.168	.328	.590	.774	.951	5
6	0	.941	.735	.531	.262	.118	.047	.016	.004	.001	.000	.000	.000	.000	0
	1	.057	.232	.354	.393	.303	.187	.094	.037	.010	.002	.000	.000	.000	1
	2	.001	.031	.098	.246	.324	.311	.234	.138	.060	.015	.001	.000	.000	2
	3	.000	.002	.015	.082	.185	.276	.312	.276	.185	.082	.015	.002	.000	3
	4	.000	.000	.001	.015	.060	.138	.234	.311	.324	.246	.098	.031	.001	4
	5	.000	.000	.000	.002	.010	.037	.094	.187	.303	.393	.354	.232	.057	5
	6	.000	.000	.000	.000	.001	.004	.016	.047	.118	.262	.531	.735	.941	6
7	0	.932	.698	.478	.210	.082	.028	.008	.002	.000	.000	.000	.000	.000	0
	1	.066	.257	.372	.367	.247	.131	.055	.017	.004	.000	.000	.000	.000	1
	2	.002	.041	.124	.275	.318	.261	.164	.077	.025	.004	.000	.000	.000	2
	3	.000	.004	.023	.115	.227	.290	.273	.194	.097	.029	.003	.000	.000	3
	4	.000	.000	.003	.029	.097	.194	.273	.290	.227	.115	.023	.004	.000	4
	5	.000	.000	.000	.004	.025	.077	.164	.261	.318	.275	.124	.041	.002	5
	6	.000	.000	.000	.000	.004	.017	.055	.131	.247	.367	.372	.257	.066	6
	7	.000	.000	.000	.000	.000	.002	.008	.028	.082	.210	.478	.698	.932	7
8	0	.923	.663	.430	.168	.058	.017	.004	.001	.000	.000	.000	.000	.000	0
	1	.075	.279	.383	.336	.198	.090	.031	.008	.001	.000	.000	.000	.000	1
	2	.003	.051	.149	.294	.296	.209	.109	.041	.010	.001	.000	.000	.000	2
	3	.000	.005	.033	.147	.254	.279	.219	.124	.047	.009	.000	.000	.000	3
	4	.000	.000	.005	.046	.136	.232	.273	.232	.136	.046	.005	.000	.000	4
	5	.000	.000	.000	.009	.047	.124	.219	.279	.254	.147	.033	.005	.000	5
	6	.000	.000	.000	.001	.010	.041	.109	.209	.296	.294	.149	.051	.003	6
	7	.000	.000	.000	.000	.001	.008	.031	.090	.198	.336	.383	.279	.075	7
	8	.000	.000	.000	.000	.000	.001	.004	.017	.058	.168	.430	.663	.923	8
9	0	.914	.630	.387	.134	.040	.010	.002	.000	.000	.000	.000	.000	.000	0
	1	.083	.299	.387	.302	.156	.060	.018	.004	.000	.000	.000	.000	.000	1
	2	.003	.063	.172	.302	.267	.161	.070	.021	.004	.000	.000	.000	.000	2

TABLE I Table of Binomial Probabilities (Continued)

n	x	p .01	.05	.10	.20	.30	.40	.50	.60	.70	.80	.90	.95	.99	x
	3	.000	.008	.045	.176	.267	.251	.164	.074	.021	.003	.000	.000	.000	3
	4	.000	.001	.007	.066	.172	.251	.246	.167	.074	.017	.001	.000	.000	4
	5	.000	.000	.001	.017	.074	.167	.246	.251	.172	.066	.007	.001	.000	5
	6	.000	.000	.000	.003	.021	.074	.164	.251	.267	.176	.045	.008	.000	6
	7	.000	.000	.000	.000	.004	.021	.070	.161	.267	.302	.172	.063	.003	7
	8	.000	.000	.000	.000	.000	.004	.018	.060	.156	.302	.387	.299	.083	8
	9	.000	.000	.000	.000	.000	.000	.002	.010	.040	.134	.387	.630	.914	9
10	0	.904	.599	.349	.107	.028	.006	.001	.000	.000	.000	.000	.000	.000	0
	1	.091	.315	.387	.268	.121	.040	.010	.002	.000	.000	.000	.000	.000	1
	2	.004	.075	.194	.302	.233	.121	.044	.011	.001	.000	.000	.000	.000	2
	3	.000	.010	.057	.201	.267	.215	.117	.042	.009	.001	.000	.000	.000	3
	4	.000	.001	.011	.088	.200	.251	.205	.111	.037	.006	.000	.000	.000	4
	5	.000	.000	.001	.026	.103	.201	.246	.201	.103	.026	.001	.000	.000	5
	6	.000	.000	.000	.006	.037	.111	.205	.251	.200	.088	.011	.001	.000	6
	7	.000	.000	.000	.001	.009	.042	.117	.215	.267	.201	.057	.010	.000	7
	8	.000	.000	.000	.000	.001	.011	.044	.121	.233	.302	.194	.075	.004	8
	9	.000	.000	.000	.000	.000	.002	.010	.040	.121	.268	.387	.315	.091	9
	10	.000	.000	.000	.000	.000	.000	.001	.006	.028	.107	.349	.599	.904	10
11	0	.895	.569	.314	.086	.020	.004	.000	.000	.000	.000	.000	.000	.000	0
	1	.099	.329	.384	.236	.093	.027	.005	.001	.000	.000	.000	.000	.000	1
	2	.005	.087	.213	.295	.200	.089	.027	.005	.001	.000	.000	.000	.000	2
	3	.000	.014	.071	.221	.257	.177	.081	.023	.004	.000	.000	.000	.000	3
	4	.000	.001	.016	.111	.220	.236	.161	.070	.017	.002	.000	.000	.000	4
	5	.000	.000	.002	.039	.132	.221	.226	.147	.057	.010	.000	.000	.000	5
	6	.000	.000	.000	.010	.057	.147	.226	.221	.132	.039	.002	.000	.000	6
	7	.000	.000	.000	.002	.017	.070	.161	.236	.220	.111	.016	.001	.000	7
	8	.000	.000	.000	.000	.004	.023	.081	.177	.257	.221	.071	.014	.000	8
	9	.000	.000	.000	.000	.001	.005	.027	.089	.200	.295	.213	.087	.005	9
	10	.000	.000	.000	.000	.000	.001	.005	.027	.093	.236	.384	.329	.099	10
	11	.000	.000	.000	.000	.000	.000	.000	.004	.020	.086	.314	.569	.895	11
12	0	.886	.540	.282	.069	.014	.002	.000	.000	.000	.000	.000	.000	.000	0
	1	.107	.341	.377	.206	.071	.017	.003	.000	.000	.000	.000	.000	.000	1
	2	.006	.099	.230	.283	.168	.064	.016	.002	.000	.000	.000	.000	.000	2
	3	.000	.017	.085	.236	.240	.142	.054	.012	.001	.000	.000	.000	.000	3
	4	.000	.002	.021	.133	.231	.213	.121	.042	.008	.001	.000	.000	.000	4
	5	.000	.000	.004	.053	.158	.227	.193	.101	.029	.003	.000	.000	.000	5
	6	.000	.000	.000	.016	.079	.177	.226	.177	.079	.016	.000	.000	.000	6
	7	.000	.000	.000	.003	.029	.101	.193	.227	.158	.053	.004	.000	.000	7
	8	.000	.000	.000	.001	.008	.042	.121	.213	.231	.133	.021	.002	.000	8
	9	.000	.000	.000	.000	.001	.012	.054	.142	.240	.236	.085	.017	.000	9
	10	.000	.000	.000	.000	.000	.002	.016	.064	.168	.283	.230	.099	.006	10
	11	.000	.000	.000	.000	.000	.000	.003	.017	.071	.206	.377	.341	.107	11
	12	.000	.000	.000	.000	.000	.000	.000	.002	.014	.069	.282	.540	.886	12

TABLE II Areas under the Standard Normal Curve

The table below gives the shaded area shown in this figure.

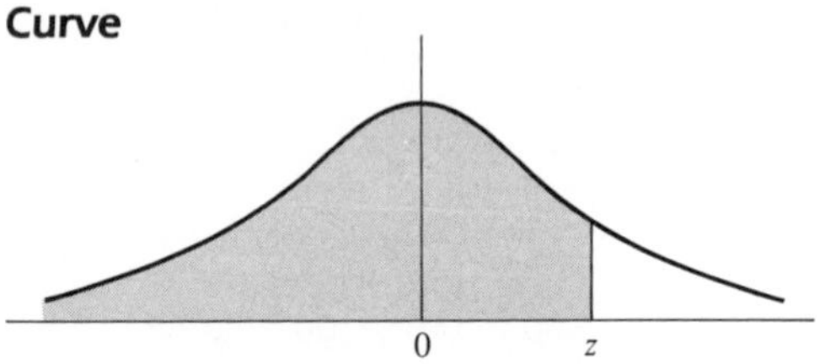

z	0	1	2	3	4	5	6	7	8	9
−3.4	.0003	.0003	.0003	.0003	.0003	.0003	.0003	.0003	.0003	.0002
−3.3	.0005	.0005	.0005	.0004	.0004	.0004	.0004	.0004	.0004	.0003
−3.2	.0007	.0007	.0006	.0006	.0006	.0006	.0006	.0005	.0005	.0005
−3.1	.0010	.0009	.0009	.0009	.0008	.0008	.0008	.0008	.0007	.0007
−3.0	.0013	.0013	.0013	.0012	.0012	.0011	.0011	.0011	.0010	.0010
−2.9	.0019	.0018	.0018	.0017	.0016	.0016	.0015	.0015	.0014	.0014
−2.8	.0026	.0025	.0024	.0023	.0023	.0022	.0021	.0021	.0020	.0019
−2.7	.0035	.0034	.0033	.0032	.0031	.0030	.0029	.0028	.0027	.0026
−2.6	.0047	.0045	.0044	.0043	.0041	.0040	.0039	.0038	.0037	.0036
−2.5	.0062	.0060	.0059	.0057	.0055	.0054	.0052	.0051	.0049	.0048
−2.4	.0082	.0080	.0078	.0075	.0073	.0071	.0069	.0068	.0066	.0064
−2.3	.0107	.0104	.0102	.0099	.0096	.0094	.0091	.0089	.0087	.0084
−2.2	.0139	.0136	.0132	.0129	.0125	.0122	.0119	.0116	.0113	.0110
−2.1	.0179	.0174	.0170	.0166	.0162	.0158	.0154	.0150	.0146	.0143
−2.0	.0228	.0222	.0217	.0212	.0207	.0202	.0197	.0192	.0188	.0183
−1.9	.0287	.0281	.0274	.0268	.0262	.0256	.0250	.0244	.0239	.0233
−1.8	.0359	.0351	.0344	.0336	.0329	.0322	.0314	.0307	.0301	.0294
−1.7	.0446	.0436	.0427	.0418	.0409	.0401	.0392	.0384	.0375	.0367
−1.6	.0548	.0537	.0526	.0516	.0505	.0495	.0485	.0475	.0465	.0455
−1.5	.0668	.0655	.0643	.0630	.0618	.0606	.0594	.0582	.0571	.0559
−1.4	.0808	.0793	.0778	.0764	.0749	.0735	.0721	.0708	.0694	.0681
−1.3	.0968	.0951	.0934	.0918	.0901	.0885	.0869	.0853	.0838	.0823
−1.2	.1151	.1131	.1112	.1093	.1075	.1056	.1038	.1020	.1003	.0985
−1.1	.1357	.1335	.1314	.1292	.1271	.1251	.1230	.1210	.1190	.1170
−1.0	.1587	.1562	.1539	.1515	.1492	.1469	.1446	.1423	.1401	.1379
−0.9	.1841	.1814	.1788	.1762	.1736	.1711	.1685	.1660	.1635	.1611
−0.8	.2119	.2090	.2061	.2033	.2005	.1977	.1949	.1922	.1894	.1867
−0.7	.2420	.2389	.2358	.2327	.2296	.2266	.2236	.2206	.2177	.2148
−0.6	.2743	.2709	.2676	.2643	.2611	.2578	.2546	.2514	.2483	.2451
−0.5	.3085	.3050	.3015	.2981	.2946	.2912	.2877	.2843	.2810	.2776
−0.4	.3446	.3409	.3372	.3336	.3300	.3264	.3228	.3192	.3156	.3121
−0.3	.3821	.3783	.3745	.3707	.3669	.3632	.3594	.3557	.3520	.3483
−0.2	.4207	.4168	.4129	.4090	.4052	.4013	.3974	.3936	.3897	.3859
−0.1	.4602	.4562	.4522	.4483	.4443	.4404	.4364	.4325	.4286	.4247
−0.0	.5000	.4960	.4920	.4880	.4840	.4801	.4761	.4721	.4681	.4641

TABLE II Areas under the Standard Normal Curve (Continued)

z	0	1	2	3	4	5	6	7	8	9
0.0	.5000	.5040	.5080	.5120	.5160	.5199	.5239	.5279	.5319	.5359
0.1	.5398	.5438	.5478	.5517	.5557	.5596	.5636	.5675	.5714	.5753
0.2	.5793	.5832	.5871	.5910	.5948	.5987	.6026	.6064	.6103	.6141
0.3	.6179	.6217	.6255	.6293	.6331	.6368	.6406	.6443	.6480	.6517
0.4	.6554	.6591	.6628	.6664	.6700	.6736	.6772	.6808	.6844	.6879
0.5	.6915	.6950	.6985	.7019	.7054	.7088	.7123	.7157	.7190	.7224
0.6	.7257	.7291	.7324	.7357	.7389	.7422	.7454	.7486	.7517	.7549
0.7	.7580	.7611	.7642	.7673	.7704	.7734	.7764	.7794	.7823	.7852
0.8	.7881	.7910	.7939	.7967	.7995	.8023	.8051	.8078	.8106	.8133
0.9	.8159	.8186	.8212	.8238	.8264	.8289	.8315	.8340	.8365	.8389
1.0	.8413	.8438	.8461	.8485	.8508	.8531	.8554	.8577	.8599	.8621
1.1	.8643	.8665	.8686	.8708	.8729	.8749	.8770	.8790	.8810	.8830
1.2	.8849	.8869	.8888	.8907	.8925	.8944	.8962	.8980	.8997	.9015
1.3	.9032	.9049	.9066	.9082	.9099	.9115	.9131	.9147	.9162	.9177
1.4	.9192	.9207	.9222	.9236	.9251	.9265	.9279	.9292	.9306	.9319
1.5	.9332	.9345	.9357	.9370	.9382	.9394	.9406	.9418	.9429	.9441
1.6	.9452	.9463	.9474	.9484	.9495	.9505	.9515	.9525	.9535	.9545
1.7	.9554	.9564	.9573	.9582	.9591	.9599	.9608	.9616	.9625	.9633
1.8	.9641	.9649	.9656	.9664	.9671	.9678	.9686	.9693	.9699	.9706
1.9	.9713	.9719	.9726	.9732	.9738	.9744	.9750	.9756	.9761	.9767
2.0	.9772	.9778	.9783	.9788	.9793	.9798	.9803	.9808	.9812	.9817
2.1	.9821	.9826	.9830	.9834	.9838	.9842	.9846	.9850	.9854	.9857
2.2	.9861	.9864	.9868	.9871	.9875	.9878	.9881	.9884	.9887	.9890
2.3	.9893	.9896	.9898	.9901	.9904	.9906	.9909	.9911	.9913	.9916
2.4	.9918	.9920	.9922	.9925	.9927	.9929	.9931	.9932	.9934	.9936
2.5	.9938	.9940	.9941	.9943	.9945	.9946	.9948	.9949	.9951	.9952
2.6	.9953	.9955	.9956	.9957	.9959	.9960	.9961	.9962	.9963	.9964
2.7	.9965	.9966	.9967	.9968	.9969	.9970	.9971	.9972	.9973	.9974
2.8	.9974	.9975	.9976	.9977	.9977	.9978	.9979	.9979	.9980	.9981
2.9	.9981	.9982	.9982	.9983	.9984	.9984	.9985	.9985	.9986	.9986
3.0	.9987	.9987	.9987	.9988	.9988	.9989	.9989	.9989	.9990	.9990
3.1	.9990	.9991	.9991	.9991	.9992	.9992	.9992	.9992	.9993	.9993
3.2	.9993	.9993	.9994	.9994	.9994	.9994	.9994	.9995	.9995	.9995
3.3	.9995	.9995	.9995	.9996	.9996	.9996	.9996	.9996	.9996	.9997
3.4	.9997	.9997	.9997	.9997	.9997	.9997	.9997	.9997	.9997	.9998

INDEX

abscissa, 21
absorption law, 391, 405
add-on loans, 347
addition rule, general, 217
  mutually exclusive events, 215
algebra of logic gates, 408
    of sets, 413
    of statements, 401
    of switches, 386
algebra, Boolean, 395
allowable manipulations, 4
  operations, inequalities, 113
    for systems, 52
amortization schedule, 378
and-gate, 409
annual percentage rate, 363
  yield, 356
annuity, 368, 373
  face value, 372
  future value, 372
  present value, 376
applications of linear equations, 7
argument of a function, 18
arithmetic mean, 354
  sequence, 359
  sum of, 362
arithmetic, Boolean, 395
associative law of switching algebra, 391
  properties of set operations, 169
augmented matrix, 64
average, 254

Bayes's formula, 237
Bayes, Thomas, 237
Bernoulli experiment, 268
Bernoulli, Jacques, 268
bimodal distribution, 295
binary operation, 166
binomial coefficients, 191
  distribution, normal approximation, 287
  expansions, 191
  probability, 271
  probabilities, table of, 274
bistable device, 384
Boole, George, 384
Boolean algebra, 395
  equation, 396

calculators, graphing, 23
cardinal number of a set, 161
  of a union, 173
Cartesian coordinate system, 21
Chebyshev's theorem, 265
circuit, dual, 389
circuits, equivalent, 388
closed Leontief model, 105
closed interval, 113
  switch, 385
coefficient matrix, 64
column matrix, 74
combinations, 183
common difference, 359
  ratio, 367
commutative law of switching algebra, 390
complement of a set, 164, 415
  of a switch, 385
complementation law, 391
composition of functions, 19
compound interest, 351, 353
computer logic circuits, 408
conditional probability, 225
consistent system, 67
constraints, nonstandard, 147
continuous probability, 277
coordinate axes, 21
counting combinations, 188
  permutations, 183
  principle, 179

Dantzig, George, 134
De Morgan's laws, 173, 391
De Morgan, Augustus, 173
DeMoivre, Abraham, 289
demand function, 26, 57
dependent events, 222
  system, 67
  variable, 15
Descartes, Rene, 21
diagonal elements, 74
discount loans, 348
discounting, 348
disjoint sets, 165
distribution, binomial, 271
  normal, 276
  probability, 247
distributive law of switching algebra, 391
  properties of set operations, 171
domain, 2
  of a function, 14
dominance law, 399
doubling period, 369
dual of a circuit, 389
  problem, 150
  Boolean, 396
duality principle, 153

effective annual interest rate, 349
elimination of a variable, 52
empty set, 161
equal matrices, 75
equality of sets, 163

equally likely outcomes, 205
equation, linear, 2, 32
Boolean, 396
equilibrium state vector, 316
equivalent circuits, 388
equations, 3
inequalities, 112
systems, 51
equivalent system, 51
event, simple, 202
independent, 222
mutually exclusive, 203, 215
expected value of a matrix game, 332
of a random variable, 255, 324
experiment, 198

face value of an annuity, 372
factorial, 181
failure, 249
feasible region, 127
finite sample space, 201
sequence, 358
set, 162
frequency, relative, 199
function, 13, 244
binomial probability, 271
probability density, 277
functional notation, 17
functions, composition of, 19
linear, 32
fundamental principle of counting, 180
future value, of an annuity, 372
compound interest, 351, 353
simple interest, 346

game theory, 324
games, nonstrictly determined, 331, 336
strictly determined, 327
Gauss, K. F., 279, 362
Gauss-Jordan elimination, 52
geometric sequence, 367
sum of, 369
graph of a function, 21
of a linear inequality, 117
of inequalities, 28
graphical method, linear programming, 129
graphical solution, system of inequalities, 121
graphing calculators, 23
grouped data, 293, 298

histogram, 248, 298
historical comments, 134, 165, 173, 279, 306, 324

idempotent law, 391
identities, 3
identity matrix, 66, 89
Boolean, 396
inconsistent system, 67
independent events, 222
variable, 15
inequality, linear, 112
infinite set, 162
initial simplex matrix, 135
installment loans, 347
installments, add-on loans, 347
interest, 344
compound, 351
simple, 345
internal consumption matrix, 102
intersection of sets, 166, 414
intervals, 112
inverse matrix, 90
isomorphism, 404, 414

Leontief model, 99
linear equations, 2, 32
functions, 32
inequality, 112
programming, 126
programming, graphical method, 129
systems, 51
loan, 344
loans, add-on, 347
discount, 348
installment, 347
logic gate, 408
logic, symbolic, 402

MAPLE, 71, 144
Markov chain, 306, 315
Markov, A. A., 306
mathematical modeling, 7
matrix, 61
addition, 76
algebra, 73
algebra, laws of, 78
games, 326
inversion, 90
multiplication, 82
payoff, 326
regular transition, 317
simplex, 135
transition, 307
maximization problem, nonstandard, 148
mean, 294, 299
arithmetic, 254
binomial random variable, 273
measure of central tendency, 255, 294
of dispersion, 296
median, 294
minimization problem, 149
mode, 294
model, probability, 199, 205
mortgage payments, 368, 377
multiplication rule, dependent events, 226
independent events, 223
mutually exclusive events, 203

NAND-gate, 409
negation of a statement, 403
nonsingular matrix, 90
nonstandard linear programming problems, 147
NOR-gate, 409
normal approximation, binomial distribution, 287
curve, 278
distribution, table of,
probability, 276
variable, standard, 283
null set, 161

objective row, 136
function, 127
odds, 212
open interval, 113
switch, 384
operations on sets, 166
optimal solution, 129
optimum strategy, 326, 333, 336
OR-gate, 409
order of a matrix, 62

ordered pairs, 14
ordinary annuity, 372
ordinate, 22
origin of a coordinate system, 21
outcomes, equally likely, 205

parallel connection, 385
parameter, 293
partial sums, 361
Pascal, Blaise, 194
Pascal's triangle, 194
payoff matrix, 326
permissible values, 2
permutations, 183
perpendicular lines, slope of, 37
pivot column, 137
  element, 136
point-slope form, 41
population, 293
present value, 355
  of an annuity, 376
principle of duality, 390, 396
probability, 199
  density function, 277
  distribution, 247
  models, properties, 208
  binomial, 271
  conditional, 225
  normal, 276
proceeds, discounted loan, 348
production matrix, 102
programming, linear, 126
proper subset, 163
pure strategy, 327

quadrants, 21

random variable, 246
  expected value, 255
  standard deviation, 263
  variance, 262
range, 298
  of a function, 14
rate, of change, 38
  of interest, 344
rectangular coordinate system, 21
recursive formulas, 359
regular Markov chain, 315
  transition matrix, 315
relative frequency, 199
row matrix, 74
  vector, 310
rule of correspondence, 14

saddle point, 327
sample, 294
sample space, 201
scalar multiplication, 77
sequences, 358, 360
  of partial sums, 361
  arithmetic, 359
  geometric, 367
series connection, 385
set builder notation, 161
sets, 160
  algebra of, 413
Shannon, C. E., 384
sigma notation, 254
simple event, 202
simple interest, 345
simplex matrix, 135
simplex method, 134
singular matrix, 90
sinking fund, 374
slack variable, 134
slope intercept form, 41
  of a straight line, 36
solution, feasible, 127
  of an equation, 2
  of a linear system, 54
  optimal, 129
solution set, 3
  for linear inequality, 112, 119
square matrix, 74
standard deviation, 263
  binomial, 273
  sample, 296, 299
  normal curve, 280
  normal variable, 283
state vector, 310, 316
statements, algebra of, 401
statistics, 293
stochastic process, 230
straight line, 39
strategy, 325
  mixed, 331
  pure, 327
strictly determined game, 327
subset, 163
sum of a geometric sequence, 369
  of an arithmetic sequence, 362
supply function, 26, 57
surplus matrix, 103
switching algebra, 386
symbolic logic, 402
system of linear inequalities, 121
systems, linear, 51
  solution by matrix inversion, 94

table of closure, 385
technology equation, 104
transition matrix, 307
transpose of a matrix, 78
tree diagram, 179, 309
trial, 198
truth value, 403

ungrouped data, 293
uniform probability model, 206
union of sets, 166, 414
universal set, 164

value of a loan, 344
variable, 15
  random, 246
  slack, 134
  standard normal, 283
  statement, 401
  switching, 386
variables, 2
variance, 262
vector, state, 310, 316
Venn diagrams, 165
vertical line test, 28
Von Neumann duality principle, 153
Von Neumann, John, 134, 325

withdrawals, 375

x-axis, 21
x-intercepts, 34

y-axis, 21
y-intercepts, 34
yield, annual, 356

zero matrix, 74
zero-sum games, 326

Interest Formulas

Simple Interest

$$I_s = PRt$$

Future Value: Simple Interest

$$S_t = P(1 + Rt)$$

Amount of Installment Payments for Add-on Loans

$$L = \frac{P(1 + Rt)}{k}$$

Proceeds of a Discounted Loan

$$P_D = P(1 - Rt)$$

Future Value: Compound Interest

$$A_n = P(1 + i)^n$$

Present Value: Compound Interest

$$P = A_n(1 + i)^{-n}$$

Future Value of an Annuity

$$V = P\frac{(1 + i)^n - 1}{i}$$

Sinking Fund

$$P = V\frac{i}{(1 + i)^n - 1}$$

Present Value of an Annuity

$$A = Q\frac{1 - (1 + i)^{-n}}{i}$$

Mortgage Payments

$$Q = A\frac{i}{1 - (1 + i)^{-n}}$$